ANTHONY
EDEN

Austen Chamberlain: Gentleman in Politics
Odyssey of an Edwardian Liberal (ed.)
British Politics since 1945: the Rise and Fall of Consensus
Simon: a Political Biography
His Majesty's Loyal Opposition
Statecraft and Diplomacy in the Twentieth Century (ed.)

ANTHONY EDEN

A Life and Reputation

DAVID DUTTON

A member of the Hodder Headline Group
LONDON • NEW YORK • SYDNEY • AUCKLAND

First published in Great Britain 1997 by
Arnold, a member of the Hodder Headline Group,
338 Euston Road, London NW1 3BH
175 Fifth Avenue, New York NY 10010

Distributed exclusively in the USA by
St Martin's Press, Inc.
175 Fifth Avenue, New York, NY 10010

British Library Cataloguing in Publication Data
A catalogue record for this book is available from the British Library

Library of Congress Cataloging-in-Publication Data
Dutton, David. 1950–
Anthony Eden: a life and reputation / David Dutton.
p.   cm.
Includes bibliographical references and index.
ISBN 0–340–56168–8 (hb)
1. Eden, Anthony, Earl of Avon. 1897–   . 2. Great Britain –
Politics and government – 20th century. 3. Conservative Party
(Great Britain) – Biography. 4. Prime ministers – Great Britain –
Biography. I. Title.
DA566.9.E28D88 1996
941.082′092–dc20
[B]   96–9223
CIP

ISBN 0 340 56168 8

Typeset in 11.5/13pt by J & L Composition Ltd, Filey, North Yorkshire
Printed and bound in Great Britain by Hartnolls Ltd, Bodmin, Cornwall

In memory of my mother

Contents

List of illustrations

14 With John Foster Dulles
    [Hulton Getty Picture Collection ©]
15 Commonwealth prime ministers meet at 10 Downing Street,
    July 1956
    [Crown copyright is reproduced with the permission of the
    controller of HMSO]
16 The Prime Minister leaves 10 Downing Street for the
    House of Commons, 12 September 1956
    [The Sport and General Press Agency ©]
17 Sir Anthony and Lady Eden at London Airport, just
    before departure for Jamaica, 23 November 1956
    [The Sport and General Press Agency ©]
18 Interviewed for Thames Television, April 1972
    [Pearson Television ©]

Preface

ROBERT Anthony Eden, first Earl of Avon, has been well served by his biographers – certainly quantitatively and arguably qualitatively. When I began work on this book there already existed at least eight English-language studies. A ninth biography appeared after I had begun my research. Among these works those of Robert Rhodes James and David Carlton are outstanding. Though I by no means agree with all the judgements and conclusions of these two writers – indeed, so different are the assessments of Eden which emerge from them that it would be impossible to do so – it has not been my purpose to supplant all of their work. Their books are best seen as the persuasive arguments for the defence and for the prosecution in relation to a career which is likely to remain controversial for many years to come.

In addition to Eden's biographers a number of authors have focused their attention on particular aspects of his long career. And, of course, the literature on the Suez Crisis of 1956 is already extensive. Against this background I early decided that I did not wish to write another conventional biography. Rather I could enjoy the luxury of building on an existing body of literature without feeling the need to cover every detail of Eden's life from his cradle to his grave. This book therefore takes the form of a series of linked essays, arranged with a sufficient sense of chronology to allow it to be read as a biography, but permitting me to explore the key themes of Eden's career without the distraction of trying to do justice to every feature of the kaleidoscopic existence of a leading politician. The dominating thread, granted the nature of Eden's career, is the study of the foreign policy of a declining great power and the role of one key player in this story. But Eden chose to make his contribution to British public life not as a professional diplomat but as a party politician and this important, but often neglected, fact looms large in some of the chapters which follow.

In writing a work of this length I have incurred a number of debts which I am pleased to acknowledge. Nigel Ashton, Philip Bell, Christopher Wheeler and Ralph White read parts or all of the typescript and made many valuable suggestions which I have done my best to incorporate in the final text. I am grateful to them all, though none, of course, bears any responsibility for those errors of fact and judgement which remain. I have received much help and kindness from the staff of the libraries and archives in which I have worked, particularly from Dr Ben Benedikz, Christine Penney and their colleagues in the Heslop Room at the University of Birmingham Library where the Avon Papers are housed. Lords Hailsham and Gladwyn kindly discussed their memories of Eden with me, while Dr Justin Nowell explained some of the intricacies of Eden's medical history. For help with other aspects of Eden's career I should like to thank Jonathan Aitken MP, Rupert Allison MP, Professor Brian Bond, Anthony Moncrieff, A. J. Nicholls, Chapman Pincher and Dr Glyn Stone. At Arnold Christopher Wheeler has acted throughout as a most efficient, understanding and helpful editor, while the expert eye of Linden Stafford has removed a host of stylistic infelicities.

I am particularly indebted to the Countess of Avon for permission to quote extensively from the papers of her late husband, the First Earl of Avon.

Transcripts of Crown-copyright records in the Public Record Office appear by permission of the Controller of HM Stationery Office.

For permission to quote from material in their possession or of which they own the copyright, I am pleased to thank the following: Alex Allan for the letters of the Baron Allan; Lord Amery for the letters and diaries of L. S. Amery and letters written by himself; Sir Colville Barclay for the letters of Sir Robert Vansittart; Lord Blake for letters written by himself; Lady Boothby for the letters of the Baron Boothby; the British Library Board for the diaries of James Chuter Ede; the British Library of Political and Economic Science for the diaries of Hugh Dalton; Robin Bruce Lockhart and the Clerk of the Records of the House of Lords, acting on behalf of the Trustees of the Beaverbrook Foundation, for the letters and diaries of Robert Bruce Lockhart; Professor Victor Bulmer-Thomas for the letters of

Ivor Bulmer-Thomas; Lady Butler for the letters of R. A. Butler; Lord
Carr of Hadley for letters written by himself; Viscount Chandos for
the letters of Oliver Lyttelton, Paul Channon for the diaries of Sir
Henry Channon; the Marquess of Cholmondeley for the letters of Sir
Philip Sassoon; the Clerk of the Records of the House of Lords,
acting on behalf of the Trustees of the Beaverbrook Foundation,
for the letters of the First Lord Beaverbrook; the Earl of Crawford
for the diaries of the Twenty-Seventh Earl of Crawford; Anne Cross-
man for the diaries of Richard Crossman; Curtis Brown Ltd on behalf
of the Estate of Sir Winston S. Churchill, copyright Winston S.
Churchill, for the letters of Sir Winston Churchill, David Higham
Associates and Cassells Ltd for the diaries of Sir Alexander Cadogan;
Lord Davies of Llandinam for the letters of the First Baron Davies;
the Earl of Derby for the letters of the Seventeenth Earl of Derby;
Piers Dixon for the diaries of Pierson Dixon; the Estate of the first
Lord Coleraine for the letters of Richard Law; Adam Fergusson for
the diaries of Blanche Dugdale; Lord Hailsham of St Marylebone for
letters written by himself; the Earl of Halifax and the Borthwick
Institute of Historical Research for the letters and diaries of the First
Earl of Halifax; Baron Hankey for letters written by himself; the
Harold Macmillan Book Trust Copyright and Archive Fund for the
letters and diaries of Harold Macmillan; the Hon. Mrs John Harvey
for the letters and diaries of Oliver Harvey; Captain James Headlam
and the Durham Record Office for the diaries of Cuthbert Headlam;
Lord Hemingford for the diaries of William Clark; Hodder and
Stoughton Ltd for the diaries of Sir John Colville; the Earl of
Home for the letters of Sir Alec Douglas-Home; Lord Jenkins of
Hillhead of the diaries of Hugh Gaitskell; Viscount Lambton for
letters written by himself; Mrs M. M. Leishman for the diaries of
Lord Reith; Little, Brown and Co. for the diaries of Beatrice Webb;
Mr H. P. R. Lloyd and the Carmarthenshire Archive Service for the
letters of J. P. L. Thomas; Macmillan General Books for the memoirs
of Harold Macmillan; the Master and Fellows of Churchill College
Cambridge for the letters and diaries of Maurice Hankey; Lord
Moran for the diaries of the First Baron Moran; Nigel Nicolson for
the letters and diaries of Harold Nicolson; Mrs Paul Padget for the
letters of Samuel Hoare; the Marquess of Salisbury for the letters of
the Fifth Marquess of Salisbury; the Earl of Scarbrough for the
letters of Roger Lumley; the Earl of Selborne of the letters of the
Third Earl of Selborne; Julian Shuckburgh for the letters and diaries
of Sir Evelyn Shuckburgh; Christie, Viscountess Simon for the letters

of Sir John Simon; Lord Strang for the letters of the first Baron Strang; the University of Birmingham for the letters of Austen and Neville Chamberlain; Sir Nicholas Hedworth Williamson for the letters of Patrick Buchan-Hepburn; Lord Woolton for the diaries of the First Earl of Woolton; and Peter Wright for the letters of Arthur Mann.

Every effort has been made to trace the owners of copyright material. If, unwittingly, any copyright has been infringed, the author offers his apology.

During the last couple of years of her life it was my mother's usual opening remark to enquire after the progress of 'the book'. Sadly she did not live to see its publication but I am happy to dedicate this work to her memory.

The role of my wife, Christine, would be difficult to encapsulate without adding substantially to what is already a long book. Tireless and expert exponent of the art of word processing, shrewd counsellor, tactful critic and, above all, ever-present comforter at those moments of frustration and despair which all authors inevitably experience – suffice it to say that without her this book could not have been written.

David Dutton
Liverpool, 1995

1

The destruction of a reputation

I can still hear Eden's voice saying 'If it could be said that by my diplomacy I had arranged this, my name would be made for ever as Foreign Secretary'.[1]

THE diplomatic correspondent, Thomas Barman, recorded an incident in Moscow during the celebrated visit of Winston Churchill and Anthony Eden in October 1944. After long negotiations with Molotov a weary Eden returned to the British embassy and sought resuscitation in a stiff whisky. As Barman recalled:

> Suddenly he stopped: 'If I give way over Lvov', he said, 'shall I go down in the history books as an appeaser?' . . . First he put the question to the Ambassador and moved on. Then he came to my colleague and put it again, and moved on. Then he put it to me. At no time did he wait for an answer. Perhaps he did not want an answer.[2]

The story is uncorroborated, but, both in Eden's concern for the verdict of posterity and in his inclination to seek reassurance from those around him, it has the ring of authenticity. Yet in 1944 Eden could at least look forward with confidence to history's judgement on his career. Three decades later it was a different story. By then Eden's reputation had become inextricably linked with one episode – the Suez Crisis of 1956. In the autumn of 1976, after two decades to reflect on Suez, Eden was relaxing at the luxurious villa of the American diplomat, Averell Harriman. There, his visitors included another well-known diplomatic correspondent, Henry Brandon. Though in his eightieth year, Eden's appearance gave the impression of good health. Tanned by the Florida sun and dressed in a green bathrobe, he 'looked almost as handsome and dashing as ever', his stride was 'springy like that of a young man'. But Eden was suffering from the cancer which had struck him over a year earlier and which would kill him within a matter of weeks. On Mrs Harriman's advice, Brandon sought to avoid Suez as a topic of conversation. Eden, however, needed little prompting:

> It soon became clear . . . that Eden was using my presence to plead . . . for justice before history, presenting his case as he saw it, offering new and dispassionate afterthoughts . . . The ghost of Suez was still stalking Eden as he was getting ready for the end and wondering about the verdict of history. In his mind his whole proud career had been scarred by a decision which misfired for lack of American co-operation.[3]

For politicians to feel concern for what historians will finally make of their actions is a natural, and not dishonourable, characteristic. Those subject to the immediate impact of their actions must hope that politicians do not allow this attention to history's verdict to blind them to the direct consequences of their deeds for their fellow citizens. But, in Eden's case, the development of a reputation is central to an understanding both of his political career and of the man himself. Eden attained high office at a sufficiently early age for the first verdicts of history to be firmly established while he was still active in public life. His evolving reputation was based on two fundamental and related facts.

The first was the role which Eden played, or was perceived to have played, in the 1930s, a decade which damaged the standing — often beyond repair — of almost every other leading British cabinet minister of that time. Eden held office for a total of just over seven years between the formation of the National Government in August 1931 and its collapse in May 1940. Yet he has no place in the cast list of Cato's *Guilty Men*, the celebrated indictment of three Prime Ministers, ten ministers, a Chief Whip and a government adviser, whose writing may have taken only a weekend but whose influence lasted for more than a generation.[4] While those around him in the government, struggling with international problems to which perhaps there were no real answers, did lasting damage to their historical reputations, Eden escaped unscathed, moving effortlessly from one public acclamation to the next. As an early biographer noted, the Chinese saying that 'he who lies down with dogs gets up with fleas' never applied to him.[5] Indeed, the more the National Government became compromised by the worsening international climate, the more remorselessly did Eden rise in popular esteem. Shortly before his appointment as Minister for League of Nations Affairs in June 1935, the *Spectator* paid this tribute:

> In the last three years, when with each month the international situation has worsened and the prospects of disarmament have become increasingly remote, and Europe is again . . . an armed camp, one man has stood out with courage and consistency for the translation of the ideals of the post-war peace system into realities. . . . at 37 he has won a position . . . that no man of comparable age has achieved in our time.[6]

To a whole generation Eden became something of an icon. With his distinguished service on the Western Front, he was the knight errant of those few survivors from the so-called 'lost generation', the most perfect gentleman in British politics.[7] William Clark, later his Downing Street press secretary, recorded of these years:

> He had been a boyhood hero. In the sixth form at Oundle in the 1930s, I thought him the bright and shining light of hope: interested in the League of Nations, which I was keen on; anti appeasement and in favour of what he called 'a better ordering of humanity's way'.[8]

It was an appeal which transcended the confines of party politics. John McGovern, the left-wing Clydeside MP, once said, 'Anthony is our idea of a gentleman. We respect him, but we shan't ever match him.'[9]

Politicians must wish that they could see the development of future events laid out before them. One might believe that Eden was possessed of this miraculous foresight. His crucial decision was to resign from the National Government in February 1938. 'Eden has today paid a big cheque into the bank on which he can draw in future,' observed Lloyd George perceptively.[10] By leaving office Eden not only confirmed a popular perception that fundamental differences divided him from Neville Chamberlain, but also succeeded in distancing himself from the policy of appeasement before it turned completely sour. 'We assumed', recalled Douglas Jay, that Eden was protesting 'against the general drift of Chamberlain's policies; and when Hitler invaded Austria . . . only a few weeks later, we naturally took this as only confirming our assumptions.'[11] With hindsight February 1938 came to be seen as a turning-point after which people began increasingly to question the Prime Minister's foreign policy. Eden's resignation was widely seen as a triumph for the pro-German group in Chamberlain's cabinet. For the rape of

Czechoslovakia that September he bore no responsibility. The following year, after Hitler's march into Prague, the government sought
belatedly to incorporate Soviet Russia into a system of collective
security. But it was not Eden who mishandled the resulting negotiations. Indeed, his offer to go to Moscow to spearhead British diplomacy was politely rejected by Chamberlain. Only when war broke
out in September 1939 did Chamberlain accept the inevitable and
restore Eden to his rightful place in the government.

To have been, until February 1938, a member of a government
from which Winston Churchill was conspicuously excluded and of
whose policies he was regularly critical inevitably set Eden and
Churchill apart. As a government spokesman Eden had, not infrequently, to rebut and reject the other's attacks. Yet, paradoxically,
Eden's association with Churchill was the other key factor in the
development of his early historical reputation. The two men were
seen to share a common objective in resisting appeasement, even if
they had tried to do so from different positions within the political
spectrum. It was at Churchill's insistence that Eden returned to
office in September 1939. Then, for nearly five years, the two men
stood together as Prime Minister and Foreign Secretary in the
struggle against Hitler. Opinion polls suggested that, after the premier, no minister stood higher in popular esteem than Eden. As early
as 1942 Churchill advised the King that, in the event of his own
death, Eden, 'the outstanding Minister in the largest political party
in the House of Commons', should be invited to form a new government.[12]

When, with fascism successfully eradicated, Churchill turned from
making history to writing it, the Eden legend of the 1930s was
enshrined in some of the most celebrated passages of Churchillian
prose. In one sense Eden, not Churchill, is the real hero of the first
volume of the great man's war memoirs. And, if the impact of *The
Gathering Storm* was as marked as that of *Guilty Men*, its eloquence
was infinitely greater. Churchill found Eden 'the most resolute and
courageous figure in the Administration';[13] there was 'not much
difference of view between him and me';[14] the resignation of this
'one strong young figure standing up against long, dismal, drawling
tides of drift and surrender' left Churchill contemplating 'the vision
of Death'.[15] Subsequent volumes of the Churchillian saga merely
confirmed the author's judgement. During the war years the two
men surveyed world problems with a single mind and Churchill paid
full credit to Eden's role in achieving victory.[16] After 1951 Eden

served again as Churchill's Foreign Secretary, enhancing his country's standing while confirming his own position as one of the world's leading diplomatists. Finally, in April 1955, it was as Churchill's chosen successor that Eden, by then happily married to Churchill's niece, replaced his political mentor in 10 Downing Street in one of the least contentious prime ministerial successions of the century.

True, it had not been a political odyssey without difficulties. By the end of the war, with his elder son killed in action and his first marriage disintegrating, Eden's private life was in disarray. That Churchill kept Eden waiting for too long was widely admitted. In the meantime the younger man's health had broken down in 1953. Yet two years later, with Eden's health apparently restored, political life seemed ready for once to bestow a happy ending on one of its practitioners. One who had worked with him closely now noted: 'It was not a Greek tragedy after all. I . . . saw A.E. get his ovation as he entered the House for the first time as P.M.'[17] After a personal triumph in the general election of May 1955, everything seemed set fair to consolidate the reputation which had been so carefully nurtured over two decades. 'By every conceivable test of history, politics and popularity', wrote James Margach, 'Eden should have been one of Britain's truly great Prime Ministers.'[18]

Less than two years later, as the now ex-Prime Minister set sail for New Zealand, this prediction, together with Eden's own hopes and reputation, lay in ruins. 'It remains extraordinary', noted Margach, 'that one who had served such a long apprenticeship in power should ultimately prove so inept.'[19] The Suez Crisis had brought national humiliation on Britain and personal obloquy upon the man who was held responsible for it. Many former supporters could never look upon Eden in the same way again. Aspects of the Suez Crisis seemed incomprehensible in terms of anything that was known about him. Indeed it was precisely because of the reputation which Eden enjoyed, and had cultivated, that Suez proved so damaging to him. The leading French protagonists of the campaign, Christian Pineau and Guy Mollet, escaped lightly by comparison. The career of Harold Macmillan actually prospered after Suez, despite his well-known enthusiasm for the venture. Nigel Nicolson, whose father was one of Eden's loyal lieutenants of the 1930s, recorded the scene as the Conservative Party met to choose its new leader: 'Without going so far as to say that it was a tragedy that a man whose reputation for integrity and diplomatic skill was unchallenged should have resigned in circumstances which threw doubt on both qualities, [Walter Elliot] almost said it.'[20]

Of course, not all Eden's admirers deserted him in adversity. A habit of trust built up over two decades was not destroyed overnight. Many believed that Eden's past record offered an adequate guarantee that, if only the secret calculations of government could be revealed immediately, his conduct would be spectacularly vindicated. For others, the best that could be said was that Suez represented an inexplicable lapse, induced perhaps by chronic ill health and overwork, which had brought an otherwise brilliant career to a premature close. Eden's admirers might almost have wished that he had died on the operating table in 1953, as he very nearly did, or at least before Colonel Nasser had nationalized the canal. Thus was born one overall interpretation of Eden's career which still has its adherents – a distinguished career irretrievably blighted by a single mistake or, as a slight variation, a great Foreign Secretary unwisely 'lured by the system into becoming Prime Minister'.[21] When Eden died in 1977, *The Times* noted that a 'long, distinguished and seemingly serene political career' had been extinguished in a 'violent storm of hostile criticism'. It could only be hoped that, 'when the dust had settled', Eden's services to the nation would stand out 'as clearly as before'.[22] Similarly, Sidney Aster concluded his short, but perceptive, biography with the thought that it was a cruel fate 'to be remembered by one failure and not by numerous achievements'.[23] Even Anthony Nutting, a junior minister who resigned in disgust over Suez, argued that Eden's career needed to be viewed in two distinct, if unequal, parts: 'whereas his record as Foreign Secretary was studded with one brilliant success after another, his brief reign as Prime Minister will be remembered only for the titanic disaster of Suez'.[24]

Yet the makings of a very different assessment of Eden's historical reputation had long existed. Some saw Suez not as an aberration but as telling proof that he had never possessed the qualities of a statesman. Not everyone had welcomed his meteoric rise in the 1930s. Jealousy was by no means the only motive of his detractors.[25] Indeed, it was characteristic of Eden to arouse widely conflicting emotions. In part, perhaps, this reflected his capacity to project entirely contrasting characteristics to different audiences. One French diplomat later attributed to Eden 'the best conference table manners *que j'ai vues*'.[26] But European observers who admired his calm negotiating skills might have been amazed by the violent bouts

6

of temper to which his personal staff were often subjected. In different situations Eden was as renowned for his outright rudeness as for his polished charm. More seriously, some believed that Eden's success owed too much to superficial qualities which disguised underlying weaknesses in his character and make-up. 'It was always to me a mystery', recalled Oswald Mosley, 'why Eden was so assiduously groomed for leadership by the Conservative Party, as his abilities were much inferior to the abilities of [Oliver] Stanley and [Harold] Macmillan.'[27] There was an unreal quality about Eden – 'so charming almost too good to be true!' – which disturbed contemporaries. 'He reminds me', noted the Liberal Percy Harris, 'of the pictures that used to appear in the back of Ouida's novels, the perfect hero, so good-looking and without a blemish.'[28] Many wondered whether Eden would have made such political strides had he not been blessed with good looks, an impeccable sense of dress and exquisite manners. Those who likened him to a Hollywood film star were not necessarily being complimentary. A Conservative elder statesman, with no cause to resent Eden as a rival, described him in 1938 as 'fussing and fidgeting, very self-conscious and blushing with handsomeness – vain as a peacock with all the mannerisms of the *"petit maître"*'.[29]

Eden's pronouncements on the League of Nations and collective security, which undoubtedly made a favourable public impression, struck a more sophisticated audience as commonplace and platitudinous. If, in the 1930s, Eden never said a foolish thing, neither did he utter very many particularly wise ones. One backbencher recorded: 'Anthony Eden was brought in to tell us about foreign affairs and made a little speech telling us nothing that we didn't know before . . . a very commonplace effort, but of course it was treated as if it was delivered to us by an oracle.'[30] So a view always existed that popular adulation had been carried too far and that Eden's reputation rested on an insubstantial basis. Was he, perhaps, praised more for what he did not do than for what he did, more for what he was not than for what he was? Certainly it did Eden's standing no harm to distance himself from those he nominally served at the Foreign Office – from whom in age he was in any case separated by a generation.

During the early 1930s Eden stood as a glittering contrast to John Simon. But such were Simon's misfortunes as Foreign Secretary that by the time he left his post in June 1935 almost anyone would have emerged well from such a comparison.[31] Within a matter of months

Samuel Hoare had also vacated the Foreign Office, discredited if not disgraced, leaving Eden to inherit his position and restore Britain's good name in Europe. But few could assert with any confidence that substantial policy differences separated Eden from the other, older and less charismatic, members of the government.

Even the presumption that Eden's resignation in February 1938 reflected a fundamental point of principle, setting him apart from his cabinet colleagues, achieved little echo in the inner circles of government. One minister, surprised by Eden's actions, recorded that 'the actual difference between him and the P.M. and the bulk of the Cabinet was that the P.M. wanted conversations to start with Italy immediately and Anthony wanted to postpone them'.[32] Dennis Bardens, whose biography, published as early as 1955, contains more perceptive comments than later writers have sometimes noted, wondered whether 'in the excitement of the moment . . . Eden was invested with an exaggerated altruism and credited with a spirit of defiance in the face of tyranny which . . . he had never shown before'.[33] Not even the Churchillian interpretation of this crisis convinced everyone. 'I have been reading the story of Eden's resignation in Churchill's memoirs,' noted Lord Beaverbrook in 1948. 'I have been told differently.'[34] Though Churchill's passage on the 'vision of death' bears the unmistakable imprint of his own pen, Eden himself was given the opportunity to comment on a draft of *The Gathering Storm* before it went to press. Points in Churchill's account about the attitude of the Chiefs of Staff to the proliferation of Britain's enemies and concerning Eden's commitment to improving Anglo-American relations were inserted at Eden's suggestion.[35] Precisely what Churchill thought of Eden at the time of his resignation, as opposed to a decade later when he produced his memoir account, is not entirely clear, but we do know that Churchill had felt doubts about Eden's capabilities at the time of his appointment to the Foreign Office.[36] While urging Eden not to mince his words in his Commons resignation speech, Churchill also felt able to append his signature to a round robin expressing confidence in Chamberlain's policies only days after Eden left office.[37] Furthermore, little in the way of a Churchill–Eden axis in opposition to government policy emerged in the period between February 1938 and the outbreak of war.

If Eden was the 'one strong young figure' in Chamberlain's administration, was this any more than a function of the 'lost generation' of 1914–18? Drew Middleton's remark that, 'had it not been for the first war, Churchill might have seen twenty such figures' provides food for thought.[38] 'Chips' Channon was more explicit: 'The War killed or ruined millions; it must have made a few people, a few young men, like Anthony Eden, who would not be where he is except that there are so few.'[39] As has been noted, Eden's remaining career was played out against the background of a reputation established in the 1930s, and in his two later periods as Foreign Secretary he enjoyed an overwhelmingly favourable press. Relative isolation from domestic politics perhaps meant that he could do little to affect the daily life of the man in the street who had few grievances to lay at Eden's door.[40] Once again, however, the image of an alternative Eden was never far away. Even during the war colleagues noted shortcomings – his eagerness to overrate his achievements, occasional indecisiveness and a tendency to immerse himself in details at the expense of a strategic understanding of the wider picture. If Eden shared Churchill's reflected glory, many doubted whether the former's brilliance would survive the latter's eventual withdrawal from public life. After the war one journalist wrote of his 'desire to debunk the halo of foreign affairs rectitude which the years have attached to [his] head'.[41]

The popular enthusiasm greeting the inauguration of Eden's premiership was not universally shared within government. Many ministers viewed serving under Eden with some apprehension. They had seen at first hand the effect which Churchill's agonizingly postponed retirement had had on his chosen successor's brittle temperament. That said, the new Prime Minister's public 'honeymoon' was of short duration. Delays over a ministerial reshuffle and a worsening economic climate led to press murmurings and backbench discontent. With the *Daily Telegraph* calling for 'the smack of firm government', the public mood had perceptibly changed six months after Eden's election triumph. By early 1956 the proportion of voters approving of his performance had fallen from 70 to 40 per cent. This change, of course, took place some time before the Suez Crisis burst on the scene. As John Major would find three decades later, it proved no easy task to follow a dominant and dominating figure into 10 Downing Street. For everyone who believed Churchill had outstayed his welcome and looked to Eden for a new beginning from a representative of a younger generation, others prepared to judge his every move against the impossibly high standards they attributed to Churchill.

*

9

The battle over Eden's historical reputation continued into his retirement, indeed it intensified to the extent that more and more of his career came within the compass of genuine historical research and debate. The defence case was led, inevitably, by Eden himself. Though physical weakness now precluded anything more than occasional excursions into active politics, Eden's literary output was impressive. By 1965 he had produced three volumes of memoirs totalling nearly 1800 pages. Inspired partly by the financial possibilities offered by his account of Suez and partly by his determination – 'in case he falls down dead'[42] – to justify on paper the most controversial episode of his public life, Eden's first volume dealt with the last phase of his political career. In doing so, while giving a detailed account of the inner workings of government so soon after the events described, Eden broke new ground in the conventions surrounding political memoirs.[43] *Full Circle* was published in 1960. Its very title showed how the author had become obsessed with the unity of his career. Far from being an aberration, Suez only made sense in the light of the lessons learnt from the earlier phases of his political life. The book's theme was 'the lessons of the 'thirties and their application to the 'fifties'.[44]

As a *pièce justificative*, however, *Full Circle* was less than satisfactory, the least impressive of Eden's three volumes. Perhaps Eden had engaged in a losing battle. His determination to conceal vital elements in the Suez story had been undermined by revelations from other sources of the murkier events of 1956. In acting as if the central charge of collusion – that Britain had had foreknowledge of and given encouragement to the Israeli invasion of Egypt – did not even need to be confronted, Eden was giving unnecessary hostages to fortune. If he genuinely believed that this secret could be carried with him to the grave, Eden was guilty of a serious misjudgement, perhaps a supreme example of wishful thinking.[45] The Americans knew the essentials of the collusion story by early November 1956.[46] Friends and former colleagues warned Eden of his mistake in ignoring the whole question. After reading *Full Circle* in draft, Lord Chandos wrote:

> I don't think there is (no doubt on purpose) any reference to the charge of collusion with the French or the Israelis. This is a matter upon which I can form no judgement, other than that complete silence may tend to provoke more controversy than if you were able to put in something which headed them off.[47]

Eden's private reply was as evasive as his published account: 'As regards the other charge against us, after much reflection I thought a detached account of events without protestation was the best way to handle the business.'[48]

Eden no doubt judged that his case would be the more persuasive if he presented the facts – or his selection of them – as dispassionately as possible. The result, however, was 'too much like a series of F[oreign] O[ffice] memoranda'.[49] His tendency was to gloss over the documents which formed the basis of his account, making a better case for himself than an objective analysis could sustain. The book is at once both curiously dispassionate and yet intensely personal. Eden makes no attempt to display himself in the round and yet, particularly in dealing with Suez, he seems capable only of seeing events from his own point of view. Selwyn Lloyd, shown a draft of the work in August 1959, noted a strong and persistent 'anti-American bias', but his main concern was for the effect the book would have on Eden himself:

> I think it will damage his own reputation so much. Many of the things said in it seem rather petty and to indicate personal malice and resentment of criticism . . . I feel that to publish the book in anything like its present form is a mistake from his personal point of view.[50]

The remaining volumes of Eden's trilogy were more impressive works. They put, as do all political memoirs, the best case from the author's point of view. Eden showed himself to have been wise about appeasing fascist dictators before the war and equally far-sighted in warning of the threat Russian dictators would pose after it. But Eden never had to face the sort of problem of omission which had confronted him over recording the collusion of 1956.[51] It was even possible now to discern elements of self-criticism. With hindsight Eden was prepared to concede that his handling of the 1936 Rhineland crisis might have been firmer: 'I think I should also admit that I should have been more responsive to what the French appeared to want to do and stiffer to Hitler. It is also rather healthy . . . to do a little penance in four hundred pages.'[52] *Facing the Dictators* and *The Reckoning* were recognized as significant contributions to the historiography of recent times, especially as they included copious quotations from Eden's own papers and previously unpublished passages from the Foreign Office archives. Though not often enthused by Eden's prose, critics appreciated the historical value of his writing. The author himself appeared increasingly obsessed by the unified consistency of his career. In part he had become convinced by the legends woven around him in earlier years. 'I confess', he wrote to one reviewer of *Facing the Dictators*, 'I am at times depressed by men's capacity to make the same mistake over and over again. The dangers of appeasement seem the hardest to learn.'[53]

Eden could not, of course, enjoy a personal monopoly of the historiography relating to his own career. During his last twenty years he kept a watchful eye on the writings of others. It was a field occupied by an assorted combination of journalists, professional historians and political contemporaries. Though he occasionally gave unattributable interviews, most enquirers were not helped, especially over the question of copyright in Eden's own writings. Not surprisingly, he opposed the introduction of the so-called Thirty-Year Rule, though he insisted that his primary concern was to protect the reputations of civil servants rather than ministers.[54] Among books which caused him concern were works by Harold Macmillan, Robert Boothby, Lord Kilmuir, Iain Macleod, Anthony Nutting, Randolph Churchill, Eugen Spier, Lord Moran, Nigel Nicolson, Hugh Thomas, A. J. P. Taylor, Harman Grisewood, Nicholas Bethell, F. S. Northedge and Maurice Cowling.[55] While avoiding the limelight of controversy himself, Eden frequently sought legal advice, and relied on friends and former colleagues to enter the arena of debate on his behalf. Others might be less discreet. When Nutting was on the verge of publishing his account of Suez, Eden heard that Macmillan had spent

> no less than four hours trying to persuade him not to be too, what he considered unfair . . . [He] implored Nutting not to publish what would make people 'spit at me, an old man in the street'. It seems hardly credible, if one did not know to what histrionics Harold can descend.[56]

Eden preferred a less direct approach. While refusing to review Nutting's book himself, he was not above suggesting 'improvements' to one written by Selwyn Lloyd.[57] When unfortunate impressions were created, as for example by Lord Swinton in relation to the motivation behind his 1938 resignation, Eden urged friends to place letters in the press:

> It had occurred to me that possibly you, who were very close to me at the time and knew all the facts, might care to write a short letter to *The Sunday Times*, saying that my differences with Neville were those

of principles about how to deal with dictators, and had nothing to do with 'amour-propre'.[58]

Similarly, a number of Eden's former colleagues, including Ted Heath, then Leader of the Opposition, expended considerable energies on his behalf trying to trace the origin of the canard that, during the Suez Crisis, he had sought powers to take over the BBC.[59]

As historians began to burrow into the details of his career, Suez inevitably became an open field for attacks on Eden. But, almost as a function of the operation of historical revisionism, few aspects of his public life escaped critical attention. This process was well advanced by the time he died. Eden's reputation as an 'anti-appeaser' was one he was particularly keen to protect. Increasingly, however, historical investigation of the 1930s placed a question mark over the very existence of any substantial opposition to appeasement within the Conservative Party. As Lord Halifax had reflected as early as 1940, 'the truth is that Winston is almost the only person who has an absolutely clean sheet'.[60] A. J. P. Taylor voiced a growing sentiment that 'Eden did not face the dictators; he pulled faces at them'.[61] After Churchill's death in 1965 those who had been close to the great war leader revealed evidence not only that his relationship with Eden had grown increasingly difficult, but also that Churchill had developed serious doubts about Eden's suitability as his ultimate successor. The controversial diaries of Lord Moran, Churchill's physician, were particularly damaging. 'It is a great relief to have charge of the F.O. instead of having to argue with Anthony,' Moran has Churchill saying in 1954 when Eden was ill, 'I can get something done.'[62]

Eden's record of achievement in his final period at the Foreign Office seemed the most solid of his career, but even this did not escape reappraisal. By the mid-1960s, as a consensus developed about the desirability of Britain joining the EEC, attention shifted to Eden's sceptical attitude towards European integration both as Foreign Secretary and as Prime Minister. Eden found a retrospective attack by Robert Boothby in the House of Lords in August 1962 particularly wounding and considered seeking legal redress.[63] When cabinet colleagues underlined Boothby's criticisms in their memoirs, Eden commissioned the young historian, Ben Pimlott, to draft an article, based on his own papers, to show that he had merely implemented agreed government policy and that the Euro-enthusiasm of some fellow ministers was essentially a later invention.[64] Finally, in the last two years of his life, Eden and his friends were engaged in a new battle as authors and television producers sought to uncover his role in the repatriation of Soviet prisoners at the end of the Second World War. Charges of callous disregard for the fate of these prisoners, made thirty years after the event and far removed from the very different atmosphere in which difficult wartime decisions had had to be taken, were difficult to refute. In any case Eden was no longer strong enough to take any vigorous part in the debate, though his former parliamentary private secretary (PPS), Robert Allan, tried to use his position as a BBC Governor to stop a television documentary on the subject.[65]

Yet in many of the books to which Eden took exception references to himself were incidental. His readiness to let them disturb his tranquil retirement is difficult to comprehend. At Eden's request several minor deletions were made and qualifying footnotes added to Nigel Nicolson's edition of his father's diaries.[66] The publication of Iain Macleod's biography of Neville Chamberlain – 'a 100 per cent defence of him, Munich, appeasement etc.' – was a matter of particular concern.[67] Eden argued that a far more lenient official attitude had been adopted towards Macleod's use of Chamberlain's diary, which referred to cabinet discussions at the time of Eden's resignation, than Eden had enjoyed in the case of *Full Circle* or *Facing the Dictators*, which he was then preparing for publication. He had, he complained to the cabinet secretary, been shabbily treated and this had caused him much hurt.[68] Macleod's book determined Eden to revise his own account 'which I shall base on the minutes and pull no punches'. He took comfort that, unless the whole Conservative Party was now 'appeasement minded, which it may be', Macleod's work would 'do the Government much harm'.[69]

In the case of Macleod, Eden felt that the author's contemporary status as a member of Harold Macmillan's government had won him preferential treatment from the Cabinet Office. Professional historians, however, fared little better. Maurice Cowling, who merely sought authority to quote from Eden's words, was a figure to 'be watched with some care'.[70] F. S. Northedge, whose general survey of inter-war foreign policy, *The Troubled Giant*, might strike most observers as unexceptionable, provoked Eden to considerable anger and a flurry of letters to such former colleagues as Lord Salisbury, Oliver Harvey and William Strang. Northedge was likely to 'start a hare the spoor of which will muddy several reputations, including mine, most unjustly'. Eden would 'gladly sue him if [he] could'.[71]

Indeed, in relation to both Cowling and Northedge, Eden sought legal advice, though in practice no grounds for action existed.

By contrast with the books to which he objected, few works of contemporary history received Eden's wholehearted endorsement. Those in this category relating to Suez were rare indeed. Only Herman Finer's *Dulles over Suez* – a 'brilliant and penetrating document' – met with approval. The reason was clear. Finer's work shifted the spotlight from Eden's own conduct and offered a sustained indictment of that of the American Secretary of State. Before he had read the whole book, Eden was freely recommending it to his friends. The first 200 pages were 'terrific'. He hoped that it could be reviewed in *The Times* by 'somebody who knew the course of these endless anxious days'.[72] Yet, even here, Eden's enthusiasm waned once he realized that his own conduct was more harshly judged in the second half of the book.

If Eden's memoir account had failed to carry all before it, the task of sustaining his reputation would be carried on, after his death, by the writing of his official biography, for which purpose his private papers were carefully guarded. Eden had ample opportunity during his twenty-year retirement to prepare his papers for their later use. His private archive is vast and Eden spent much time augmenting it with documents secured from the patient and co-operative staff of the Cabinet and Foreign Offices. The presumed needs of his chosen biographer were clearly uppermost in Eden's mind. In addition to numerous 'notes for my biographer' which he continued to dictate until shortly before his death, offering new or renewed justifications for and explanations of past actions, many documents – often of an apparently trivial nature – were highlighted for the biographer's attention. If Eden was not intending actually to write his own biography, he was none the less determined to influence the way it was composed.

For many years the task of producing the definitive account of Eden's career was entrusted to his personal friend, the distinguished historian, Sir John Wheeler-Bennett. Though Wheeler-Bennett did not live to carry out his mission, there are indications from his other writings that he would have produced a work to meet with Eden's approval. Here, the overall impression is of an 'effective, indefatigable, patient [and] brilliant negotiator'.[73] Wheeler-Bennett's Eden was one who, at the announcement of Chamberlain's visit to Hitler in September 1938, left the Commons Chamber 'pale with shame and anger'.[74] It is also clear that Wheeler-Bennett used privileged access to Eden's papers to defend him against charges that he had thwarted cabinet colleagues in their efforts to draw Britain closer to Europe in the early 1950s.[75] Wheeler-Bennett had known Eden for a number of years and had worked in the Foreign Office during the war. He was of that generation which looked upon the Eden of the 1930s as 'the Golden-Boy, slim handsome charming. . . . Though he did not succeed, we knew it was no fault of his, and he retained our confidence.' Wheeler-Bennett even attributed his own determination to chronicle the inter-war years to Eden's 'courageous struggle for sanity in diplomacy'. His career as a historian was 'born of a desire to place on record the story of what may be called the Eden ethos'.[76]

In choosing a biographer only five years his junior and whose health was not robust, Eden knew that his plans might well need to be changed. After Wheeler-Bennett's death in 1975 the mantle of responsibility for completing Eden's official biography fell on A. J. Nicholls of St Antony's College, Oxford, who had originally intended to work on the project as Wheeler-Bennett's assistant, as in the case of the same author's *The Semblance of Peace*. Nicholls's enthusiasm quickly waned, however, and with his own political sympathies moving increasingly to the left he withdrew from the task in April 1976. By the autumn the baton had passed to Martin Gilbert whose credentials and ongoing performance as Churchill's authorized biographer suggested that Eden's career would ultimately receive its due deserts. But Gilbert's commitment to the vast Churchill project precluded any chance that he would be able to turn his attention to Eden in the foreseeable future. Eventually it was the versatile Robert Rhodes James who brought the work to fruition, though his commitments as an MP delayed the appearance of the biography until 1986, nearly ten years after Eden's death.

Rhodes James, whose earlier writings had apparently accepted the mounting consensus about Eden's shortcomings, took up the cause of his subject's rehabilitation with gusto, refighting – and in some cases winning – Eden's old battles. The book has a clarity and elegance of which Eden himself would have approved. Few people could read an often moving account without feeling enhanced sympathy for and understanding of the man portrayed. Indeed, it was in the drawing of Eden the man that James was most effective. Eden emerges – as he had appeared in his early days in public life – as a sensitive, sympathetic figure of high integrity, honourable almost to the point of folly.[77] But in transferring the argument on to an essentially human plane James often failed to confront existing verdicts which it was his task to challenge. Eden's biographer was too good a historian to attempt a complete whitewash. But, as in Eden's own memoirs – though with a lighter touch and considerably more skill – events are portrayed through Eden's eyes, while his opponents are lightly and sometimes unfairly dismissed. The Nasser presented by James is essentially the megalomaniac who stalked Eden's mind for the last two decades of his life rather than an Arab nationalist boasting a considerable record of achievement within Egypt. Thus, while Eden's conduct and predicament over Suez are portrayed with understanding and sympathy, most of the charges built up against him since 1956 look as damning as before. Compassion cannot fully compensate for a neglect of critical analysis.

In part James was the prisoner of the genre of official biography. He who is 'authorized' is also, consciously or otherwise, constrained. The author based his account largely on Eden's own archive and generally also accepted the emphasis which Eden's careful arrangement of his papers had been designed to create. James derived considerable advantage from the permission he received to quote cabinet papers relating to 1956, a year before these were officially released. He thus planted a first marker – on Eden's behalf – in a renewed round of historical debate. None the less, in other areas he seemed wilfully to turn his back on readily available evidence for no other reason than that it threatened to tell him things he wished not to know. James's decision to ignore the detailed diary of Evelyn Shuckburgh, Eden's private secretary from 1951 to 1954, is the most obvious example of his selectivity.[78]

But perhaps the least convincing part of the book was the final chapter entitled 'Victory, 1957–77'. 'There is nothing in Anthony Eden's life', James wrote, 'more remarkable and admirable than the manner in which he rebuilt his shattered reputation and health after this year of catastrophe.'[79] In his retirement Eden certainly came to enjoy a happiness and tranquillity largely denied him hitherto. The devoted care of his wife, Clarissa, helped ensure that he survived, in reasonable health, for longer than might have been expected early in 1957. With patent sincerity Eden recorded his private thoughts in the late summer of 1966: 'Reflected on how happy we should be in retirement with Alvediston and Villa Nova, if only we could be let alone! I would only ask five years of it, if gods would allow.'[80] Yet the suggestion that these same years saw a rehabilitation of Eden's standing, particularly in relation to Suez, flies in the face of much of the evidence presented above. Though some of the passion aroused in 1956 died down and the experience of the Six Day War of 1967 caused many to re-evaluate the Suez episode, this did not fundamentally affect the general thrust of historiographical opinion.

Little over a year before his death, Eden made a further revealing diary entry. 'Life is a comedy to those who think, a tragedy to those who feel.'[81] Eden unquestionably figured among those who feel and there are certainly elements of classical tragedy about his career. There seems little doubt that he died feeling that he had been, and would continue to be, judged harshly by his fellow men. 'He was consumed deep down', wrote one who saw him shortly before his death, 'by the sorrowful knowledge that history would not treat him with kindness.'[82] Seen in this context the sustained criticism which characterized David Carlton's brilliantly destructive biography, published four years after Eden's death, is less surprising than it might at first appear, even though the book understandably caused pain to Eden's surviving family and friends. Any fresh attempt to examine the career of this exceptionally complex man must tread a wary path. Knowledge of the tragic finale of 1956 provides a strong temptation to interpret the earlier episodes of his political life with one eye firmly on the last. Much that has been written of Eden as Foreign Secretary has sought to find the fatal flaw which will somehow make sense of the events of his premiership. In fact Eden's career before he reached 10 Downing Street is of sufficient intrinsic importance to merit analysis in its own right. Yet to divorce consideration of one part of Eden's career from another would be equally misleading. Not only did Suez do catastrophic damage to his overall standing at the bar of history. He himself was forever conscious of the essential unity of his career, even as he lived through it. Increasingly the prisoner of his past, Eden felt compelled to live up to his own reputation.

Yet if he was fearful of history's ultimate verdict, his anxiety was perhaps misplaced. History, almost by definition, will never reach a definitive conclusion on Eden's role in public affairs and the debate about his standing in British and international politics will continue for a long time yet. What follows is an attempt to reassess the main themes of Eden's career, using above all his own private papers and in the light of the vast amount of available scholarly literature relating to his political life. The focus is on the period 1931 to 1957 when he was at or near the centre of British politics.

Eden's early life has been well documented elsewhere.[83] Born in 1897 into the minor aristocracy, he grew up under the shadow of a domineering and eccentric father. As a boy 'he was the quietest, most studious of them all, preferring his books to his father's horses', and, with a skill which greatly assisted his early career, 'showed a real genius for expressing the right sentiment at the right time'.[84] After Eton, Eden fought with distinction on the Western Front, winning the Military Cross. He then secured a First in Oriental Studies at Oxford and entered parliament in 1923 for the safe seat of Warwick and Leamington. He rose rapidly and as PPS to the Foreign Secretary, Sir Austen Chamberlain, from 1926 to 1929 began that close association with foreign affairs which characterized his entire public career. He also won the patronage of Stanley Baldwin, whose brand of consensual Conservatism he much admired. Indeed, Eden remained loyal to Baldwin when the latter's fortunes were at a low ebb in the years of Labour government after 1929, and when a change in the Conservative leadership seemed a distinct possibility.[85]

And it was to Baldwin and Chamberlain that Eden owed his appointment as Parliamentary Under-Secretary at the Foreign Office at the formation of the National Government in August 1931, a decisive step which saw him overtake some of the more fancied Conservatives of his generation on the ladder of ministerial advancement:

> [Chamberlain] told me there was a chance I might go to F.O. That he had spoken strongly to Reading [the new Foreign Secretary] and that S.B. had agreed to his doing so. He hoped something would result but S.B. had given away so much to the Liberals it was impossible to say. He – S.B. – apparently greeted my name with more enthusiasm than any other. The F.O. in a national govt. with the S of S in the Upper House is higher than I hoped for and I do not expect that I shall get it.[86]

But get it he did. At just thirty-four years of age Eden was ready to make his mark.

2

'One strong young figure'

> Mr Churchill more than once remarked to me that, though he himself had not thought so at the time, he realized that he had been fortunate to be out of the Government for so long in the thirties. He explained that when one is a member of an administration one has inevitably to yield in opinion to some extent for the sake of the team.[1]

D URING an idle moment early in the Second World War, members of the Foreign Office Political Intelligence Department 'discussed the politicians who were criminally responsible for war and should be hanged on lamp-posts of Downing Street'. There were not many candidates. Most agreed that 'sinner No. 1' was Sir John Simon, though supplementary votes went to Stanley Baldwin, Samuel Hoare, 'Labour lunatics', Lord Beaverbrook, Geoffrey Dawson and, of course, Neville Chamberlain.[2] Of those indicted Simon and Hoare were successively Foreign Secretary during the crucial years 1931–5. Throughout this period Anthony Eden was the second-ranking minister in the Foreign Office. Far from being tainted by association, however, Eden managed greatly to enhance his political standing. In 1931 he was virtually unknown outside a narrow political circle. Just over four years later Eden was himself Foreign Secretary at the early age of thirty-eight, widely seen as the coming man of British politics and enjoying a place in popular esteem unmatched by any of his cabinet colleagues. The public 'saw in him the embodiment of the highest principles in a world of depraved and cynical power politics'.[3] This period, then, was vital to the development of both Eden's career and his long-term historical reputation. Though historians have now largely abandoned earlier crude distinctions between 'appeasers' and 'anti-appeasers', it still seems unlikely that Eden's later fortunes would have prospered in the way they did had he not distanced himself from the doings of his ministerial colleagues during his first years in office. To be with 'the appeasers', but not of them, proved to be a not insignificant achievement.

Soon after the formation of the National Government, and Eden's appointment as Parliamentary Under-Secretary at the Foreign Office, the world was confronted by the first of what came to appear a series of major challenges to the international order, which culminated in the outbreak of a second great war eight years later. In mid-September the Japanese army in the Chinese province of Manchuria seized the city of Mukden, initiating a diplomatic crisis which rumbled on until 1933, but which was later imbued with far greater significance than anyone had seen at the time. With hindsight it became easy to link the opportunity lost in 'the wastes of Manchuria' with the ignominy incurred on 'the beaches of Dunkirk'.[4] If Japanese aggression had been stopped by a united League of Nations, asserted one later critic, 'it could have had no successors'.[5] Indeed, it was the challenge to the League's authority which makes the Manchurian episode important in any chronicle of Eden's career, since it was above all as a doughty champion of that institution that he now proceeded to forge his reputation. Looking back, it was easy to blame the Foreign Secretary, Simon, for all that had gone wrong. Lord Robert Cecil, who spent half his long life championing the cause of internationalism in general and the League in particular, pronounced that the members of the government were resolute about only one thing in foreign affairs, and 'that was to do nothing'.[6] Eden could be safely excluded from this criticism. The 'Manchurian precedent', he later wrote, 'had fateful implications'. But he had had 'no direct part' in the unfolding of those events by which Japan drove her 'war-planes and weapons through the Covenant', except for occasionally representing the government on the League's Committee of Twelve.[7]

Even so, Cecil might have been surprised by the tone of contemporary entries in Eden's diary. Cecil was 'too much an intolerant, and not over scrupulous, League purist' for Eden's liking, his 'League fanaticism' a 'menace to British Imperial interests'.[8] But such surprise would have been mitigated by an examination of Eden's contributions to parliamentary debates on the League in the 1920s. From his earliest days he had adopted a realistic rather than idealistic attitude towards its potentiality. To be called upon to give heaven-sent judgements or to formulate impeccable decisions, he had declared in 1926, was to ask too much of it. The League's greatest benefit should be the more humble one of creating an opportunity for international statesmen to meet and exchange opinions. Far more harm had been done to the League by people with their heads in the clouds and their brains in their slippers than by its most inveterate enemies.[9] Eighteen months later Eden gave an even clearer picture of his views:

> I think we should refuse to . . . tie our hands in directions where we cannot see what the future may hold for us. I do not believe we can maintain peace . . . by handing out a hotchpotch of pledges . . . I believe that the policy of 1925 [i.e. Locarno] is in the long run better and wiser than the policy of 1924 [i.e. the Geneva Protocol].[10]

Such words were a timely reminder that Eden's first mentor in politics was Austen Chamberlain and not Cecil. Like Chamberlain Eden knew that the burden of implementing the high-minded idealism of League enthusiasts would always fall inequitably upon the shoulders of Britain. As Eden put it:

> The responsibility of intervention upon any nation called upon to fulfil it would fall in the first instance most severely upon this country . . . I do not think anyone can seriously suggest . . . that in these conditions the Protocol does not mean new and far-reaching commitments for this country.[11]

The whole purpose of Chamberlain's diplomacy in the later 1920s had been to define the extent and limits of Britain's commitment to and involvement in European affairs and there is no evidence that his young parliamentary secretary dissented from this objective. Even when Chamberlain and the Conservatives lost power, Eden was keen to stress that the League Covenant was satisfactory as it was and did not need the ministrations of well-meaning zealots. 'Attempts which are made from time to time to block up holes, imaginary or real, and to stiffen up the machinery, are not likely to have very happy results.'[12]

In the circumstances it is scarcely surprising that Eden – like the government as a whole – reacted cautiously to news of the Manchurian crisis, occurring as it did in a remote part of the globe with no obvious British interests immediately involved. Such inclinations were undoubtedly confirmed by the coincident crisis of the national economy and the parlous state of Britain's defence forces after more than a decade of the Ten Year Rule – which had founded defence spending on the assumption that the country would not have to face a major war for at least ten years. But Eden's statement that his own involvement in the crisis was at the margin seems substantially correct. The evidence suggests that his own views were in line with the ideas he had enunciated a few years earlier and thus with the thinking of the leading members of the government. This meant accepting that Britain was in no position to risk involvement in a major confrontation on the other side of the world, particularly with a country like Japan, which Britain wished to remain a friendly power.[13] Even economic sanctions against Japan were likely to hurt Britain at a time when the national economy needed no further blows. 'We with vital interests at stake', he insisted, 'must continue to moderate the zeal of those with none.'[14] When Sir Arthur Salter called for the economic coercion of Japan, Eden judged his memorandum 'academically valuable but practically useless'. 'Collective offensive action', he added, 'which sanctions imply – even if desirable, is, we know, unobtainable.'[15]

From the standpoint of his later reputation, it was fortunate that Eden's views were confined to the corridors of power rather than displayed in the public arena.[16] Criticism of government policy – particularly in retrospect – fell largely, if unfairly, on the Foreign Secretary. It was Simon's thankless task to front British policy as, for example, with his celebrated speech to the League Assembly in December 1932. Here he sought to prevent the League being carried away on an anti-Japanese bandwagon – whose enforcement would fall largely on Britain's faltering shoulders – and upon the basis of a League report whose findings were more evenly balanced between Japan and China than the majority of delegates wished to concede.[17] The fact was that Britain had few options but to avoid a Far Eastern war for which she was wholly unprepared. But it was easy to argue that Simon had betrayed an institution which, as a Liberal, he should have upheld. According to one critic he did 'more than any man since 1918 to destroy the prestige of the League by his scarcely concealed support of Japan in her rake's progress of imperialistic crime'.[18] Through no effort of his own, Eden benefited from the simple fact of not being Simon.

In one sense Simon's appointment as Foreign Secretary had been something of a setback to Eden's career aspirations. Between the formation of the National Government in August 1931 and the election in October the post had been held by Lord Reading whose peerage enabled Eden to act as the ministry's spokesman in the Commons. Even so, there is no evidence that Eden initially found Simon's appointment unwelcome in personal terms. The new Foreign

Secretary was widely regarded as among the most intellectually distinguished of contemporary politicians. By the time, however, that he wrote his memoirs, Eden could confirm the, by then, widely accepted perception of a considerable gulf between himself and Simon. *Facing the Dictators*, published in 1962, was a far more adept and successful exercise in personal propaganda than *Full Circle*. At the same time Eden played a not inconsiderable part in confirming Simon's already unenviable reputation. As Prime Minister, Ramsay MacDonald had 'made a mistake in appointing Simon to a position for which he was miscast by temperament and training'.[19] Having read the early chapters of the book in typescript, Oliver Harvey judged that 'Simon shows up in his worst colours and I'm afraid HMG will stand condemned by history because of his lamentable tergiversations and retreats. . . . Simon did more than anyone to devalue Locarno.'[20] Reviewing the book with greater detachment than Harvey could probably muster, Roy Jenkins confirmed that Eden emerged from it as 'the best minister of a dismal decade'.[21] 'From the start of the Disarmament Conference through the Abyssinian conflict and the Spanish Civil War to the beginning of appeasement under Chamberlain he was almost consistently right.'[22] Yet it is interesting that almost two decades before the publication of *Facing the Dictators* Eden had begun to contemplate a book of reminiscences about his early career, commissioning the young Cambridge historian F. H. Hinsley to produce a draft. The result was not what Eden had in mind. 'General impression of book', noted his literary secretary, 'is that A. E. supported [Simon] . . . to a degree greater than I thought true.'[23] This work was never published.

Ironically, it was Simon's immersion in Far Eastern affairs which gave Eden his first real chance to develop a national, and indeed international, reputation, and which began to distance the junior minister from his master in the eyes of contemporary observers and the majority of subsequent opinion. As early as 1924 the League Assembly had determined to hold an international disarmament conference. That conference finally convened in Geneva in February 1932. Just before it opened, Simon suggested to MacDonald that his own lengthy absence from London might be inadvisable and proposed using the services of the former Foreign Secretary, Austen Chamberlain, who had retired from front-bench politics after the 1931 general election, as a permanent British delegate to the conference. Nothing came of this suggestion. Over the following months, therefore, Simon had to conduct an exercise in shuttle diplomacy, taking overall responsibility for both the Disarmament Conference and the ongoing Far Eastern crisis. He found the combined burden 'a grind'[24] – a situation scarcely improved by his deep scepticism concerning the likely outcome of the disarmament negotiations and his marked distaste for Geneva itself. It was, he told Baldwin, a dreadful place, 'well worthy of Calvin'.[25]

As 1932 came to a close and with little progress in either of his major foreign policy preoccupations, Simon again raised the issue of his workload. His preference now, he told MacDonald, was for a younger man to join the cabinet as Assistant Foreign Secretary with standing duties at Geneva. He would be 'perfectly ready to see Anthony Eden in that position'.[26] This time Simon's promptings found a more sympathetic ear. In the course of 1933 Eden, though not securing the suggested cabinet rank, was allowed to take over more and more of Simon's responsibilities at Geneva.

Few episodes in the diplomatic history of the 1930s proved so barren of achievement as the World Disarmament Conference. The detailed account of its arcane debates has largely been ignored by recent historians.[27] Yet it was in this unpromising environment that Eden's reputation and political prospects began to flourish. As Robert Vansittart, the Permanent Head of the Foreign Office, later put it, 'it was all dreary, all ingenious, its vain details deserve no chronicle, but helped Eden to make a name'.[28] Here he 'won his spurs'.[29] Two factors are important. In the first instance the conference gave Eden his first opportunity to display those negotiating skills which were to become a mainstay of his career as a whole. Eden worked hard. He knew his brief. But his personal demeanour was perhaps more important. The man who was capable in private of violent displays of temper was, in the conference chamber, the embodiment of patience and finesse. 'I admired Mr. Eden very much during these difficult discussions', recorded one who worked closely with him at this time. 'Although the rough and tumble methods of Herr Nadolny [the German representative] must have been extremely distasteful to him, he never lost his head or his temper.'[30] Whereas Simon tended to antagonize and alienate through his – usually unintended – displays of intellectual arrogance, Eden's skill was to conciliate and appease. Spain's representative spoke for many:

After the Madame Tussaud wax works which was Austen Chamberlain, after the insular, enigmatic and arrogant Simon, there was at last a

man alive who spoke to all comers as man to man and who needed by
no means to smile in order to be open and friendly. . . . Clear, definite,
efficient, he was always moderate, not merely 'willing to listen' but
an excellent and intelligent listener of what was said, no matter who
said it.[31]

In the second place Eden managed to convince all at Geneva – and
the majority of those watching in Britain – that he was more sincerely committed to disarmament than were Simon and the government as a whole. His age was a clear advantage. 'I unhesitatingly gave
my support to Eden,' recalled Harold Macmillan, 'who seemed to
embody all the aspirations of the war generation.'[32] His idealism and
commitment to peace were seen as the products of his war experience. A man who had fought and had two brothers killed in the last
war had every reason to strive to avoid another world conflict. 'How
much might have been saved in the future', judged the military
adviser to the British delegation, 'if these two young men [Edouard
Daladier and Pierre Cot] and Mr. Eden, all of the war generation, had
been left to work out a settlement with Germany before it was too
late.'[33] Notwithstanding his pronouncements in the 1920s and the
line he had taken over Manchuria, Eden thus became associated in
the popular mind with a more wholehearted commitment to the
internationalism of the League than was attributed to the government of which he was a minister. As the League still occupied a high
place in public esteem, Eden was increasingly seen as the only
leading politician in tune with popular sentiment on this issue.[34]

Eden had been quietly critical of Simon's performance from the
very beginning of the Disarmament Conference. He was convinced
that Simon had no real expectation that anything tangible would
emerge from the discussions. As early as the end of January 1932
Eden recorded that the Foreign Secretary was 'anxious not to go to
Geneva. I think really because he has little hope of disarmament
conference.'[35] But it was above all Simon's methods which exasperated Eden. 'The handling of this country's disarmament policy', he
noted in July, 'has been weak and timorous from the first.'[36]
Notwithstanding Eden's mounting irritation, it is not at this stage
easy to isolate significant policy differences between the two men.
Indeed, Simon was responsible for launching one of the conference's
more promising initiatives, for which he won Eden's guarded praise.
This was the suggestion to consider 'qualitative' disarmament, that is
the limitation of armaments, not by numbers, but by the complete

elimination of certain types of weapons, particularly those which lent themselves to 'offensive' warfare. 'We can hope for little progress at this conference', Eden told Baldwin in May 1932, 'except by means of qualitative disarmament.' He continued: '[Simon] took the lead in originally putting forward the proposal, and . . . there is general appreciation here both for the manner and the matter of the speech in which he did so.'[37]

Eden's increasing exasperation with Simon's conduct may be traced in his diary entries for 1932. But there are few signs of the younger man trying to suggest alternative lines of policy. Moreover, relations between them remained, to all outward appearances, good. In November Eden made a point of telling Simon of the warm reception given by the Commons to MacDonald's clear statement of support for his work as Foreign Secretary.[38] Almost certainly Eden's diary – together with its later use in his memoirs – gives a distorted picture of the true relationship between Simon and his deputy. As Eden himself once wisely wrote, albeit in a different context, 'a diary that is regularly kept . . . is apt to be but a peevish record of daily irritations'.[39]

It was Simon's evident lack of enthusiasm for his task, his inherent caution – which Eden interpreted as procrastination – and, as Eden saw it, his inability to stand up for Foreign Office policy inside the cabinet which irritated the younger man. The impatience of youth was a factor here. But Eden – not himself a member – also failed to appreciate the delicate balance of power within an all-party cabinet where senior figures such as MacDonald and Neville Chamberlain sought to promote their own powerful views on foreign policy, and where the service ministers were also bound to defend their corner. 'It is not very easy', noted Simon with feeling, 'for a For. Sec. to satisfy a P.M. whose forte is foreign affairs.'[40] Not that Eden's complaints lacked all substance. Simon's constitutional inability to reach a quick decision had been noted throughout his career. As Lloyd George once memorably remarked, 'Simon has sat on the fence so long that the iron has entered into his soul.'[41] His complex intellect could always see all sides to a question, but this very attribute denied Simon the certitude needed to make up his mind. His tendency towards equivocation was exacerbated by the insecurity of his governmental position. He was a Liberal National within an administration dominated, notwithstanding the person of the Prime Minister, by the Conservatives. At all events Simon never seemed to carry the weight which his high office merited. As Eden recorded:

> He will not fight for his own policy. He expects the Cabinet to find his policy for him. That they will never do. They want to be told. The only result of present procedure is F.O. pushed into the background, which is not good either. . . . Poor Simon is no fighter. Nothing will make him one.[42]

Eden was inevitably annoyed that proposals upon which he had himself worked hard inside the Foreign Office were not presented with sufficient force by Simon at the cabinet table:

> He has never fought for his own hand . . . The policy is as good as can be expected in the circumstances and it now remains for Simon to go for it. Anyway the ink wells at the F.O. are dry and if the Cabinet will not have it Simon should ask them to send someone else to Geneva.[43]

It was not possible to do more than 'give a man material to fight with: no one else can give him gutts [sic]'.[44]

Eden certainly brought enthusiasm to the British involvement in the Disarmament Conference which had been patently absent hitherto. Alexander Cadogan, the senior Foreign Office official dealing with the conference, voiced his approval: 'He seems to me to have a very good idea of what is right and what is wrong, and if he thinks a thing is right he goes all out for it, hard, and if he thinks a thing is wrong, ten million wild horses won't make him do it.'[45] Very quickly Eden's star began to rise. Next to Arthur Henderson, noted the *Manchester Guardian*, he was Britain's best representative in Geneva for many years.[46] His performance made him 'the blue-eyed boy of the National Govt'.[47] 'I am proud of my protégé,' purred Austen Chamberlain.[48] Less inclined than Simon to dance to the tune of the service ministries,[49] Eden was clearly keen for Britain to take the lead in the disarmament discussions. In February he proposed that the 'only way' to make real progress was to produce 'a convention complete in all its details and lay it on the table and ask that the Conference should accept it'.[50] Early the following month he warned that the conference was doomed unless definite proposals were put forward. The only country in a position to do this was Britain. A breakdown would merely provide Germany with the excuse to rearm. Once again Simon proved a less than effective advocate:

The S of S began by giving the substance of the sum of that last minute which you wrote him. Either he didn't put it very well, or the understanding of his colleagues was not very good, as they seemed to have the impression that you were anxious to fling the convention on the table at once.[51]

In the event it was agreed that MacDonald and Simon should both travel to Geneva. Whether they would use Eden's draft convention remained unclear. But the latter succeeded in concealing his disappointment: 'My telegram today will tell you how opinion here has reacted at the news of your impending visit. . . . The effect has been excellent, the Conference sadly needing such a tonic, and I have heard no criticism anywhere.'[52] Eden finally persuaded MacDonald that the conference 'must have some meat'. The Prime Minister then presented the convention on 16 March in a speech which 'was criticized for rambling and ranting, but which did the job'.[53]

Eden's commitment to disarmament certainly struck a chord with the popular mood, but it is pertinent also to place his views within the spectrum of political opinion as a whole. At this time Eden was very much at odds with some of those with whom his later reputation would be linked. This became apparent when he faced the task of answering Winston Churchill in a parliamentary debate on 23 March. The latter had attacked the whole premise upon which the disarmament negotiations were based. 'As long as France is strong and Germany is but inadequately armed', insisted Churchill, 'there is no chance of France being attacked with success, and therefore no obligation will arise under Locarno for us to go to the aid of France.' But to press France to disarm would merely encourage Germany, whose internal politics were giving increasing cause for concern. Eden responded that without French disarmament it would be impossible to secure the necessary period of European appeasement. The reaction of the House suggested that most members were with the minister rather than his distinguished backbench colleague.[54] In the wake of Eden's performance Neville Chamberlain commented approvingly: 'That young man is coming along rapidly; not only can he make a good speech but he has a good head and what advice he gives is listened to by the Cabinet.'[55] Indeed Churchill's views were probably closer to those of Simon than Eden. In cabinet in May the Foreign Secretary described the international situation as 'definitely disquieting in relation to the increasingly militant attitude adopted by Germany'.[56]

Despite his outward composure even Eden became exasperated by the lack of progress at Geneva where the Germans were 'proving extremely unhelpful'. It was, he said, 'rather like a 1917 campaign in Flanders: we can only make such progress as we may in the mud between the pill-boxes, and leave the strong points to be attacked at the last – and, as in Flanders, the pill-boxes are occupied by Germans!'[57] He believed that the British government needed to make further concessions. A hostile critic

> might even thus sum up our Draft Convention in the form in which it is now taking shape: that we maintain our position at sea; that we gain a parity in the air we had not previously enjoyed; . . . that on land we shall possess the most effective striking force in Europe.[58]

But Simon seemed more conscious than Eden of the potential dangers awaiting the unwary. It was important to avoid being manoeuvred into a position where Britain appeared to agree that Germany be allowed a measure of rearmament. Eden should give no such suggestion to Nadolny and ensure that the French were involved in any developments:

> Then France and others will make it plain that they will not consent to rearmament of any sort. What I dread is that we should seem to concede a little bit in this direction only to find that Germany demands more, and the Conference breaks down after we have, by our concession, lost the overwhelming public support which would be given to a stiffer attitude.[59]

The Foreign Secretary was now unwell and in need of a break from his duties. Eden helped try to persuade Simon to go on a two-month cruise to restore his health.[60] 'If matters continued as they were, we would in effect have no Foreign Secretary, only the appearance of one, which would be worse than none.'[61] Simon's performance seemed less impressive than ever, especially to those such as Eden who were expected to carry more and more of the department's workload:

> Simon told me he could not take questions Monday, would I? . . . It eventually transpired that there was a question on bombing that he did not want to answer because he could not express approval of government policy though he has urged me to often enough and has done little enough against it. Not very noble. He added: 'I shall certainly feel ill again by then. Indeed I feel my illness creeping

upon me already. It will certainly be bad on Monday.' Makes one wonder whether the whole thing is not a sham.[62]

Increasingly the perception gained ground of a widening breach between the policies of the two men. In August Harold Laski commented in the *Daily Herald*: 'To play second fiddle to Sir John Simon while the latter is losing the confidence of all reasonable progressives is a difficult essay in loyalty.'[63] When both men were at Geneva together the contrast was stark. One minister warned Baldwin of Simon's mounting unpopularity. It was, he said, painful to have to listen to criticism of the Foreign Secretary not only from foreigners, but also from Dominion delegates and members of Britain's own League staff. Eden, by contrast, was 'the real power here and as long as he is here he gives the French and others some sense of confidence'.[64]

Yet for all his enthusiasm, by the time that Eden assumed chief British responsibility at the Disarmament Conference, the chances of a successful outcome had probably passed. On 30 January 1933 Hitler had become German Chancellor. Thereafter, any agreement reached would not have been worth the paper upon which it was written. This probably intensified Eden's inclination to criticize the earlier management of the negotiations and to postulate the idea of missed opportunities which never really existed. This became apparent when the conference reached a crisis in the autumn. By this stage Eden himself had become far more conscious of the German menace. In late August he listed the 'order of German ambitions' – Austria, the Polish Corridor and her former colonies – suggesting that it would be 'easier to say no at the first than at the last'.[65] Then, on 14 October, Simon presented a plan to the conference secretariat for progressive general disarmament. By lunchtime that day it was clear that the Germans were going to leave the conference and give notice of their intention to quit the League. Despite the coincidence of timing, Germany's decision had been taken earlier and was not simply a response to Simon's initiative. For Eden, however, here was proof of the Foreign Secretary's incompetence. He would not, he said, like to have Simon's conscience about 'the earlier part of last year when Brüning was still in power. We missed the bus then and could never overtake it.'[66] Yet a disarmament agreement in 1932 would have required a more far-reaching commitment to French security than the British were able to give. In later years Eden clearly recognized this fact, describing his contemporary criticism of Simon as 'a sweeping and youthful judgement'.[67] Even his official biographer concludes that it is 'very debatable' whether more vigorous British conduct in 1932 would have averted subsequent failure.[68]

What is striking is the similarity of response by Simon and Eden to Germany's withdrawal from the conference. The Foreign Secretary had already advised the cabinet of the consequences of failure. Time, he warned, was on the side of those in Germany who were preparing to reverse the verdict of the last war:

> If, after all this talk, no Disarmament Convention is reached, Germany will, in law, continue to be bound by . . . Versailles, but I do not see how she is going to be made to observe it. She will claim to be free to rearm without any conditions, and once the new regime in Germany settles down and becomes more respectable, there will be many people who will say that there is a good deal in the German claim that she has waited fourteen years for equality and cannot be expected to wait any longer.[69]

Once the blow had fallen, Simon remained convinced that conciliation was the only course to pursue, even though Germany's latest action had merely created new obstacles. Britain should still try to meet any German claims which were reasonable and just. Only then would it be possible to 'strengthen the peace structure of Europe which has been so shaken by recent events'.[70] While arguing that Hitler was an impediment to any agreement and that the Nazi Party preached a doctrine which regarded preparation for war as 'a noble ideal pervading every aspect of national life', Simon persuaded the cabinet that it remained important to maintain contact with Germany both because her assent was essential to any future disarmament convention and so as to ensure that she did not commit herself to unilateral rearmament or a denunciation of Versailles.[71]

On the same day Eden expressed comparable sentiments in a letter to Lord Irwin. Though he did not anticipate results from renewed contact with Hitler in the immediate future, 'If . . . we go on with our work, Hitler may be sufficiently impressed to be anxious, if not to take part in it officially, to keep in touch – may be ultimately to sign.' Success might depend upon further British concessions, including dropping existing reservations about so-called police bombing. 'If Germany does sign, the price of any concession we make should surely be well repaid.'[72]

It fell to Eden to prepare the ground for further negotiations with Germany. At meetings with the French ambassador, Corbin, on 31

October and 10 November he urged the French to adopt a more flexible attitude towards German demands.[73] Eden also wanted a quick British response when Hitler made known his specific demands for rearmament:

> What frankly troubles me . . . is the fear that if we do not despatch our telegram to Berlin soon, we may be too late. The German Government may decide on some new development which would change the situation . . . Then, indeed, we should be in difficulty, for we could not show any convincing evidence of our activity since the Disarmament Conference adjourned.[74]

Simon's reaction was more cautious. He recognized that Germany was the one power which was really indifferent whether or not there was a disarmament agreement. She would therefore set her demands high and, if they were refused, would be none the worse off because she 'will do what she intends to do just the same and she does not think that anybody will physically stop her'. Simon was moving towards the conclusion that no progress was possible unless Britain offered 'some kind of assurance' to France on the question of security. 'Those who wish the end must provide the means.' Nothing else would move the French from their existing position and nothing else was 'ever going to stop the Germans'.[75]

In so far as Eden's relations with Simon deteriorated towards the end of 1933, this had less to do with policy questions than the press campaign to which the Foreign Secretary was now subjected. *The Times* called for a new British initiative while criticizing Simon for having granted too many concessions to France. Other newspapers called for Eden's inclusion in the cabinet and, in the case of the *Daily Mail*, his elevation to the Foreign Secretaryship. In the wake of this assault from Fleet Street, Eden found his chief much agitated. 'He could not possibly go [to Geneva] without me. . . . He quite realised that he was being sent to be sacrificed etc. etc. I tried to calm him.'[76] In the event Eden became Lord Privy Seal on the last day of the old year. The position had previously been held by Baldwin, but for Eden it did not carry cabinet rank, though it involved admission to the Privy Council. This limited promotion was well received by the press, but not by the Foreign Secretary. Simon was 'friendly but I think not too happy at the way my appointment had been taken'.[77] Eden was still not summoned to attend the cabinet committee on disarmament. He feared that Simon's jealousy was 'so rampant' that it was not worth raising the matter with him. Not surprisingly, speculation over Simon's future continued unabated. Whenever Eden now visited Geneva on his own, some saw this as a slight to the Foreign Secretary, even though the latter had probably settled details of the British representation himself. 'These impressions', concluded Eden, 'are created by his vacillation.'[78]

Eden's star thus continued to rise almost as a function of Simon's decline, despite the absence of significant policy differences between them. A Conservative backbencher commented on the younger man's qualities: 'Anthony Eden, they tell me, can make up his mind . . . and give advice clearly – hence his success and advancement. He also possesses a pleasant manner, is nice looking, and does the right things in the right way – what more is required?'[79] Be that as it may, Eden remained part of a Foreign Office consensus which still saw its primary goal as the securing of a disarmament convention to satisfy both France and Germany. Indeed *The Times* had noted that Eden's appointment as Lord Privy Seal was taken in Berlin as a sign of British determination to bring disarmament negotiations back to Geneva as soon as possible.[80] It was a consensus which Churchill would challenge in the Commons in March when he spoke of the illusion of seeking an arms limitation agreement.[81] A British memorandum on disarmament was completed on 25 January and submitted to Germany four days later. Disarmament was to be phased over a ten-year period with the major powers beginning the immediate destruction of their heaviest classes of tanks and artillery. Meanwhile Germany would be granted a limited measure of rearmament so that the goal of equality could be achieved within seven years.

To ascertain how the leading European powers reacted to these proposals, it was agreed that Eden should visit Paris, Berlin, Rome and Paris again. The press offered fulsome support to Britain's young ambassador of peace. Eden's reaction to his visits to these capitals is instructive. Before leaving on 16 February with William Strang, Lord Cranborne and Robin Hankey, his new private secretary, Eden described his pilgrimage as no more than an 'outside chance', but one which was worth taking.[82] He held out few hopes for his visit to Paris, fearing that the government of Gaston Doumergue, hastily formed after the Stavisky riots had left Paris reeling amidst allegations of governmental corruption, would be unable to give its attention to Britain's disarmament proposals. He formed an instant dislike for the French Foreign Minister, Louis Barthou, whom he found 'bristly and foxy . . . a nasty old man at heart', but judged Marshal Pétain a 'sane and charming soldier'.[83] Perhaps, however, he was disposed to respond favourably to any sign of conciliation offered by Hitler. In Berlin the Führer convinced Eden that he would uphold the Treaties of Locarno notwithstanding his contempt for the diktat of Versailles. Hitler seemed ready to accept some form of supervision – about which Simon had been sceptical – if an acceptable convention were signed. Additionally, he appeared ready to take account of French susceptibilities, especially in relation to paramilitary formations.

To an extent which is scarcely echoed in his memoir account, Eden clearly joined the ranks of those who had fallen under Hitler's spell.[84] 'He is a surprise,' he reported to Baldwin. 'Without doubt the man has charm.'[85] To his wife he was even more effusive: 'Dare I confess it? I rather liked him.'[86] Perhaps more importantly Eden seemed to have little understanding of the internal realities of the German regime: 'My general impression here is that Hitler is infinitely better than his entourage. Something may be done with him; nothing with Bülow and company.'[87] Eden found it hard to believe that Hitler wanted war, especially as the Nazi government had too much to do internally for the next five years to be planning external conquest. The common bond of service in the last great war appeared to draw Hitler and Eden together. 'Poor man . . . he was badly gassed by us and blind in consequence for three months.'[88] Eden judged that Hitler's proposals offered a genuine basis for a convention and that the German Chancellor had 'made an earnest effort to come to meet us'.[89] From the British embassy in Berlin Basil Newton concluded that Eden's visit had been a distinct success. Eden and Hitler liked one another and were impressed by each other's good intentions.[90]

The Foreign Office on the other hand reacted less enthusiastically to Eden's discussions with the German Führer. The meeting occasioned what was probably the first significant *policy* divergence between Eden and Simon. The latter wrote that Eden's enthusiastic espousal of Hitler's plan had 'put us in a position of great embarrassment. The proposals contained in it are such as we could not possibly ourselves put forward and sponsor.'[91] Receiving this rebuke, Eden confided his exasperation with Simon to the privacy of his diary. The Foreign Secretary was 'not only a national but an international calamity'. It was 'really hopeless to attempt to work for such a man'.[92] But Simon was speaking for the Foreign Office – indeed the government – as a whole. Hitler's demand for the immediate legalization of a German air force was viewed as particularly unacceptable.[93] Even MacDonald noted that 'we should not allow Germany to dump its confidences upon us in order to use us for its own policy'.[94] Many years later, as he drafted his memoirs for this period, Eden seemed to admit his error:

> Perhaps I attach too much importance to Hitler's offer to me . . . because . . . I much doubt if Hitler and Musso would ever have held to any agreements we made with them. If you think I have, don't hesitate to tell me, but I had a feeling we would have gained if we could have tied Hitler up and he had then broken his word.[95]

Such reasoning has its merits, but it is a long way from Eden's attitude in 1934.

From Berlin Eden moved on to Rome. He found Mussolini 'genuinely anxious' to help and adopting 'a very realistic attitude'. His doubts still focused on Paris to which he was due to make a return visit to stress British solidarity with the French government:

> If they had a strong government and one prepared to face the realities of the present European situation, I believe myself that they would seize Hitler's offer and make use of it, but one can hardly expect such courage or initiative from a makeshift government, and this, I fear, is all that Doumergue's cabinet really is.[96]

Eden's second visit to Paris proved as disappointing as he had anticipated. The French government's domestic preoccupations left it in no better position to comment on the British memorandum than on his first trip nearly a fortnight earlier. None the less, and notwithstanding the private disagreement over Hitler's proposals, Eden's public standing seemed to have been enhanced by his European tour. *The Times* adopted an optimistic tone. The problems of disarmament seemed less intractable; the German attitude had been clarified; and the British memorandum would serve as a basis for further discussion.[97]

Overall, however, it is important to stress that Eden remained at one with the British government, and with Simon himself, on the broad lines of policy. As 1934 progressed, both men became increasingly concerned about Germany's long-term goals but each was convinced that Britain should still strive for an armaments convention, the price for which might have to be a firm British guarantee of French security. This was the message Eden presented to the Foreign Secretary on his return from Paris.[98] On 14 March Simon urged the cabinet that, to keep the Disarmament Conference in being and avoid a return to competitive rearmament, Britain needed to make further concessions to Germany and give 'additional undertakings' to France. In effect he was moving away from the goal of disarmament *per se* towards that of an agreement on rearmament which would at least try to put a ceiling on what was inevitable.[99] Britain now accepted that Germany was the 'ultimate potential enemy' against whom her long-term defence policy should be directed. Simon took some comfort from the belief that Germany was more likely to strike in Eastern and Southern Europe than in the West.[100] For Eden the saving grace was that time was on Britain's side: 'Internally Germany is probably in for a difficult time in the next twelve months and I do not myself believe that a military menace from that quarter is imminent. We have probably five years, at least, during which many things may happen.'[101] In the long term, however, a British commitment to the continent might be the only solution.

The weeks following Eden's European tour were not easy. Barthou was unimpressed by the British proposals, and on one occasion went out of his way to humiliate Simon. The meeting of the Disarmament Conference in late May proved in effect to be its last session. Everyone seemed to think that Eden had done all that was humanly possible. Simon joined the chorus of praise: 'This is just to send you my warm congratulations and thanks for your achievement. . . . It is a good conclusion: France appeased but restrained – Germany reproved but invited – Britain taking the lead but avoiding the pitfalls. Well done.'[102] In the public perception Eden now stood – however unjustifiably – at some distance from his senior colleagues, a separation which would be vital for the development of his subsequent reputation. Gilbert Murray of the League of Nations Union insisted that Eden was 'practically the only member of the government in whose work for disarmament my people have any confidence'. It would be 'very difficult ever to convince the general peace forces in this country that the government as a whole worked sincerely for disarmament'.[103]

Eden soon had further opportunities to consolidate his growing standing. On 9 October Barthou and King Alexander of Yugoslavia were assassinated by a Croatian terrorist. Momentarily this incident seemed to threaten European security by rekindling latent rivalries in the Balkans. Eden had modified his earlier hostile view of the French Foreign Minister, especially as Barthou had proved amenable to compromise with Britain during talks in London in July over the idea of an 'Eastern Locarno' pact. Eden felt 'very sorry' at his assassination:

> His courage and his wit at least were attractive, and having at length realised the necessity for Anglo-French friendship, I believe that he would have proved less parsimonious than most Frenchmen in the give and take necessary to make such friendship effective.[104]

At the League Eden assumed the role of *rapporteur* and, devising soothing formulae, defused a menacing situation. His solution was anodyne – a public declaration of the responsibility of all countries to stamp out terrorism and a condemnation of Hungarian neglect – but appropriate to the occasion. For this he was again widely praised. The Foreign Secretary told the Commons that he took 'particular satisfaction' in dwelling upon the important part which Eden had played.[105] Even in his diary he paid tribute to Eden's 'adroitness' in framing and carrying unanimously the resolution of the League Council.[106] One observer suggested that 'there was no other man in Europe who could have done it'.[107] This was an exaggeration, but one to which an increasing body of opinion subscribed.

Simon and Eden were less in harmony over arrangements for the plebiscite on the future of the Saar. Under the provisions of Versailles the inhabitants of this region had the right to determine their own destiny after fifteen years under League jurisdiction. The vote was to be held on 13 January 1935. By late 1934, however, there were doubts, as a result of increasing Nazi agitation, whether order could be maintained during the transitional period which, most observers expected, would be the prelude to a resumption of German sovereignty. Eden later claimed that it was he who first argued that Britain should sponsor an international police force during the period of the plebiscite and that this policy was pushed through the cabinet with the support of Baldwin but against the opposition of several ministers including the Foreign Secretary on 3 December.[108] By contrast, Simon recorded – 'as some foolish or grudging commentators . . . seem to suppose the whole thing was a brilliant improvisation at Geneva (where Eden carried it through splendidly)' – that the initiative had been his. The idea of an international force originated with a paper he wrote for the cabinet on 16 November; he developed the idea in a letter to Eden a few days later; and a complete plan was then presented to the cabinet.[109] Certainly the cabinet record confirms that Simon raised the question of an international police force on 28 November.[110] In the circumstances it is instructive to turn to the account of a third party. According to Neville Chamberlain, Simon 'very timidly and half heartedly' brought forward a proposal for an Anglo-Italian force, but seemed ready to withdraw it when it was suggested that it might be inappropriate. The idea would never have gone ahead 'if I [Chamberlain] had not urged it so strongly'.[111] It was another case of Simon's old weakness – his inability to stick with a policy and champion it against all opposition – which made Eden increasingly contemptuous of Simon, though he was still usually able to conceal the strength of his feelings from his chief.

Eden's successes at Geneva seem to have encouraged him to believe that the League could play a more active role in preserving the peace than he had once thought. After three years in office he was coming more closely to fit the image of himself which had, somewhat misleadingly, existed throughout this period. By the end of 1934, he later recalled, British authority and that of the League stood higher than at any time in the life of the National Government.[112] At the time he noted that Britain's offer of troops to help supervise the Saar plebiscite had 'been a real tonic for the League and may have far-reaching consequences'.[113] The year 1935 would provide further scope to enhance his credentials as a League man.

Before 1934 was out, however, there was renewed speculation about the Foreign Secretary's place within the government. His position was under threat for political rather than policy reasons. Despite his outstanding intellect, Simon was perceived to be the weakest link in a cabinet which was by no means confident of receiving the British people's renewed endorsement at the election widely anticipated for 1935. Eden believed that the government could win again – providing it took the right steps while there was still time:

> The country is perhaps a bit bored with us, with some of us in particular. If, however, we face this fact and reorder our ranks accordingly, we should, I believe, be able to secure a decent majority when the election comes. Much depends, however, upon the thoroughness of our own overhaul.[114]

Undoubtedly Simon's removal was central to Eden's thoughts. Indeed, his diary entries show that he was discussing the possibility of the Foreign Secretary's replacement as early as November.[115] A month later Eden confided to Baldwin that he 'could not go on as at present' because he was interfered with so much in his work at Geneva. The two men appear to have discussed the possibility of Baldwin himself taking over the Foreign Office, at least for the remainder of the present parliament.[116] Neville Chamberlain also speculated on a move from the Exchequer to succeed Simon but doubted whether MacDonald and Baldwin would bring themselves to sack Simon, whose status as a Liberal National remained important to the government's 'national credentials'.[117] The Chief Whip favoured Chamberlain's appointment, failing which he would 'come down on Eden every time'.[118]

*

It was by no means evident to British policy makers in early 1935 how best to proceed in relation to the perceived German threat. As Eden noted, the preferred option of 1934 was probably no longer available. The French 'think, I suspect, that German rearmament has already gone so far that agreement at any reasonable level is past paying for'.[119] In the absence of a clear policy the government fell back on the idea of renewing personal contacts with Hitler to explore the possibility of German participation in a series of multilateral security pacts, including one of mutual assistance against aerial attack. After considerable discussion it was agreed that Simon should be accompanied by Eden, with the latter going on alone from Berlin to make further visits to Moscow, Warsaw and Prague. It would be up to Eden to allay those fears of the Eastern European powers which would undoubtedly be aroused by the discussions in Berlin. Once again Eden was irritated by Simon's indecisiveness: 'It is really the old story. J. S. cannot make up his mind whether to go on to Moscow or not. He is afraid of missing something if he doesn't and afraid of getting something if he does.'[120] Chamberlain confirmed that Simon was 'jealous' and did not like Eden to go alone to any foreign country.[121] Eden was at least relieved to be accompanying the Foreign Secretary to Berlin, for he feared that Simon – his political fortunes at a low ebb – was too keen to 'score some "success"'. 'The chances are', he added, 'this can only be done by double-crossing our friends.'[122] Meanwhile the Soviet ambassador in London took comfort that only Eden would proceed to Moscow. The Lord Privy Seal 'represents the Conservative Party and is an "English gentleman" – qualifications which Simon lacks'.[123]

Before the visit to Berlin took place, however, the British government published a Defence White Paper on 4 March which contained some plain speaking on German rearmament. Hitler took offence and insisted that the mission must be postponed, ostensibly because of a cold he had caught in the Saar. Though he later argued that the British visit should now have been abandoned – 'in other words, we ought to have returned the diplomatic cold' – Eden seems at the time to have agreed that it should still go ahead, even though Hitler had now announced the existence of the Luftwaffe.[124] Finally, the British delegation arrived in Berlin on 25 March. From the outset Eden seemed to take a stronger line towards the Germans than did Simon: 'Dinner at Embassy and rehearsed tomorrow's doings. Fairly happy about it, though at each rehearsal J. S. modifies the firm opening and I have to put it in again.'[125] Hitler's interpreter noted that throughout the talks with the German Chancellor Eden seemed more sceptical than Simon.[126] Eden's contemporary diary account lends substance to this conclusion:

> Only thing Hitler wants is Air Pact *without limitation*. Simon much inclined to bite at this, and to suggest separate conference on this. I had to protest and he gave up the idea. Total result of visit for European settlement very disappointing. Simon toys with idea of letting G. expand eastwards. I am strongly against it. Apart from its dishonesty it would be our turn next.[127]

Recording his thoughts *en route* to Moscow, Eden concluded that the basis for a general European settlement, in which he had certainly believed a year before, no longer existed. Britain's only option now was to join with other members of the League in reaffirming their faith in this institution.[128] Eden was moving towards the conclusion that Germany's military threat could only be checked if Britain, France and Italy came together in a common front, groping in other words towards that concept of collective security with which his name is so generally associated. His memorandum, discussed by the cabinet on 8 April, showed how much his thinking had evolved since his first meeting with Hitler in 1934:

> if we refuse to be scared . . . by Germany's growing demands, if we resist the temptation to accept everything that Germany asks for as a basis for discussion . . . if for a moment we can cease to be an honest broker and become the honest facer of truths, then I am confident that there is no call to view the future with alarm.[129]

Taken in isolation these words, faithfully repeated in Eden's memoirs, are the very stuff of 'anti-appeasement'. But Eden had taken some time to reach this position and it was one from which he would again deviate in the years ahead. Nor would it be fair on Simon to suggest that his views were the complete antithesis of Eden's. Simon agreed that the general result of the Berlin conversations had been disappointing.[130] He recognized that Hitler's ambitions were 'very dangerous to peace in Europe' and that the natural consequences of German policy might be 'terrible beyond conception'. It was all 'pretty hopeless'.[131] The Berlin venture had done little to restore the Foreign Secretary's fortunes. 'I don't say . . . that Simon was in any way to blame', conceded Chamberlain, 'but I did regret that the position was not more thoroughly explored in some respects.'[132]

Eden had already concluded that the Soviet Union could play a significant part in any overall calculations of the European scene, if only because Germany was constantly citing the Russian threat.[133] His visit to Moscow – making him the first British minister to meet Stalin – thus assumed considerable importance in his own mind. But Simon's instructions to Eden before his departure for Moscow suggested that he was not keen to include the Soviet Union in any armaments discussions for the time being.[134] In what was to become one of his recurrent themes of the next few years, Eden saw the need to reckon with the intensely suspicious nature of the Soviet regime. In Moscow he was impressed by Stalin, while recognizing that dictator's fundamental ruthlessness. His most difficult moment came when the Foreign Affairs Commissar, Maxim Litvinov, suggested that the British government did not believe in German aggressiveness. Eden demurred, but had to admit that his government had not 'hitherto wished to believe badly of German intentions'.[135] It would have been reasonable to make a mental note to include himself in this mild rebuke.

Not surprisingly in view of the continuing evolution of his views, Eden attached great importance to the forthcoming conference at Stresa with France and Italy. 'I don't believe the international situation to be so difficult', he insisted, 'if we will pursue a straight and firm course in support of the collective peace system.'[136] But Eden himself played no part in these proceedings. Returning in poor weather from his tour of Eastern Europe, he suffered mild heart strain and was ordered to rest for two months. After lengthy cabinet discussions it was decided that MacDonald should accompany Simon to Stresa. From his sick-bed Eden viewed with disappointment their refusal at the conference – in accordance with the cabinet's decision – to go beyond British obligations under Locarno. He considered that a real opportunity had been missed to construct an effective barrier against German aggression. Any such barrier, however, would almost certainly have crumbled in the face of Italian designs on Abyssinia – an episode which would complicate Eden's relations with Simon's successor.

In the last months of his Foreign Secretaryship Simon seemed uncertain which way to turn. Eden found him 'clearly confused as to what to do next'.[137] So, while Simon sometimes appeared ready to follow Eden's option of collective security, he still refused to abandon the continuing quest for agreement with Germany. With Eden out of action, British policy in the second quarter of 1935 therefore followed the twin but fundamentally incompatible tracks which led both to the Stresa Front, which seemed to commit Britain, France and Italy to a common stand against German aggression, and to the Anglo-German Naval Agreement. Though there is no evidence that Eden objected forcefully to this bilateral pact with Germany,[138] he did seem to have recognized the fallacy behind the idea of making concessions in the vain hope of satisfying Hitler's voracious appetite. Simon's capacity to see all sides of an issue was, in one sense, a virtue, but it was, thought Eden, 'sad' that his 'capacity to see Germany's side only results in a desire for words or deeds which will only be read as a sign of fear in Germany . . . and an encouragement to ask for more'.[139] Soon after returning to work Eden delivered a forceful speech in Fulham, which contained a scarcely veiled warning of Britain's commitment to a 'collective system' against those who sought by 'power politics to break up the peace'. His words provoked annoyance in Berlin, but won the backing of Sir Robert Vansittart.[140]

By the spring of 1935 a cabinet reshuffle was imminent, if for no other reason than MacDonald's desire to hand over the burden of the premiership to Baldwin. It was generally assumed that Simon would be the principal casualty of any reorganization. The succession at the Foreign Office was a matter of understandable interest and concern. Something like an Eden group was beginning to emerge in the Commons, a body of largely younger members anxious that their man should be the beneficiary. There were a number, wrote Roger Lumley, ready to back Eden 'in any way in which they can be useful, because they feel that you are the only person who has the right ideas about foreign policy and who can steer us through the present situation'.[141] William Ormsby-Gore confirmed that Simon's 'number

[was] really up at last' and that reconstruction was now assured,[142] while Philip Sassoon found Baldwin 'quite fit to *deal with problems*'. 'I hope he will', he continued, 'and in the way we want him to! I am sure you will get it.'[143] A considerable head of steam thus developed behind Eden's candidacy, supported by sections of the press. In view of his hard work over the government's India Bill, however, there was another strong candidate for Simon's post. 'It is a race between Sir Samuel Hoare and Anthony Eden for the Foreign Office,' judged 'Chips' Channon.[144] Victor Cazalet thought the job should go to Eden but 'I suspect Hoare will be there.'[145] Though there is evidence that Baldwin seriously considered naming Eden, he had abandoned the idea by late May, before finally deciding on Hoare. Eden would 'probably fill the post of the Wandering Minister'.[146] If Eden had supporters, he already had opponents, many of them senior figures in the government and the Conservative Party, jealous perhaps of the precocious rise of a politician still in his thirties.[147]

Baldwin was well aware that Eden disliked the idea of continuing with junior status at the Foreign Office, but it was at Hoare's request that dual responsibility was retained. It was, thought Hoare, essential that Eden should keep his present position. 'If there was trouble over Abyssinia or Hitler, it would all be blamed on the exclusion of Eden and oblivion of the League.'[148] Eden was coaxed into acceptance by membership of the cabinet with the status of Minister for League of Nations Affairs. Hoare later recorded that – anxious to give the younger man 'the fullest possible scope for his great talents' – he and Eden had no difficulty in working out an appropriate division of responsibilities to avoid creating a dyarchy within the Foreign Office.[149] At the time, however, the new Foreign Secretary clearly appreciated the difficulty of distinguishing between League questions and those relating to foreign affairs generally.[150] The danger existed that two men, charged with responsibility for advising the cabinet on foreign policy, could present diametrically opposing opinions.[151] As the *Manchester Guardian* noted, if Eden were to be Minister for League Affairs in any real sense, he would usurp most of Hoare's responsibilities.[152] Austen Chamberlain, who had personal, and largely unfavourable, experience of a similar division of duties in the 1920s, was sceptical of the new arrangement. 'If it works well', he concluded, 'it will only be because of the personal characteristics and relationship of Eden and [Hoare].'[153] Just as potentially damaging as genuine divisions inside the Foreign Office was the perception of them among the government's critics. As early as 11 July the leader of the Labour Party argued that, whenever Eden spoke, 'I can always see throughout his speeches that there is an ideal to which he is working, but I can see nothing in the speech of [Hoare] but a temporising and a parleying with all the forces of disorder in the world.'[154] In fact, however, at least until the crisis of December, the relationship between Eden and Hoare worked as well as could be expected. Neville Chamberlain noted that they were co-operating 'admirably' and that Eden found the difference from Simon's regime 'incredible'.[155] Eden confirmed that 'happily we see eye to eye in every respect'.[156]

*

Hoare's Foreign Secretaryship was dominated by a single issue and one which exercised a profound impact on Eden's career. Benito Mussolini had never disguised his ambition to see Italy become a world power. After 1931 internal problems drove him increasingly towards an overtly aggressive foreign policy. The immediate object of his expansionist ambition was the ancient kingdom of Abyssinia which had, against the odds, prevailed against an earlier thrust of Italian imperialism forty years before. Eden was closely associated with this developing crisis, managing – as with earlier episodes – to escape with his reputation unscathed, indeed enhanced, while those of his senior colleagues suffered grievous damage. After the Wal-Wal incident, a frontier skirmish in which more than a hundred Ethiopians were killed by Italian troops in December 1934, a crisis became increasingly apparent. To his credit Eden quickly saw the extent of Mussolini's intentions. 'If we do not maintain a firm front towards the Italians about this incident', he warned, 'it will only be the forerunner of others.'[157] Italy aimed at the absorption of Ethiopia 'morsel by morsel'.[158]

At this stage, Simon too seemed to favour a strong line, but his failure – and that of MacDonald – to raise the matter with Mussolini at Stresa, despite being accompanied by a Foreign Office specialist on Abyssinian affairs, caused Eden much concern.[159] On his return Simon warned the cabinet that a British-sponsored initiative against Italy would lead to that power leaving the League, to the pleasure of Nazi Germany.[160] Almost certainly Simon was influenced by Vansittart's view that the loss of Abyssinia – or at least a part of it – was an acceptable price to pay to keep Italian friendship in a world dominated by the German menace, notwithstanding the damage likely to be done thereby to the League. Such reasoning increasingly set Eden apart from Vansittart despite his growing appreciation of the German threat. Indeed, Eden's failure to see the relative dangers posed by the two fascist dictatorships in an appropriate perspective was and remained a fundamental weakness of his understanding of the international scene.

Eden found Simon's warning in early May 1935 to the Italian ambassador, Dino Grandi, of the likely British response to Italian aggression 'nothing like strong enough' in its failure to emphasize British obligations under the League Covenant.[161] He later complained that Grandi was bound to conclude that the British were 'troubled and uncertain in our course'.[162] But at the time Eden had little idea of how the League's will could be made to prevail, being concerned above all with the political difficulties of defying public opinion on this issue:

> Opinion here has grown steadily in support of the League during the recent difficult European situation, so that there is now an overwhelmingly powerful volume of opinion . . . in support of the League as the one means of maintaining peace. . . . If this analysis is right, you can readily picture what the reactions will be should our Italian friends be so unwise as to take steps in Africa which would damage, and may be destroy, the League's authority.[163]

Eden arrived in Geneva on 19 May and adopted a much firmer line with the Italian delegate, Baron Aloisi, than Simon had with Grandi. He remained, however, pessimistic about a successful outcome: 'There is I suppose a remote possibility that we shall be able to bring enough pressure upon Mussolini to stop him in his mad adventure. But how again this is to be done I have not the least idea at present.'[164] Winning Simon's backing for his insistence that Italy should accept arbitration for all Italo-Abyssinian incidents since the previous December, and also seeming to establish a common front with France, Eden secured agreement for two draft resolutions. These provided no long-term solution but made it 'harder for Mussolini, in the eyes of the world at least, to proceed to extreme measures against Abyssinia'.[165] Simon concurred: 'This does not secure a peaceful issue, but it gives peace a chance and saves the League from being flouted.'[166]

Back in England Eden still did not see the way ahead with any clarity. Accepting the need to help Mussolini save face, he saw little room for manoeuvre beyond assisting the Duce to secure his rights from the Italo-Abyssinian Treaty of 1928. 'I find it difficult', he concluded, 'to know what is the next step to take.'[167] One man with a clearer line of policy was Vansittart. On 8 June the Permanent Under-Secretary noted that Italy would have to be bought off 'in some form or other, or Abyssinia will eventually perish'. Eden had already implicitly rejected this approach when, five days earlier, he had endorsed the opinion that Britain had to 'stick to League principles and stand the racket'.[168] With a new Foreign Secretary installed on 7 June, it was Vansittart's views which prevailed. On 19 June the cabinet decided that a solution might be found if landlocked Abyssinia were offered access to the sea at the port of Zeila in British Somaliland in exchange for territorial concessions to Italy in the Ogaden desert. 'I was not enthusiastic about the proposal,' Eden later recalled, 'still less when it was suggested that I should be its sponsor in Rome.'[169] Be that as it may, he went along with it.

After passing through Paris and attempting to justify the recent Anglo-German Naval Agreement to a sceptical Pierre Laval, Eden met Mussolini in Rome on 24 June. The Duce made it clear that, from an Italian point of view, a peaceful resolution could only be obtained through the satisfaction of claims which were so wide that no British concessions to Abyssinia could form sufficient compensation. It seemed that nothing short of full Italian control over Abyssinia would satisfy Mussolini. It also appeared that the earlier solidarity with France was on the point of collapse. A minor historical controversy has developed over whether Eden felt personally affronted by his reception from the Italian dictator. Eden denied this to be the case, both in a Foreign Office minute a few months later and in his memoirs. Mussolini may not have been overtly rude to his visitor, but it still seems likely that, as one of Eden's biographers has noted, 'there was a complete failure of personal chemistry between the two men which, added to Mussolini's political rebuff, did at least begin the process by which Eden was to develop an obsessive hatred of the Italian leader.'[170] Any discomfort Eden experienced was compounded by the necessity to explain the British offer and its refusal to the Commons on his return. Thereafter policy and personality became increasingly entangled in his assessment of the Italian problem.

Over the following weeks British policy drifted with little sense of direction. The cabinet set up a sub-committee to consider the possible imposition of sanctions against Italy – but more as a means of postponing a decision than as a statement of future intent.[171] In his memoirs Eden suggested that 'we had reached a climacteric in the thirties; the time had come to take a stand.' 'I date our Abyssinian failure from these weeks,' he continued, 'its consequences stretched far into the future.'[172] Eden wanted to tell Laval that Britain would be prepared to fulfil her League obligations if others would do the same, but the cabinet decided that Hoare should continue to exercise diplomatic pressure.[173] Meanwhile Eden stressed the importance of not appearing to support the Italian case – 'our moral position will then be destroyed'.[174] At the cabinet on 24 July there was general agreement on the advantages of playing for time, as the longer a catastrophe could be avoided the greater would be Italy's realization of the difficulties confronting her.[175] On occasion, however, Eden seemed as ready as the government as a whole simply to allow the situation to develop while keeping fingers crossed. On 30 August he wrote:

> So far as I have any glimmering of hope at all, it is the possibility that Italy's financial difficulties will increase rapidly..., that she will not be able to launch her campaign as soon as she anticipates, that she may find the Council of the League stiffer than she had calculated, and that all these factors may yet make a settlement possible, if not before the conflict breaks out, at least before it has assumed serious proportions.[176]

Overall, British policy remained, as defined by the cabinet at the end of July, 'to see that the emergency did not develop to the point where the question of [the] fulfilment [of British obligations under the Covenant] arose'.[177] The government therefore strove to arrange three-power discussions in Paris in mid-August between Britain, France and Italy. Eden was not optimistic and was worried that Britain, at the expense of her good name, might engage in an attempt to make the Abyssinians accept 'an unjust and unworthy settlement'.[178] In the event, the conversations in Paris ended in deadlock, while revealing mounting French apprehension at the prospect of an irreversible breach with Italy, with all that this might mean for the European balance of power.

In an important passage in his memoirs Eden states that on 20 August Hoare told him that there would be a wave of public opinion against the government if it repudiated its obligations under Article 16 of the Covenant, which allowed for military sanctions against an aggressor.[179] There is no other record of this conversation, but no reason either to assume that it did not take place. Hoare's remarks must, however, be taken in the context of the contemporary domestic political environment. The government faced an imminent general election, about whose outcome it was by no means confident. Such evidence as was available, notably the results of the Peace Ballot announced at the end of June, suggested popular commitment to and faith in the League which was now scarcely matched among the governing élite. As Hoare explained to a meeting of ministers on 21 August, he and Eden, 'after considering the position *at home* and abroad, were clear in advising . . . that it was nothing less than essential to follow the regular League of Nations procedure in this crisis'.[180] Conversations with leading opposition figures had confirmed his view that there would be strong public feeling against the government if it reneged on its League obligations. But Hoare was careful to include the proviso that 'France would go as far as we were prepared to do'.[181] He spelt out his thinking even more clearly in a letter to Neville Chamberlain:

> I see myself the making of a first-class crisis in which the government will lose heavily if we appear to be repudiating the Covenant. When I say this, I do not mean that I have changed my views since we both discussed the question in London. . . . Our line, I'm sure, is to keep in step with the French and, whether now or at Geneva, to act with them.[182]

In similar vein Hoare told Sir George Clark in Paris that Britain must 'play out the League hand in September', but that if it appeared that there was no basis for collective action and, in particular, that full French co-operation was not forthcoming, then the impossibility of imposing sanctions would have to be recognized.[183]

If, therefore, there was an element of cynical political calculation in Hoare's apparent readiness to embrace the League with greater ardour than hitherto, this was something which Eden ought fully to have understood. In particular, it seems strange if he did not realize that the British position might be modified, either once the election was over, or if French co-operation for a League-based solution broke down. It is against this background that Hoare's speech to the League Assembly on 11 September needs to be assessed. Eden later claimed that this speech surprised and indeed alarmed him since the wide-ranging nature of Hoare's commitment to the League might impose intolerable obligations on Britain in the Far East, where China and Japan were in a state of undeclared war.[184] Yet it is difficult to believe that Eden saw Hoare's stance as incapable of subsequent modification, especially as he had been present at conversations with Laval on the previous day when the French Prime

Minister's resolve was clearly weakening. Subsequent cabinet meetings showed that Britain was still trying to avoid the issue of
sanctions and pinning her hopes on a negotiated settlement, acceptable to both Italy and Abyssinia. On 24 September the cabinet
recognized the serious consequences of backing away from Hoare's
apparently firm stance 'from the point of view of domestic policy no
less than from that of foreign policy'. As, however, no immediate
decision was called for on sanctions, 'unless the burden was to fall
on all the States members of League of Nations, we should not
commit ourselves'.[185] A week later Hoare made abundantly clear
the sort of solution he really had in mind. If Italy, finding itself in
difficulties, seemed disposed to compromise, he 'hoped that the
powers would not be too intransigent and would be willing to
consider a settlement which, without destroying Abyssinian independence, would give Italy some satisfaction'.[186]

Such a solution remained, however, elusive, especially when
Mussolini turned down the compromise proposals of the Committee
of Five, which had been set up by the League Council to handle the
whole problem. On 22 September Eden, despairing of Laval's
reliability, advised Hoare that effective League action was now
dependent on a firm British lead. But Eden had been away from
London for almost a month and, wilfully or otherwise, was increasingly out of step with the subtleties of the cabinet's thinking. Even
after Mussolini began military operations in October, Hoare persisted
with a policy of caution which was, superficially at least, at odds with
his wholehearted public endorsement of the League. The cabinet was

> reminded that the policy had been to play the Geneva hand to the
> full, but the question was raised as to whether it had not been played
> as far as was safe. But the view of [Hoare] was strongly supported that
> the moment had not yet come for a change in policy. It was urged,
> however, that until the situation with France was cleared up we
> should not apply any new sanctions apart from the maintenance of
> the arms embargo to Italy.[187]

Eden's line at Geneva was clearly in advance of government thinking
and Hoare was authorized to send him a message advising a more
prudent line. Eden was urged to go as slowly as possible and,
bearing in mind the uncertainty of Laval's position, avoid taking
the initiative.[188]

In his memoirs Eden suggests that the continued search for a
compromise was now a course of 'doubtful wisdom' – 'it looked

like an all-out win for one side or the other'.[189] Such reasoning did not, however, deter him from interrupting his electoral campaign – polling was scheduled for mid-November – and going with Hoare to Geneva for further talks with Laval. Here compromise was certainly on the agenda with Eden keen only to draw a distinction – not altogether clear in view of Mussolini's ongoing military campaign – between a territorial exchange promoted by Britain and France and inviting the League to grant a mandate in Abyssinia as a reward for Italian aggression.[190] Eden preferred a 'frank exchange of territory' to an 'Italian protectorate'.[191] Providing any settlement was acceptable to Italy, Abyssinia and the League, he was now ready to envisage a large cession of Abyssinian territory including Tigre province and the Ogaden, and on this basis Maurice Peterson, head of the Foreign Office's Abyssinian department, was authorized to continue negotiations in Paris.[192] Meanwhile, 'rather against Eden's inclination', it was agreed to postpone any decision on oil sanctions.[193]

*

In view of the political crisis which the Hoare–Laval Pact occasioned, as well as the confirmed place it came to occupy in the demonology of appeasement, it is hardly surprising that many cabinet ministers, Eden included, later distanced themselves far more clearly from what was, essentially, a government policy than their contemporary conduct justified. None the less it was, Eden later revealed, 'the one thing for which he blames himself'.[194] On 2 December the cabinet authorized Hoare, on his way to Switzerland via Paris for a much needed holiday, to press 'on by every useful means with discussions with the countries concerned with a view, if possible, to a peaceful settlement'.[195] Eden evidently recognized that important discussions might ensue in Paris, since he later made a point of recalling his 'warning to S. H. . . . before he left . . . to beware of Van's advice there and S. H.'s reply "Don't worry, I shan't commit you to anything"'.[196] When Hoare brokered his deal with Laval, he clearly anticipated Eden's support. The proposals were 'the best that we can get . . . better than might have been expected in the circumstances'. Indeed, he believed Eden would see that the terms came 'well within the framework of the Committee of Five and the formula that we agreed to in London'.[197] Not surprisingly in view of his consistent attitude over several months, it was Laval's support which Hoare most prized. 'We are in fact relieved at having brought the French so far and so solidly with us.'[198] If Eden was, as has been claimed, 'astonished and appalled' by the proposals, it seems unlikely that he conveyed the strength of his feelings to the cabinet on 9 December.[199] His reservations were on questions of detail. In particular he secured support for the principle that Hoare's proposals should be presented simultaneously to the Abyssinian emperor and the Italian dictator and argued successfully against any postponement of the planned meeting of the League's Committee of Eighteen, responsible for the organization of sanctions. Rather than presenting his *own* opposition to the pact, Eden suggested that there were 'features' of the proposals which would prove very distasteful to some members of the League.[200]

Many years later Eden claimed to be puzzled as to how Hoare could have thought that the proposals agreed with Laval could be reconciled with his own speech to the League in September.[201] As has been shown, however, this largely missed the point. Enough had happened between September and December, to which Eden had been party, to show that Britain had moved on from a simple commitment to the principles of the Covenant, if this had indeed ever been the essence of government policy. It is broadly true, as Hoare claimed, that the pact followed the lines of the scheme drawn up by the Committee of Five in September, though the Hoare–Laval proposals considerably extended the concessions to be made to Italy.[202] Indeed a telephone conversation between Eden and Hoare after the cabinet meeting of 9 December suggests that the former was concerned by the scale of the concessions rather than the principle of rewarding Italian aggression. He did not seem 'much worried'. 'The only part of the scheme he disliked', Hoare recorded, 'was the big economic area in the South.'[203]

Little had changed when the cabinet met again on 10 September, the minutes recording that the terms on offer were the best which could be obtained from Italy from the Abyssinian point of view.[204] By the time of its third meeting in three days on 11 September, however, the mounting tide of public and political indignation was taking its toll. Growing uneasiness on the government's own benches in the Commons was probably the crucial factor. Eden now voiced the hope that he would not be expected to champion the proposals in detail at Geneva. He reminded his colleagues that British foreign policy was still based on the League. Forced to choose between aggression by Italy and the League's collapse, he would consider the former a lesser evil. Though Eden was reminded of the fundamental principle that Britain should not contemplate entering hostilities without the assurance of French support, opinion had begun to move against Hoare and his pact.[205] Hastening to Geneva, Eden reported that the impression created by the Paris proposals was 'even worse' than he had anticipated.[206] His tone to the Committee of Eighteen showed that he was actively distancing himself from the pact. 'Without actually repudiating parentage', Eden showed that the government now 'looked upon their child without affection and even with distaste.'[207]

The majority of the cabinet had come to the conclusion that continued espousal of the pact would be politically disastrous. The more unpopular it became, the more easily Eden found flaws which had not, it seems, been so apparent to him on first examination. The comparison with the proposals of the Committee of Five 'looks all right on paper', but the intention of the Committee 'was never to give the Italians as much as this'.[208] A serious misjudgement had occurred but, as Chamberlain stressed, despite some misgivings, in the first instance 'nobody [had] suggested rejection'.[209] Realizing that it was a case of either sacrificing Hoare or destroying the government, Simon gave a candid assessment of what had happened. 'A mistake was made. We are all in it because, in very sudden and difficult circumstances, we confirmed the Paris plan.'[210] The rest of this story is well known. On 17 December the cabinet agreed that the pact could not be defended. It fell to Eden to make the best of a bad job, returning to Geneva with the argument that Britain and France had merely been trying to find a formula to restore peace, but had never sought actively to sponsor this particular solution if it proved unacceptable to the parties actually concerned. From the apparent wreck of Hoare's career, Eden emerged the almost inevitable beneficiary. While Hoare tendered his resignation, Eden was in Geneva where a Canadian diplomat noted him 'brilliantly displaying the charm and diplomatic skill of a chosen man moving upwards'.[211] Though Baldwin seems to have bought off backbench criticism by hinting to Austen Chamberlain that he would be recalled to his old post, Eden duly emerged as Foreign Secretary on 22 December 1935.[212] He was still only thirty-eight years of age.

One or two eyebrows were raised. There were suggestions that some of the 'mud' of the Hoare–Laval débâcle was 'sticking' to Eden, while the Daily Express observed that he had been closely associated with the discredited plan and yet had gained high office because of its rejection.[213] But these were minority opinions. The Daily Herald expressed a more general point of view. The appointment of anyone else would 'be deplored and resented by the country. Of all concerned in this miserable affair, he only has emerged without discredit.'[214] Eden was now the undoubted star of the National Government. When, in the autumn of 1935, he received the freedom of the borough of Leamington, there were 25 000 applications for 3000 seats. Being number two minister at the Foreign Office for more than four years had been an inherently dangerous predicament, granted the difficulties in winning laurels for the conduct of British diplomacy at this time. But Eden had used the position to advantage. His subordinate status was widely interpreted to convey distance rather than association. When attacking the government as a whole in a speech in October 1935, the former Labour Chancellor, Philip Snowden, could still pay 'a warm personal tribute to Mr. Eden'.[215] Over Hoare–Laval, as over so much that had gone before, the popular perception of Eden's role suggested a far greater divergence between his views and those of the government than was justified. If the deal negotiated in Paris epitomized Hoare, and indeed the majority of his colleagues, then Eden was seen to stand for Geneva and an altogether higher code of conduct. 'To the average League supporter', argued the *Spectator*, 'he symbolizes not merely British policy but the ideal League policy.'[216] Not being Hoare in 1935 proved as valuable as not being Simon had been in earlier years.

Eden, in fact, had become the beneficiary of the image – myth would be too strong a word – which had grown around him since 1931. It would be unfair to suggest that he had been completely at one with his governmental colleagues over the direction of British foreign policy. But the fact that he was now in a position to take over the Foreign Office hardly suggests that he had been fundamentally out of step hitherto. The record reveals few of the divergencies which the public believed to be an everyday occurrence. Sometimes, perhaps, Eden had failed to bring his views to the attention of his colleagues with sufficient force. This was a problem which bedevilled much of his ministerial career. The trained diplomat tended to make a virtue of glossing over difficulties with the balm of conciliatory language. His subordinates and friends got a clearer picture of his dissatisfaction with aspects of government policy. But the fact remains that no issue in this period brought Eden seriously close to resignation. Divisions over the conduct of British foreign policy in the 1930s were never as clear-cut as its early historiography tended to suggest. Very few were those who appreciated from the outset that Hitler's regime was not one which could be treated according to the normal canons of diplomatic practice and that its advent had transformed the nature of international politics. 'Something went very wrong with us in the thirties', wrote William Strang two decades later, 'and the responsibility was general and not to be attributed to individuals.'[217] Much as he might regret it, Eden bore his share of this collective guilt.

3

Eden and the dictators

I fully realised we had to gain time. N[eville] C[hamberlain] however sincerely believed that Hit[ler] and Muss[olini] were negotiable. This was the real difference between us.[1]

A NTHONY Eden always recognized that his performance as Foreign Secretary during the 1930s would be crucial to his ultimate historical reputation. Though the clear-out of the Chamberlainite old guard after May 1940 was not as complete as was once supposed,[2] much of the internal dynamics of post-war Conservativism was determined by a rather simplistic assessment of the politics of the 1930s. To have been on the 'right' side of the appeasement divide was an important factor in determining political fortunes for many years afterwards and it was not entirely accidental that four of the five men who successively led the Conservative Party after Neville Chamberlain could all claim the credentials of 'anti-appeasement'.[3] The reality, of course, was never that straightforward. Historians have long since abandoned the simple categorization of 1930s politicians as appeasers and anti-appeasers. Appeasement, in so far as it is even a useful term to encapsulate the foreign policy of pre-war governments, contained countless variations, subtleties and shifts of emphasis. Most who later laid claim to unsullied virtue were inclined at the time to embrace some aspect of it. As early as 1941 Oliver Harvey, then Eden's private secretary, judged that all parties were 'equally responsible' for the policy.[4] But for many years the illusion remained more important than historical reality. Not only foresight but also a higher standard of morality could be attributed to those who had dissented from the government's line. What divided Eden and Chamberlain, asserted John Wheeler-Bennett, was the 'degree of sacrifice which could legitimately be made by Britain in the cause of peace without bringing her to the position where her word and her bond were doubted in the councils of Europe'.[5] Yet for more than two years in the 1930s Eden bore direct ministerial responsibility for British foreign policy, no longer constrained by subordinate status to John Simon and Samuel Hoare. His strong point was that he resigned in February 1938, before events in Europe brought British policy increasingly into disrepute. His problem, on the other hand, was that the quest for a settlement with the dictator powers by no means began in the spring of 1938.

Ironically, it fell to Eden, during his second spell as Foreign Secretary, to initiate the process upon which a more scholarly assessment of pre-war foreign policy could be based. With the Second World War still in its early stages, the government began to consider the possibility of publishing its diplomatic documents following the precedent of the earlier *British Documents on the Origins of the War* relating to the events of 1914. Eden's somewhat equivocal attitude towards early publication reflects in part the ambiguity of his own position in the 1930s. After discussing the matter with his immediate predecessor, Eden decided in early July 1941 against publication. But he then changed his mind and decided that publication would, on balance, be to the national advantage.[6] By November, however, he had 'suddenly developed fresh qualms'. Eden was 'anxious about the effect on his reputation and fears he may look like an appeaser too'.[7] Sir Alexander Cadogan was more forthright: 'Does A. realise that *he* is responsible for the great and tragic appeasement – not reacting to German occupation of the Rhineland in 1936? How lucky he is – no one has ever mentioned *that*! And that was the turning point.'[8] The Prime Minister's reaction was more charitable. It was, thought Churchill as he reviewed the documents relating to German colonial claims, a 'pity that Mr. Eden's name should be mixed up directly with appeasement'.[9] Yet, far from accepting historical reality with its shades of grey, Eden sought in the years ahead to portray the dividing lines of the 1930s and his stance in relation to them in ever more stark and clear-cut terms. 'Appeasement' became a blanket term of abuse, separated from his own outlook by an unbridgeable divide. The lessons of the 1930s and their supposed application to later decades developed into a near obsession,[10] and he bitterly assailed those who offered even a moderately charitable interpretation of the rationale behind appeasement.[11]

Yet if Eden as Foreign Secretary to Baldwin and Chamberlain never found his policy as out of step with the desires of his governmental colleagues as he tended to imply, this is not to suggest that his position within the administration was a noticeably comfortable one. In his first weeks in office he had to beat off the attempts of Sir Warren Fisher, the head of the Civil Service, to claim a veto over ambassadorial appointments.[12] After a particularly difficult cabinet meeting, he later recalled, Baldwin took him aside and explained that in any cabinet there were twenty-two would-be Foreign Secretaries and only one Minister of Labour.[13] Youth probably counted against him and he never in his first spell as Foreign Secretary carried the weight which his office implied. Senior figures such as Chamberlain and Simon belonged to an older generation, while Eden seemed on the whole not to cultivate the friendship of those younger ministers who were generally more sympathetic to his own outlook. Perhaps, as one Italian diplomat surmised, Eden remained at this stage of his career the PPS of Sir Austen Chamberlain.[14] 'Not impressive in Cabinet,' judged Hoare at the time of Neville Chamberlain's accession to the premiership in May 1937.[15] In fact the surviving documentation suggests that Eden was not particularly effective in expounding his views and fighting his department's corner at the cabinet table. Not for nothing did Cadogan once compare Eden to John Simon.[16] As a result, in so far as Eden and his ministerial colleagues did find themselves at odds, this was something of which the Foreign Secretary and his immediate circle were altogether more conscious than the rest of the cabinet. 'The majority of the Cabinet are against A. E.', asserted Harvey nearly four months before Eden's resignation, but the surprise with which most ministers greeted Eden's departure hardly suggests that this was a judgement they would have shared.[17]

At no time in his career did Eden cultivate political friendship for its own sake. As one who worked for him later put it, he felt 'a strong affection for a far narrower circle of friends than most men'.[18] It was probably as a reaction to relative isolation among his peers that Eden placed particular value upon the friendship, advice and encouragement of a small group of ministers and officials of his own generation within the Foreign Office. Of these the most important were his junior minister, Lord Cranborne, J. P. L. Thomas, who became his PPS in May 1937, and Oliver Harvey, from January 1936 Eden's private secretary in the Foreign Office. Eden gave what many saw as undue weight to the views of these men, which they repaid with unwavering admiration and support. Critics maintained that this entourage was able to play upon the Foreign Secretary's vanity, but all remained close friends of Eden until their deaths. By contrast, Vansittart and Cadogan, who successively served as Permanent Under-Secretary, never exercised comparable influence. Though the former came to believe that he and Eden shared much common ground on the German question, the Foreign Secretary 'could not stand it being thought that any opinion he uttered in the Cabinet was really the voice of Vansittart'.[19]

Public reaction to Eden's appointment in December 1935 was overwhelmingly favourable, particularly among those under fifty. It was, asserted the *New Statesman*, 'the best Christmas present the PM could have given us'.[20] That praise should have come from such a quarter emphasized that Eden's appeal was not restricted to the Conservative Party. His enthusiasm for the League of Nations enabled him to symbolize the emotions and aspirations of many on the centre and left of the political spectrum. Eden's promotion, suggested Henry Channon, was 'a victory for "The Left", for the pro-Leaguers'.[21] The appointment, insisted the *Spectator*, 'demonstrates the will of the Government, and still more the will of the people, that the foreign policy of Great Britain shall be maintained not only in words, but in deeds'.[22] In this sense the nation seemed to have chosen its own Foreign Secretary and Eden was probably the only figure who could now help the government recover from the discredit of the Hoare–Laval Pact. Such disquiet as there was came from the Conservative right. 'I expect the greatness of his office will find him out,' warned Churchill. 'Austen [Chamberlain] wd. have been far better and I wonder why he was overlooked.'[23]

But, whether they rejoiced at Eden's appointment or not, few could fail to appreciate his difficult inheritance. The problem of Italian aggression against Abyssinia remained unresolved and Eden, though he had earlier been involved in looking for a compromise, was now compelled to seek a League 'victory', which would be seen to thwart Mussolini's ambitions. The danger, however, clearly existed that the loss of Italian friendship would result in driving that country into Germany's arms. At the same time Eden needed to restore confidence in the government both at home and abroad and to repair Anglo-French relations.

*

To his credit Eden placed the German problem at the top of his agenda. One of his first acts was to circulate to the cabinet a collection of reports received from the Berlin embassy since Hitler's assumption of power. 'I had by this time', he later recalled, 'occasionally used the

word "appeasement" in a speech or minute for the Foreign Office
in the sense of the first meaning given in the Oxford English
Dictionary, "to bring peace, settle (strife, etc.)".' This, he insisted,
was a different concept from the policy subsequently pursued by
Neville Chamberlain of seeking to pacify by satisfying demands.[24]
But it seems beyond question that at this stage Eden truly believed
that it was possible to negotiate with Hitler. His aim, he told Harold
Nicolson, was to avoid another war, to which end he was prepared to
make great concessions to satisfy German appetites providing they in
return would sign a disarmament treaty and rejoin the League. His
idea was to work for a package deal for the following three years and
then 'suddenly to put it before the League'.[25]

Inside the cabinet room Eden was more specific. In a covering
memorandum, entitled 'The German Danger', circulated with the
embassy reports, he concluded that the Führer aimed at 'the destruction of the peace settlement and the reestablishment of Germany as
the dominant power in Europe'. Such ambitions would be achieved
through the complete militarization of the German nation and by
economic and territorial expansion to absorb all Germans presently
citizens of neighbouring states. Hitler sought new markets and new
fields for German migration, together with control of raw materials
which Germany lacked. Though this analysis failed to define the
likely extent of German expansion, it is not one from which most
historians would dissent. Of particular interest, however, are the
conclusions which Eden drew from his own exposition. The first
was that Britain had to hasten and complete her own rearmament.
The second was that the government should examine the possibility
of 'some *modus vivendi* – to put it no higher' with Nazi Germany
which would be both honourable and safe for Britain while lessening
European tension.[26] These twin goals of rearmament and the quest
for an agreement remained central to Eden's policy throughout his
Foreign Secretaryship. If this was appeasement through strength, it
was appeasement none the less. Though a settlement would 'not
easily be realised', Eden had not concluded that the Third Reich's
ideological fanaticism placed it beyond the pale of diplomatic negotiations.[27] Significantly, Vansittart, another whose anti-appeasement
credentials would later shine brightly, was thinking along the same
lines and concluded that advantage lay in making one serious
attempt to come to terms with Germany. Eden did not dissent. His
proviso was that agreement should be achieved not by offering
Germany 'sops' which would only whet the appetite they were

designed to satisfy, but by making proposals for a 'final settlement'. Precisely what Britain might offer as her part of this deal Eden was not at this stage ready to define, but he seemed willing to consider colonial restitution, financial help for the German economy, special trading concessions in areas such as the Danube basin and even the abandonment of the Rhineland demilitarized zone.[28] The experience of Italy's behaviour, which Eden attributed in part to her economic distress, made the notion of economic appeasement particularly attractive: 'Our purpose being to avoid war, it should follow that we should be wise to do everything in our power to assist Germany's economic recovery, thereby easing the strain upon the German rulers, and making an outbreak less likely.'[29] A cabinet committee was established to investigate the matter further. 'Thank heaven we are moving that far at any rate,' recorded Lieutenant-General Pownall, assistant secretary to the Committee of Imperial Defence. 'Vansittart's memorandum gave up the demilitarized zone as a lost cause and strongly urged the handing back of colonies as well.'[30] It could thus be argued that Eden was prepared for a more systematic quest for an agreement with Germany than anything attempted since Hitler had come to power.

In the following weeks Eden gave every indication that he was serious about this mission. On 15 February 1936 his junior minister, Lord Cranborne, saw the counsellor at the German embassy, Prince Bismarck. He insisted that the British government attached particular importance to cultivating friendly relations with Germany and that it was Eden's belief that close collaboration between Britain, France and Germany was essential for European peace.[31] The remilitarization of the Rhineland, forbidden under both the dictated Treaty of Versailles and also the 1925 Treaty of Locarno, of which Germany had been a willing signatory, was likely to be high among German desiderata. By the time that Cranborne spoke to Bismarck, however, Eden was already contemplating offering Germany such a concession, notwithstanding the implications it would have for the strategic balance between Germany and France:

it seems undesirable to adopt an attitude where we would either have to fight for the zone or abandon it in the face of a German reoccupation. It would be preferable for Great Britain and France to enter betimes into negotiations . . . for the surrender on conditions of our rights in the zone while such a surrender still has got a bargaining value.[32]

Not surprisingly, Eden was unenthusiastic about the ratification of the Franco-Soviet Pact of May 1935 since this might give Germany the pretext for a unilateral denunciation of Locarno. He also declined to give a direct answer to Flandin when the French Foreign Minister enquired about how Britain would react to a German remilitarization. As the zone was constructed 'primarily to give security to France and Belgium', Eden argued, it would in the first instance be up to those two governments to decide what price they were prepared to pay for its maintenance.[33] Though he did not rule out the possibility of Hitler embarking on further foreign adventures, this prospect represented not a bar to Anglo-German negotiations but an 'additional reason for coming to terms quickly'.[34] On 24 February Eden wrote to the British ambassador in Germany of the 'supreme effort' which would soon be made to reach an understanding with Germany, and in a further paper submitted to the cabinet's Foreign Policy Committee suggested a package which included a new Air Pact, negotiations for remilitarization and recognition of Germany's special interests in Central and Eastern Europe. Precisely what implications such concessions might have for the sovereignty of Czechoslovakia and Poland remained unclear.[35]

In public Eden managed to create a somewhat different impression. When asked in the Commons on 12 February to give an assurance that the Locarno provisions would be carried out in the event of a flagrant violation of the demilitarized zone, he insisted that the government would honour its obligations.[36] His maiden speech as Foreign Secretary a fortnight later restated his well-known claim to be a champion of peace through collective security, but contained a statement which, as an early biographer noted, 'might be used in textbooks as an awful example of a platitude':[37]

> It would be equally idle to deny that that is an anxiety which we must all share on the Government Bench, an anxiety which is not minimised, though it is mitigated, by reflection that the course which this country pursues in the next year or two may well be a decisive factor on events.[38]

By early March the likely imposition of oil sanctions against Italy made it probable that Mussolini would take Italy out of the League and denounce her commitments under Locarno. In this situation the cabinet met to consider Britain's obligations under that treaty. With Baldwin's support Eden secured agreement that the best solution was to embark on diplomatic action to prevent the issue of Britain's Locarno obligations arising in the first place. To avert a 'very difficult situation' the government should 'come to some agreement with Germany'. Eden was authorized to follow up his earlier approaches and begin formal negotiations with Germany for an Air Pact.[39] In the light of the cabinet's decision Eden duly met the German ambassador, von Hoesch, on 6 March and received a favourable initial reaction to the idea of an Air Pact. At a further meeting the following morning, however, von Hoesch announced that the Germans were already in the process of restoring full German sovereignty to the demilitarized zone. The Germans in fact had seized the key prize which Eden had hoped to use as his chief bargaining counter in forthcoming negotiations. Almost certainly he was taken by surprise. He had not paid sufficient attention to Germany's repeated insistence that the Franco-Soviet Pact, ratified in February, was incompatible with French obligations under Locarno.[40]

This was, Eden later recalled, 'the most carefully prepared example of Hitler's brazen but skilful methods' to date.[41] Even so, the Foreign Secretary determined that Germany's action should not be the occasion of armed conflict. His reaction was one of irritation at the way the Nazis had handled the matter. The saving grace, he believed, was that von Hoesch's announcement was accompanied by a fresh series of German proposals, designed to confirm 'their unchangeable longing for a real pacification of Europe between states which are equals in rights and equally respected'. These included negotiations for a new demilitarized zone, twenty-five-year non-aggression pacts between France, Belgium and Germany, pacts between Germany and the countries of Eastern Europe, an agreement on limiting air strength and, tantalizingly, the re-entry of Germany to the League. As this list contained several of the goals which Eden had had in mind over the previous weeks, he could scarcely ignore the German offer, even though remilitarization had been secured by unilateral action rather than negotiation. Yet his reaction was based on a curious contradiction. On the one hand what had happened raised grave doubts about Germany's good faith:

The myth is now exploded that Herr Hitler only repudiates treaties imposed on Germany by force. We must be prepared for him to repudiate any treaty even if freely negotiated (a) when it becomes inconvenient, and (b) when Germany is sufficiently strong and the circumstances are otherwise favourable for doing so.[42]

Few would now dispute this judgement and, as Eden later concluded, if Hitler was not to be believed over Locarno – which he had previously made a point of differentiating from the hated Treaty of Versailles – he could not be believed in anything.[43] But at the same time the Foreign Secretary persuaded the cabinet that Germany's power for mischief meant that it was still in Britain's interests to conclude with her 'as far-reaching and enduring a settlement as possible whilst Herr Hitler is in the mood to do so'.[44] He seemed to believe that there were short-term advantages in securing an agreement even if Hitler might tear it up at any moment. The cabinet now authorized Eden to inform the Commons that, while one of the main foundations of Western European peace had been cut away, it was Britain's manifest duty to replace it if peace were to be secured:

> It is in that spirit that we must approach the new proposals of the German Chancellor. We must examine them clear-sightedly and objectively, with a view to finding out to what extent they represent a means by which the shaken structure of peace can again be strengthened.[45]

If Eden had earlier ruled out 'sops' to Germany, this did not now prevent him from persuading Baldwin to set up a committee under the Earl of Plymouth to consider a colonial mandate as a *quid pro quo* for Germany's return to the League.[46]

Whether, as Eden hoped, the crisis would pass without conflict was, of course, not entirely in Britain's control. As he had already conceded, any response to a German violation of the demilitarized zone was primarily a matter for France and Belgium. Indeed, what had happened in the Rhineland was as likely to precipitate a crisis in Anglo-French as in Anglo-German relations. Not surprisingly, Eden summoned the French ambassador as soon as his interview with von Hoesch on 7 March was over. Urging the need for frank Anglo-French discussion on the German memorandum, he expressed confidence that the French would do nothing to make the situation more difficult. 'We must not close our eyes to the fact that a *"contre partie"* was offered [by Germany] and that that would undoubtedly have a very considerable effect on public opinion.'[47] He took comfort that Flandin, by calling for a meeting of the League Council, seemed not to be treating the matter as one of 'flagrant' violation necessitating an immediate response under the terms of Locarno. Overall Eden gave the impression of 'a man who was asking himself what advantages could be drawn from a new situation, not what barriers should be put up against a hostile threat'.[48]

Accompanied by Lord Halifax, Eden now went to Paris to win the backing of France and Belgium for his policy. He found Flandin outwardly inflexible, calling for sanctions including, in the last resort, military action. The British ministers responded that, although Britain would not join in military action against Germany, she was ready to consider a renewal and reinforcement of her Locarno obligations. Vansittart had already warned that 'if we can *not* manage France, there may be a *débâcle* at Geneva'. The dangers of allowing Germany to drive a wedge between Britain and France were, the Permanent Under-Secretary argued, extremely great, not least for Eden himself:

> The acid test of Germany's sincerity will come during [the next] five years. On form up to date, she can *not* be relied upon to be as good as her word. . . . If the wedges were driven, and then Germany proved false when grown to full strength, I fear that your position would be very difficult. And you have at least a quarter of a century more of politics ahead of you.[49]

It is now generally accepted that France's acquiescence in Germany's Rhineland coup was the result less of British pressure than of the refusal of the French General Staff to take unilateral action.[50] In his memoirs Eden was keen to emphasize evidence that the French government and people were far from united in favour of a forthright approach.[51] But in his report to the cabinet on 11 March he seemed genuinely concerned that the French might indeed take military action. The possibility of war was real. The British line of condemning Germany and then developing a constructive policy to re-establish a European settlement had, he stated, no chance of acceptance. France and Belgium both believed that if the German challenge was not taken up, a much more formidable situation would arise two years later, when war would have to be fought under very unfavourable conditions. If France and Belgium resorted to military measures, Britain would have to decide whether she intended to fulfil her treaty obligations – 'and we should be in an impossible position if we refused'. During the ensuing discussion Duff Cooper, the Secretary for War, seemed to cast doubt on Eden's assumption that there were advantages in concluding an agreement with Germany now with the possibility of a conflict delayed until Britain's rearmament plans had been completed. As Cooper pointed out, 'though we should have reconditioned at any rate to some extent our small forces, yet by that time Germany would have 100 divisions and a powerful fleet. We should not relatively, therefore, be in a better position.' None the less, the cabinet accepted Eden's proposal. Britain should inform Germany that she was still anxious for a peaceful settlement but hoped that Germany could contribute to that end by agreeing to withdraw all forces from the Rhineland above those needed for a symbolic occupation.[52]

Over the following week of negotiations, which left Eden 'haggard with exhaustion', the French position became progressively more flexible.[53] Agreement was reached on 19 March. Britain would join a mutual agreement with Belgium and France against further unprovoked aggression. The promise of staff conversations should negotiations with Germany fail to bear fruit, which provoked some opposition inside the British cabinet, induced Flandin and Van Zeeland, the Belgian Prime Minister, to accept remilitarization as a *fait accompli*. Eden's attitude was governed throughout by 'the desire to utilise Herr Hitler's offers in order to obtain a permanent settlement'.[54] To the extent that the situation had been defused without a major breach in Anglo-French relations, Eden had cause for satisfaction, though he scarcely merited the praise which Vansittart showered upon him:

> You have extricated your country . . . from a position in which it might have been either dishonoured or isolated, or forced into dangerous courses. . . . We have *got* to be cautious, and not be carried away prematurely, or we may pay for it with our national existence. . . . The future will bear you out right enough – and you have such a long personal future still.[55]

Whatever Eden had achieved had been at the cost of a considerable shift in the European strategic balance. As Cranborne pointed out, without the demilitarized zone, France could not effectively fulfil her obligations in Central and Eastern Europe which would ultimately have to be liquidated.[56]

Eden's Commons speech on 26 March was certainly a considerable personal triumph, 'the speech of his life' according to Neville Chamberlain.[57] He announced that Britain's aims were to avert the danger of war, to create conditions in which negotiations could take place and to bring those negotiations to a successful conclusion which would involve Germany's return to the League.[58] It was his contention, at the time and later, that the French and Belgian governments did not have sufficient public backing to allow for the effective use of force. Being democracies they could not have acted without that backing, even had they wished to. Similarly, opinion in Britain, famously epitomized in the remark of Eden's taxi-driver that 'Jerry can do what he likes in his own back garden, can't he?',[59] did not believe that Germany's 'crime' was sufficiently grave to justify a military response. This was true enough though, as Harold Macmillan tellingly argued, 'it is hard to see how this differs from the defence, two years later, of the Munich settlement'.[60] In addition, at least one of those who later figured among Eden's close associates believed that the account given in *Facing the Dictators* unduly minimized the support in the Commons and the country for a firm response to Germany's coup.[61] Almost certainly, however, opinion in all three countries merely confirmed a conclusion which Eden had already reached – that military action was out of proportion to what Germany had done.

Be that as it may, the Rhineland later assumed an important place in that mixture of fact and fantasy which made up the early historiography of appeasement. As the pace of German aggression escalated, March 1936 was widely seen as the last time that Hitler could have been stopped without war. Later comment assumed that an immediate military response would have been a relatively simple operation, limited perhaps to 'police action'. This, it was argued, would have been successful and might have led to the fall of Hitler. 'If the French government had mobilized,' wrote Churchill with the authority of a man who had been proved 'right' about the Nazi menace, 'there is no doubt that Hitler would have been compelled by his own General Staff to withdraw, and a check would have been given to his pretensions which might well have proved fatal to his rule.'[62] Almost certainly, however, Germany could only have been thwarted if the democracies had been prepared for military conflict. As Philip Bell puts it, 'the opportunity open to France was to stop Hitler *by war*, not without war'.[63] The idea that Hitler would give up the Rhineland without a fight was not, as Chamberlain recognized at the time, 'a reliable estimate of a mad dictator's reactions'.[64]

Even so, as the man with direct responsibility for British policy at this time, Eden understandably came to feel vulnerable about his conduct over the Rhineland crisis. An allied move against Hitler, he conceded more than twenty years later, 'would have been the right thing to do and many millions of lives would have been saved'.[65] Unusually, his memoir account conveys a note of self-reproach – the

German occupation was 'an occasion when the British and French governments should have attempted the impossible'.[66] But, if Eden's conduct in 1936 is open to criticism, it is less in his failure to propose military action than in his conviction that Britain should still aim for a political agreement with Germany, even if the latter might break that agreement at any convenient time. Such a strategy inevitably implied that the time would come when Britain would have no alternative but to fight to stop Hitler's onward march. Though Eden wanted Britain to use whatever time might thus be gained to strengthen her armaments, it was by no means certain, as Duff Cooper had reminded the cabinet on 11 March, that any postponement of conflict would be to Britain's advantage. The cabinet minutes from the spring of 1936 onwards do reveal more urgency about the need for rearmament. But British industry faced long-term problems in its shortage of skilled labour and plant capacity to produce the materials of war.[67] Lord Weir had already warned ministers, including Eden, that the rearmament programme could not be completed in five years without affecting the export trade, unless quasi wartime controls were imposed on British industry – a condition which, Eden seemed to accept, was politically impossible.[68]

The ultimate outcome of the Rhineland crisis produced little cause for satisfaction from a British point of view. The German government lost interest in its own proposals for a European settlement once the danger of a French military response had passed. On 1 April Ribbentrop delivered a revised version of Hitler's earlier peace proposals, elaborately set out under nineteen sub-headings. Eden took up a suggestion from Flandin for the presentation of a questionnaire to Hitler to determine the latter's intentions, which was finally delivered on 7 May. It represented, as Eden privately admitted, 'all that had survived of our obligation to march into the Rhine'.[69] To the cabinet he justified the questionnaire on the grounds that it would buy time for rearmament, though there was some concern inside the Foreign Office that its conciliatory language might only encourage Hitler to strike again.[70] The Germans' response, however, was repeatedly postponed until the whole matter was finally buried by Hitler in a Reichstag speech at the end of January 1937. Meanwhile Anglo-French staff talks went ahead, but were limited to logistical questions and represented only a small step towards a military alliance. Eden himself seemed exasperated by French behaviour. 'Our French friends are really difficult to understand and to tolerate with the patience that they no doubt deserve,' he noted after thwarting a final French attempt to impose sanctions upon Germany.[71] British policy remained to seek agreement with Germany, however improbable this now was. Far from the remilitarization marking the end of Hitler's ambitions, his mind was turning to the 'next item on Germany's menu'. 'Miracles', commented Vansittart on 17 March, 'do still come about, and we must continue to hope and work for one.'[72]

*

In the eyes of cabinet ministers such as Neville Chamberlain, the increasingly worrying behaviour of Nazi Germany provided the most telling of reasons for coming to terms with Mussolini's Italy. Granted the circumstances in which he had succeeded Hoare, this would never have been an easy path for Eden. It was important for his credentials as a League man that the Italian dictator's ambitions should be thwarted.[73] Furthermore, he never shared the acute fears of the majority of the cabinet that Germany and Italy might come together in an anti-British alignment. In fact, his unwillingness to contemplate the dangers posed by a nightmare coalition of Britain's three potential enemies – Germany, Italy and Japan – remained one of Eden's most important differences with his colleagues throughout his Foreign Secretaryship. As he explained in early March 1936, a *rapprochement* between Italy and Germany was unlikely, partly because of Italy's interest in maintaining Austrian independence, partly because it would embitter Italy's relations with Yugoslavia and partly owing to the 'well-known unreliability of Italian policy, of which Herr Hitler was well aware'.[74] It may fairly be argued that neither Hitler nor Mussolini was appeasable, in the sense of offering the chance of a long-term, reliable settlement with Britain. But Eden's greater willingness to chase the illusion of a German agreement is one of the least impressive features of his record in the 1930s. In all probability he had concluded as early as his conversations with Mussolini in June 1935 that the Italian dictator could not be trusted and that concessions to him were therefore pointless.[75] Eden's references to Mussolini – the 'anti-Christ' – display a depth of animosity never matched in his attitude towards Hitler. As he once told Oliver Harvey, he always had to be 'particularly careful' to prevent his personal prejudices in regard to Mussolini from colouring his political attitude 'too much'.[76] Almost certainly Eden failed in this self-imposed objective.

For a while, in the first weeks of 1936, it seemed that the League's will might indeed prevail, with the Italians unlikely to conquer Abyssinia before the rainy season set in. On 17 January Eden declared that aggression must not succeed and that Britain would always be found on the side of collective security.[77] By the time that the cabinet met on 26 February, however, there were signs that the Italians could yet secure victory before the weather turned against them. Eden now argued that 'on balance' the League ought to agree to impose oil sanctions.[78] He seems not to have been fully confident that such a measure would be successful, especially as American oil exports to Italy had already reached three times their normal level but, short of actual war, an embargo was the only weapon the League had left.[79] With two ministers insisting that their dissent be recorded, Eden was asked to avoid taking the lead at Geneva and to carry out the government's policy while attracting as little publicity to himself as possible.

When Flandin, who wanted another attempt at mediation, put obstacles in the Foreign Secretary's path – in particular a demand for support in maintaining the demilitarized zone – Eden had little option but to make a speech in which he took the lead over sanctions to an extent that the cabinet had not wanted. But the way in which Britain, in deference to French wishes, acquiesced in delaying further meetings of the sanctionist states suggested that this was little more than a gesture. 'If it is applied', Eden argued, 'it will be applied less by reason of its probable effectiveness, than as a means of demonstrating the determination of Members of the League to persist in the policy upon which they have embarked.'[80] Thereafter the Rhineland occupied centre stage and Eden had not succeeded in imposing oil sanctions before the Abyssinian emperor fled to Palestine in May. Eden was still hoping in April that the Italians might be defeated by the weather, but, as Vansittart commented, 'it has got to rain *very* hard – and very *quick* – and very *long!*'[81] The idea of closing the Suez Canal to thwart Italy's imminent triumph had considerable support inside the Foreign Office, but Eden failed to speak up for this plan at the cabinet on 22 April – precisely the failing, from a departmental viewpoint, which he had so often held against Simon.[82]

Though Eden found it difficult to contemplate reaching a settlement with Italy – 'the British public is not prepared to sit round the table on friendly terms with the aggressor for many a long day – nor can I blame them'[83] – the twin setbacks over the Rhineland and Abyssinia provoked a reappraisal of British policy within the

Foreign Office and inside Eden's own mind during the spring of 1936. From the CID secretariat Pownall noted that Eden had been 'talking to Hankey about "Collective Security". At last the light is dawning! It has failed in the "perfect case" of Italy, it will never work for a German crisis.'[84] By 22 April Eden was ready to share his doubts with the cabinet as to whether the League could prevent a German putsch against Austria.[85] The following day the Conservative Foreign Affairs Committee showed that it too was taking stock of a changed situation. Eden received a report on its proceedings from his friend and PPS, Roger Lumley. Everyone present had agreed that the time for a showdown of some kind had arrived. Right-wing members argued that this should come about through a frank recognition that sanctions had failed. It was necessary to accept Italy's *fait accompli* in Abyssinia and let her return to the negotiating table, so that all energies could be directed towards resisting Germany. Lumley reported considerable support for abolishing the sanctionist Article 16 of the League Covenant and for the notion that any reorientation of policy must be based on attempting nothing which it was beyond Britain's capacity to perform. He ended by insisting that Eden, having been 'so very pro-League', was the only person who could now credibly advocate 'a realistic policy'.[86]

Four days later Eden and Chamberlain agreed that it was time to divest the League of its power to impose sanctions, leaving it as no more than 'a moral force and focus'. Thereafter peace would depend on a system of regional pacts to be registered and approved by the League.[87] In cabinet on 29 April, Eden posed three fundamental questions: what future was there for the League, granted its failure over Abyssinia; what future was there for sanctions; and what policy should be pursued towards Germany? With such radical thinking in the air, the cabinet accepted Eden's suggestion of a new committee to consider the future direction of British policy.[88] This committee met the following day and heard Eden admit that collective security was likely to be less rather than more successful in future. The Foreign Secretary found the meeting 'quite helpful', noting a general consensus that some modification of Article 16 would be necessary and that there must at the very least be 'some stocktaking' by the League after the 'hapless affair' of Abyssinia.[89]

It is against this background that the decision to lift sanctions against Italy must be viewed. There was some danger that Eden, having been so associated with the League in the public mind, might suffer a personal reverse if he could not adapt to changed circumstances. 'The League is discredited,' noted Thomas Jones in early May. 'Eden has played his hand out.'[90] But politicians' minds had moved faster than those of the general public and there was an equal danger that Eden would lose popular credibility if he were now seen to renounce his League credentials. In the event matters worked out better than he might have expected. Inside the government there was a growing conviction that attention should be focused on Germany. This was reasonable enough, granted that the cabinet's Defence Requirements Committee had determined as early as 1934 that Germany was 'the ultimate potential enemy against whom our long-range defence policy must be directed'. Eden recorded a meeting with Baldwin on 20 May. The Prime Minister was characteristically vague but at least clear that he wanted 'better relations with Hitler than with Musso'. When asked by Eden how this was to be achieved, Baldwin could only reply, 'I have no idea, that is your job.'[91]

Not that Eden was any longer sanguine of an accommodation with Germany. Reports reaching the Foreign Office painted a gloomy picture of Europe's future. The best that Eden could do was to insist that this made no difference 'to our intention to probe and explore Herr Hitler's offers and to construct, if possible, something reliable out of them'.[92] By early June he was still more pessimistic: 'We must continue to aim at the general agreement. . . . It may be that it is unattainable, but it is we who must prove this, by making every effort to attain it.'[93] Eden was disappointed that Hitler had still not responded to the list of questions sent to him in early May. This failure to reply suggested that 'his intentions were bad'.[94]

Be that as it may, logic suggested that, if the diplomatic focus was to be directed towards Germany, little would be achieved by policies which continued to incur Italian hostility. Eden had told the Commons as early as 6 May that sanctions had failed.[95] At the cabinet three weeks later majority opinion favoured lifting sanctions and Eden found himself, along with Neville Chamberlain, in a minority that 'felt tenacious'. It was, he conceded, a 'desperately difficult' decision and, not for the last time, he admitted that personal factors might be colouring his attitude. 'To some extent one is inevitably influenced by dislike of Italy and her methods.'[96] Certainly Eden's conditions for Italian negotiations contained reservations which he was less inclined to impose when it came to Germany. While accepting that the fall of Addis Ababa on 5 May had left the sanctions policy in ruins and that the quest for a German settlement made continuing Anglo-Italian friction undesirable, he remained reluctant to grasp the sanctions nettle. In response to reports that Mussolini was ready to negotiate a Mediterranean security agreement in exchange for the lifting of sanctions, Eden minuted:

> There is a touch of blackmail about this, and we are not in a mood to be blackmailed by Italy. Some constructive contribution by her is also called for and there is no evidence of it here. If Mussolini thinks he has only to beckon and we will open our arms, he is vastly mistaken.[97]

Two days later he conceded that it was no longer practicable to secure the withdrawal of Italian forces through sanctions and that he would therefore be circulating a memorandum proposing their termination.[98] By 10 June he was 'rather veering' to the view that, if sanctions were to be removed, there was something to be said for Britain taking the initiative.[99] On that same day, however, it was the Chancellor, Neville Chamberlain, who seized that initiative.

When, therefore, Chamberlain, speaking to the 1900 Club, described the continuation of sanctions as 'the very midsummer of madness', he was scarcely bouncing Eden into a policy of which the latter disapproved. Rather Chamberlain seems to have concluded that Eden, having accepted the intellectual argument, needed to be prodded by a public statement which it was easier for the Chancellor than for the Foreign Secretary to make. Thus Chamberlain noted his belief that Eden was 'entirely in favour of what I was going to say' even though he was 'bound to ask me not to say it!'[100] In practice this worked to Eden's advantage, since it enabled British policy to be changed without undue damage being done to his personal status as the upholder of the League.[101] Eden was no doubt annoyed by Chamberlain's initiative – though characteristically 'as nice as possible about it' – but, as Vansittart pointed out, 'the idea to do the big thing (and the right thing) was entirely spontaneous and yours'.[102] It was Cranborne, who had a knack of holding views erroneously attributed to Eden, who emphasized that the raising of sanctions would be 'bad, both for the future of the League and the government and of our Mediterranean position in the future'.[103] It was therefore entirely consistent that Eden himself should have proposed abandoning sanctions at the cabinet on 17 June and announced that policy to the Commons the following day. He did his best to salvage some credibility for the League by insisting that the Italian conquest of Abyssinia should not be formally recognized. Such a step would be 'bitterly unpopular at Geneva'.[104]

*

The failure of British policy and the need to work out an alternative strategy were apparent at an important cabinet meeting held on 6 July 1936. Eden reported that recent meetings of the League Council and Assembly had been the most depressing he had attended. The international situation held out a permanent risk of some dangerous incident arising or even the outbreak of war. The alternatives were to work for a new Locarno Treaty – which would 'smash the League' – or to go back to the universality of the Geneva Protocol of 1924, which the country was unlikely to support. The only other option was something on the lines of the French proposal for a series of regional obligations combined with economic sanctions to be imposed on an aggressor by those states which were not members of that particular regional arrangement. The government needed to decide on future policy in relation to the League, Eastern Europe and the former German colonies. As Hitler would never commit himself as regards the whole of Europe, there was strong support for the idea that policy might be framed 'on the basis that we could not help Eastern Europe'. The objects of policy were defined as trying to secure world peace and keeping the country out of war. If Britain had been strong enough and public opinion 'better instructed', it might have been feasible to guarantee the whole of Europe, East and West, but this was not the case. Here, surely, were the beginnings of the policy which culminated in the Munich settlement just over two years later, of which Eden would, on occasions, be so critical. In the telling words of the cabinet secretary, the *'present* policy' – that is the policy for which Eden bore ministerial responsibility – 'of yielding to Germany on every occasion' merely encouraged Hitler to pursue his aggressive designs. But when the question was asked whether the Führer ought to be told that there was a definite limit beyond which Britain would not let him go, it was pointed out that for the moment at least Britain had 'neither the means nor the heart' to stop him.[105] The contemporary assessment of the deputy editor of *The Times* that 'the government have no foreign policy at all' was perhaps not far off the mark.[106]

If Eden did not have enough problems with which to concern himself, the summer of 1936 saw a new and unlooked-for development on the international horizon. Just three days after the League agreed to end sanctions against Italy, news arrived of a military revolt in Spain against the existing Popular Front government. Civil war soon resulted. Eden's policy was on this occasion clear from the outset, though less high-minded than once imagined. He believed that Britain's interests lay in the defeat of both extreme factions and the emergence of a middle-of-the-road Spain amenable to British pressure. More importantly, the state of British rearmament made it imperative to keep the conflict localized and not allow it to widen out into a general European war.[107] Many myths soon grew up to the effect that the war in Spain represented a microcosm of a forth-coming European-wide struggle between democracy and totalitar-ianism, and similar myths once attached to Eden's role in the war's diplomacy. According to one writer, he

> did not escape the effects of the emotional storm which swept over a whole generation of Englishmen and women; he saw the Spanish Civil War, in its simple primary colours, as a conflict between absolute right and absolute wrong in which the dictator should at all costs be totally defeated and democracy totally defended.[108]

Eden's thinking always remained more firmly rooted to the ground than this implies. He was also once credited with being the motive force behind the policy of non-intervention – a policy into which he was supposed to have pushed the reluctant French Popular Front government of Léon Blum, whose sympathies inevitably lay with the elected government in Spain. Blum did visit London on 23 July – but to discuss Locarno rather than Spain – and there is no evidence that Eden exerted any particular pressure upon him. Almost certainly French policy was determined, as over the Rhineland, by French considerations and, in particular, by fear that any partisan approach to the Spanish situation risked opening up fissures within the French body politic itself.[109]

Thus Eden had less responsibility for 'shaping the international character of the initial phase of the Spanish Civil War' than was once supposed.[110] This is not to deny that non-intervention suited British interests, domestic and foreign, better than any other policy that was realistically available.[111] By the end of July Eden was support-ing at least an informal embargo on the export of arms to Spain, not least because he did not wish to see resources diverted from Britain's own rearmament programme.[112] But non-intervention was based upon a practical calculation of Britain's best interests rather than high-minded idealism. Cranborne probably expressed Eden's own thinking:

Non-Intervention is not a moral principle. It is a device to prevent the war spreading. If that can be achieved, all other considerations pale into nothingness. The agreement is probably being broken by both sides, by everyone, in fact, except us. But any agreement is better than none.[113]

From Eden's point of view the main effect of the war and of Mussolini's defiance of the Non-Intervention Agreement – which all the powers had accepted by early September 1936 – was to increase his misgivings about the quest to restore good relations with the Italian dictator. Eden's success in negotiating a twenty-year agreement with Egypt in late August enabled Britain to concentrate her attention on Italian ambitions in the western Mediterranean.[114] He wanted to warn Mussolini that any alteration to the *status quo* in that area would be a matter of the greatest concern to Britain.[115] During the autumn Eden alternated between a readiness to 'let bygones be bygones' in relation to Abyssinia and a reluctance to accede to the Italian wish for the formal recognition of their conquest.[116] The cabinet discussed Italy on 4 November, when Samuel Hoare, restored to ministerial rank as First Lord of the Admiralty, pointed out that, despite rearmament, Britain would remain unprepared for conflict for some time to come. This meant that British policy would have to proceed 'very quietly' and the first objective should be to remove Italy from the list of countries with whom Britain might have to reckon. Eden did not dissent, though he suggested that Hoare's position might be mistaken for 'flabbiness'. But he warned that recognition of the Abyssinian conquest would be 'a very difficult gesture to make'. Something should be done to remove Italian suspicions of British intentions, but in doing so 'we must not break with the League of Nations'.[117] In view of the policies Eden had pursued earlier in the year, this was a curious conclusion to reach. Yet he insisted that Italian policy would always be opportunistic and that Britain must be on her guard against 'increasing the dictator's prestige by our own excessive submissiveness'.[118] Not surprisingly, Leo Amery, listening to Eden in the Commons, found his attitude towards Mussolini 'grudging and sniffing'.[119]

If not as blatantly as Italy, Germany too soon violated the Non-Intervention Agreement in support of Franco's rebel forces. But Eden did not allow this breach of an international agreement to divert him from the ongoing quest for an Anglo-German settlement.

He was, if anything, more hopeful about the international outlook than for some years past.[120] The basis for his optimism was a series of reports suggesting that Germany's economic position was steadily worsening and that this would force Hitler to seek assistance from Britain and France. The important point to note is less Eden's insistence that he would not seek to ease Germany's economic difficulties without a political return – 'she must be given *nothing* except she changes her political ways' – than his continuing belief that there was a realistic chance of a settlement.[121] In the last months of 1936 a growing divergence began to emerge on this question between Eden and some of his senior officials, including Vansittart, whom he was already planning to replace as soon as the opportunity offered. Figures such as Vansittart and Wigram had concluded that Germany would only make an agreement in return for the recognition of its political and economic supremacy in Eastern Europe.[122] So, while Eden was ready to ask Germany to define her desiderata in the colonial sphere, Vansittart preferred to 'keep Germany guessing' and avoid the impression that Britain would consider a colonial deal. The Permanent Under-Secretary now saw negotiations as little more than a delaying tactic – 'holding the situation till at least 1939' – to give British rearmament the chance to reach a position of strength.[123] If, therefore, Eden was marginally more cautious about negotiating with Germany than some cabinet colleagues,[124] he was now altogether keener on 'appeasement' – albeit with conditions – than several of his professional advisers.

Eden's attitude towards the dangers – or lack of them – posed by a potential coalition of enemy states also tended to separate him from key figures in the Foreign Office. When news arrived in November that Germany and Japan had concluded an Anti-Comintern Pact to oppose international communism, Laurence Collier judged this development to be definitely 'inimical' to British interests. Eden, on the other hand, refused to be alarmed, insisting that the principal target of this agreement was Soviet Russia rather than Britain.[125] This was true enough, but suggested a limited appreciation of the expansionist dynamism of the aggressor states. The contrast between Eden and Vansittart is instructive:

> Van came in and talked somewhat hysterically about this alliance being directed against us and not Russia. I fear that he is not balanced and is in such a continual state of nerves that he will end by making would-be aggressors think the more of us as a possible victim![126]

In the closing months of 1936, however, Eden found himself forced to bow to ministerial and service opinion that Britain needed to reduce the number of her possible enemies by responding to Mussolini's offer of a 'gentleman's agreement' on the Mediterranean. This agreement, concluded on 2 January 1937, noted the mutual interest of both countries in the maintenance of peace and affirmed that there was no incompatibility between their respective interests in this area. The two powers committed themselves not to alter the *status quo* in the Mediterranean. Though this stipulation offered some comfort in relation to Italian ambitions in Spain, the agreement fell short of Eden's aims. Less than a month earlier he had minuted that 'an agreement that only adds one more scrap of paper to the world and changes nothing in fundamentals is not worth so much effort'.[127] Yet despite the absence of tangible concessions by the Italian government Eden seemed reasonably content. 'This is very satisfactory,' he minuted. 'It is particularly satisfactory that we should have obtained our formula regarding Spain, and I am glad that we were firm about this.' Confident that there were few advantages in Britain being seen to be too keen to court Italian friendship, he added: 'We shall lose nothing in Italian eyes by continuing to *nous faire valoir.*'[128]

Eden's mood soon changed. On the very day that the signature of the agreement was announced, 3000 Italian troops landed in Cadiz. Though the 'gentleman's agreement' did not specifically deal with non-intervention, Eden felt let down and his attitude towards the possibility of a future accord with Italy was permanently affected. His initial inclination was to put as brave a face as possible on what had happened:

> At least we have given nothing away to Italy. It remains to be seen whether what we have gained will prove of any material value. Time alone will show and nothing would be more foolish than openly to attempt to pull Mussolini away from Hitler.[129]

On reflection, however, Eden decided to call a halt to German and Italian aggression in Spain by the strict application of the Non-Intervention Agreement to volunteers. After a discussion with Baldwin from which he concluded that he had the latter's backing, Eden arranged a meeting of ministers on 8 January to consider a complicated plan to employ the Royal Navy to supervise all approaches to Spanish ports. His attitude towards the outcome of the civil war had clearly undergone a significant change of emphasis.

The character of the future government of Spain, he now believed, was 'less important to the peace of Europe than that the dictators should not be victorious in that country'. If the present situation continued, moderate influences in Germany would be eclipsed and she would undertake 'other adventures' against, for example, Czechoslovakia. To be firm in Spain was to gain time and 'to gain time is what we want'. Hoare opposed what was effectively a plan to blockade the whole Spanish coast and Eden was understandably surprised by Baldwin's complete failure to support him. Without the Prime Minister's backing Eden's cause was lost and the final decision to prohibit the flow of British volunteers and merely to request other nations to follow Britain's lead represented a considerable rebuff. Whatever the technical problems relating to his scheme, this was perhaps the moment of his Foreign Secretaryship at which Eden tried, albeit unsuccessfully, to make his strongest stand against the dictator powers.[130] Not surprisingly, the outcome left him feeling 'tired and depressed'.[131]

The cabinet assembled for its first scheduled meeting of the year on 13 January. Eden noted that 1937 might be a critical year because of Germany's declining economic prospects. It was not clear whether she would follow a policy of co-operation or one of foreign adventure. Eden argued that two camps existed inside the Reich – the army, the Foreign Office and Schacht's Finance Ministry which were cautious, and the Nazi Party which was inherently aggressive. It was Britain's task to restrain the latter and vigorous rearmament would be of the greatest possible assistance. She should be firm 'but always ready to talk'.[132] After the cabinet meeting broke up, several ministers discussed the possibility of a British visit to Hitler, but Chamberlain and Eden were agreed that this would be 'very dangerous without invitation and clear purpose'.[133] Even so, the Foreign Secretary had clearly not ruled out the possibility of further negotiations. This may reflect the influence of Alexander Cadogan, who had returned from China in 1936 at Eden's invitation to take over the position of Deputy Under-Secretary, with the clear expectation of succeeding Vansittart – an expectation so far thwarted by Eden's inability to prise the latter from his post. Cadogan had dined with Eden on 10 January, using the opportunity to 'preach to him to get in touch with Germany and not leave her "in Coventry"'.[134]

Eden used a Commons speech on 19 January to appeal to Germany for a policy of co-operation, making 'the way smooth for peace and prosperity'.[135] While dilatory negotiations dragged on for a five-power conference to replace Locarno, Eden probably placed more faith in behind-the-scenes discussions on the possible return of some former German colonies. This reflected his conviction that Germany's financial problems offered scope for economic appeasement. Schacht had held informal talks with Blum as long ago as August 1936 and pressure now mounted for a meeting between Schacht and Frederick Leith-Ross, the government's Chief Economic Adviser. Eden insisted that a colonial deal must be part of a general agreement extending to Europe. 'There must', he wrote, 'be a definite change of political orientation in Germany, and Germany's undertakings and assurances must not therefore be merely "eyewash".'[136] Eden indicated what a general settlement might involve at a meeting of the cabinet's Foreign Policy Committee in mid-March. He wanted a new treaty to replace Locarno, guarantees for the countries of Eastern Europe, Germany's return to the League and an agreement on arms limitation.[137]

The first months of 1937 offered the Foreign Secretary few encouraging signs. Hitler's Reichstag speech on the fourth anniversary of his accession to power spoke ominously of the rights of national minorities in Eastern Europe and of the need to crush Bolshevism in Spain. In addition, discussions on colonial restitution could not make progress in the face of a French refusal to consider the cession of Togoland and the Cameroons.[138] By mid-March Eden had to admit that the chances of an agreement were very small. German policy was committed to the destruction of the principles of collective security and in particular to weakening France's ties with Britain and her allies in Eastern Europe.[139] The importance of British rearmament appeared 'daily more apparent',[140] but Eden was increasingly cautious about any British commitments in Central and Eastern Europe. In a major address to his constituents in Leamington in November 1936 he had emphasized that British arms would only be taken up in self-defence, in defence of the Empire and in fulfilment of specific treaty obligations: 'nations cannot be expected to incur automatic military obligations save for areas where their vital interests are concerned'.[141] Speaking in Bradford a month later, Eden seemed to modify this position, arguing against 'comfortable doctrines' that Britain could 'live secure in a Western European glass-house'.[142] But it was the Leamington speech which increasingly encapsulated his views. In March he noted a distinction between 'those who accept my Leamington definition of our commitments . . . and those who want to go further and undertake in advance to fight for Czecho-Slovakia, etc.', a pledge which, according to Eden, would result 'in a considerable division in our own public opinion'.[143] Such thinking suggested that the gap which had been apparent a few months earlier between Eden and some of his officials was narrowing once more. The Foreign Secretary was near to adopting the policy of 'cunctation' – keeping Germany guessing until rearmament had created a British position of strength.[144]

*

By the time therefore that Chamberlain replaced Baldwin as Prime Minister in May 1937 there were few signs that Eden was out of step on the key issues of foreign policy with the administration as a whole. There were differences of emphasis, but not more striking than those which have always existed and will always exist in cabinet government. Eden was noticeably cooler than some of his colleagues towards the appeasement of Italy, confident that Britain had the capacity to deal with any Italian threat to British interests.[145] But his attitude was coloured by a personal antipathy towards Mussolini, based on a perception of the Italian dictator's untrustworthiness. Eden was no longer optimistic about an agreement with Germany. He had not ruled this out, but was not prepared to indulge in unilateral concessions to secure a merely temporary diplomatic success. In addition, he placed increasing importance on the support which rearmament could give to British policy, but without showing much understanding of the economic and political constraints which conditioned Treasury thinking on this issue. But only over his proposed blockade of the Spanish coast had Eden suffered a significant cabinet reverse, while as regards the central approach towards the civil war – non-intervention – the government had pursued a policy with which he was in entire agreement. The in-coming Prime Minister had little reason to suppose that he and his Foreign Secretary would find themselves at odds.

4

Eden and Chamberlain

I think that, on the facts, an ordinary man ought to have stayed.
A, being what he is, was right to go.[1]

NOTWITHSTANDING the tendency of several ministers to interfere in foreign policy – a problem to which Duff Cooper, the Secretary for War, had drawn the cabinet's attention[2] – by early 1937 Eden enjoyed a considerable latitude in the direction of British diplomacy. This was partly because it was an area of government in which the Prime Minister had never felt comfortable and partly as a result of Baldwin's poor health and preoccupation from the autumn of 1936 with the constitutional implications of the King's affair with the American divorcee, Wallis Simpson. Eden's contact with Baldwin that August consisted of 'one letter and one telephone . . . and these about my spending a weekend with him in S. Wales'.[3] But his later remark that he 'looked forward' to working with a Prime Minister who would give him 'energetic backing' was more the result of Baldwin's failure to support him in January 1937 over the Spanish embargo than it was an indictment of the outgoing premier's conduct overall.[4]

Eden seems to have been entirely happy that Chamberlain should be Baldwin's successor. He had the 'makings of a really great Prime Minister if only his health held out'.[5] But Eden's subsequent comment that he and Chamberlain were 'closer to each other than to any other member of the Government' reflected more his failure to cultivate close friendships among his colleagues than any particular warmth between him and the new Prime Minister.[6] None the less there is no evidence that either man anticipated difficulties in their ministerial relationship, still less that it would end in Eden's resignation, and Chamberlain would surely not have retained Eden's services had he been aware of any significant divergence of views. The two men had markedly different personalities and twenty-eight years separated them in age, but these factors were as likely to forge a complementary and successful partnership as they were to lead to disharmony. As Oliver Harvey remembered Eden saying, 'Neville and I will be able to get a move on'.[7]

Nevertheless Chamberlain was bound to adopt a different style of leadership from his predecessor. Though very much the same age as the lethargic Baldwin, he was active and energetic and rightly saw that foreign affairs would be at the forefront of the government's and his own agenda for some time to come. Samuel Hoare had already urged him to make his new government as unlike the old as possible: 'you cannot compete – and you certainly would not wish to – with Baldwin's slipshod, happy-go-lucky quietude.' Hoare also warned that the Foreign Office was 'so much biassed against Germany (and Italy and Japan) that unconsciously and almost continuously they are making impossible any European reconciliation'.[8] This was an exaggeration but it may have helped convince Chamberlain that it was his duty to adopt an active role in the making of British foreign policy. At all events the new Prime Minister soon concluded that the quest for European-wide appeasement could usefully begin with Mussolini's Italy, not because he any more than Eden overestimated that country's strength – 'if only we could get on terms with the Germans I would not care a rap for Musso'[9] – but because the increasing closeness of Italian–German relations made this a sensible starting-point in the drive for a lasting European peace. Such tactics might cause problems for Eden, who in January 1937 had written that 'nothing would be more foolish than openly to attempt to woo Mussolini away from Hitler'.[10] On the other hand, at no time before his resignation in February 1938 did Eden abandon hope of what was for Chamberlain always the more important aim of securing a settlement with Germany.

If Eden's differences with Chamberlain in May 1937 over Britain's likely enemies were of emphasis rather than principle, there existed greater scope for disagreement in the Foreign Secretary's perception of her friends, real and potential. During 1936 Eden had become keen that the United States should play a more active role in European affairs. Precisely when this strand of thinking first developed is not altogether clear, though recent research suggests that American neutrality and non-membership of the League may have been a consciously limiting factor in Britain's quest for an agreement with Mussolini over Abyssinia in late 1935.[11] An agreement with the United States could be invaluable in easing Britain's mounting strategic dilemma. But the aspiration of involving America in British interests and commitments would not be easy to achieve. Eden faced a legacy of nearly two decades of mutual suspicion and even antagonism in political, economic and naval relations. Within his own ministerial lifetime the experience of the Manchurian crisis had shown how difficult it was to co-ordinate a common Anglo-American response to acts of aggression. But while others, including Chamberlain, tended to conclude that it was futile to count on American support, Eden may have believed that the United States would have taken action in the Far East had Britain shown herself more ready to collaborate. Since that time the principle of neutrality had been enshrined in American law, but talks with the American ambassador, Judge Bingham, early in 1937 encouraged Eden to believe that President Roosevelt was dissatisfied with the existing neutrality legislation and would seek to amend it at an appropriate time.[12] Speaking in Aberdeen two months later Eden described partnership with the United States as a 'great stabilising factor, the influence and authority of which was of evident advantage to mankind'.[13] That same month Britain's ambassador in Washington offered corroborative testimony that American attitudes were softening: 'never in history have Anglo-American relations been so friendly and cordial as now, except during the eighteen months when the two countries were associated together in war.'[14]

Eden took up this theme at the opening session of the Imperial Conference of dominion leaders in May 1937, painting a glowing picture of Anglo-American relations. But his assertion that the United States was prepared to work closely with Britain seems to have exaggerated American willingness for any sort of partnership in the international arena.[15] Talks between Ambassador Lindsay and Secretary of State Cordell Hull in the spring indicated that the United States might be willing to enter a trade treaty with Britain, but on its own terms, and Eden was no more ready than Chamberlain to dismantle the system of Imperial Preference worked out at Ottawa in 1932 to secure such an accord.[16] America's help could be particularly effective in policing the western Pacific, increasing Britain's bargaining power in the Far East and allowing her to concentrate her naval resources in the Mediterranean, and it was for such an agreement that Eden strove during the remainder of 1937 without, it must be admitted, any particular success. Its implications for world peace might, he believed, prove incalculable. One idea was for Eden or Chamberlain himself to visit Washington. This might lead to a useful exchange of views and have a salutary effect upon the European situation, but Chamberlain remained sceptical and the plan was dropped.[17] The outbreak of hostilities between China and Japan in July seemed to Eden to provide an obvious focus for Anglo-American co-operation. On 20 July he suggested a joint approach to bring the fighting to an end and to offer the mediation of the two countries. Only such an initiative, he argued, could curb the expansion of Japan's influence in China.[18] 'We must lose no opportunity of co-operating with US Govt,' he minuted a few days later.[19] Over the following months, however, it became clear that the State Department opposed the imposition of any type of sanctions against Japan. By September Eden admitted that there was no sign of any desire in Washington to improve Anglo-American co-operation.[20]

But if Eden had been thwarted for the time being he had not lost all hope of drawing the United States into Britain's foreign policy orbit and he remained alert to any signs that the American mood was beginning to change. Chamberlain, by contrast, was altogether more cautious. 'I tried to get them to come in on China and Japan', he assured his sister in late August, 'but they were too frightened of their own people, though I believe if they had been willing to play there was an off-chance of stopping hostilities.'[21] Not surprisingly, Eden welcomed Roosevelt's famous 'quarantine' speech of 5 October as indicating a readiness to consider closer American involvement in world affairs.[22] The President declared that world lawlessness was spreading and suggested that 'when an epidemic of physical disease starts to spread, the community . . . joins in a quarantine . . . to protect the health of the community'. Eden saw in this evidence that the United States was abandoning its 'psychological withdrawal from the world stage'.[23] But Roosevelt had said very little – a 'characteristic mixture of vague rhetoric and moral uplift'[24] – and Chamberlain's reaction was altogether understandable: 'In the present state of European affairs . . . we simply cannot afford to quarrel with Japan and I very much fear, therefore, that after a lot of ballyhoo the Americans will somehow fade out and leave us to carry all the blame and the odium.'[25] Though there were as yet few signs of actual conflict, here lay the seeds of a dispute which would do as much as anything to bring the Eden–Chamberlain partnership to breaking point.

The gap between Prime Minister and Foreign Secretary became clearer as the nine-power Brussels Conference on the Far East convened in early November. Eden announced that he would travel 'not only from Geneva to Brussels, but from Melbourne to Alaska' to secure full American co-operation. Yet America's mood scarcely matched his own. Roosevelt indicated that any attempt to pin the United States down to specific commitments or even to an explanation of the 'quarantine' speech would be 'objectionable and damaging'.[26] But Eden still managed to conclude from conversations with Norman Davis that, if efforts at conciliation in the Far East came to nothing, the President would be able to do 'something more'.[27] He remained concerned not to rebuff any sign that the United States was stirring from its isolation:

> Any suggestion that we are lukewarm in the matter of joint action might fatally impair the good will of President Roosevelt, and we should be made to appear once more as having rebuffed an American offer of co-operation as in the case of Manchukuo.[28]

To suggest that Britain would follow any American initiative was reasonable enough. The problem was that the Brussels Conference produced none.

Meanwhile Eden's attitude towards European problems continued along paths which he had marked out before Chamberlain's accession to the premiership. Throughout the spring and summer of 1937 he resisted calls for the granting of *de jure* recognition of Italy's conquest of Abyssinia. As he minuted, 'I can imagine nothing more gratifying to [Mussolini] than the picture of a British Foreign Secretary "touting" for signatures to secure the expulsion of Abyssinia in order that conversations may begin with Italy.'[29] This was not the sort of language Eden tended to use in relation to Germany. In a memorandum in early April he argued that Mussolini's ambitions were ultimately incompatible with British interests. While, therefore, he did not rule out a short-term arrangement, any longer-term settlement would depend on Britain securing such military superiority as would allow her to dictate terms to the Italian dictator.[30] A little later Eden was telling his military colleagues that Italy posed a greater immediate danger than did Germany and that 'trouble was more likely to arise in the Mediterranean than elsewhere in Europe'.[31] Yet he still took a dim view of Italy's military capacity and rather discounted the menace of a Rome–Berlin axis. Indeed, 'Germany cannot feel comfortable if she has to face the world with only a weakened and untrustworthy Italy as an ally.'[32] By contrast, Eden saw continuingly hopeful signs emanating from Berlin, sharing the common misconception that Hitler could be separated from more radical elements within the Nazi movement.[33] 'Nothing could be more beneficial for the interests of H. M. G.', he noted in early June, 'than an improvement of Anglo-German relations even if it only proved temporary – Spain is a possible field for our collaboration.'[34] As the summer of 1937 proceeded, therefore, Eden, undeterred by repeated setbacks, continued to work for the opening of Anglo-German negotiations.

That Eden had not abandoned all hope of a German settlement was clear at the Imperial Conference in May, where he returned to the theme of economic appeasement, designed 'to restore Germany to her normal place in the Western European system'.[35] He also told the Commonwealth Prime Ministers that Germany, unlike Italy, had co-operated 'loyally, efficiently and zealously' with the Non-Intervention Committee set up to supervise the policy. Yet this statement was made barely a month after the Luftwaffe had destroyed the Spanish market town of Guernica. Eden's wide-ranging statement to the conference on 19 May was in line with ideas he had been developing over the last six months. He spoke of Britain's agreed defence priorities, as outlined in his Leamington speech the previous November, stressed Britain's commitment to extensive rearmament and underlined the importance of drawing in the United States to contain the ambitions of aggressor states. His emphasis upon Britain's formal treaty obligations had obvious implications for the status of the League in British thinking and for the future prospects for Central and Eastern Europe. If Britain were to declare her readiness to fight for Czechoslovakia or Austria, this would take her 'far beyond where the people of this country were prepared to go. There could be no greater danger than for the Government to declare themselves in favour of a policy which did not command the general support of public opinion at home.'[36] In effect, this meant that Articles 10 and 16 of the Covenant, which together spelt out the obligation to defend the territorial integrity of member states if need be by force, had been abandoned. As the South African delegate noted, 'both Chamberlain . . . and Eden agreed . . . that this was the only sound course to follow'.[37]

By the summer the first signs appeared of a significant gap opening up between the Foreign Secretary and his Prime Minister. Eden urged that the Committee of Imperial Defence should accept a new assumption – that Italy could not be considered a reliable friend and should 'for an indefinite period be regarded as a possible enemy'. At the cabinet on 14 July he proposed a show of strength in the

Mediterranean as the best way to moderate Italian behaviour, but encountered the Prime Minister's opposition.[38] Chamberlain was irritated that Eden's preoccupation with Italy tended to cloud perceptions of the greater danger posed by Germany. It was the latter, the premier insisted, which presented the biggest threat to Britain and which should have priority in defence preparations. Yet, contrary to the impression given in his memoirs, it was Eden who now took the lead in suggesting a fresh approach to Italy, addressed directly to the Duce. He told Chamberlain on 16 July that he intended to draft 'a personal letter to Mussolini. Such a course is perhaps excusable in that I have met Mussolini and that he has written to me direct – not lately, but in the early stages of the Abyssinian dispute.'[39]

When, therefore, the Prime Minister pre-empted Eden and himself took the initiative, the latter's objections can scarcely have been on grounds of policy, but rather the result of Chamberlain's unorthodox methods. As irritation at the intrusion of others in his departmental affairs would become a feature of Eden's three periods at the Foreign Office, it is reasonable to suppose that he felt considerably put out by Chamberlain's actions, even though it would also have been entirely characteristic for him to have concealed the strength of his feelings from the Prime Minister. Lasting damage may have been done to the Eden–Chamberlain relationship of which Chamberlain would have been largely unaware. As over the ending of sanctions in 1936, Chamberlain seems to have been motivated by Eden's lack of urgency in executing a policy upon which the two men were in theory agreed. He did not show Eden his letter before sending it to Mussolini, 'for I had the feeling that he would object to it'.[40] Chamberlain recognized that this would cause problems, but was keen to restore relations as quickly as possible – 'there is no desire on my part to take credit away from the F. S. and I shall try now to put him in the foreground again'.[41] Like Prime Ministers before and since, Chamberlain was becoming irritated at the slow and cumbersome ways of the Foreign Office, though at this stage his criticism was more generally directed at the office itself, rather than its political head. 'But really that F. O.!' he noted in October. 'I am only waiting for my opportunity to stir it up with a long pole.'[42]

From Eden's point of view the damage was less easily repaired. 'We should . . . go slow about Anglo-Italian *rapprochement* and talk as little as possible in the press about Abyssinia,' he now argued. The Italians would try to push Britain into concessions in advance of negotiations.[43] Quite possibly Eden would in any case have imposed conditions upon the opening of negotiations had the plan of a direct approach to Mussolini been left in his hands. As it was, he now wanted to link the question of better Anglo-Italian relations with the problem of continuing Italian intervention in Spain. This became in his mind a fixed precondition before negotiations could begin. 'By all means let us show ourselves ready to talk', he told Vansittart, 'but in no scrambling hurry to offer incense on a dictator's altar.'[44] This was the sort of memorable phrase which, taken in isolation, does much to sustain Eden's credentials as an opponent of appeasement. Not surprisingly, and unlike his letter to Chamberlain of 16 July, it duly finds its place in *Facing the Dictators*.[45]

With Eden absent on holiday, a Foreign Office meeting chaired by Lord Halifax decided in favour of negotiations. Yet concessions to Italy aroused Eden's moral scruples in a way that comparable gestures towards Germany did not. 'I am very reluctant to recognise *de jure* conquest of Abyssinia', he informed Halifax, 'and really do not think I could bring myself to any kind of approval of what Italy has done, though I share your desire to be realist.'[46] Halifax seemed to appreciate that the breach was widening. 'This letter from Anthony disquiets me,' he warned Vansittart. Eden's attitude was 'dangerously divergent from what the P.M. contemplated'.[47] Two meetings between Eden and Halifax produced few results, prompting the latter to warn Chamberlain that Eden would try to stall any forthcoming negotiations with Italy.[48] Eden even seemed 'sensitive' to the idea of including recognition as part of an overall package. In truth, as Halifax now appreciated, the Foreign Secretary was doubtful 'about getting anything out of the Italians. He dislikes and mistrusts their general make-up.'[49]

Almost certainly Chamberlain found his Foreign Secretary's scruples tiresome. As far as he was concerned, Italy was only an annoying distraction, blocking the way to the more substantive 'appeasement' of Europe – the general settlement to which Eden had so often referred. Encouraged by Mussolini's reply to his letter, Chamberlain saw the way forward to 'reduce our anxieties in the Mediterranean'. Even then 'the key of the situation is in Berlin and that is not so easily handled'.[50] The simmering conflict between Prime Minister and Foreign Secretary came into the open at the cabinet meeting on 8 September. Eden's preconditions for opening negotiations had become more exacting. *De jure*, he argued, could only be granted as part of a general European settlement rather than a 'nefarious bargain' between Britain and Italy. Chamberlain insisted that the opportunity opened up by his letter to Mussolini must be seized and stressed the urgent need to reduce the number of Britain's potential enemies. It was important to 'do everything we possibly could to recover the better atmosphere of the early summer'. A compromise was reached when it was agreed that the Italian Foreign Minister, Ciano, should be invited to the forthcoming conference at Nyon, without mention of the possible future discussion of the Abyssinian issue.[51] But Eden showed that he did not take as pessimistic a view as the Prime Minister of the international outlook when he sent Chamberlain an assessment of the relative strength of the British Navy and that of other states, together with a comparison with the position existing in 1914. His conclusions were optimistic. Chamberlain found some merit in Eden's analysis, but added that 'the proposition that our foreign policy must be, if not dictated, at least limited by the state of the Nat. Defences remains true'.[52]

Something of a battle of wills was developing between the two men, but Chamberlain seemed to underestimate the likely impact of their encounters upon Eden's always brittle temperament. 'I must say A. E.'s awfully good in accepting my suggestions without grumbling', he noted in September, 'but it is wearing to have always to begin at the beginning again and sometimes even to rewrite [the Foreign Office's] despatches for them.'[53] It seems unlikely that Eden took as relaxed a view of the situation as this. He believed that 'P.M. *au fond* had a certain sympathy for dictators whose efficiency appealed to him . . . and that he really believed it would be possible to get an agreement with Musso by running after him'.[54]

Eden had strengthened his hand by securing agreement for an international conference at Nyon to deal with Mediterranean problems. This duly met on 10 September and represented a considerable success for the Foreign Secretary. Both at the time and in his memoirs Eden argued that Nyon demonstrated the democracies' capacity to make a stand against aggression. The nine participating powers agreed a plan to protect shipping routes with naval patrols in specified areas, and made arrangements for instant counter-attack. 'It was more than time that the democracies should make themselves felt in Europe,' he reported at the conclusion of the conference.[55] According to Eden, Nyon showed that the best way to deal with the nightmare coalition of Germany, Italy and Japan was not 'continually to retreat before the three of them' but to counter-attack 'the weakest member . . . in overwhelming force'.[56] But recent historians have played down the extent of Eden's achievement, with one even arguing that the conference 'fits more comfortably with Chamberlain's search for appeasement, than Eden's opposition to it'.[57] Certainly the Nyon discussions were limited to technical matters and the practical effect of the decisions reached was limited. Strikingly, Eden took credit for avoiding the creation of an 'Anglo-French-Soviet bloc on an ideological basis' and received the cabinet's congratulations for securing the agreed objective 'that the Conference should not have an anti-Fascist tendency'.[58] Be that as it may, Chamberlain, who was privately worried lest Nyon damaged the prospects of an Anglo-Italian *rapprochement*, soon regained the initiative.[59] On 29 September he persuaded the cabinet that Britain's ties with France, strengthened at Nyon, should not impede early talks with Mussolini.[60]

*

The last months of Eden's Foreign Secretaryship are littered with clashes with the Prime Minister which, taken cumulatively, make his resignation appear more or less inevitable. Individually, none seems sufficiently grave or clear-cut to warrant his departure, though in later years – and particularly in his memoirs – Eden gave the impression that the issues in dispute were more stark than was the case. But to such considerations must be added Eden's mounting irritation at Chamberlain's interference in his departmental affairs, his characteristic reluctance – perhaps inability – to articulate the nature of his policy differences to the Prime Minister and his cabinet colleagues as clearly as to the figures in his immediate entourage, and the capacity of that entourage to influence his thinking and convince him that resignation had become an increasingly probable outcome.[61] As early as mid-October Harvey recorded that Eden was 'criticised and thwarted by half his colleagues who are jealous of him and would trip him up if they had half a chance'.[62] If this was also the Foreign Secretary's reading of the situation, it was probably not one shared by the cabinet as a whole.

Early in October Eden authorized a cautious start to negotiations with Italy. But when Italy then joined the Anti-Comintern Pact and abandoned the League of Nations his doubts were understandably confirmed. For Chamberlain, on the other hand, these events seemed only to justify an even greater effort at reconciliation. To the Prime Minister's annoyance Eden criticized Italy in a Commons speech on 1 November. His remarks, thought Chamberlain, would throw Germany and Italy 'together in self-defence when our policy is so obviously to try to divide them'.[63] A message from Chamberlain, urging that Eden should say nothing to upset the dictators, was not delivered. According to Harvey it was becoming imperative for Eden to have a frank talk with the Prime Minister. The latter must 'either support AE or AE must resign and the Government would then fall'.[64]

For the time being Eden chose instead to raise again the question of rearmament. Judging that the British population was ready for sacrifices to meet the challenge of the dictators, he argued that it was time to supplement home-based production with purchases from abroad. But as on other occasions Eden showed no appreciation of the argument which dominated Chamberlain's Treasury-minded thinking, that the strength of the British economy would be the fourth arm of defence in a long war, an arm which excessive rearmament expenditure at too early a stage would only weaken. Indeed, Eden confessed that he was 'very ignorant of the financial aspect of all this question'.[65] When the two men did meet on 8 November their conversation went 'very badly'. Chamberlain complained that the Foreign Office never made genuine efforts to get on terms with the dictators while Eden countered with renewed criticism of the rearmament programme. For a while, rumours circulated that Eden had resigned.[66] A good financial position, he argued, would be small consolation if London were laid flat because the RAF was inadequate to protect it.[67] There was much to be said for this line of argument, but another year would pass before the Exchequer lost its preeminence in cabinet discussions.[68]

Much has been written – not all of it accurate – of Lord Halifax's visit to Germany in November 1937, at the invitation of the German Air Minister Hermann Göring. 'My own recollection', Eden later wrote, 'is that, when I first heard of this proposal I was not eager, but saw no sufficient reason to oppose it.'[69] After the war, however, Halifax, in the belief that 'if legends can ever be overtaken by accurate history, it is worth recording the truth', sent an account of his mission to be placed in the Foreign Office files. According to this version of events it was Eden who suggested that there might be value in Halifax's going to Germany:

> I pooh-poohed the idea, but he stuck to it and said he was going to speak to Neville Chamberlain and we could all have a word about it when he had done so. This, in due course, we did and as a result of their joint exhortations to me to take the opportunity I went.[70]

Eden's later misgivings were partly the result of the promptings of his Foreign Office staff and partly a consequence of the way the visit was arranged, especially when it became clear that Halifax would be going to Berchtesgaden to see Hitler. The Foreign Secretary was also upset by the tendency of the press, under Downing Street influence, to play up the importance of the mission. A heated meeting on 16 November ended with Chamberlain advising Eden, who was suffering from influenza, to go home and take an aspirin.[71] Entries in Harvey's diary reveal doubts about the wisdom of Halifax's visit, but also show that Eden's entourage still 'favoured [an] approach to Hitler' and envisaged a deal over Czechoslovakia involving autonomy for the German-speaking Sudetenland.[72] So while Eden might not have gone as far as Halifax in suggesting to Hitler that Danzig, Austria and Czechoslovakia were all questions where 'possible alterations in the European order . . . might be destined to come about with the passage of time', it was not unreasonable that, when the cabinet reviewed the visit, Eden 'expressed great satisfaction with the way the Lord President had dealt with each point in his conversations with the Chancellor'.[73]

At this same cabinet meeting Eden reverted to his ambition of securing Anglo-American co-operation in the Far East. Undeterred by the failure of the Brussels Conference to produce anything more than an anodyne deprecation of the use of force, he recommended a further approach to the United States in view of the Japanese seizure of Chinese customs vessels off Shanghai. Chamberlain consented, without enthusiasm, to a request that the Americans should join staff conversations and agree to send ships to the Far East if Britain also did so.[74] One American observer judged that Eden hoped that an Anglo-American compact in the Far East would immobilize the European dictatorships and thereby eliminate the danger of a hostile coalition between Germany, Italy and Japan.[75] Not surprisingly, the Americans rejected all such ideas. If the United States would not play the role for which Eden had cast her, Britain's other obvious ally remained France and Prime Minister Chautemps and Foreign Minister Delbos visited London on 28 November. Significantly, the French ambassador in London noted no evidence to support current rumours of strained relations between Eden and Chamberlain.[76] The British ministers seemed reluctant to offer guarantees for the future safety of Austria and Czechoslovakia. Their overriding concern was that a settlement of Germany's grievances should be reached without recourse to force. The British and French delegates agreed that there was no objection in principle to handing back Germany's former African colonies, providing this was part of a 'general settlement', a point which Eden reiterated to Ribbentrop in early December.

Eden spelt out his current thinking in a memorandum for the Committee of Imperial Defence (CID). This paper showed the influence of Laurence Collier, a Foreign Office official who was most sceptical of the chances of an Anglo-German agreement. It would, Eden argued, be a mistake to try to detach any one member of the German-Italian-Japanese bloc by acquiescence in plans for expansion. All three had aims which were inimical to British interests. It would be safer, for the time being, to tolerate the present state of armed truce. Acknowledging that Britain might have to accept further *faits accomplis* at the hands of the aggressor powers, the paper concluded by advocating the 'unheroic policy of so-called "cunctation"'.[77] More positively, Eden emphasized the value of the French alliance and argued that it must be a constant peacetime aim to increase the likelihood of United States armed support in case of war. Chamberlain, by contrast, doubted the wisdom of counting c.1 potential allies. He thought they would only enter a war when it was too late to change the course of events.[78] Even so, on the central question of Anglo-German relations any gap between the two men remained relatively narrow. Chamberlain was more optimistic of an agreement, Eden less so. Not that Chamberlain was prepared to accede to every demand Hitler made without some form of *quid pro quo*. A letter written at this time to his sister shows that he was contemplating an overall settlement involving Germany's return to a reformed League and an agreement on disarmament.[79] The most significant difference was Eden's contention that a settlement with Germany would have to await a significant strengthening of Britain's military capability.[80] This made him doubtful whether diplomacy could compensate for military deficiency. Foreign policy, he stressed, should be seen as operating in conjunction with and not in substitution for military strength.[81]

The problem was whether the level of rearmament which Eden thought necessary could ever be achieved within the prevailing climate of economic orthodoxy which suggested that defence expenditure needed to be strictly limited. In December the cabinet discussed the interim report of Sir Thomas Inskip, Minister for Co-ordination of Defence, on 'Defence Expenditure in Future Years'. Asserting that £1500 million was the most that could be afforded over the period 1937–41, Inskip proposed drastic revisions in the plans of the service ministries. The most severely hit was the army and it was significant that he succeeded in relegating to fourth position in the scale of Britain's defence priorities – 'which can only be provided after the other objectives have been met' – co-operation in the defence of the territories of any allies in war. Even the air force was subject to restrictions. The quest for parity with Germany, Inskip suggested, need not apply to bombers.[82] It was his central contention – and one shared by Chamberlain and the Treasury – that 'the provision of adequate defences within the means at our disposal will only be achieved when our longterm foreign policy has succeeded in changing the present assumptions as to our potential enemies'. This was perhaps the most succinct justification of the policy of appeasement ever produced.

Eden was concerned that British strategy seemed too ready to adopt a defensive posture and expressed 'some apprehension' at the inability to 'help our allies on land'. But he had to admit that the report's central argument was 'irresistible' and that the overall position was quite different from that in 1914. During the ensuing discussion Halifax pointed again to the danger posed to Britain by a possible coalition of three hostile powers. The conclusion he drew, 'and he had no doubt that it could be shared by the Foreign Secretary', was that this threw an immensely heavy burden on diplomacy and that Britain should therefore make every possible effort to get on good terms with Germany. There is no evidence that Eden dissented. Indeed, he reacted by arguing that the first task was for Britain and France to agree on a colonial policy, presumably in terms of what Germany might be offered in exchange for a European settlement.[83]

*

As 1937 drew to a close, therefore, there was evidence that Eden was not entirely happy with all aspects of government policy. But there were few signs of yawning chasms of principle dividing him from the rest of the cabinet. He took advantage of a letter to Chamberlain on the last day of the old year to stress that 1938 would be a difficult time internationally, not least because British rearmament remained incomplete. But his tone was scarcely that of a man who had concluded that his position was becoming untenable:

> This letter . . . is really to say thank you . . . for your unvarying kindness and help to me this year. I really find it hard to express how much I have appreciated your readiness at all times to listen to my

problems and help in their solution, despite your many other pre-occupations.[84]

Significantly Eden wrote also of 'elements of encouragement' on the international horizon, 'first and foremost' of which was co-operation with the United States, which was making 'real progress'. What accounted for Eden's optimism was Roosevelt's decision to send a naval representative, Captain Ingersoll, to London for secret talks with the Admiralty. The Foreign Secretary interpreted this as a significant breakthrough, paving the way for a joint naval demonstration in the western Pacific, and rearranged his Christmas holiday to enable him to join in the discussions. It was, he said, what he had been working for for years.[85] But Washington saw the talks as an opportunity for an informal and non-binding exchange of information and plans rather than the prelude to a display of naval force.[86] Not surprisingly, they produced only limited results with agreement on broad strategic co-operation, respective spheres of action and the reciprocal use of facilities.[87] Interestingly, Ingersoll reported that Eden was 'more interested . . . in immediate gestures to impress the Japanese than he was in long-term future planning'.[88] One historian has found it difficult to ascribe Eden's confidence in the outcome of these staff conversations to 'anything more than wishful thinking'.[89] But while he must have been disappointed by the immediate results of Ingersoll's mission, Eden's determination to foster good Anglo-American relations remained undimmed.[90]

Chamberlain's priorities lay elsewhere. Indeed, he had no wish to see his efforts at European appeasement complicated and perhaps thwarted by high-sounding, but probably ineffectual, American gestures. Before government ministers went their separate ways for Christmas, he again raised with Eden the problem of Italy, suggesting that the question of *de jure* recognition was becoming deadlocked and asking him to suggest a way out.[91] Eden and the Foreign Office produced two plans. One was for the granting of *de jure* recognition 'on its merits'; the other for making it part of a bargain with the Italians. Rather surprisingly, in view of the importance he had sometimes placed on not giving concessions without securing reciprocal gains of equal value, Eden preferred the former option. To grant Mussolini *de jure* recognition as part of a deal now seemed to him 'rather sordid' and likely to be 'dangerous to our reputation'.[92] This attitude allowed Chamberlain to assume the mantle of the man who would not make unilateral concessions to dictators – 'I should be afraid that [Mussolini] would regard our action as dictated by fear and that he would not abate one jot of his demands or give one tittle of what we wanted' – and to point to the inconsistency of Eden's stance:

> We are . . . in much the same position here as in the proposed German conversations. . . . What we were aiming at was a general settlement to which every country must make its contribution and we felt that, providing there was reasonable ground for supposing that we might achieve a general settlement, we should have little difficulty persuading our people that it was worthwhile paying the price for it. So now with Italy. . . . Everyone must make their contribution and we shall be prepared to make ours, provided we are satisfied that we can count upon a full response from others.[93]

Seizing upon Eden's notion that Britain must maintain her 'moral position', the Prime Minister insisted that the only way to do this was to make *de jure* recognition part of a 'general scheme for appeasement in the Mediterranean and the Red Sea'.[94]

Chamberlain had hit upon the fundamental flaw in Eden's opposition to opening conversations with Italy. The Foreign Secretary's attitude reflected less an objection in principle to the appeasement of the dictators than a perception, based largely on personal instinct, that, as regards Mussolini, this was an unprofitable line to pursue. When it came to Germany he seemed keen to build upon Halifax's talks of the previous November. It was important, the cabinet's Foreign Affairs Committee heard, 'if we are really anxious to prevent the hopes created by the recent conversations from evaporating, that there should be no long delay'.[95] But against Chamberlain's assertion that the problems of opening conversations with Italy and Germany were fundamentally the same, Eden scribbled the curious comment 'except for the difference between the peoples, which is important'. 'There seems to be a certain difference', he wrote to Chamberlain from France, 'between Italian and German positions in that agreement with the latter might have a chance of a reasonable life, especially if Hitler's own position were engaged, whereas Mussolini is, I fear, the complete gangster and his pledged word means nothing.'[96] As late as 31 January he added, 'As you know, I entirely agree that we must make every effort to come to terms with Germany. The Italian conversations are rather a different matter.'[97] Eden's jaundiced conclusions about Mussolini turned out to be entirely justified.

But his altogether more favourable assessment of Hitler brings his judgement into question.

Even so, it was by no means inevitable that the question of Anglo-Italian relations would produce a breach between Eden and Chamberlain. The former insisted that he had 'no closed mind' about talks or the Prime Minister's programme for them. It was, he insisted, a question of priorities. The big issues for 1938 would be Anglo-American co-operation, the assertion of 'white race authority' in the Far East — which objective would be assisted by the racial impediment to any closer German–Japanese *rapprochement* — and relations with Germany. 'To all this Musso is really secondary.'[98] This apparent narrowing of any differences was accompanied by a renewed effusion of personal friendship and understanding on Eden's part:

> I do hope that you will never . . . feel that any interest you take in foreign affairs, however close, could ever be resented by me. I know, of course, that there will always be some who will seek to pretend that the Foreign Secretary has had his nose put out of joint but that is of no account beside the very real gain of close collaboration between Foreign Secretary and Prime Minister, which, I am sure, is the only way that foreign affairs can be run in our country.[99]

The fact that the two men seemed still to be speaking with one voice regarding Germany helps to make sense of the elevation, at the beginning of 1938, of Sir Robert Vansittart to the high-sounding, but largely impotent, post of Chief Diplomatic Adviser to the government and his replacement as Permanent Under-Secretary by Alexander Cadogan. Vansittart had long been recognized as among the leading critics of Germany inside the administration. His removal, thought Churchill, would be represented 'as a victory for the pro-Germans in England' and would 'arouse the suspicion of the French'.[100] If this were true, it owed as much to Eden as to anyone else. Indeed, Eden had been determined to remove Vansittart from his power-base for some time, having offered him the Paris embassy as early as September 1936.[101] Vansittart refused 'as he [saw] through Anthony's motive' and not even Baldwin's intervention could persuade him to change his mind.[102] Eden renewed his efforts early in 1937, again without success.[103] By the end of the year, however, Vansittart had reluctantly agreed to move sideways, deceived perhaps into thinking that his new position would be one of real power.[104] Vansittart had never been an easy colleague.

Many found his views too extreme and his excitable temperament, which seemed sometimes to border on instability, was not best suited to complement Eden's own. 'Van had the effect of multiplying the extent of Anthony's natural vibrations,' noted Chamberlain.[105] Even so, Eden's determination to exclude him from the policy-making process would have been surprising if the evidence did not point to his own continuing commitment to improving relations with Berlin.[106] Predictably, Vansittart came to look upon Eden as not only 'second rate but also a dirty dog'.[107]

While Eden was still in France Chamberlain received details of Roosevelt's so-called 'peace initiative'. The President proposed to launch an appeal leading to a world conference to discuss general issues such as disarmament, raw materials and the principles of international conduct. 'It was Neville's judgement and methods in that business', Eden later recalled, 'which made separation between us inevitable.'[108] Chamberlain's reaction was entirely predictable. The Brussels Conference and Roosevelt's failure to follow up the quarantine speech with anything tangible had made him, more than ever, sceptical of the contribution which America could make to the containment of European aggression. Nothing in Roosevelt's latest pronouncement made him change his mind. As John Simon dismissively put it, the President's message was the 'outline of a general scheme for a better world', whereas Britain was confronted by 'concrete pressing anxieties and dangers'.[109] Chamberlain did not reject Roosevelt's proposal outright, but stressed the danger that the presidential initiative would cut across the government's own plans for talks with Italy and Germany. He therefore asked Roosevelt to postpone his appeal 'for a short while'.[110] Chamberlain's true feelings were more forthright. He found the President's proposals 'fantastic' and likely to excite the derision of Germany and Italy.[111] For Eden, on the other hand, the American initiative was altogether more important – more important than any plans for conversations with the European dictators. He was not as naïve as to imagine that Roosevelt's proposals represented in themselves any substantial advance in Anglo-American co-operation, but was anxious that Britain should not alienate American opinion. Every favourable sign from across the Atlantic should be greeted with open arms. 'The Foreign Office had worked for it for a year past.'[112] As Eden's research assistant later put it, the American proposal was jejune and would almost certainly have led to nothing. But even a one in a thousand chance that it might induce a tentative American step away from neutrality justified Britain in playing Roosevelt's game.[113]

As important as their respective assessments of the Roosevelt initiative was the way in which Chamberlain replied to the President without attempting to consult his Foreign Secretary. Though Eden was on the French Riviera, the presidential timetable did allow for him to be contacted had the will to do so existed. Almost certainly Chamberlain hoped to kill off the American plan before Eden could express an opinion.[114] The Foreign Secretary was summoned back to England by his staff and found Cadogan and Harvey awaiting him with the full story of what had happened. Harvey's advice may be assessed from his diary entry: 'the more I think of Roosevelt's proposals the more convinced I am that we must clinch with them and co-operate for all we are worth in spite of the risks they contain.'[115] On 16 January 1938 Eden visited Chamberlain and tried to persuade him not to snub the President's initiative. He argued that it represented a more important development than the proposed talks with the dictators which were by no means sure to succeed. It was 'almost impossible to overestimate the effect which an indication of United States interest in European affairs may be calculated to produce'.[116] Cadogan thought that Eden was exaggerating 'as much one way as PM does on other'.[117] Eden certainly contemplated resignation and there is evidence that his entourage played upon his sensitivity and vanity to raise the stakes and turn the dispute into an issue of heroic proportions. Cranborne argued that the difference was 'of the most fundamental kind',[118] while, according to Harvey,

> There is no room for compromise on the question whether America is more important to us than Italy or even Germany. . . . The time has come for a changeover from the old generation to the new; this issue has revealed the chasm which divides us; if the democracies have been falling back it is because, unlike the totalitarians, they have not yet faced this fact. The country is aware and uneasy but it cannot *do* anything unless it is given a lead. The lead must come from you.[119]

Harvey suggested that the Foreign Office anticipated 'no real progress' towards a general settlement to emerge from the talks with Germany and Italy.[120] If this was so, then either Eden was out of step with his own department or – at least as regards Germany – he had been misleading the Prime Minister and his cabinet colleagues. Perhaps – and this is by no means impossible or untypical of his career as a whole – Eden's close circle more truly reflected attitudes commonly attributed to the Foreign Secretary than he did himself.

The disagreement between Chamberlain and Eden led to four meetings of the cabinet's Foreign Policy Committee. Harvey argued that Eden could risk taking a strong line – 'they couldn't let you go on this!'[121] Shaken by the Foreign Secretary's threat to resign, Chamberlain agreed to a telegram on 21 January urging Roosevelt to go ahead with his initiative after all. In the meantime no move would be made towards Italy for at least a week. It was, judged Cadogan, a great victory for Eden.[122] But his closest confidants were more sceptical. Harvey, Cranborne and his PPS, J. P. L. Thomas, all believed that fundamental differences had been revealed and that Eden would be better off outside the government. Cranborne, in particular, thought that Eden had been worn down in the Foreign Policy Committee and might have committed himself too far as regards *de jure* recognition.[123] But Eden's negotiating position had never been as strong as Harvey implied. The secrecy attaching to Roosevelt's initiative would have made it extremely difficult for Eden to have resigned and given a convincing public explanation or to have disclosed the extent of the Anglo-American co-operation he had been seeking in the Far East. At all events, the crisis of January 1938 must be seen as a turning-point in Eden's career. The President himself seemed to lose interest in his own scheme and little more was heard of it. There is not even much evidence that Roosevelt felt rebuffed or discourteously treated by the British Prime Minister.[124] But Eden's was not the sort of temperament which could now put these matters aside as so much water under the political bridge. Permanent damage had been done to his relationship with Chamberlain, leaving him vulnerable to any renewed signs of disagreement, even if the issues then at stake were not of fundamental importance.[125]

For the time being, however, relations appeared to have been repaired. The Prime Minister's close adviser, Horace Wilson, insisted that Chamberlain's 'one wish' was to get back to their old co-operation, while Eden reassured a sympathetic MP that relations were 'perfect'.[126] As regards Britain's international position Eden's views had not changed. He believed neither that Britain's allies were so weak nor that her 'coalition' of enemies was so strong as they were portrayed. 'I cannot help believing', he wrote to Chamberlain on 31 January, 'that what the Chiefs of Staff would really like to do is to reorientate our whole foreign policy and to clamber on the bandwagon with the dictators, even though that process meant parting company with France and estranging our relations with the United States.'[127] But the question of conversations with Germany and particularly Italy would not go away. Though Roosevelt had expressed misgivings about the possible recognition of the Italian conquest of Abyssinia, the crisis meetings of the Foreign Affairs Committee suggested that Eden was still not, in principle, as opposed to the quest for a *rapprochement* with the dictators as his advisers seemed to believe. Indeed Eden made clear his readiness to prepare for negotiations with Germany and did not exclude an initiative towards Italy, providing only that *de jure* recognition was to be 'an integral part of measures for world appeasement'.[128] By the end of January he was convinced that *de jure* recognition must be dependent on an agreement to withdraw Italian 'volunteers' from Spain. Such an accord, thought Harvey, would constitute 'real appeasement'.[129]

Eden's terms for an Anglo-German agreement would have been demanding.[130] But any difference with Chamberlain on this point related less to the practicability of seeking a settlement than his perception – probably unfounded – that the Prime Minister was prepared to 'throw colonies away as a sop apart from any general settlement'.[131] At the cabinet on 9 February Eden actually urged that conversations should begin before Hitler's Reichstag speech on 20 February, an occasion which the Führer might use to express disappointment at the lack of progress since Halifax's visit to Berlin.[132] Eden's memorandum on the instructions to be given to Britain's ambassador, Sir Nevile Henderson, on the opening of talks suggested that Britain did not regard the frontiers of Central and Eastern Europe as fixed for all time, though she would not condone any forcible change in the international status of any country.[133] That particular issue came to the fore more quickly than anticipated, as Germany prepared to incorporate Austria inside the Reich. Eden was probably thinking along the same lines as Harvey, whose instinct was 'not to take this too tragically'. The ban on *Anschluss* had been 'wrong from the start' and was 'perhaps the weakest point in our postwar policy'.[134] Eden's reaction to Austria's predicament was similar to that which the government would display in the spring of 1938 as it contemplated its inability to do anything about the fate of Czechoslovakia. He 'did not want to put himself in a position of suggesting a resistance which we could not, in fact, furnish'.[135] To Phipps in Paris Eden was more direct. The government was 'considering what steps might be taken to bring about a measure of appeasement which would include *inter alia* Austria'.[136] As one writer has concluded, 'the official papers show that Eden pursued the accepted line on Germany to the last'.[137]

Italy was a somewhat different matter. Eden's doubts about Mussolini's good faith prevented him from pursuing a consistent course. When Chamberlain tried to arrange a meeting with the Italian ambassador, Eden dug 'his toes in'. Cadogan was understandably exasperated:

> This is all wrong – he has agreed in principle to talks with Musso, and nothing has happened to change situation fundamentally. . . . A is silly on this question – he doesn't like the medicine, and makes no bones about it; but then he seems to agree to take it, and uses every excuse – clutches at it – to run out.[138]

Eden was annoyed to learn that Chamberlain had used the good offices of his sister-in-law, Lady Ivy Chamberlain – 'a kind and well-meaning woman with little intelligence and in consequence too great a conceit of herself'[139] – to establish a direct channel of communication with Mussolini. Lady Ivy had encouraged the Prime Minister to believe that Eden represented a considerable personal impediment, at least in Italian eyes, to any improvement in relations.[140] Though Eden was largely ignorant of it, Chamberlain had developed a network of contacts, bypassing the Foreign Office, some passing through the shadowy hands of Sir Joseph Ball, a former MI5 officer who had continued his clandestine activities in the service of the Conservative Party.[141] In the long term Lady Ivy's amateur initiatives were probably devoid of diplomatic significance, but their revelation undoubtedly played upon Eden's sensitivities and helped convince him that his position was being systematically undermined. J. P. L. Thomas had already reached the conclusion – shared, no doubt, with Eden himself – that a group of senior ministers including John Simon, Samuel Hoare, Kingsley Wood, the Health Minister, and Lord Swinton, the Secretary of State for Air, were purposefully working to transfer the control of policy from the Foreign Office to 10 Downing Street.[142]

Eden raised the question of Lady Ivy's 'unofficial diplomacy' with Chamberlain on 8 February and received a fulsome apology.[143] Once again a difficult situation appeared to have been overcome. A meeting with Dino Grandi, the Italian ambassador, even left Eden hopeful that Italy was now anxious to extricate herself from Spain, with the result that 'it may be possible to achieve useful Anglo-Italian conversations after all'.[144] Rumours of a split again circulated in the press but a meeting on 11 February seemed to clear the air. 'I saw Anthony on Friday morning', wrote Chamberlain, 'and we were in complete agreement, more complete than we have sometimes been in the past.'[145] This may have exaggerated the degree of harmony, but Prime Minister and Foreign Secretary had certainly been most successful in concealing their disagreements from their cabinet colleagues. 'There is no foundation for it,' noted Duff Cooper of press suggestions of a breach over Italian policy.[146] Even Simon thought that Eden was 'rapidly approaching the P.M.'s point of view'.[147]

It was therefore against a background of relative calm that the final crisis of Eden's Foreign Secretaryship broke. Chamberlain would not be deflected from the opening of talks with Italy. 'Don't let us', he cautioned Eden, 'in our over-anxiety not to be over-eager, give the impression that we do not want to have conversations at all.'[148] The disagreement now seemed to turn upon subtly differing assessments of the international situation. Simon pinpointed the issue at stake:

> There seems to be no doubt that the turn of events makes Italy anxious to withdraw from intervention in Spain. Chamberlain moreover is most anxious . . . to establish Anglo-Italian relations on a better footing, securing the withdrawal of Italian troops from Libya, the ending of hostile propaganda in the Middle East and a friendly understanding in the Red Sea area in return for a recognition of Italy's conquest of Abyssinia. The Foreign Office is not equally enthusiastic, holding that Italy is on the verge of financial collapse and that we ought not to purchase her friendship by such concessions. . . . We cannot prepare to fight Germany, Japan and Italy all at once: let us at any rate get rid of *one* potential enemy.[149]

Eden judged that Mussolini was now in a very uncomfortable position. The Duce had commitments in Abyssinia and Spain, 'neither of which is turning out well', and an uncertain friendship with Germany which had Austria 'as the first item on the list of intended victims'. And Austrian independence was the issue which had helped keep Germany and Italy apart in the early 1930s. In such a situation 'we have nothing to gain by showing ourselves over eager'. If Mussolini really wanted a settlement, Eden argued, the opportunity to liquidate his commitment in Spain offered an excellent chance to prove his sincerity.[150] Thus the more the Austrian situation deteriorated, the more urgent in Chamberlain's eyes the question of the Italian negotiations became; and the more Eden wanted to delay precipitate action.[151]

Intelligence reports suggested that Italy had agreed to the Nazi take-over of Austria in return for Hitler's recognition of the Mediterranean as an area of exclusive Italian influence. Fortified by Harvey's advice that Spain would have to be 'cleared up' before there could be any question of *de jure* recognition, Eden was therefore more than usually sceptical of Italian good faith when a meeting was arranged between Chamberlain, Grandi and himself for 18 February.[152] Eden tried first to avoid a joint meeting with the Prime Minister and then, when this failed, wrote to advise that Grandi should not be given any commitment about immediate conversations 'until we have had time to go carefully into all the implications'.[153] Chamberlain, more than ever convinced that Britain must seize the moment if there were to be any chance of preventing the consolidation of the Rome–Berlin axis, recognized Eden's stalling tactics for what they were. He now confided to Horace Wilson his determination 'to stand firm, even though it meant losing my Foreign Secretary'.[154] It was significant that the Prime Minister thought the issue might be a resigning matter.

Events now moved rapidly to a crisis. Chamberlain, Eden and Grandi met in the cabinet room at 11 a.m. The ambassador succeeded in convincing Chamberlain of the unique opportunity for a wide-ranging Anglo-Italian agreement in which the question of Spain would pose no particular difficulty. Eden was more sceptical. As he later recalled:

> G[randi], who was a very skilful diplomat, did his stuff admirably. Whenever he paused N. C. encouraged him. He sat there nodding his head approvingly while Grandi detailed one grievance after another. The more N. C. nodded, the more outrageous became Grandi's account until in the end it would almost seem that we had invaded Abyssinia.[155]

Eden registered his dissent sufficiently forcefully for the ambassador to describe him and Chamberlain as 'two enemies confronting each other, like two cocks in true fighting posture'.[156] After Grandi had left, Prime Minister and Foreign Secretary discussed the situation alone, with Eden still advising caution and refusing to agree that Grandi could now be told that Britain would enter formal conversations. 'Upon this, Neville Chamberlain became very vehement, more vehement than I have ever seen him, and strode up and down the room saying with great emphasis, "Anthony, you have missed chance after chance. You simply cannot go on like this."'[157] Both parties realized that this meeting had been decisive. 'I had no idea', wrote Chamberlain, 'until the 18th, that it would come to a break, but after the repeated efforts of the F.O. to stave off conversations and to prevent my seeing Grandi, I knew the time had come when I must make my final stand and that Anthony must yield or go.'[158]

It was agreed that the dispute would have to go to the cabinet on the following day. Ministers were taken by surprise at the gravity of the situation.[159] 'So little was I aware of what was going on', recorded Halifax,

> that I passed a note to Sam Hoare . . . to ask what was the purpose of [the Prime Minister's] rather boring lecture on history. To which he replied that he did not know, but that he anticipated rather serious difference between the Prime Minister and Foreign Secretary.[160]

Eden scarcely helped his case by stating that the difference was on a point of method. Simon seized on this admission, arguing that the whole cabinet had accepted the general principle of Italian conversations for the previous six months. 'The question was one of timing, on which opinion might differ but where a conclusion could be adopted by us all without any sacrifice of principle.'[161] On the merits of the case, as they understood them, and scarcely helped by Eden's own exposition, even those members generally sympathetic towards him now lined up behind Simon and Chamberlain. No one could see that the issue under discussion was a matter for resignation.[162] It was agreed that the cabinet should meet again the following day, Sunday 20 February. Meanwhile, Foreign Office officials discussed the conditions upon which Eden might be prepared to stay at his post, but, as the Foreign Secretary himself was now arguing that the international situation precluded all hope of Italian conversations for the time being, the chances of a compromise did not look good.[163] Halifax called round at the Foreign Office to see if anything could be done but sensed that the mood of the Eden camp – 'the corner of a boxing ring where the seconds received back the pugilist and restored his vitality by congratulations and encouragement' – was now fixed. 'I felt at once that the atmosphere, emanating I thought mainly from Bobbety [Cranborne], was very much pro-resignation.'[164] Oliver Stanley probably hit the nail on the head:

Eden 'has been through Hell to make up his mind, and he's damned well not going to unmake it'.[165]

The final cabinet meeting of 20 February was perhaps best and most objectively recorded by Halifax:

> The talk went on for a long time, partly argument, partly appeal, without any clear result emerging, and I eventually ventured to say that I thought that the situation that had been created was both unfair to the Cabinet and, in a sense, unreal. Unfair because they had till this moment had no intimation of this alleged fundamental difference, and unreal because it was precisely the difference of viewpoint . . . that was of value to the Cabinet, as giving them the best of both minds. Was it not the case that this difference of temperament had been aggravated by smaller incidents?[166]

Eden spoke of 'fundamental differences of principle'. In his own mind these no doubt now existed, but Simon challenged the description and said that as a member of the Foreign Policy Committee he had been quite unconscious of any such divergence. Inskip and Halifax agreed.[167] No compromise proved possible and, 'in spite of every sort and kind of appeal and ingenious formula', Eden confirmed that he would have to resign.[168] Halifax among others tried to rationalize Eden's behaviour:

> I cannot help thinking that the difference on the actual timetable of the conversations was not, and never has been, the principal difference. I suspect it has been the cumulative result of a good many different things: partly subconscious irritation at Neville's closer control of foreign policy; partly irritation at his amateur incursions into the field through Lady Chamberlain, Horace Wilson and his own letter to Mussolini; partly Anthony's natural revulsion from dictators . . .; and partly . . . his excessive sensitiveness to criticism of the Left.[169]

Significantly, Halifax added a final explanation – that Eden was overstrained and that his judgement was not at its best. Harvey and Thomas later complained that it was Simon who deliberately spread the rumour that the Foreign Secretary was unwell and unable, physically, to continue at his post.[170] But throughout his career Eden displayed a chronic tendency to overwork himself to the point of exhaustion and Halifax's suggestion that 'the thing had got out of proportion, and he was no longer seeing it straight', seems worthy of consideration.[171] Malcolm MacDonald, the Dominions Secretary, who dined with Eden on 19 February confirmed this interpretation, while even Harvey admitted that Eden was 'quite exhausted'.[172]

*

Just forty years of age, Eden had quit one of the highest offices of state. The note which Chamberlain passed to him at the cabinet on 20 February, expressing sorrow that they had come to the parting of the ways, seems to make a nonsense of any notion that the Prime Minister had consciously worked to remove Eden from his post.[173] On reflection, however, Chamberlain confessed to relief that the cabinet's peacemakers had been unsuccessful. 'A[nthony] would never have been able to carry through the negotiations with any conviction and in his hands they might well have failed.'[174] Yet the precise reason for Eden's departure remains elusive. More than twenty years later, after reading draft chapters of *Facing the Dictators*, the Foreign Office official Harold Caccia still felt that Eden had not made his motivation entirely clear.[175] The present account has not sought to minimize the differences which separated Prime Minister and Foreign Secretary. But neither has it accepted the image of the heroic anti-appeaser projected by Churchill and others. Speaking in the Commons after Eden's resignation, Churchill claimed that 'all over the world . . . the friends of England are dismayed and the foes of England are exultant'.[176] There was an element of truth in this – but more because of the perception than because of the reality of Eden's attitudes and policies. For most of his Foreign Secretaryship far less divided him from the rest of the government than many outsiders imagined. He and Chamberlain 'agreed wholeheartedly on the most important issue in foreign policy: that Britain could and should avoid war with Germany'.[177]

It is true that the idea of making concessions to Mussolini came to stick in Eden's throat. Yet he entertained greater hopes of success with Germany and, while less willing than some to court that country's friendship, he never placed Germany or Hitler himself beyond the diplomatic pale. As late as the autumn of 1937 Eden would have welcomed direct negotiations with Hitler or his Foreign Minister to discuss the international situation.[178] If the crude distinction between 'appeasers' and 'anti-appeasers' has any historical validity then the latter were surely those few who decided that the nature of Nazi ideology precluded normal diplomatic intercourse, and certainly the granting of concessions. Austen Chamberlain, Eden's old mentor, was perhaps one of the few. 'Are you going to discuss revision with a Government like that?' he had enquired as early as April 1933. 'Germany is afflicted by this narrow, exclusive, aggressive spirit, by which it is a crime to be in favour of peace and a crime to be a Jew. This is not a Germany to which we can afford to make concessions.'[179] By contrast, in a debate in November 1937 Eden had insisted that 'we will join no anti-communist and no anti-fascist bloc. It is nations' foreign policies, not their internal policies, with which we are concerned.'[180] As for his perception that the Führer was more 'appeasable' than the Duce, this appears an embarrassing misjudgement, best explained by the 'complex between him and Musso'.[181]

The historian can discern a series of differences of detail and emphasis between Eden and Neville Chamberlain from the summer of 1937. The importance of any differences was probably exaggerated in Eden's mind by a clash of personalities and a mounting perception of prime-ministerial interference.[182] Some of these differences were never brought fully to the attention of their ministerial colleagues, partly because both sides strove to keep their antagonism muffled and partly because of Eden's tendency to cloak his feelings in the language of diplomatic ambiguity. Among members of his immediate circle he seemed more capable of articulating clear alternatives to Chamberlain's policies and found his uneasiness enhanced by what one historian has called 'the vapourings of his coterie'.[183] This made it difficult for Eden to escape the charge of inconsistency. 'When I talked to him', reflected Chamberlain, 'he always appeared to agree, but the moment we seemed really to be getting to close quarters there was always some reason why it was necessary to draw back.'[184] Hankey confirmed that he was 'always chopping and changing, blowing hot one day, cold the next'.[185]

None the less, by February 1938 a clear but narrow divergence had emerged. Chamberlain was convinced that the opportunity had to be seized to improve Anglo-Italian relations in order to disturb the Rome–Berlin axis and to put Britain in a stronger position for future negotiations with Germany. Eden could see nothing to be gained from such a strategy and placed mounting emphasis on the need to strengthen Britain's rearmament programme as the only satisfactory basis for talks with Germany. At the cabinet on 16 February he again called for the pace of rearmament to be quickened, despite Simon's warning that existing commitments would not only place a terrible strain on the national finances, 'but could not be increased without financial disorganisation to an extent that would weaken the resistance of the country'.[186] Whether Britain could

safely have increased her armaments expenditure in the period
1936–8 remains a matter of debate. But it is worthy of note that, a
year after Eden's resignation, Winston Churchill paid tribute to the
National Government's success in authorizing massive rearmament
without doing serious damage to Britain's credit.[187] There must also
be doubts as to whether much was to be gained by allowing time for
rearmament so as to negotiate from a supposed position of strength.
As Thomas Jones pointed out, 'this assumes that we can catch up
with Germany – which we cannot – and that Hitler takes no dramatic steps in the meantime which is unlike Hitler'.[188]

Yet behind specific differences over potential enemies there lay a
significantly different reading from Chamberlain's of Britain's overall
global predicament. On the one hand Eden never took as alarmist a
view of the three-power enemy coalition. On the other he placed
greater reliance than the Prime Minister on Britain's potential
friends and allies.[189] Though assessments must in part depend on
the 'might have beens' of history, there is at least a case that Eden's
strategic judgement was not as sound as Chamberlain's. 'Is it in truth
at all possible that this Tokyo–Rome–Berlin combination will hold
together so strongly as to come into a war against us?' asked Eden in
January 1938.[190] Such a military coalition did of course come into
being in December 1941. In the same letter Eden insisted that 'the
French army is absolutely sound' – a widespread view which the
events of May and June 1940 rapidly disproved. Eden's assessment
of America is more difficult to evaluate. 'To me nothing matters so
much in the world today as the steady growth of Anglo-American
understanding,' he wrote less than a month before his resignation.[191] This judgement seems sound in the light of the role which
the United States ultimately played in the victory over the dictator
powers; less so against the evidence that it took a direct Japanese
attack on American territory and a foolhardy declaration of war by
Hitler to cement the Anglo-American military alliance. In the shorter
term Eden's efforts to secure a partnership in the Far East seem to
have been wildly optimistic. Even if it had been achieved, professional service opinion would not necessarily have rejoiced. 'If at the
present time, and for many years to come,' wrote Lord Chatfield, 'we
send a Fleet to the Far East, even in conjunction with the United
States, we should be left so weak in Europe that we should be liable
to blackmail or worse.'[192] Likewise, it seems unlikely that the
Roosevelt initiative carried the importance which Eden imagined.
As one writer has argued:

It is hard to avoid the conclusion that Roosevelt launched his plan
. . . partly to mitigate the impression created by the negative nature
of his response to a direct British appeal for joint action [in the Far
East], and also to avoid the necessity for a blunter refusal of any
further British requests by distracting international attention with a
scheme which would sound impressive while obviating the need for
unacceptable U.S. involvement.[193]

Though it was once customary to criticize Chamberlain, with his
background in domestic administration, for his ignorance of
diplomatic affairs, it seems that he had a firmer grasp of the inter-
relationship between political, military and economic factors than
his Foreign Secretary.

Indeed, perhaps the weakest element in the 'Edenite' interpretation of the Chamberlain–Eden conflict is its failure to give the Prime
Minister any credit for his own insights and understanding. As has
been suggested, Chamberlain was less committed to a policy of one-
sided concessions than Eden, particularly in later years, tended to
claim. Remarks made to a correspondent in January 1938 could
easily have come from Eden himself. 'Until we are fully rearmed',
warned the Prime Minister, 'our position must be one of great
anxiety . . . we must adjust our foreign policy to our circumstances.'[194] Their shared consciousness of British weakness led
Chamberlain to press for agreements; Eden by no means ruled
them out, but preferred to keep Germany guessing. In the last resort
both men were in error to the extent that Hitler, with his fanatical
long-term aims and his determination to pursue them, was not
appeasable. Against this point, questions as to whether any opportunity existed in 1938 to wean Mussolini away from his brutal
friendship and bring Italy back into the Stresa alignment seem
entirely secondary. But there were no easy solutions to the diplomatic
and strategic dilemmas of the 1930s and Eden's claims to an understanding of them are not aided by the denigration of Chamberlain's.
David Dilks, who found himself working for Eden while carrying
out an extensive study of Chamberlain's career, tried to present the
former with a balanced view. The documentation he had examined
gave Dilks a somewhat higher opinion of Chamberlain's policy than
he had held before. Chamberlain's view of the dictators may not
have been consistent or fully justified, but he was less deceived by
their protestations of good faith than many imagined. Dilks feared,
however, that Eden would regard his conversion as treachery.[195]
Almost certainly, Eden did.

5

The wilderness months

Eden resigned and made his career by doing so.[1]

A NY chronicle of Anthony Eden's political career is essentially a
story of the exercise of power. For an unusually large proportion of his public life Eden sat on the government front bench, most
of it as a cabinet minister. From September 1931 until January 1957
he was out of office for two periods only, totalling just under eight
years. By any criterion this is a record of success. Yet spells in
opposition can be as important as periods of government in determining the development of a politician's career. This was certainly
the case with Eden between his resignation in February 1938 and his
return to Chamberlain's government eighteen months later. Lord
Blake has suggested that in leaving office Eden had reason to believe
that he had now sacrificed all prospect of rising to the top in public
life.[2] Such a view seems difficult to sustain. Within days of the
resignation taking effect, as shrewd a judge as Maurice Hankey
was speculating that 'before long he will come back in another
office' and that one day 'he may become P.M.', though his enumera-
tion of the qualities which might take Eden to Downing Street –
'good looks, good address in Parliament and on the platform, a nice
wife, money and, above all, the divine gift of popularity' – were not
necessarily those Eden might have ascribed to himself.[3] In the event
Eden's success in distancing himself from the government's foreign
policy before crucial developments in Austria and Czechoslovakia in
1938 was a vital factor in his subsequent political ascent. Yet in
other ways these eighteen months worked to his considerable disadvantage. Through his behaviour in this period Eden convinced
many that he lacked the qualities of leadership which they had
hitherto attributed to him. By the time that he returned to office,
the internal dynamics of the Conservative Party had dramatically

changed. When it came to choosing Chamberlain's successor, Eden was, in the words of one of his own supporters, 'out of it'.[4] If, therefore, Eden's resignation secured his future career, the next year and a half also ensured that it would be the best part of two decades before he reached the summit of his political ambition.

The problem of the succession to Neville Chamberlain had been a live question from the beginning of his premiership. He was, after all, sixty-eight years old at the time he took over and, though vigorous for his age, was unlikely to lead the government through more than one general election. Baldwin seems to have sensed this. 'Chamberlain will be your leader for a short time,' he is said to have told Duncan Sandys. 'Then Anthony will have a long reign.'[5] The main argument against Eden was perhaps his youth − but this difficulty would be overcome providing Chamberlain's tenure was not too brief. Apart from Eden, the cabinet contained few serious contenders for the succession, though several ministers probably nurtured private ambitions. Halifax's peerage seemed an insuperable obstacle, while both Samuel Hoare and John Simon aroused strong hostility on the government's own benches. Neither had been a conspicuous success as Foreign Secretary and Simon was not even a Conservative. Outside the government there was always Churchill − but the historical inevitability of his finest hour in 1940 tends to obscure the extent to which he had excluded himself from serious consideration in earlier years. At the time that Eden became Foreign Secretary, Churchill was sixty-one years of age and widely regarded as a political failure. His idiosyncratic conduct during the abdication crisis scarcely fostered any hopes of rehabilitation. It was generally assumed that his talents would only be called upon again in the event of a major upheaval in British and European politics.[6]

For some time after his resignation Eden's prospects of being the next Prime Minister still appeared bright. In April the Labour MP Josiah Wedgwood predicted that he would probably succeed Chamberlain before the next election, due in 1940 at the latest.[7] In the spring of 1939 a Gallup Poll suggested that 38 per cent of those asked nominated Eden as their preferred successor to Chamberlain. A mere 7 per cent opted for Churchill.[8] Yet, within political circles faster than among the public, Eden's standing was being steadily eroded. When war came his moment had passed − at least for the predictable future.

The period has an added significance for Eden's career and historical reputation. The weakest point in his claim for rectitude in the

1930s – that he had merited the accolade of 'anti-appeaser' – was that he had managed to remain a member of the government for more than six years before his resignation. It was always possible, and indeed widely believed, that Eden had been fundamentally at odds with his colleagues over the central issues of foreign policy, but had chosen to fight his corner from within the government rather than assume the more problematical position of a backbencher. Churchill, after all, has written: 'I knew well what his difficulties were with some of his senior colleagues . . . and with his Chief, and that he would act more boldly if he were not enmeshed.'[9] But the test of his credentials would surely come once Eden was freed from the restraints of ministerial responsibility. Looking back after thirty years, however, Eden seemed to appreciate that his conduct out of office would not necessarily speak for itself. In an attempt to forestall criticism he explained his perhaps over-cautious approach for the benefit of his biographer:

> In part this was dictated by my belief that those who agreed with me were not numerous enough in the Party to have any hope at all of commanding a majority. Chamberlain's position and that of the old men who buttressed him . . . was then unassailable. I considered that the only consequence rebellion could achieve would be to weaken the Government to the advantage of the Labour opposition. There was no advantage in this . . . their call for disarmament at so late an hour was the worst of all policies.[10]

Precisely what consequences Eden anticipated flowing from his resignation remain an open question. But it seems likely that he was influenced by the report from his friend Paul Emrys-Evans of a meeting of the backbench Conservative Foreign Affairs Committee held on 17 February 1938. Evans recorded a 'unanimous feeling' that the time was 'most inopportune' to open discussions with Italy with *de jure* recognition of the conquest of Abyssinia on the agenda. A speech by Churchill praising Eden – 'no figure has emerged on the political stage since the war equal in quality, equal in strength, equal in power of self-restraint' – but also calling for stronger action by the government, was well received, though it later incurred the wrath of the whips.[11] At the very least, therefore, there were grounds for believing that Eden's departure would have considerable repercussions within the Conservative Party. According to Duff Cooper's PPS, the situation in the Commons would be 'hopeless' with more than a hundred of the government's usual supporters likely to withhold their backing.[12] Might such turmoil have led even to the fall of Chamberlain's government? Across the Channel the shock-waves caused by Eden's withdrawal were such as to cause the Prime Minister, Camille Chautemps, to contemplate a change of government in France.[13]

Eden discussed the possibility of the government's fall with Oliver Harvey two days after his resignation. Harvey recognized that the course of events was difficult to predict. Either Eden's first speech from the back benches might produce 'an immediate reaction from the country' which could overturn the government, or there would be no immediate reaction, in which case Eden should sit back, let Chamberlain make the running and then attack him when the occasion arose.[14] His subsequent behaviour suggests that this was advice which Eden followed. Even so, a week later Eden still considered that the government had been mortally wounded. He thought it would 'stagger along more and more discredited, politics getting much more bitter, losing by-elections and all support from the "floating vote", and further resignations coming later'.[15] But Eden's difficulties were twofold. On the one hand he had no wish to be accused of manipulating the crisis created by his resignation to the advantage of the Labour opposition. This would scarcely enhance his prospects of leading the Conservative Party in the not too distant future. As the *Spectator* warned, only one thing could undo Eden – excessive zeal by the opposition in their endeavour to make party capital out of his resignation.[16] In the second place, but for much the same reasons, Eden made no effort to encourage other resignations in the wake of his own. Should such a rallying cry have failed to overturn Chamberlain, the accusation of disloyalty would be hard to escape and Eden might succeed only in burning the boats which were supposed ultimately to take him to the party leadership. If, therefore, Eden entertained any hopes that his resignation would lead to the government's collapse, such a process would have to occur of its own accord.

In fact, no cabinet minister followed Eden's lead, although Malcolm MacDonald, Walter Elliot and Lord de la Warr were at one time expected to do so.[17] In later years Eden became irritated by suggestions that he had dissuaded his colleagues from resigning. 'All we did was to tell these personages that they must make up their own minds. They none of them had the slightest intention of giving up office.'[18] Yet nearer the time Eden admitted having given no warning of his impending resignation. 'It took them by surprise and he says on this he let them down,' recorded one well-disposed MP.[19] Harvey believed that if Eden had not been so engrossed with his departmental work and had had time to see more of his colleagues out of hours, more would have followed his lead and left the government.[20] In the event those ministers who were generally most sympathetic to Eden were as surprised by his withdrawal as any others. Though half wishing to follow his instinct rather than his reason, Elliot judged that it would have been dishonest not to accept Chamberlain's assurance that entering conversations with Italy in no way committed Britain to further concessions. Eden's departure on a point of procedure did not justify upsetting the government.[21] W. S. Morrison and Oliver Stanley were 'equally amazed' at what had happened, while MacDonald was clear that Eden had made a mistake. The Foreign Secretary had agreed as to the necessity of talks with Germany and Italy and his disagreement with Chamberlain had turned only on questions of method.[22] Had the dispute amounted to a point of principle, MacDonald would have resigned, but he could 'scarcely break up the National Government on so small an issue'.[23]

If Eden had been looking to damage the administration, the most obvious opportunity existed in his two speeches of explanation – to the Commons and to his constituents – delivered within a week of his resignation. In the event the chance passed Eden by. The atmosphere in a crowded House was one of tense excitement. Expectations were high that Eden's speech would prove devastating and Simon was booed when he started to answer Foreign Office questions.[24] Only hours before Eden rose to make his speech, Churchill sent an urgent note of advice. The retiring minister should pull no punches. His cause was the cause of England.[25] Eden, however, adopted the sort of diplomatic and circumlocutory style for which he was often criticized. Hampered by an inability to refer to the secret Roosevelt initiative, he couched his statement in generalities leaving most of his audience little wiser than before he spoke:

> I am certain in my mind that progress depends above all on the temper of the nation, and that temper must find expression in a firm spirit. That spirit, I am confident, is there. Not to give voice to it is, I believe, fair neither to this country nor to the world.[26]

Though Harvey predictably found Eden's statement 'first rate', others were less enthusiastic.[27] According to Leo Amery the speech revealed 'good feeling and sincerity' but left the impression that there was really no issue to resign on – 'unless he objected fundamentally to discussing with Italy at all'.[28] Nicolson concurred that Eden did not 'really make a good speech', while Harold Macmillan later conceded that it had not made the issues behind the resignation any clearer.[29] Geoffrey Dawson concluded that he could have made a better case on Eden's behalf himself.[30] Others proceeded to do so. By general consent Cranborne's Commons statement was more impressive. He adopted a combative tone and spoke of the dangers of surrendering to blackmail. Later in the debate Churchill predicted that Eden's resignation might constitute a milestone in history but, if this were the case, most listeners could only assume that the member for Epping was privy to facts of which they were unaware. Even Chamberlain seemed inadvertently to help Eden's case. His dismissal of the League in his winding-up speech at least opened up a tangible gap with his former Foreign Secretary which all could understand. Yet in the eyes of many the man for whom this parliamentary occasion had been created had failed a first and important test. 'During the two days' debate', noted Duff Cooper, 'one could feel opinion veering steadily towards the Government.'[31] In the division only one Conservative, Vyvyan Adams, voted against the government. The twenty or so abstainers contained few big names and scarcely had the makings of a government in waiting.

There remained the occasion afforded by Eden's speech at Leamington on 25 February. In the government camp Channon sensed apprehension lest he were too bitter, but thought Eden too shrewd a politician to burn his boats.[32] Eden's task was not an easy one. The point behind his resignation was 'not a clear issue' and it would be difficult to explain it to the country.[33] But Eden sought to avoid controversy as far as possible, even allowing the Cabinet Secretary to vet his speech in advance. The result was a performance which won even Chamberlain's praise for its 'dignity and restraint'.[34] Eden was glad that it met with Halifax's approval. He had wanted to be 'fair to all concerned'.[35] The feeling inside the Carlton Club was that the speech was 'a very quiet affair, with no revelations or claims for further publication'.[36] Even Harvey accepted that it was not a 'fighting' speech.[37]

The impression is clear that in the first days after his resignation Eden was scrupulously careful not to be seen to rock the boat. An emergency session of the executive of the League of Nations Union and protest meetings among branches around the country suggested a groundswell of opinion in his favour. There was talk of a postal ballot on the question, 'Do you approve of Mr. Anthony Eden's stand for good faith in international affairs and will you support his demand for the re-establishment of peace and security through the League of Nations?' But Eden and Cranborne did not want their resignations used to challenge the government and, on behalf of the Union, Gilbert Murray agreed to respect their views and prevent LNU meetings getting out of control.[38] Many years later Eden rationalized his attitude by arguing that his purpose had been to convert the party rather than break it up.[39] Be that as it may, the effect of his restraint was that a potentially serious crisis for the government soon evaporated. Behind the scenes Chamberlain's party machine worked hard to limit any damage. Sir Joseph Ball, anxious that the occasion of Eden's resignation should be presented as a technicality, took 'certain steps privately' to undermine his case in the Conservative press.[40] Meanwhile the whips strove to conceal the extent of his support in the Commons and the subsequent resignations of the officers of the Conservative Foreign Affairs Committee passed off quietly.[41] A week after Eden left the cabinet Simon noted 'an immense swing-over to the P.M.'s side in all quarters where the vaguest idealism does not obscure realities'.[42]

*

In his memoirs Eden asserts that he resigned because he could not agree with the foreign policy which Chamberlain and his colleagues wished to pursue. His intention, for the time being, was to bide his time. 'When', he suggested, 'the attempt to negotiate with Signor Mussolini and Herr Hitler failed to produce results of value would be the moment to point the moral and try to influence British foreign policy.' Yet, in slightly contradictory vein, Eden adds that he was determined to use his greater freedom to multiply his warnings to the government and nation.[43] In the event, once his two speeches were out of the way, Eden went abroad, so that, in his own words, 'while the clouds were gathering over Austria, I remained in the South of France, reading the French newspapers and listening to the Vienna wireless'.[44] Indeed, Eden made no further public pronouncements for two months until the Royal Society of St George invited him to propose a toast at their banquet on 26 April. Eden, writes David Carlton, 'practically hibernated for several months'.[45] Thomas found him in France 'at the top of his form. Six sets of tennis per day and scarlet in the face with sun. My feet are blistered, my wind has gone, but A is ruthless.'[46] Such a characteristic did not, however, extend to the world of politics.

Had Eden wished to show that the government's policy towards Germany was misguided, the Austrian *Anschluss* in March offered an obvious occasion to make political capital and rally to his cause those disillusioned backbenchers whose confidence in the administration had already been shaken by his resignation. His own position, however, might have been untenable. His reluctance to open conversations with Italy had hardly been designed to encourage the latter to help Britain maintain Austria's independence. Significantly, Churchill had spoken on 22 February in favour of concessions to Italy in return for the re-establishment of something like the Stresa Front to protect Austrian integrity.[47] Certainly, Chamberlain believed that the *Anschluss* might have been avoided if he had had Halifax and not Eden at his side at the time of his approach to Mussolini in 1937.[48]

As at the time of his resignation Eden chose to adopt the line of 'wait and see' while the Austrian crisis unfolded. One friendly backbencher suggested that events were arguing Eden's case more forcefully even than he could do himself.[49] Another believed that his silence — and absence — created a very desirable feeling of uncertainty and alarm in government circles.[50] The editor of the *Yorkshire Post*, on the other hand, warned Eden to be ready to return to England at a moment's notice. If the government were shown to have no policy to meet the new situation, Arthur Mann would consult Churchill and endeavour to draw out Baldwin. A word from the latter 'would probably turn enough Conservatives . . . against the Government to cause its defeat'.[51] Channon found the Commons 'humming with intrigue' with 'insurgents' rushing about 'very over-excited' about Eden's possible return to government.[52] In the event Chamberlain responded with greater purposefulness than many had expected. In a parliamentary statement on 24 March he enumerated those obligations which might lead Britain to use force and left open the possibility that Britain's longstanding friendship with France might result in coming to the latter's aid in defence of Czechoslovakia. The Edenite Mark Patrick summed up the situation as it affected his mentor:

> The nett result of all this is that the situation which would have followed an 'Isolationist' statement yesterday has not arisen. I think we are entering upon the lull I forecast in my last letter; and therefore my arguments for your immediate return fall to the ground.[53]

Not surprisingly, nothing that merits the title of an 'Eden Group' emerged in the spring of 1938, though a few sympathizers began meeting together after his return from the continent.[54] During his first months of exile Eden's closest confidants were his loyal junior ministers, Cranborne and Thomas, his equally devoted private secretary, Oliver Harvey, and the former Prime Minister, Stanley Baldwin, who had after all been his mentor for most of the decade. Baldwin's advice seems to have been crucial in determining Eden's cautious line.[55] Despite occasional correspondence and conversations between them, it would be misleading to list Churchill among Eden's intimates. As Robert Rhodes James admits, Eden did not yet know Churchill very well or wholly trust his judgement.[56] The feeling was probably mutual. When rumours circulated of an intrigue to bring back Eden, Churchill was at pains to reaffirm his loyalty to Chamberlain.[57] Drawing up the outline of a possible future government of National Safety in May, Eden was ready to concede no more than the Ministry of Supply to Churchill with, of course, himself as Prime Minister.[58]

If the situation had been other than this, Eden might have been more wary of the older man as a possible rival for the ultimate succession to Chamberlain. But, as a journalist of the time recalled, it was Eden whom most people thought of as the next Prime Minister.[59] Furthermore, Churchill's stance on issues other than foreign policy was likely to alienate precisely those younger, more progressive Tories upon whom Eden's future hopes were pinned. As R. A. Butler saw it, Eden might 'become a new focus for a political movement among the younger and more sentimental members of the House'.[60] 'I am sure you are right,' Arthur Mann assured Eden. 'If the Conservative Party has any hope for the future it needs to go "left centre" – your course.'[61] Close association with Churchill could thus have been a political liability to Eden. 'Don't be worried, my darling,' wrote Harold Nicolson revealingly to his wife, 'I am not going to become one of the Winston brigade. My leaders are Anthony and Malcolm [MacDonald].'[62] Those who followed Eden rather than Churchill tended to be more mindful of the possible implications in their constituencies, and therefore for their political futures, of any overt signs of revolt against one of the most disciplined party-political regimes of the century. Throughout 1938, therefore, the two men struck markedly different poses. Boothby later contrasted Eden's 'frigid disapproval' of government policy with the 'tremendous salvoes' from Churchill's heavy guns.[63] Ironically, it may be argued that it was Eden's reticence which gave

Churchill his opportunity to re-emerge as a potential Prime Minister – at the former's expense.

It was a sign of Eden's determination not to kick away the political ladder by which he had already risen so far and so relatively fast that his return to government was frequently discussed. Channon was writing of such a possibility only four days after the resignation was announced, no doubt having the recent precedent of Samuel Hoare in mind.[64] In April Halifax made soundings about Eden becoming First Lord of the Admiralty, while on the latter's first visit to the Commons after his resignation speech Chamberlain himself made a tentative approach with probably the same office in mind.[65] Though Eden rebuffed both gestures, this perception of his position by the government's two leading ministers is not without significance and Eden maintained regular contact with Halifax throughout the year. The new Foreign Secretary still regarded his predecessor as a 'Prime Minister in the making'.[66] For the time being, however, Eden assured his supporters that he would not allow Chamberlain to profit by his return to the cabinet. His intention was to make a few big speeches on such topics as 'Democracy and Young England' in which he would largely avoid questions of foreign policy and indicate that he stood for the post-war generation against the old guard.[67] His restraint appeared to pay dividends with Thomas reporting that his support had grown and not diminished. 'The reason is that you "have behaved alright when you might have smashed the party . . . and we shan't forget it".'[68] By April, Harvey noted, Eden was 'much more reconciled and accustomed to idea of No 10. I think he feels it inevitable now.'[69]

Dramatic gestures therefore had no place in Eden's plan of campaign. When the independent-minded Duchess of Atholl suggested that he should follow her lead and resign the Conservative whip, Eden merely referred her to the statement of intention contained in his resignation speech.[70] Once he began to take a more active role on the public platform, his caution was very apparent. Only occasionally did his speeches contain anything which could be construed as overt criticism of the government's policy. For the most part he contented himself with platitudinous assertions offering no open challenge. There was 'nothing too offensive to dictators', commented Harvey after one performance. Eden had to avoid becoming a symbol for war against Germany and Italy, while distancing himself from the 'truckling attitude of the present Government'.[71] Typical was a speech to his constituents after the May crisis of 1938 in which

British firmness had appeared to thwart Nazi designs on Czechoslovakia. 'If appeasement is to mean what it says', proclaimed Eden, 'it must not be at the expense either of our vital interests, or of our national reputation and our sense of fair dealing.'[72] Such sentiments were sufficiently high-minded to be unexceptionable, yet for those who wished to read between the lines there was enough to suggest that the government's policy was not reaching such high standards.

It seems likely that Eden had engaged upon a more subtle approach to undermine Chamberlain's position than a frontal assault on the government's foreign policy. In April Harvey advised him not to make any 'purely foreign affairs' speeches for the time being. He should aim for 'higher and broader' speeches than the Prime Minister's. Eden should constitute himself as the spokeman of those floating voters whom the government had now lost and thereby emerge as an alternative leader.[73] The most common theme in Eden's public utterances was a rather vague call for a policy of national union, presumably involving the establishment of a national government on a broader basis than Chamberlain could command. He elaborated this idea in private correspondence. If Britain were to live through the next few years, other than on sufferance, a united national effort was needed on a scale comparable to that of the dictator powers. For this, a national leader was called for and, added Eden, 'N.C. has none of the necessary attributes'.[74]

Advised that an economic downturn might lead to worsening business conditions, rising unemployment and smaller revenue receipts, Eden purposefully widened the scope of his speeches to embrace social and economic issues. If rearmament expenditure was to be met in such circumstances, argued Ronald Tree, this might necessitate 'an all-round tightening of the belt' with Chamberlain 'temperamentally quite incapable' of putting over such an appeal to the electorate.[75] Thomas Jones, a confidant of both Baldwin and Lloyd George, arranged for Eden to visit Durham and South Wales to see the human misery and physical decay caused by unemployment. The task was to advance Eden's 'political and economic education'. Jones organized a two-weekend conference in November on unemployment and related problems, and urged upon Eden the need for 'a big forward social policy at home'. After his long immersion in diplomacy, Eden had much to learn. 'I think we succeeded in making him feel completely baffled by the complexities of the problem,' noted Jones.[76] Yet while such activities might better prepare Eden for the premiership, precisely how Chamberlain's grip on power was to be broken remained an open question. Rumours of the government's collapse continued into the summer without anything tangible arising. Harvey speculated on the development of events. Though Eden was the one man who could lead an all-party team,

> the difficulty of course that all see is how A.E. is to succeed N.C.? It is impossible to foresee how circumstances will fall out but I feel sure S[tanley] B[aldwin] will play a vital part in securing the succession. A.E. will have difficulty with his own party unless S.B. weighs in.[77]

Conservative Central Office was under no illusions about what lay behind Eden's platform campaign and advised Chamberlain to counter Eden not on foreign policy but on questions such as higher old-age pensions to nullify the younger man's appeal.[78]

In avoiding too overt a challenge to the government, Eden escaped the charge of disloyalty but ran the parallel risk of dissatisfying those who looked to him for a lead. Duncan Sandys, a young MP more usually associated with the Churchill camp, wrote on behalf of those who believed strongly in the 'modern progressive Conservatism' of which Eden had spoken and who were now ready to follow 'wherever you choose to lead'.[79] Such men felt fewer inhibitions than Eden about attacking the government openly. If Eden was serious about national unity, his failure to move the argument further forward was bound to disappoint others. 'He sticks chiefly to safe phrases about national unity,' complained the *New Statesman*. By June Amery surmised that Eden was reconciled to Chamberlain's policy and had 'made it up with him'.[80] Nicolson became increasingly exasperated by Eden's 'flabby formulas', while Victor Cazalet concluded at the end of 1938 that Eden had not got 'very much to say at the present time except a good many platitudes about democracy etc. with which we all agree'.[81] Even Eden concluded that 'we need an inspiration and we are not getting it' – without seeing the irony of his own words.[82]

While Eden concentrated his efforts on the public platform, parliament remained the forum in which the major foreign policy issues were discussed. When Chamberlain brought his Italian agreement before the Commons in April, Eden gave serious consideration as to whether he should speak. It was after all a development of the issue which had precipitated his resignation. Cranborne advised that any criticism Eden made of the terms of the agreement would be quoted all over the country by the opposition and would only embarrass and alienate government supporters. 'So probably silence is golden', he concluded, and 'in that case it might be wiser to stay away from the House for the debate itself.'[83] Churchill hoped that he and Eden might concert their forces, but Eden, 'greatly relieved not to be officially engaged in this orgy of friendliness to Mussolini', argued that he was bound by his promise of February not to hamper the course of the Italian negotiations. Any criticism he made would be attributed to pique, so it was best to leave Chamberlain 'a clear run for his frolics', even though many would resent his absence.[84]

In the event it was the ambitions not of Italy but of Germany which determined the pattern of international diplomacy during 1938. As Eden later recorded, the German invasion of Austria in March 'should have pointed the danger for Czechoslovakia, even to the obtuse'.[85] Yet during the early stages of the Czech crisis Eden made no significant interventions or comments. His past record suggested that he had few quarrels with the government's policy in Central Europe. As Foreign Secretary he had already urged reforms on the Czech President and Foreign Minister to improve relations with their German-speaking minority.[86] On 22 April Eden discussed with Halifax his own idea of a reformed Czechoslovakia guaranteed by Britain, France and Germany.[87] Though this might in its details have been very different from the solution which Chamberlain championed as 'peace with honour' five months later, it does suggest that Eden was at one with the thinking which determined government policy from the spring onwards – that the existing frontiers of the Czech state were untenable.

Only gradually did Eden start playing a more prominent part in national life. By late May he had decided upon regular attendance at the Commons, though not to speak. This Harvey judged a wise strategy since events were moving rapidly in Eden's favour and it was best not to appear to precipitate them.[88] Eden remained in close contact with Baldwin who advised that the country would not put up with Chamberlain for much longer.[89] His speech to his constituents in Leamington in early June was well received and renewed talk of a party split by the autumn.[90] He now proposed to make each speech 'rather hotter' than the last.[91] Still, however, there was a side to Eden which insisted on treading warily, anxious to point to the dangers of German armed strength, but reluctant to assume the role of the 'irreconcilable opponent of the Dictators'.[92] He believed that the government had lost a lot of credibility and that there would soon be great pressure for cabinet reconstruction. This would create 'a pretty embarrassing period for me'.[93]

Chamberlain, however, displayed more resourcefulness than Eden had expected. His decision to seize the initiative over Czechoslovakia by sending out Lord Runciman as a mediator between the Czech government and the Sudeten Germans seems to have taken Eden by surprise. Chamberlain had 'completely regained the initiative over Anthony,' concluded Baffy Dugdale. 'Who would have thought a year ago that nobody knows or cares where Anthony Eden is, and that all eyes should be on Runciman!'[94] Indeed, as the Czech crisis came to a head in early September, Eden was on holiday in Ireland. His absence worried government critics. Robert Boothby thought he had 'sunk himself'.[95] But on 8 September Eden returned to London. With Baldwin still advising him to concentrate on domestic issues, Eden seemed uncertain what line to take, especially as no concessions made by the Czech government had proved sufficient to satisfy the Sudetendeutsch. 'Full of impressions of industrial conditions and need for revival of Disraeli Tory democracy', Eden was alarmed at the international situation and thought that 'we should publicly endorse next Czech offer.'[96] After his return from Ireland he visited Lord Halifax and pressed for a more definite warning to Hitler to the effect that in the event of a conflict over Czechoslovakia, and France being drawn in, Britain would not stand aside. He left the meeting fearful that the government would still let Czechoslovakia down.[97] Yet two letters written to him that same day encapsulated Eden's dilemma. On the one hand Cranborne reported difficulties in his own constituency – 'I am likely to be stoned. It shows how wise you were to be more moderate.'[98] But a former MP sent a blunt warning of the consequences of inaction on Eden's part:

> During the last few months I have met many people who are beginning to lose confidence in you. Your stock has fallen appreciably. I wish to God it wasn't so, but there it is. People say 'if he really believes in collective security; if he is a staunch supporter of the League . . . why doesn't he come out and fight in the open? Is he lying low, waiting for another job? Does he expect the rank and file of the Tory Party to forsake the party machine and invite him to lead them when Chamberlain retires?'[99]

Eden's behaviour, up to and during the Munich Conference at the end of the month, suggested that he was only too conscious of these conflicting pressures. Captain Liddell Hart, who had recently been acting as his adviser on defence matters, found him 'very wrought-up and anxious', pacing up and down during their conversation.[100]

In a letter to *The Times* on 12 September Eden underlined the dangers involved in a lack of clarity and firmness towards Germany. But he still considered that there was scope for a negotiated settlement between the Czech government and its German minority.[101] Once again, however, Eden's initiative was overtaken by the course of events. On 15 September Chamberlain flew to meet Hitler at the Führer's mountain retreat at Berchtesgaden. Eden anticipated 'one more surrender' which the country would be invited to call 'peace'.[102] Three days later, as the details of Chamberlain's deal with Hitler became known, he seemed determined to attack the government in the Commons but, after further discussion with Halifax, admitted that he might himself have followed the same course. Significantly, he made it clear to his supporters that he had no wish to lead a revolt or to secure resignations from the government.[103] Yet such a passive attitude was bound to irritate many – as Eden clearly realized. Chamberlain's solution might see him hailed as a triumphant peacemaker, but 'surely there ought to be someone influential to explain in debate just exactly what he has done'.[104] The Liberal leader, Archibald Sinclair, wanted Eden to come out into the open, while Labour's Hugh Dalton spoke bitterly of Eden's failure to give a lead – 'What is the bloody fellow doing?'[105]

When Chamberlain was forced at a second meeting with Hitler to make further concessions at the expense of the Czechs, Eden urged the cabinet, through Halifax, to reject these new terms. It was perhaps his most resolute move during the crisis, but he could scarcely do otherwise having just warned his constituents about the dangers of continued retreat.[106] The majority of the cabinet followed Eden's line and for a few days the country balanced delicately on the edge of war. Tradition has it that, when Chamberlain announced to an excited House that he would accept an invitation to a third meeting with the German Chancellor, Eden was among a small band who refused to join in the outburst of spontaneous relief and emotion which greeted this statement. This may have been the case, but it did not stop Eden from writing to Chamberlain on the eve of his departure for Munich to wish him Godspeed and good fortune. Such behaviour suggests that he was as ready as the premier to explore every last avenue for a peaceful resolution of the crisis.[107]

What upset Eden most about the terms finally negotiated at Munich was less their content than the fact that they were imposed upon the Czechs as a diktat. This point was communicated to Halifax by telephone.[108] But Eden eschewed more overt opposition. With the Prime Minister again in conclave with Hitler, the anti-appeasement Focus group held a lunch on 29 September. Churchill suggested sending a telegram to say that, should Chamberlain impose further onerous terms on the Czechs, he would face serious opposition in parliament. The telegram was to be signed by a cross-party group consisting of Churchill, Attlee, Sinclair, Lord Lloyd, Robert Cecil and Eden. But the plan collapsed when Attlee and Eden refused to participate. Eden reasoned that such a gesture would be interpreted as hostile to the Prime Minister, by any criterion a curious excuse.[109] When Eugen Spier came to publish an account of this incident, Eden insisted that the recollection of his own actions was incorrect.[110] His official biographer attributes the story to the spite of Lady Violet Bonham Carter, admittedly no friend of Eden's, but as the events described are confirmed by others including Harold Nicolson this defence is less than compelling.[111] Lady Violet may have exaggerated the extent of Churchill's dismay – 'sitting alone in his chair immobile, frozen like a man of stone. I saw tears in his eyes' – but not significantly so.[112] Dining later that day with members of the Other Club, which he had helped form back in 1911, Churchill was 'in a towering rage and deepening gloom'. His failure to persuade Eden to support his initiative greatly upset him.[113]

Parliament debated the Munich settlement in early October. Even before this Baffy Dugdale concluded that Eden was 'ratting'.[114] This may have been unfair, but some of those closest to him were still anxious not to align themselves with Churchill and a 'cabal who were notorious for plots against the Government'.[115] Eden's Commons speech was overshadowed by more forceful contributions from colleagues such as Churchill and Richard Law. He praised both Runciman and Halifax for their efforts for peace and noted that 'we all owe [Chamberlain] a measureless debt of gratitude for the sincerity and pertinacity which he has devoted to averting the supreme calamity of war'. Eden included his now usual call for a national effort by a united nation. Overall, however, his opposition was voiced in 'friendly tones'.[116] His most telling criticism was to expose the similarity of the Munich territorial concessions to those rejected under the one-sided Godesberg proposals with which Hitler had confronted Chamberlain on 22 September. His purpose, he later wrote, was to get the Commons to think hard 'and in this I hope I was successful'.[117] Amery was impressed by a speech of studied moderation, while Nicolson found it superb. Such reactions were not, however, universal. According to Harvey, Eden's friends were 'disappointed, not to say shocked'.[118] Even more revealing is the fact that Eden himself was greatly impressed by Chamberlain's own speech on the third evening of the debate and, as a result, very nearly voted in the government lobby before deciding to abstain.[119] He and Amery first discussed the situation with a group of younger members who had all agreed to abstain. It was because none of these wanted to go back on their decision that Eden and Amery ultimately withheld their support.[120]

Eden's attitude to the settlement was probably encapsulated in a letter to Baldwin at the end of September. His words scarcely suggested that he would have been prepared to go to war rather than accept the terms of the agreement. Munich, he argued, had at least given Britain time 'at the expense of the unhappy Czechs'. But Chamberlain's primary task remained to broaden his government. If he would now invite Liberal and Labour co-operation, he would be in an impregnable position while impressing the world with the country's resolute intentions.[121] In the light of the Commons debate Samuel Hoare concluded that there was no basis for a working agreement between Churchill and the government. As regards Eden, on the other hand, he advised Chamberlain to 'get him back if and when you can'.[122] Eden himself seemed ready for a reconciliation. A week after the Munich debate he had a further long interview with Halifax where he emphasized the agreement he felt for 90 per cent of Chamberlain's speech. Halifax took up Eden's cause and tried to persuade the Prime Minister that this was the right psychological moment for a government of national unity.[123] Chamberlain, however, remained unconvinced. He explained the situation in terms which, in later years, would no doubt have gratified Eden:

> I fear the difference between Anthony and me is more fundamental than he realises. At bottom he is really dead against making terms with dictators and what makes him think it is possible to get unity is my insistence on the necessity for rearmament and the news that I didn't like Hitler personally. He leaves out or chooses not to see . . . that the conciliation part of the policy is just as important as the rearming, and I fear that if he were again a member of the Cabinet, he would do what he did before, always agree in theory but always disagree in practice.[124]

*

The Czechoslovakian crisis acted as the catalyst which brought the so-called Eden Group of dissident MPs into existence. There is no reason to doubt Eden's own account that a number of his supporters had begun meeting during the summer, although as late as the end of July he was claiming a reluctance to join any group, being 'less embarrassing to everybody' when he was on his own.[125] The decision to abstain in the Munich debate widened the circle of those who associated themselves with the former Foreign Secretary. In particular Munich compelled Eden and his erstwhile adversary, Leo Amery, into co-operation – though Boothby later wrote of a separate 'Amery Group' to which he claimed allegiance.[126] The Eden Group of 'Glamour Boys', as they were contemptuously labelled by the party whips, was an informal gathering of about twenty MPs without fixed membership or rigid policy positions. In addition to Eden, Cranborne and Thomas, its more prominent members included Amery, Mark Patrick, Duff Cooper – who resigned from the cabinet over Munich – Paul Emrys-Evans, Richard Law, Harold Macmillan and Harold Nicolson.[127] It was thus largely Conservative in composition and attracted the younger, more progressive elements within the party. The group maintained a separate identity from the smaller number who surrounded Churchill, though there was some overlap and Macmillan and Edward Spears, a close friend of Churchill, sometimes acted as intermediaries between the two groups.[128] Indeed Churchill was deliberately excluded because 'he would dominate our proceedings and would associate us with courses we did not want to follow'.[129]

The Eden Group usually met once a week, generally at the home of Ronald Tree in Queen Anne's Gate. This convenient location boasted a division bell enabling members to return to the Commons to vote when necessary. It was also subject to telephone surveillance by Sir Joseph Ball, who effectively ran a private intelligence service from Conservative Central Office.[130] Yet such anxiety on the part of Central Office may have been excessive. The private newsheet, the *Whitehall NewsLetter*, founded by Victor Gordon-Lennox, diplomatic correspondent of the *Daily Telegraph*, served as a mouthpiece for the group's views.[131] But, apart from the Munich debate, the group exercised little parliamentary impact where its members' contributions were hardly disloyal. Ironically, the group, which continued in existence when Eden returned to government in September 1939, was probably more influential once deprived of his membership. Eden's own tactics remained essentially unchanged. As Leo Amery recalled:

Eden and I attempted to follow up the [Munich] debate by a couple of articles in the *Sunday Times* outlining our conception of a national policy on which men of all parties might agree, and developed our ideas in our speeches over the next few months.[132]

Speeches in Cardiff and Southampton in mid-October reiterated Eden's well-rehearsed theme that an acceleration of rearmament would depend on an all-party national effort. His reception left him heartened that the country was ready to respond.

If, however, the Munich crisis led to a broadening of the opposition front against Chamberlain's policy, that process did not last long. Eden was important to any such development since his appeal had never been restricted to the Tory Party. When he resigned from the cabinet, Harvey had argued that the government would inevitably forfeit the confidence of the left, while Eden's repeated calls for national unity seemed to confirm his readiness to cross the lines of party demarcation.[133] Yet his behaviour after Munich tended to contradict such notions. The Commons debate had made it clear that there could be no effective parliamentary challenge to Chamberlain unless Tory dissidents joined forces with the Labour Party. Macmillan was in the forefront of efforts to establish such links. Hugh Dalton and some other Labour leaders showed interest as did Churchill. But for Eden national unity meant exactly that and not a minority – if cross-party – grouping designed to attack the government. For him there was always a clear distinction between criticism and disloyalty. So, while ready to speak out on 2 November against the government's Italian policy, Eden drew the line at anything which smacked of conspiracy. His reluctance to get involved in Macmillan's initiative proved decisive since more dissident Conservatives regarded Eden as their leader than any other figure. When Dalton pressed for a meeting with the Tory rebels, it became clear that Duff Cooper would not come without Eden and Eden would not come at all.[134] The scheme was left to fizzle out. As Dalton recorded: 'Macmillan is much disappointed with Eden and does not know what he is playing at. . . . Many of the Tory critics, however, are followers of Eden and will not move further or faster than he.'[135] Similar reasoning determined Eden's reaction to Amery's call for a campaign to introduce National Service. This would best come, he countered, as the consequence of the formation of a truly national government.[136]

Eden's tactics inevitably raised suspicions as to his motivation. His refusal to share a platform with Attlee and Sinclair led Mrs

Dugdale to conclude that he aimed 'to be P.M. of the Tory Party'.[137] Another judged that his wish was to be recalled to the cabinet 'for being good and causing no trouble'.[138] This was an oversimplification of Eden's position. Yet political calculation was never far away as entries in Harvey's diary show. 'He obviously looks at things', noted Amery, 'a little more from the point of view of their effect on the political situation and on himself than I am apt to do, probably rightly.'[139] In his own mind Eden had ruled out returning to office except as part of a wider reconstruction. He discussed his dilemma with the ever loyal Harvey. There were no easy options:

> What should AE do? Should he break away from the Party and lead a crusade in the country? Or should he stay just inside the Party, pressing re-armament? Too firm a stand now might force the PM to have an immediate election which he might win. There must be an election within a year – a policy of attrition from within, damaging speeches from the backbenchers, may be more effective in breaking the hold of the Party machine and securing a more easy change-over from the PM's regime to a wider Government. I gather balance of opinion of AE's supporters in the House of Commons is in favour of less heroic course.[140]

While some questioned his motivation, Eden's tactics left others disappointed or alienated. One step he refused to take was to support anti-government candidates in the series of by-elections held that autumn. At Oxford the Master of Balliol, A. D. Lindsay, standing against the official Conservative, Quintin Hogg, praised Eden in his election address but failed to persuade him to come to speak on his behalf.[141] Vernon Bartlett, who successfully contested Bridgwater, tried charm and flattery but to no avail. No one else could possibly rally the country, 'not Winston, whose vindictiveness in old days is partly responsible for Hitler'.[142] Bartlett won applause on the hustings for the mere mention of Eden's name, but was repeatedly faced by the question of why, if Eden was all that Bartlett said he was, he still declined to say a word in the candidate's support.[143] If, therefore, Bartlett's victory was 'for the Eden policy', it was a triumph to which Eden himself contributed nothing.[144] But Eden could always find the best of reasons for restraint. A memorandum drawn up by his brother, Timothy, gives an interesting insight into his hesitations:

> I have been thinking about what you told me with regard to the risk which this country may run if you succeed in turning out Chamberlain

and putting yourself and others in his place – the risk that, from the German point of view, you will have substituted for a statesman obviously friendly to the German people one who is considered by them to be hostile, and that, therefore, Hitler will be in a stronger position to cultivate a spirit of hatred towards England than he is now, and to translate that spirit into action before we are ready to defend ourselves.[145]

Small wonder, perhaps, that Dalton was ready to draw a contrast between Churchill and 'the Edens and other gentlemanly wishy-washies'.[146]

It is difficult to escape the conclusion that Eden failed to sense that the tide had finally begun to turn against Chamberlain and that the stance taken at the time of his resignation might no longer be appropriate. To be, as Eden saw it, cautious but statesmanlike ran the risk, in the eyes of others, of appearing hesitant and irresolute. The Munich crisis had temporarily enhanced Eden's prestige as the symbol of opposition to Chamberlain's policies, but such credentials could easily be lost. A much heralded speech at the Queen's Hall in November was widely considered a flop. It might have been one 'delivered by Lord Halifax' and left the general feeling that 'he is not a leader'.[147] Liddell Hart, who had helped prepare the speech, was dismayed to discover that Eden had removed most of the more telling passages and it was left to Violet Bonham Carter, moving a reluctant vote of thanks, to save the day with a forceful indictment of the government.[148] One particular difficulty was to foresee the circumstances in which Eden's vision of a national government would become a reality, since leading Labour figures were unlikely to serve in a cabinet led by Chamberlain and containing Simon and Hoare.[149] A poll in the *News Chronicle* in late November suggested that 40 per cent of those questioned would favour a new political grouping of MPs from all parties under Eden's leadership. At the same time, however, 49 per cent claimed to be satisfied with Chamberlain's premiership.[150]

Some who were closest to Eden began to show their concern. Cranborne concluded that 'Anthony must go further now than he has up to the present' and try to define the basis of the unity of which he so often spoke.[151] A note of criticism even crept into Harvey's diary, particularly when Eden claimed that the gap between himself and the government was narrowing and proved it by voting with the government over the question of creating a Ministry of Supply.[152] Churchill found him 'very shy at present', while Nicolson wanted Eden to be 'just a little less cautious'. The youth of the country was waiting for a stirring lead.[153] The Duchess of Atholl, about to sacrifice her career by standing as an independent in Kinross, insisted to Eden that 'a lot of people' would join a national opposition once a lead was given.[154] But perhaps the most astute political advice came, uncharacteristically, from his own brother:

> Something positive and definite for the future will have to be proposed by you very soon, if you are to maintain your position as leader on foreign affairs in this country. This is the sort of lead for which people all over England are longing and waiting, and until it is given, there is . . . little chance of turning out or even of completely reforming the present government.[155]

Such criticism seems to have had some effect and a speech Eden delivered in the Commons on 10 November was more forceful than many earlier efforts, though it left listeners uncertain whether he was trying to split the government, angle for a new coalition or lay down a marker for a new centre party.[156] A follow-up speech in Oxford also contained more substance, leaving the political world 'agog to know what he means to do – found a new party or capture the Conservative Party'.[157] But in December Eden accepted an invitation to visit the United States, where convention precluded any attack on the British government from a visiting British politician. An expectant meeting at the Waldorf Astoria heard Eden warn of the German danger, while 'refraining carefully from incorporating into it any explicit criticism of Mr. Chamberlain's policy'.[158] This meant a return to 'right non-committal platitudes' though, according to the British ambassador, the trip was certainly a success.[159] By the end of 1938 there was renewed talk of Eden's return to the cabinet, with the Party Chairman convinced of the advantages of such a step.[160] Eden wondered whether he might be offered Inskip's job as Minister for Co-ordination of Defence and, according to Harvey, 'rather fancied the idea'.[161]

Little had changed in Eden's outlook by the New Year – or in the effect which his tactics were having on those who awaited a lead. He had, thought Vernon Bartlett, missed the boat by not coming out definitely against the government and was now regarded with suspicion by the left.[162] Gordon-Lennox confirmed that Eden was thinking more of political tactics than of strategy and was 'lying pretty for the leadership of the Party'.[163] A forceful speech by

President Roosevelt left Harvey wishing that Eden could 'speak out as bluntly as that'.[164] It was perhaps indicative of his capacity to sit on a fence that colleagues picked up differing impressions of Eden's attitude towards a return to government. Churchill thought him resolved to stay outside, but Harvey found him very anxious to get back 'simply because he itches to be back at work and pushing things along in view of what he knows to be the country's danger'.[165] In all probability Eden hardly knew his own mind. A further talk with Halifax led him to expect an approach from the government. He wanted Baldwin's advice, but seemed to believe that a reconciliation was possible 'for in fact appeasement is dead'. His own views and those of the Foreign Secretary were 'much closer than they were'.[166] As the perception grew that Halifax's influence was increasing at the expense of that of the Prime Minister, Harvey concluded that the gulf between the government and the Eden Group had been much reduced. 'Indeed, as regards Germany, the two points of view are now very near.'[167] Cranborne agreed. 'I really do think that we are all coming closer together,' he assured Halifax.[168] In March 1939, three days before the Germans entered Prague, Chamberlain took pleasure in telling his sister that Eden lost no opportunity to let him know how cordially he approved of government policy.[169] But Chamberlain still held the reins of power sufficiently tightly to determine the composition of the cabinet. He had no intention of taking Eden back except upon his own terms. Eden was in a predicament of his own creation, torn between trying to convince the government of his loyalty and satisfying the expectations of his followers. 'In short Our Anthony is in a dilemma from which he would very much like me to extract him.'[170]

Be that as it may, the belief that a return to office was now a distinct possibility seems to have determined the tone of Eden's Commons speech in the wake of the Prague coup. He denounced the German action in somewhat stronger terms than had the Prime Minister but avoided any attack on Chamberlain personally, even though Prague represented the end of the latter's post-Munich illusion of lasting peace. Eden's remedy, as so often before, was to call for a new and more broadly based National Government.[171] He 'was not hostile to the Government', insisted Channon, 'and he took me by the arm at one moment'.[172] Eden found Halifax, whom he was now meeting regularly, sympathetic to his call for a widening of the government and the Foreign Secretary promised to raise the matter with Chamberlain.[173] Eden even began considering which of his followers should accompany him on his return to the cabinet while Cranborne, emphasizing Halifax's mounting influence, concluded that the government was 'now doing just what we would wish'.[174] By 24 March Eden was, according to Nicolson, convinced that his group should come out – violently if need be – for a three-party government with industry, wealth and manpower organized on a war footing. A meeting of the group three days later decided to table a Commons motion to this effect.[175]

Once again, however, the government reclaimed the initiative by offering guarantees to Poland and Romania. In private Eden gave the credit to Halifax for this strengthening of British resolve. But it was entirely characteristic that Eden went out of his way in a speech on 31 March to play down suggestions of splits within the cabinet and followed this up with an almost fulsome letter of support to Chamberlain.[176] When the Commons debated the country's new commitments on 3 April Eden argued that this policy had united the nation behind the government and that this would be of immense value in subsequent foreign negotiations. He mentioned the desirability of Russian support, but defended the government's decision to put Poland first, even though Liddell Hart had already advised him that the guarantee was strategically impracticable without a prior Soviet commitment.[177]

By the spring of 1939, therefore, Eden could scarcely be regarded as a government critic. He had even accepted reality and withdrawn his opposition to the recognition of Franco's victory in Spain. No one moment can be identified at which his status as the 'Prime Minister in waiting' passed irrevocably to Winston Churchill, though it was at about this time that his close confidant, J. P. L. Thomas, gave a prescient warning: 'I have a feeling . . . that among some of our supporters in the country Winston's speeches are a bit dwarfing yours as he has been very good of late.'[178] By May Eden himself had come to see Churchill as a possible premier.[179] It was a sign of changing times that his parliamentary performances no longer created a mood of expectation. As Channon recorded: 'It was a snubbing experience for him since at one moment there were only fourteen members in the Chamber. A few filtered in to hear him, but at no time, although he spoke for half an hour, was the House crowded or even interested.'[180] If Churchill and Eden had appeared to be coming together at the time of Munich, they were now once more drifting apart. Indeed, Eden, doubting whether Chamberlain could 'last physically much longer', seemed ready to contemplate a government headed by Halifax in which he would return to his old post at the Foreign Office.[181] When in mid-April the Commons debated the Italian invasion of Albania, Churchill's contribution upset Chamberlain while that of Eden prompted 'a friendly note and he replied in equally cordial terms'.[182] Significantly Harvey, who had advised Eden against this particular speech, feared that the latter overestimated his chances of an early return to government. Eden 'really feels sorry for the P.M., not understanding why he should not welcome his help'.[183]

To his credit Eden became increasingly aware of the importance of an alliance with the Soviet Union and was critical of the dilatory manner in which the cabinet approached the question. The government's conduct of the negotiations lacked 'boldness and imagination'. The proposal for a tripartite pact with France and Russia should have been grasped with enthusiasm.[184] More clearly than the Prime Minister, Eden saw the weakness of Britain's bargaining position. The country's need of Soviet support was hardly matched by anything Britain had to offer in return. 'I think that in the end we shall give Russia all she is asking for. There is no doubt that we could have got better terms . . . had we not been so hesitant in the early stages.'[185] He was struck by the government's reluctance to entrust these delicate negotiations to a figure of appropriate stature – in marked contrast to Chamberlain's earlier readiness to talk personally with Hitler – and in June he offered his own services in this capacity. 'An hour's talk between principals may be worth a month of writing.'[186] Eden had, after all, the benefit, not shared by any member of Chamberlain's government, of having met Stalin in 1935. But the Prime Minister remained wary, seeing in Eden's initiative 'a means of entry into the Cabinet and perhaps later on the substitution of a more amenable P.M.!'.[187]

Chamberlain discussed the question of cabinet reconstruction with the proprietor of the *Daily Telegraph* in early July. Interestingly, the two men agreed that Eden's inclusion – 'while he has a following in the country' – was not of the same consequence as that of Churchill.[188] It was perhaps the growing realization that Chamberlain had no intention of restoring Eden to high office which prompted his group into renewed activity in the summer of 1939. Significantly, the dissidents decided that both Churchill and Eden must be taken into the cabinet as an earnest of the government's determination not to contemplate another Munich. Macmillan acted as liaison with the *Daily Telegraph* as efforts began to organize a press campaign along these lines. But though the *Telegraph* seemed ready to co-operate, the message came back that there were doubts about Eden's 'toughness'. In the past he had been too influenced by a misguided sense of loyalty and Churchill now seemed the better bet as a potential national leader. The *Telegraph* campaign began with an editorial arguing for Churchill's recall, but containing no specific reference to Eden.[189] Soon afterwards a Low cartoon in the *Evening Standard* showed both men on the steps of 10 Downing Street with another figure entitled 'Public Opinion' knocking at the door and carrying a newspaper emblazoned with the headline, 'The Need for a Stronger Government'. As so often Low had encapsulated a growing public mood, though on this occasion his artistry could not capture the change which had come over the relative positions of these two leading outsiders. Eden hardly helped his own cause by spending an increasing amount of his time with the Territorial Army which he joined as a major in his old King's Royal Rifle Corps. Preparing for the battlefield was admirable in itself, but inevitably gave an impression that he was abandoning the political arena. By July even an Eden admirer concluded that he had lost the ear of the country, though in any government which included Churchill 'it would be politic to include Anthony as well'.[190]

As the government gradually modified its stance over the Russian negotiations, Eden concluded that it was 'pursuing the right course in international affairs'. Over reconstruction, however, Chamberlain remained 'as stubborn as ever' and Eden started turning his mind to a possible autumn election and the line he would then take. It was unlikely that he could give full support to a government constituted as at present and invite the electorate to endorse it for a further five years.[191] Yet this left open precisely what attitude Eden would adopt, particularly as it was becoming more difficult to find policy issues upon which to differentiate himself. Since resigning Eden had sought to avoid giving the impression that he was motivated by other than questions of policy. Cranborne suggested that, as the government had now 'adopted the policy for which we then pressed', it would be possible to offer full support while reserving the right to determine their attitude on foreign policy questions in the light of events. If such an approach failed to satisfy the party managers, 'we shall just have to go on without them'.[192] Neither man was convinced that Chamberlain would make no further concessions to the dictators, but their cautious attitude had merely boxed the Edenites into a corner. As Cranborne put it:

Should there be further evidence of government defeatism in their relations with foreign countries, as for instance the Tokyo negotiations, you will probably feel that it is your duty to . . . take a lead against it, at whatever cost to party unity. But unless there is some concrete evidence that the government's deeds fall short of their words, I do not personally quite see what . . . would be the grounds of your attack.[193]

In all probability an inherent sense of party loyalty would always hold Eden back. When rumours spread that the government might do a deal over German claims to Danzig, Nicolson spoke for a growing number when he summed up the problems created by Eden's equivocal stance:

If only Anthony Eden would now come out in rage against this subversive attempt, we should be safe. But Anthony does not wish to defy the Tory Party and is in fact missing every boat with exquisite elegance. We drift and drift and pass the rudder into other hands. I am much depressed.[194]

The Eden Group met frequently in the second half of July. The possibility of an autumn election continued to cause concern since, if no further international crises had then occurred, Chamberlain might be able to stigmatize the group as warmongers. Preparations were put in hand to collect material to present the Edenite case as effectively as possible. 'My real anxiety is that we should not be caught napping.'[195] By mid-August Eden was pondering whether the group should stand as independent Conservatives or even seek to create a new party. In asking 'what should our relations be with Winston', he showed that divisions still existed between the two groups of Conservative dissidents.[196] In fact Eden refused to join Churchill in opposing Chamberlain's adjournment of the Commons at the beginning of August. Eden 'rings Winston upon the telephone and begs him not to be too violent'.[197] It was almost as if each decisive step in Eden's thinking prompted a counter-move of indecision in the mind which had produced it.

It was, in any case, probably too late for the political realignment which Eden envisaged. Cranborne was obliged to reiterate the points he had made a month earlier. The government's guarantees to Eastern Europe, the formation of a Peace Front and the adoption of conscription ruled out any breakaway on grounds of policy. And while such a move might be justified by the Edenites' continuing lack of confidence in key members of the administration, including

Chamberlain himself, their speeches had tried to emphasize the essential unity and determination of the country from the cabinet downwards. Indeed, in his last Commons speech before parliament rose, Eden paid tribute to the forthright nature of Chamberlain's remarks. Something like another Munich would transform the situation. But, short of this, 'I do not see on what we should base a sudden transformation from nominal supporters to virulent opponents'.[198]

The Eden Group did not meet again until parliament was recalled on 24 August. By then, in the wake of the Nazi–Soviet Pact, events were moving inexorably towards a crisis. On 1 September German troops marched into Poland. The famous Commons debate the following day, when Chamberlain appeared to hesitate over a declaration of war, offered Eden a final opportunity to seize the initiative. Cooper and others urged him to intervene, but he declined to do so.[199] His attitude seemed to encapsulate his behaviour over the previous eighteen months. It was thus left to Arthur Greenwood, Labour's Deputy Leader, to 'speak for England'. Chamberlain saw, as probably he always had, that the coming of war would necessitate Churchill's return to government, and it was the latter's intervention which prompted an invitation to Eden as well. The younger man's 'influence with the section of Conservatives who are associated with him, as well as with moderate Liberal elements', made him a necessary reinforcement.[200] But the offer of only the Dominions Office, without a seat in the new War Cabinet, left Eden in 'great perplexity' and uncertain whether he should accept.[201] A sense of duty, coupled with Chamberlain's assurance that he could be a 'constant attender' at the War Cabinet, ultimately determined his response.[202] Eden had returned – if not triumphantly – to the centre stage, or something like it. But Chamberlain remained, for the time being at least, firmly in command, while Churchill had emerged, unofficially but unquestionably, as the second man in the government.

*

How is the year and a half which Eden spent outside the National Government to be interpreted? 'Though I knew that I had figured high in opinion polls for Prime Minister before and after my resignation in 1938', Eden recalled in his retirement, 'I had never seriously given that office a thought.'[203] Much that has been recorded above contradicts this assertion. Almost certainly Eden saw himself as Chamberlain's most likely successor. Not for several months after his resignation did Eden need seriously to consider the possibility that Churchill would be the eventual beneficiary of Chamberlain's downfall. But he had to be careful not to alienate the ranks of Conservative MPs whose support he would ultimately need, but who, in large numbers, remained solidly behind the Prime Minister. Such an interpretation makes sense of Eden's withdrawal from Westminster soon after his resignation and the particularly muted tone he at first adopted. It also helps explain why he turned first to Baldwin for advice. 'S. B. is the one man who may one day be able to turn the scales for him,' noted Harvey.[204] The idea may have been that Baldwin should return as a stop-gap premier with Eden his heir-apparent.

Eden was careful, perhaps too careful, to confine his opposition to grounds of policy – not always well explained – and high principle, so that he should not be dismissed as merely factious or self-interested. Increasingly, however, changes to government policy, especially after Munich, narrowed the ground upon which Eden could legitimately differentiate himself from those in power. He increasingly took refuge in the call for a broadly based government, a legitimate enough ambition, but one which dulled through endless repetition. Too much seemed to depend on Chamberlain eventually taking action, whereas those close to Eden saw that he himself needed to take a stronger line. 'You cannot afford to leave it to reason and the facts,' warned Law in July 1939:

> It's obviously quite impossible for you and Winston to take the active leadership of a movement to put yourselves into this Government. It's quite possible . . . for you to put yourself at the head of a movement to turn the Government out, and offer an alternative Government. . . . It means attacking the P.M., and that, you will say, will only rally the Party to him. That is what will happen. . . . You've got to pass through that stage and you will pass through it very quickly. . . . The country is crying out for leadership; it hardly knows its own need because, so far, there is no leader to evoke a response.[205]

The very man who called for a truly national government shied away from open collaboration with those who shared his misgivings about Chamberlain's capacity for the premiership. As David Dilks put it, when working as Eden's research assistant in the 1960s, Eden was careful not to prejudice his campaign for a national government and accelerated rearmament. Whereas men such as Duff Cooper and Macmillan proposed a more courageous course, believing, it seems, that it was possible to turn Chamberlain out, Eden preferred to work on Halifax to try to broaden the base of the existing administration.[206] Such conduct was honourable enough, though not perhaps the stuff of which dynamic leadership is made. Indeed, Eden's period out of office is as important in revealing his shortcomings as a politician as it is in highlighting his views on the central questions of foreign policy. 'I am afraid he has not got it in him to make a stand,' concluded Mrs Dugdale. 'If only his politics were like some of his tennis strokes down the centre of the court!'[207]

But Eden almost certainly shared with Chamberlain – and not with Churchill – the belief that war with Germany could still be avoided. Cranborne, often better at articulating Eden's thoughts than he was himself, put it in these terms as late as February 1939:

> Appeasement must no longer be regarded as an alternative to rearmament. Rearmament must be regarded as the preliminary – and the essential preliminary – to appeasement. . . . I am quite sure that our proper line is to rearm with ever increasing resolution and wait for the Dictators to come to us, and not to go running after them.[208]

This probably summed up Eden's thinking at this time, and for much of the 1930s. It distanced him from the government, though not to the extent which he later claimed. Be that as it may, Eden's career had now taken a decisive turn, if not quite in the direction he might have wished. For the next eight months he served as a popular Dominions Secretary. At the formation of Churchill's government in May 1940 he was promoted to the War Office, but still without a formal seat in the War Cabinet. With hindsight Eden came to look upon his seven months in this post as the most rewarding of his career as he took responsibility for a massive expansion of the British army and the creation of the Home Guard. Churchill 'maintained that I shall never do anything in my life to equal my . . . administration at War Office'.[209] Only in December 1940, when the sudden death of the British ambassador in Washington enabled Churchill to reshape his government, did Eden revert to his old position as Foreign Secretary. Exactly five years after his first appointment to this same post, he returned to a major office of state.

6

Eden and the United States

I have always tried to be careful with the Americans. They are
sometimes a little difficult to deal with.[1]

B Y the end of his life it would probably have been fair to describe
Anthony Eden as an anti-American. The Suez Crisis provides an
obvious explanation. To his dying day Eden believed, with considerable justification, that Britain's policy and his own career had been
wrecked by the United States' refusal to offer the backing which a
loyal ally could rightly expect. The United States was not interested,
he wrote in 1957, in working with its allies or in supporting them.
'Its only interest lies in superseding them.'[2] American policy during
Suez had been motivated by 'spite and injured dignity'.[3] Eden's
particular scorn was reserved for John Foster Dulles. It was, he once
noted, 'easier to deal with Ribbentrop or Laval' than the American
Secretary of State.[4] But he readily extended his contempt to the
American nation as a whole. It annoyed him that his successor in
Downing Street saw the re-establishment of the fractured Special
Relationship as the primary goal of his government's foreign policy
and he wrote scathingly that Macmillan's diplomacy would make
Britain 'the 49th state' of the Union.[5] Macmillan himself intervened
to try to tone down the blatant anti-Americanism of the first draft of
Eden's memoirs, *Full Circle*.[6]

Twenty years earlier things had been very different. As was argued
above, Eden's failure to secure Neville Chamberlain's support for the
notion that the allegiance of the United States should be a primary
objective of British foreign policy had been as important as any issue
in compelling his resignation in 1938. Yet despite Eden's insistence on
another occasion that he was an 'Atlantic animal',[7] the crisis of 1956 is
not a sufficient explanation of the transformation of his attitude. In
a revealing diary entry Eden once traced, with thinly veiled
resentment, the stages by which Britain had become increasingly

dependent upon the United States. 'I have seen the phases of U.S. advance', he recalled:

> First in 1914–1918 war when so large a part of our life and wealth was consumed. I was attached in April 1918 to the New York division as it arrived and trained with it for some weeks. . . . The growing wave of U.S. forces must have made Germans hopeless of victory. But their share in fighting and casualties was small.[8]

Suez exacerbated an anti-American strand in Eden's make-up, but it by no means created it. Such feelings were present during the Second World War.

By the time that Eden resumed responsibility for his country's foreign policy in December 1940, the role of the United States had become a critical factor in Britain's very survival. The fall of France in June dramatically transformed the immediate and long-term outlook, making it difficult for many inside the British government to see how victory could now be secured. The War Cabinet had concluded on 27 May that Britain did have a fighting chance of survival even if France fell, but that prospects of final victory were largely dependent on increasingly active American co-operation.[9] Two months later the Foreign Office confirmed that the future of the Empire was 'likely to depend on the evolution of an effective and enduring collaboration between ourselves and the United States'.[10] To his credit Winston Churchill had, from the outbreak of hostilities, sought to bring the United States as far into the British war effort as deeply ingrained American isolationist sentiment allowed. In September 1939, when only First Lord of the Admiralty, Churchill seized upon Roosevelt's suggestion that the two men should start up a correspondence. With Chamberlain's blessing Churchill began to cultivate a personal contact which he believed essential to Britain's survival as a great power.[11]

Though the actual relationship between the two men began more tentatively than Churchill implies in his memoirs, and was never as warm on Roosevelt's side as on Churchill's, Anglo-American collaboration remained central to the latter's strategic thinking throughout the conflict and beyond. 'My whole system', he told Eden in 1942, 'is based on friendship with Roosevelt.'[12] Nothing should be done in planning the post-war world, Churchill insisted a year later, to prejudice the 'natural' Anglo-American Special Relationship.[13] This inevitably involved subordinating purely British interests when necessary. There were 'various large matters in which we cannot go further than the United States are willing to go'.[14] That there might be a downside to Churchill's single-minded preoccupation in terms of Britain's increasing subservience to a considerably greater power, and the destruction of that independent status in world affairs which Churchill defined as a primary war aim, has long been apparent and did not require revelation by 'revisionist' Churchillian historians. But, particularly after June 1940, the British Prime Minister was, realistically speaking, confronted by a limited scope for manoeuvre, for which some of his critics have shown insufficient regard.

As Foreign Secretary in wartime Eden had to work within parameters set by Churchill. There could be no return to the sort of regime which he had enjoyed under Baldwin five years earlier. 'In wartime', Eden later wrote, 'diplomacy is strategy's twin',[15] and the figure of the premier was the dominant one in both. Churchill, like Prime Ministers before and since, considered himself blessed with an assured grasp of the central questions of diplomacy and was little inclined to give way to the supposed professionals of the Foreign Office. He had 'studied Europe for forty years. The Foreign Office were always on small points like chrome and ships in Sweden.'[16] This situation could be exasperating for Eden. Against Churchill's plea in September 1943 that he should 'try to have a little confidence in my insight into Europe', Eden wrote, 'He might have a little in mine!'[17] In the last resort, however, it was something Eden had to accept. This was particularly the case both in the priority given to Anglo-American relations and in Churchill's determination to keep such matters in his own hands. Significantly, the Special Relationship was forged first and foremost at the heads of government level. Nor was it easy for Eden to develop a parallel relationship with his American opposite number. Roosevelt's Secretary of State, Cordell Hull, did not cover the same range of business as Eden and was often kept woefully ignorant by the President of crucial developments in the war.[18] Eden's position was therefore always somewhat anomalous within the Anglo-American equation. None of this, however, meant that he accepted Churchill's vision of the Special Relationship without qualification or without seeking to modify it when the need arose.

Of Eden's wartime partnership with Churchill much has been written. Nearly forty years ago one observer counselled caution upon those tempted to penetrate further into their relationship:

The complementary qualities of character and age and mind, the pride and indulgence of the old prophet and the devotion and patience of the chosen successor, have been so often and so justly celebrated that further description seems doomed to be as stale and insipid as a retelling of some Bible story or familiar Shakespearean plot.[19]

More recently, however, historians have discovered complexities in their wartime association which neither fully revealed in his published memoirs. Despite an age-gap of twenty-three years relations between Eden and Churchill were certainly characterized by mutual affection and an inherent loyalty on Eden's side which, notwithstanding his undoubted ambition, precluded any serious threat to the Prime Minister's position. But Lord Strang overstated his case when he insisted that Churchill's differences with Eden were, 'in general, emotional rather than rational, a matter of degree rather than principle'.[20] Eden was no yes-man and serious disagreements did occur. Though his immediate entourage was more likely to view the extent of his disgruntlement than was the Prime Minister, it often fell to Eden to represent his department's longer-term view against Churchill's short-term military objectives.[21] In such situations Eden by no means always came off second best. Those who observed from the fringes of power often underestimated Eden's capacity. 'He has many of the necessary attributes of a Statesman', conceded Sir Henry Pownall, then Chief of Staff, South-East Asia Command, 'but he lacks an essential one – *guts*. Like all the other ministers he is completely overawed by the PM and would never put up a fight against him in the way that CIGS is continually having to do.'[22] With tenacity and patience Eden was fully prepared to take Churchill on when he believed that the situation demanded it and often got his way in the long run. If he was more inclined to use charm than bluster to win his case, this in no way lessened his achievement. As one perceptive writer concluded, 'if Eden had been nothing but a wet blanket, Churchill would not have stood him for long. As things were, Churchill took from Eden a great deal of opposition and frustration over his pet ideas or stubborn prejudices, and still liked him and trusted him.'[23]

In terms of the underlying goals of Britain's diplomacy Eden and Churchill were fundamentally at one. In November 1942, by which time it had become legitimate to contemplate the post-war world, Eden wrote that 'the aim of British policy must be, first, that we should continue to exercise the functions and responsibilities of a world power'.[24] It was a conclusion with which Churchill could readily have associated himself. Furthermore, both men – and indeed the overwhelming majority of their political contemporaries – took it for granted that the British Empire would survive the conflict and form an essential component of the country's international status in the post-war world. Where the two men parted company was over the advantages to be gained from an American alliance.

Eden, like his principal Foreign Office advisers and the majority of the War Cabinet, believed that too close an association with the United States could only compromise national autonomy. Despite his own claims to American ancestry,[25] Eden looked on the Anglo-American relationship without the emotional – some would say sentimental – element which Churchill invariably infused into it. He saw the potential value of Churchill's close partnership with the President, particularly in view of his assessment of the chaotic workings of the US government, but could be far more critical – even dismissive – of Roosevelt than Churchill ever was.[26] Though not wishing to see America retreat into its traditional patterns of isolationism once the war was over, Eden was also concerned that a pre-eminent United States would overshadow Britain in areas where the latter had hitherto held sway. If Anglo-American co-operation was desirable, undue subservience to the United States was not. He would have been horrified to read Leo Amery's complaint that he 'live[d] in terror of anything that could conceivably offend a single American'.[27] As a result Eden was far readier than Churchill to see in the United States, as he saw in the Soviet Union, a potential post-war rival. In a revealing note written in 1944, he warned that, if Britain lost Arab goodwill in the Middle East, 'the Americans and the Russians will be on hand to profit from our mistakes'.[28]

At a distance of more than fifty years it might appear that Eden was setting out to resist the inevitable. In the context of Britain's war effort, however, his stance seems more reasonable. Until July 1944 the British Empire had more men engaging the enemy across the globe than did the United States.[29] Moreover, Eden's perception of Britain's post-war role was not based simply on an overestimation of her inherent strength. Rather than placing Britain's eggs in a single basket, he believed that an Atlantic partnership would need to be balanced by her position in Western Europe, particularly in association with a revived France, and, if possible, by continuing friendship with Russia.

In his dealings with the United States Eden at least enjoyed the advantage of a popular following among Americans second only to Churchill. He was remembered as the handsome young politician who had had the courage to resign from Chamberlain's government in protest against appeasement.[30] That image was enhanced by a successful visit to the United States in the spring of 1943. As the British embassy reported, Eden's presence 'created a picture of a serious and honourable young statesman which replaced the image of the glamorous but lightweight character, which had previously obtained in considerable sections of the press here'.[31] When it was suggested that Eden might leave the Foreign Office to concentrate on his duties as Leader of the Commons, Robert Sherwood, Director of Overseas Operations in the Office of War Information, protested that the effect in America would be deplorable. Eden was 'the one Englishman who is completely trusted by the United States'.[32] But Americans also saw Eden as someone more likely than Churchill to exert a decisive impact on the eventual peace settlement and were not always at ease with his ideas.[33] As Harry Hopkins, Roosevelt's trusted and influential confidant, put it as early as 1941, 'I would not like to see [Halifax] have much say about a later peace [and] I should like to have Eden say less'.[34]

*

While America remained neutral during the first two years of the war, Eden was keen, as he had been in 1937–8, to persuade her to take as active a role as possible in defence of British interests. As France neared defeat his hope was that Roosevelt might inspire continued resistance by at least breaking off relations with Germany and he urged Churchill to use his influence with the President in this sense.[35] Churchill held back from such a step and France duly capitulated, though it is doubtful whether an American initiative would have had any effect. But out of the fall of France there developed the seeds of a significant disagreement between Eden and the American President. Roosevelt believed that France had now forfeited all claims to great-power status – a conclusion which Eden, with specifically British interests to the forefront of his analysis, was loath to accept.[36]

In his first public speech as Foreign Secretary Eden emphasized the similarity between British and American thinking on war aims as a way of countering the arguments of American isolationists that Britain's real ambition was to restore the imperial edifice for her own selfish interests.[37] Granted his belief that the goal of social improvement at home should be the concomitant of a new post-war international order, this was fair enough. Behind the scenes, however, Eden provided ample evidence to confirm the misgivings of those Americans who had no wish to fight a war for British national interests. In a departmental minute on the famous destroyers-for-bases deal, the new Foreign Secretary warned that

> by allowing U.S. to establish important military garrisons in most of West Indian islands and Newfoundland we have struck a grievous blow at our authority and ultimately I have no doubt at our sovereignty, in all these places. The consequences upon our position in the American Continent . . . are likely to be most unhappy for those who believe in the value of the British connexion. We have received in return 50 old destroyers of which 2 have so far . . . been made seaworthy.[38]

Though Eden, unlike Churchill, was keen early in 1941 to let the Americans know the full extent of British shipping losses as the best way of securing active support in the Battle of the Atlantic, his concern at the possible encroachment of the United States upon areas of traditional British interest remained unchanged:[39]

> Incidentally do we want to see US bases established, say, at Auckland and in Fiji, at Takoradi and Trincomalee? Some of these are a far cry for US, others are not, and I would not happily contemplate a whole-sale extension of US bases throughout the British Commonwealth.[40]

Eden's thinking was further revealed when, in August 1941, Churchill had his first meeting with Roosevelt at Placentia Bay, Newfoundland. From this conference there emerged the celebrated Atlantic Charter, based on a British draft produced by Alexander Cadogan but phrased to reflect a largely American approach to the future peace. Believing that Roosevelt meant to 'keep out of the war and dictate the peace if he can', Eden judged that the President had bowled Churchill 'a very quick one'.[41] At a time when getting American support to sustain the British war effort was a top priority for most British leaders, Eden's private secretary, Oliver Harvey, confirmed that he was 'rather isolationist' about the Americans coming into Europe, hoping that Britain could remain 'the predominant partner'.[42] Indeed, it 'chilled [Eden] with Wilsonian memories' to see an American President 'talking at large on European frontiers'.[43] The Foreign Secretary was moving towards that healthy scepticism of American intentions – the view, as Chamberlain had once put it, that 'it is always best and safest to count on nothing from the Americans but words'[44] – which had characterized the Foreign Office's outlook for most of the 1930s and of which he had earlier been so critical.

One area where Eden still hoped for positive American collaboration was the Far East, where overall British policy remained to keep Japan neutral. Though he broadly accepted this strategy, Eden was concerned that Britain should not adopt a supine attitude herself nor be seen to restrain any American initiatives against Japan. In November 1940 the exiled Dutch government in London had requested an assurance of British military assistance should Japan attack the Dutch East Indies. Eden consistently pressed for such an assurance from the War Cabinet.[45] 'I suggest', he noted the following September, 'that a display of firmness is more likely to deter Japan from war than to provoke her to it.'[46] When Harry Hopkins visited London early in 1941, Eden tried to impress upon him that Britain was taking a 'very strong line' with Japan and would stand for 'no nonsense'. As Hopkins reported, 'Eden is very anxious that we find a way to emphasize our determination to prevent Japan from making further encroachments. He believes that if we take a positive line towards Japan we might make them pause before attacking Hong Kong.'[47] When in the spring the United States moved part of their fleet from the Pacific to the Atlantic Ocean, Eden was alone in the Defence Committee in opposing this move and stressing that the US Pacific Fleet was the only barrier against a Japanese advance to the south.[48]

More than his colleagues Eden urged the importance of not antagonizing the United States and of encouraging any initiatives she might take against Japan. The danger of a misunderstanding similar to that which had occurred between John Simon and his American opposite number, Henry Stimson, in 1932 during the Manchurian crisis was, he argued, real. 'I cannot conceal . . . the danger inherent in our lagging behind the United States Government in dealing with Japan, *a fortiori* in our actually attempting to dissuade them from firm action.'[49] In July, after lengthy discussions, Eden won cabinet support for what was potentially a high-risk strategy. The government in effect committed itself to back whatever line America decided upon. The danger existed that Japan might be provoked into an attack on British territory at a time when Britain was still militarily weak in the Far East, fully engaged against Germany and Italy, and lacking any firm commitment from the United States.[50]

As Eden explained:

> President still cherishes the idea that we can deal with Japan when
> we are free from our European preoccupations. . . . [I]t might be well
> to suggest that even in these days prevention is better than cure and
> that the best prevention is fear on the part of the Japanese of
> immediate war with the United States if they go too far.[51]

At the same time he hoped that Churchill could use his influence
with Roosevelt. While appreciating the President's constitutional
difficulty about giving an undertaking to help any foreign power
without Congressional authority, he hoped that Churchill could
secure a private assurance that, if the British or Dutch became
involved in hostilities with Japan, Roosevelt would seek Congressional authorization to give all possible help.[52] By September he
judged that the Japanese had begun to hesitate, a situation brought
about by 'the contemplation of the forces that may possibly confront
them'. 'Our right policy', he concluded, 'is, therefore, clearly to keep
up the pressure.'[53]

Eden recognized that, as a result of commercial pressure, the time
would soon come when Japan would either have to reach some
understanding with Britain and America or, at the risk of war
with both countries, break out of her economic strangulation.[54]
His preference remained to keep Japan out of the world conflict,
but he was still convinced that strong American action would deter
rather than incite aggression.[55] By contrast Churchill, for all his faith
in Roosevelt, was at this stage more wary than Eden of provoking a
war with Japan which left the United States neutral. The Prime
Minister wanted a 'Jap-American war' rather than a 'Jap-British war
which the Americans might or might not enter'.[56] In the event, of
course, Churchill had his way and the United States became a co-
belligerent following the Japanese attack on Pearl Harbor on 7
December 1941. Precisely what would have happened, at least in
the short term, if Japanese aggression had been directed solely at
British and imperial possessions remains open to debate. There is at
least an argument that Eden's line could have led to disaster – the
nightmare scenario which had haunted British policy makers in the
1930s of a war against three major enemies without the compensating support of the United States. In part Eden was putting into
practice the lessons of earlier years with, it must be admitted,
greater clarity than he had always seen at the time. This became
evident when Lord Hankey, as chairman of the War Cabinet's Sub-

Committee on Supplies to the Allies, urged the importance of staving off hostilities with Japan until the last possible moment. In a revealing minute Eden noted: 'The old appeasement again. Of course we do not want to fight Japan but I fear that Lord H. will never learn that to be gentle with aggressors does not avoid hostilities. His part in pre-war foreign policy was most unhappy.'[57]

*

It was above all over the question of France that Eden's attitude towards the United States became most strained. To begin with he shared Roosevelt's view that the shame of June 1940 effectively precluded France's restoration as a great power for the foreseeable future.[58] Nor did Eden immediately jump to the cause of de Gaulle as the expression of a continuing French will to resist Nazi aggression. John Colville recorded the remark, at one moment of particular exasperation with de Gaulle in the middle of 1941, that Eden hated all Frenchmen.[59] Indeed, at this stage it was largely left to Churchill to urge an end 'to the cold-shouldering of General de Gaulle and the Free French movement'.[60] For some months the Prime Minister's entreaties to shift emphasis from the Vichy regime to de Gaulle went unheeded.[61] By the summer of 1941, however, a revised Foreign Office policy was coming into place, of which Eden emerged as the doughty champion.

Increasingly, the restoration of a strong France – though it might take many years to achieve – was seen as a primary war aim. Britain's problems in post-war Europe might derive from a resurgent Germany or an untrammelled Soviet Union. In either case the attractions of a strong, friendly power on the other side of the Channel were only too apparent. As Eden once put it, France was a 'geographical necessity'.[62] To have any chance of revival France needed leadership and within the Foreign Office it became clear that such leadership could only come from the admittedly awkward, but always formidable, personality of Charles de Gaulle. William Strang, one of the architects of this Foreign Office policy, later summed up the general's importance:

From the time of his establishment in London . . . down to the time when he entered a liberated Paris and duly became the head of a recognised Provisional Government, [the Foreign Office] fostered the Free French movement, accorded it progressive degrees of recognition, and bore with what patience and understanding they could the storms that he let loose upon them. This was, above all, Eden's policy.[63]

By August 1941 it was Eden who, while still wondering whether de Gaulle might be 'crazy', now urged Churchill to give the general the benefit of the doubt and make an effort to accommodate his idiosyncrasies.[64] Prime Minister and Foreign Secretary increasingly reversed their earlier positions with Churchill now complaining that Eden was too ready to sever relations with Vichy – 'you seem prone to cut threads that may never be rejoined'.[65] By the end of the year Eden had concluded that the French resistance movement was overwhelmingly behind de Gaulle. When Eden returned from Moscow in December 1941 he began to press for the inclusion of the Free French in the proposed Declaration of the 'United Nations'. They were 'in every sense an ally' with forces collaborating with the British and Americans in many theatres, particularly New Caledonia.[66]

Eden faced not only a reluctant Prime Minister but an adamant President and Secretary of State. In general the Americans adopted an understandably more sceptical attitude towards France's great-power pretensions than did the Foreign Office. 'The British', Roosevelt noted in 1943, 'wanted to build up France into a first class power, which would be on Britain's side.'[67] This was an accurate assessment of Eden's policy, but there were obvious reasons why the United States might be reluctant to subscribe to such an objective. Roosevelt's ideas about the post-war world centred on the concept of four great powers – the United States, the Soviet Union, Britain and China – and it was not until the final months of conflict that he reluctantly allowed – at British prompting – a revived France to intrude into this vision.[68] Additionally, it took the United States a long time to concede that a post-war France headed by de Gaulle was the only viable option. Indeed, there was a case for supposing that some modification of the existing Vichy regime might best serve American interests, providing it would transform its wartime subservience to Nazi Germany into a post-war subservience to the United States.[69] But American thinking on such matters was never divorced from considerations of character and personality.

As de Gaulle consolidated his position, some American attitudes began to soften. Secretary for War, Henry Stimson, concluded that the general would have to be recognized in the wider interests of Anglo-American relations. On this issue, however, Roosevelt and Hull were of one mind. 'President's absurd and petty dislike of de Gaulle blinds him,' Eden recorded in March 1944.[70] Hull's hostility went even deeper. 'He hates de Gaulle with such fierce feeling', noted Stimson, 'that he rambles into almost incoherence whenever we talk about him.'[71] Eden concluded that Hull had 'an obsession against [the] Free French which nothing [could] cure'.[72] On some issues the President was willing to ride roughshod over the views of his Secretary of State. On this matter, however, Hull's feelings confirmed Roosevelt's own prejudices.[73]

America's entry into the war only compounded Eden's problems. On the one hand Britain now seemed certain to prevail in the struggle against the aggression of Germany, Italy and Japan. But, on the other, Eden was reluctant to concede that Britain's new ally could call the tune in terms both of shaping grand strategy and of constructing the future peace. As he later explained:

> I accepted . . . that the United States must in time become the dominant partner in Anglo-American councils. In 1942 this was not so. Her wealth and productive capacity were not yet at full stretch and her armed forces were only beginning their training and deployment. Our effort, on the other hand, after more than two years of war, was nearing its peak.[74]

In any case Eden held an increasingly poor opinion of the President's judgement and of his likely impact on the final peace settlement. After reading a report of a typically rambling Rooseveltian statement on the future of India, he commented that it was a 'terrifying commentary on the likely Roosevelt contribution to the peace, a meandering amateurishness lit by discursive flashes'.[75] Though Eden was readier than many governmental colleagues to accept the terms of the Lend–Lease agreement lest Britain was offered worse terms at a later date, his uneasiness about America's role in the war and its consequences for Britain remained unchanged.[76] In conversation with Robert Bruce Lockhart in February 1942, Eden complained that the Americans seemed very imperialistic in their outlook, that while they were ready to criticize others they were themselves extremely sensitive to criticism, and that their performance in aircraft production left much to be desired.[77] While Churchill's carefully cultivated friendship with Roosevelt seemed – at least in the Prime Minister's eyes – to be paying dividends, for Eden the partnership between the two war leaders tended to exclude him from key decisions of wartime diplomacy and increased his mounting irritation at the way Churchill ran the war. Early in February he spilled out his frustration into the pages of his diary:

> It is impossible to understand what P.M. and Roosevelt arranged as to Pacific Council, nor do either seem to know. Anyway they say

different things and R's announcement today of two Pacific Councils is in flat contradiction to what Winston told us and the Dutch who are, excusably, becoming increasingly exasperated.[78]

A particular problem from Eden's point of view was to reconcile Britain's new American alliance with her existing relationship with the Soviet Union, which had been at war with Germany since the previous June. Eden saw that the Soviet Union's contribution to the defeat of Germany could hardly be ignored and that her ambitions for the post-war world would have inevitable implications for Britain. Though Eden conceded that any choice confronting Britain between the United States and the Soviet Union would have to be decided in favour of Anglo-American co-operation, this did not prevent his decision in April – in the face of American opposition – to pursue an Anglo-Soviet agreement.[79] He was annoyed by Roosevelt's decision in February to approach Stalin directly over the future of the Baltic States. This was the first of several occasions, Eden later recalled, when the President moved out of step with Britain in the illusory hope of securing better results with Stalin through a unilateral initiative.[80] His contemporary diary account was more revealing: 'I did not feel that we could just hand over to Americans conduct of Anglo-Soviet negotiation nor do I believe that Roosevelt in the least understands difficulty he is up against.'[81] It annoyed Eden that the United States seemed unwilling to treat Britain as an equal. In June he complained that while Britain had kept the Americans fully informed of the progress of their talks with the Russians, including submitting the draft of a proposed treaty to Washington before the Russians were allowed to see it, Britain was being told virtually nothing of the talks between the United States and the Soviet Union. These talks had produced a communiqué referring to the need for a second front in Europe that year, 'a matter which at least concerns us as much as the Americans'. 'If we are to work harmoniously together', Eden concluded, 'they should reciprocate.'[82]

But it was France which still provided the main focus for friction between Eden and the United States. Continuing American ties with Pétain's Vichy regime inevitably created difficulties for Britain's relations with the Free French. The United States sought to maintain contact with Vichy in the hope of bringing Pétain round to an anti-German stance. Eden was sceptical of this objective and aware of the difficulty of selling it to a British public which held the Marshal and his regime in low esteem after the capitulation of 1940. But he agreed to go along with the American policy, 'provided the Allied war effort will be helped thereby'. Of this, however, he became increasingly doubtful. Vichy had shown no wish to help the Allied nations, while America seemed too ready to give in to Vichyite blackmail.[83] The danger was that Britain's relations with the Free French – and thus with the likely leadership of post-war France – would be compromised by futile efforts to bring about a change of heart inside the Vichy regime.[84] Britain, insisted Eden, had been responsible for building up de Gaulle inside France and could scarcely drop him now. Though his failings were well known, the general had to be given credit for keeping the French flag flying at Britain's side since the catastrophe of 1940.[85] A latent disagreement between Eden and Churchill on this issue rumbled on through the spring and summer, with the latter motivated as much by his concern not to step out of line with Roosevelt as by his own animosity towards de Gaulle. While Eden tried patiently to persuade Churchill to adopt a more sympathetic attitude towards the Free French leader, Churchill still insisted that there was 'nothing hostile to England this man may not do once he gets off the chain'.[86]

Eden broadcast to the French people on Bastille Day 1942. The full restoration of France as a great power was, he declared, a fixed war aim and also a practical necessity if post-war reconstruction was to be undertaken 'within the framework of that traditional civilisation which is our common heritage'.[87] Such words were for public consumption. But they were not out of line with his more private comments. A year later, on 13 July 1943, Eden was to be found arguing that, in order to contain Germany after her military defeat, Britain's treaty with the Soviet Union would need to be balanced by an understanding with France in Western Europe. Significantly, he added that 'in dealing with European problems of the future we are likely to have to work more closely with France even than with the United States' – a judgement Churchill would never have accepted. Though Britain should aim to concert its policy towards France as far as possible with Washington, there were, thought Eden, 'limits beyond which we ought not to allow our policy to be governed by theirs'.[88] Indeed, the Americans 'knew very little of Europe and it would be unfortunate for the future of the world if U.S. uninstructed views were to decide the future of the European continent'.[89] Such a remark could have come from the pen of any one of a host of sceptical Foreign Office officials over the previous two decades.

Fear that the United States might otherwise step into a policy vacuum persuaded Eden and the Foreign Office to pay increasing attention in the second half of 1942 to British plans for the post-war world. In such matters Eden spoke for the traditional views of his department, notwithstanding his own reputation as a champion of internationalism. His ideas about Europe's future indicated the sort of balance-of-power approach which was certain eventually to run foul of the idealism of American liberals. Eden's interest in post-war questions was, noted Harvey in October, increasing steadily. The Foreign Office had prepared two important papers, one proposing a Four-Power Plan for the running of the world after the war and the other dealing with colonial and Far Eastern questions in the sense of placing such areas under regional councils with responsibility for defence and economic matters.[90] Churchill, his mind focused almost exclusively on the immediate problem of actually winning the war, was not receptive to such ideas. When Eden left the Four-Power Plan at Chequers for the Prime Minister's perusal, it encountered the latter's celebrated rebuff: 'I hope that these speculative studies will be entrusted to those on whose hands time hangs heavy, and that we shall not overlook Mrs. Glass's Cookery Book recipe for Jugged Hare − "First catch your hare." '[91] For Churchill the important thing remained to do nothing to endanger the Anglo-American alliance. Britain should therefore support the United States to the limit of her ability in the Pacific war. 'A successful joint war against Japan would form a very good background for collaboration about the settlement of Europe, the British Empire, India and other things like that.'[92] Eden, by contrast, was as sceptical about America's conduct in the Far East as he was about its policy towards France.

American behaviour in the Asian theatre threw a spotlight on to the United States' attitude towards the future of the British Empire. 'F.D.R. little if any sympathy for the British Empire or for the French Empire,' Eden recalled. 'In his heart probably hostile to both.'[93] The Foreign Secretary shared a growing perception that American moralizing about the iniquities of imperialism and the almost morbid interest which Roosevelt and Hull took in plans for its dismemberment served merely as a cloak to conceal their own selfish ambitions.[94] If the United States believed that Britain was intent upon restoring the colonial *status quo* in the Far East, she herself seemed determined to foster American interests through an informal imperialism based on her client state in China. In this respect American policy was, as Eden put it on another occasion, 'exaggeratedly moral, at least where non-American interests are concerned'.[95] The United States' attitude amounted to giving away other people's property 'to an international committee on which America will be one of three or more'.[96] He felt that Roosevelt's view of China's potential for great-power status was unrealistic. The Chinese war effort, he told the cabinet, was largely maintained by the vitalizing leadership of Chiang Kai-shek whose ability to sustain morale was dependent upon the support he received from Britain and the United States.[97] But in the last resort Eden had little option but to accept the American lead. In return, however, he expected to be consulted in relation to an area where Britain still had substantial interests. 'Several times', he noted,

> the United States Government have, with the object of saving time, conceded to the Chinese points of great importance to His Majesty's Government, before our Ambassador . . . could be in a position to discuss them with the Chinese Government. . . . We are dealing with questions of the greatest importance to our commercial, financial and shipping interests.[98]

Eden was particularly annoyed when Roosevelt wrote to Chiang about the use of Chinese troops in Burma without first discussing this matter with London. 'I regard this', he wrote, 'as almost intolerably impertinent. . . . We might as well not exist.'[99]

Not surprisingly, therefore, Eden reacted angrily when Churchill poured liberal quantities of cold water on his attempts to protect specifically British interests in the post-war world. With a trip to the United States in the offing, Eden was adamant that he would not go 'without any policy in his pocket'.[100] His response to Churchill was more strongly worded than usual and indicated his determination to stand his ground on an issue of fundamental importance. His plan had not, he insisted, been a vague project for an indefinite future. His aim was to set in place a policy which would carry Britain into the years of peace. It was a 'bad business to have to live from hand to mouth when we can avoid it, and the only consequence of so doing is that the United States makes a policy and we follow, which I do not regard as a satisfactory role for the British Empire'.[101] Continuing developments in the triangular relationship between Britain, America and France showed how difficult Eden's task would be.

As Eden recognized, the Allied invasion of North Africa – Operation

Torch — which began on 8 November 1942, might transform 'the whole perspective of the French problem. We may hope as a result ... for a wider Fighting French organisation and one which will show more statesmanship than the present quarrelsome clan.'[102] In the interests of speed the Allies needed to seize every possible political advantage in North Africa and to minimize opposition from the existing Vichyite forces there. It was to this end, and after the failure of a planned coup in Algiers, that the American General Clark signed an agreement on 13 November to recognize Admiral Darlan, Minister of Marine in the Vichy government, as head of the civil administration in North Africa in exchange for his co-operation with the Allies. As Eden had attended a meeting in London on 17 October with Churchill, Smuts, Clark and Eisenhower, at which the possibility of an arrangement with Darlan was discussed in general terms, he could scarcely distance himself from the deal altogether. But his misgivings rapidly increased. 'Just before conference at 11 a.m.', he recorded on 15 November, 'telegram came thro' from Eisenhower announcing an agreement with Darlan. Didn't like it a bit, and said so.'[103] Certainly Eden was not proposing to switch support from the Free French, about whose exclusion from Torch he had felt considerable unease, seeking to secure compensation for de Gaulle with the transfer of Madagascar to the general's authority once the North African landings had taken place.[104]

Lord Cranborne, Eden's closest friend in politics, warned of the possible consequences of the Darlan deal. It would give the worst impression in France and other occupied countries of what would happen after the war and make Allied complaints against war criminals sound very hollow.[105] At the same time Eden managed to convince de Gaulle that Churchill's readiness to follow this particular American policy was distasteful to the Foreign Secretary personally. The general found Eden 'deeply disturbed' and 'moved to the point of tears'.[106] The reaction of the Commons where Eden was heckled on this issue emphasized the domestic political dangers inherent in the Darlan affair. Churchill and Eden confronted one another over the telephone on 20 November. Though Eden warned that Darlan would 'make rings, diplomatically, round' Eisenhower, Churchill appeared unmoved. At one moment in the 'shouting match' he revealingly declared that Darlan was not as bad as de Gaulle. 'That man hates us and would give anything to fight with Germans against us.'[107] The argument continued at the cabinet without resolution. Seeing that he was making little progress,

Eden modified his line of attack. He made it clear to Churchill that the Darlan agreement must not become permanent. 'If we do not eliminate Darlan *as soon as the military situation permits*', he insisted, 'we shall be committing a political error which may have grave consequences not only for our good name . . . but for the resistance of the oppressed people for whose liberation we are fighting.'[108] At Churchill's request Eden agreed to let matters rest for the time being, but Cranborne continued to urge that the British position should be clearly established in Washington. 'Otherwise, we shall reach a position where we have one client and they another, and then ultimately we shall have a serious row.'[109] The Americans, Eden replied, were becoming 'increasingly tiresome' over the Darlan affair 'and almost everything else'.[110] On this occasion, however, fate was on Eden's side. On Christmas Eve 1942, Darlan was assassinated. At least one difficulty in Eden's relations with the United States had been removed. 'I have not felt so relieved by any event for years.'[111]

*

Darlan's murder removed one particular irritant. But it did little to bring Eden and the US government into line over the future of France. America's enthusiasm to be rid of de Gaulle remained unabated throughout 1943 and her championship of General Giraud as a possible alternative leader of post-war France inevitably caused further difficulties for Britain in general and Eden in particular. In many ways de Gaulle was his own worst enemy. His reluctance even to meet Giraud in Casablanca in January drove his supporters inside the Foreign Office to the verge of despair. As Cadogan recorded,

> He wouldn't budge, so we had no alternative but to inform P.M. This, I should think, is the end of the Free French movement. Roosevelt will say to P.M. 'Look at your friend: this is how he behaves.' And Winston will have to agree with him – and shed de G. A *great* pity. But in point of fact, would de G., being what he is, *ever* collaborate with *anyone*?[112]

Churchill warned Eden that de Gaulle's continued obstinacy would lead to his removal from the leadership of the Free French, as a condition of continuing British support for that movement.[113] De Gaulle ultimately relented, but was scarcely in a co-operative mood upon arrival in Casablanca. With so little upon which to base French pretensions, his tendency none the less was to behave like Stalin with 200 divisions behind his words.[114] Churchill was obviously worried that Britain's continuing association with de Gaulle risked damaging his carefully fostered relationship with Roosevelt. 'I beg you', he wrote to Eden, 'on no account to allow our relations with the United States to be spoiled through our supposed patronage of this man who is also our bitter foe and whose accession to power in France would be a British disaster of the first magnitude.'[115]

Eden, though he had earlier confessed that 'he always [felt] a sympathy with this man in his difficulties', was not blind to de Gaulle's failings.[116] But he remained convinced that de Gaulle was the necessary price Britain had to pay for the re-creation of a strong France. Though in all his experience Eden had never had dealings with anyone more difficult to cope with, there were still points in his favour. The first was that de Gaulle stood out among all available alternatives as a vital and dynamic personality committed to democratic republicanism. In the second place de Gaulle was never likely 'now or hereafter' to collaborate with the Germans. The Nazi virus, argued Eden,

has penetrated very deeply into political and industrial circles in France and what we have most to fear after the war is the establishment of a government of this type, which . . . will look to Germany rather than to ourselves. The Americans may look with equanimity on such a prospect, but we cannot do so.[117]

Eden was thus in a difficult position when he visited Washington in mid-March to initiate discussions on a range of post-war issues. At a personal level the trip was an undoubted success. According to the British ambassador, the Foreign Secretary never put a foot wrong. 'Nothing could have gone better in all ways . . . and I do believe that the contacts he made will have enduring value.'[118] The talks themselves, however, revealed the potential for serious rifts in the alliance. France was the most difficult item on the agenda. At the end of Eden's visit Roosevelt told the press that there had been agreement on 95 per cent of the matters under discussion. When asked what the remaining 5 per cent related to, Harry Hopkins answered 'Mostly France'.[119] Eden had to proceed with the utmost caution, knowing that Churchill's views were nearer to those of the Americans than to his own. As Harvey put it:

A. E. is between the hammer and the anvil. For British interests we are all convinced that we should work with de G. . . . The P.M. is passionately persuaded that de G. is our enemy, that he will work against us and even after that he means to bedevil Anglo-American relations.[120]

In discussions with Roosevelt and Hull, Eden held his ground on the French issue. By this time he was looking to the establishment of a single French authority, which the Allies could recognize as a *de facto* administration, provisionally exercising sovereignty over parts of France and over the whole French Empire, pending the establishment of a permanent government chosen by the French people.[121] By contrast Roosevelt and Hull preferred to keep their options open and deal separately with different individuals in various parts of the French Empire. Eden got the impression that the President wanted to hold the 'strings of France's future in his own hands so that he could decide that country's fate'.[122] Hull became somewhat heated, as he sought to justify his country's policy towards Vichy.[123] It was one of Eden's especial talents to know how to defuse such situations, but his private thoughts reflected a different picture. Richard Law found him in a 'constant state of irritation with the Americans'. Eden could not help 'entertaining some doubts about the cooperation being realised, having regard to the difficulties existing in the United States'.[124] He was exasperated by Roosevelt's chaotic ways of doing business. It was like a 'mad house' and he confessed to feeling 'more at home in the Kremlin' than in this atmosphere of 'confusion and woolliness'.[125] More importantly, his doubts about Roosevelt's understanding of international affairs were intensified. The President's ideas about the future map of Europe, including the projected creation of a new state of Wallonia out of parts of Belgium, Luxembourg, northern France and Alsace-Lorraine, struck Eden as unrealistic. His later assessment was sarcastically condescending:

> Roosevelt was familiar with the history and geography of Europe. Perhaps his hobby of stamp-collecting had helped him to this knowledge, but the academic yet sweeping opinions which he built upon it were alarming in their cheerful fecklessness.[126]

Though the main bone of contention, France was not the only issue to strain Anglo-American harmony during Eden's fortnight in Washington. His wide-ranging discussions covered both the conduct of the war and the construction of the post-war world. According to Roosevelt's Chief of Staff, the Foreign Secretary proved an effective spokesman for the British point of view, having a better understanding of his country's general political policy than was the case with many American leaders and knowing 'what Britain wanted'.[127] As regards the Far East, Eden found Roosevelt insistent that China be treated as a major player on the post-war stage. Eden was sceptical, doubting whether China could stabilize itself without a revolution, and disliking the idea of the Chinese 'running up and down the Pacific'. American suggestions that Britain might agree to the surrender of Hong Kong as a goodwill gesture were typical of the double standards which Eden had come to expect from the US government. He drily noted the absence of similar gestures from the Americans themselves.[128] One American observer concluded that the British were 'going to be pretty sticky about their former possessions in the Far East'.[129] Differences of opinion over relations with the Soviet Union, simmering for more than a year, also came to the surface. Eden recognized that Stalin, whatever his other territorial claims, would insist on the absorption of the Baltic States. Roosevelt, while accepting that this might prove to be the case, was wary about the possible impact on American public opinion and still hoped to use this concession as a bargaining counter in negotiations with the Russians.[130] Not surprisingly, Eden's report to the War Cabinet on 13 April emphasized both the achievements of his visit and the difficulties encountered.[131]

Churchill, who followed Eden to Washington in May, proved more susceptible to American pressure than his Foreign Secretary. Roosevelt subjected Churchill to a barrage of complaints about de Gaulle's intolerable behaviour. With few signs of reluctance Churchill telegraphed to London to suggest that the time had come to eliminate de Gaulle as a political force. The cabinet was summoned for 9 p.m. on 23 May. Eden recorded that all present were against Churchill's proposal and 'very brave about it in his absence'.[132] Foreign Office drafts had already been prepared and Eden and Attlee, as Deputy Prime Minister, replied to Churchill that this was scarcely the moment, with de Gaulle and Giraud finally showing signs of coming together, to ditch the former. They warned that the French National Committee and even the Free French fighting forces were unlikely to function effectively if de Gaulle were removed.[133] Churchill's response showed that his first consideration remained the solidity of the Atlantic alliance:

I have no intention of marring my relations with the President by arguing in the sense of your various telegrams. The only channel for such discussions must be the Foreign Office and the State Department.

I have given you my warning of the dangers to Anglo-American unity inherent in your championship of de Gaulle.[134]

In the event, Churchill agreed that no action should be taken pending the outcome of talks between de Gaulle and Giraud and asked Roosevelt not to push matters to the point of a quarrel with Britain.

Meanwhile Eden laboured long and hard in Algiers in early June, not only to bring the two French generals together, but also to reconcile Churchill and de Gaulle. His efforts appeared to meet with success. De Gaulle and Giraud agreed to a French Committee of National Liberation under their joint presidency, while Churchill entertained the new committee to lunch and telegraphed an enthusiastic report to Roosevelt.[135] De Gaulle, however, seemed intent on his own destruction. His continuing attempts to sideline Giraud inevitably revived American animosity. De Gaulle had done 'much to shake all confidence in him,' noted Eden only a week later.[136] By 18 June Roosevelt was again demanding the general's head on a charger. 'We shall be hard put to it', Eden recorded,

> to keep in step with Americans – or rather pull them into step with us
> – over the French business and not commit some folly which will give
> de Gaulle a martyr's crown or control of the French army or both.
> F.D.R.'s mood is now that of a man who persists in error. It has all that
> special brand of obstinacy like Hitler at Stalingrad.[137]

Faced by this renewed American démarche Eden and Churchill took up predictable positions. Over the next month relations between Foreign Secretary and Prime Minister moved via 'heated controversies and late-night wrangles' to a point of crisis. Far from giving way, Eden raised the stakes by arguing that it was time for Churchill to recognize de Gaulle's committee and to get Roosevelt to do the same.[138] At the Defence Committee on 8 July Churchill used a message from the President to justify his own refusal to contemplate recognition and to launch another tirade against the general. Eden countered that the Americans had mishandled the French problem from the beginning. Their treatment of de Gaulle would only make him a national hero.[139] Four days later the two men went over the same ground in a lively exchange lasting until after 2 a.m. Harvey heard that the 'broadsides were pretty hot'.[140] They agreed to put their respective views on paper, but the Foreign Office draft, stressing that a strong France was of vital interest to Britain if not necessarily to the United States, caused Churchill to remark that he and Eden might be 'coming to a break'.[141] Though it was eventually agreed that neither memorandum should be submitted to the cabinet, Churchill's own paper merits quotation in highlighting the difference of perception over the American alliance:

> I am resolved never to allow de Gaulle or his followers to cloud or mar those personal relations of partnership and friendship which I have laboured for nearly four years to develop between me and President Roosevelt by which, I venture to think, the course of our affairs has been most notably assisted. I must ask my colleagues to face this position squarely, as it is fundamental so far as I am concerned.[142]

Those who believed that, in the last resort, it was always Eden who gave way to Churchill could not have been familiar with this episode.[143] Within a week the Prime Minister seemed to have 'swallowed [Eden's] thesis whole', and was now ready to press Roosevelt, albeit unsuccessfully, on the issue of recognition. Eden preferred a different interpretation – that of 'asking Americans to face up to the realities of our situation'.[144] At all events, by the end of July Churchill defined his goal as 'the recognition of the Committee in its collective capacity and its eventual inclusion with the same status as that of other refugee Governments in the array of the United Nations'. If de Gaulle would play his part in restoring the confidence of Britain and America, he, Churchill, was ready to play his.[145]

*

Eden's concern that British interests in post-war Europe were not best served by undue subservience to the United States was not restricted to the case of France. During 1943 it became clear that the Italian military effort was on the point of collapse and much Foreign Office attention was given to the surrender terms which would need to be imposed. On this issue Eden was more likely to receive the backing of his Prime Minister who, unusually, often found himself at loggerheads with Roosevelt – though in late July Eden feared that Churchill was once again becoming too willing to dance to the tune of his American piper.[146] On this matter Eden found the Americans 'both obstinate and foolish'.[147]

Over relations with Portugal, nominally Britain's oldest ally, Churchill adopted his more characteristic posture. As the war turned in the Allies' favour, the attraction of holding facilities in the Azores for use in the Battle of the Atlantic became more apparent. Churchill favoured claiming such facilities under the threat of force, but Eden preferred a more diplomatic approach based upon reference to

Portugal's alliance with Britain. The Foreign Secretary found American intervention into these delicate negotiations singularly unhelpful:

> They wont accept any agreement between us and Portugal unless they have equal share. This after position had been explained to them and accepted by them months ago. I told [J. G. Winant, US ambassador in London] that his people were impossible and that they had better go and attack Portugal themselves if this was what they wanted.[148]

In the overall context of the war, this was a relatively minor matter, but Eden's attitude epitomized his mounting distaste for American policies and methods. To Churchill he wrote: 'It is important that the Americans should realise that modern Portugal . . . is not a second Guatemala, from whom anything which the Americans desire can be obtained by threats or bribes.'[149] Eden's tactics were time-consuming but ultimately successful. An agreement with Portugal was signed in August and British entry into the Azores fixed for 8 October.

Eden's disagreements with the United States could not be attributed to a lack of personal contact. In August he had further meetings with Roosevelt and Hull in Quebec. France and the French National Committee again loomed large. Hull's attitude had not softened in the least and he remained unresponsive even to Eden's reminder that Britain had to live just twenty miles from France. The American Secretary of State tried to exploit the differences which had clearly existed between Eden and Churchill by noting that the latter's telegrams to the President had failed to expose disagreement with US policy.[150] Eden could only conclude that Hull was motivated by a personal hatred of de Gaulle.[151] With Roosevelt refusing to go beyond recognizing the French Committee as administering those territories it actually held, both parties had to accept that they would go their own way on this issue. Further difficulties arose over colonial questions. Hull referred to American proposals to guarantee independence to colonial territories within a measurable period. In the face of Hull's persistence Eden had to admit his distaste for these proposals which failed to take account of the diversity already existing within the British Empire.[152] With the conference over, Roosevelt confided that he could have made more progress – particularly over France – with Churchill, but for Eden's presence.[153]

From Eden's point of view a clear danger inherent in Churchill's Special Relationship with the American President was its possible impact on Britain's relations with the Soviet Union. Always sensitive to the deep suspicions ever present in Soviet foreign policy, Eden would ideally have preferred to place Britain's dealings with the two countries on an equal footing. In practice this was scarcely possible, but he saw no reason to give Stalin unnecessary grounds for complaint. He feared, therefore, that the Quebec meeting would 'infuriate U[ncle] J[oe]' and began to press upon Churchill the case for tripartite consultation. There had been a tendency, he argued, for Britain to agree matters first with the United States and then present the results to the Soviets, and 'there are signs that the Russians resent this procedure'. His idea was for a consultative body consisting of ambassadors Winant, Maisky and himself. 'Even if this proved of no very great practical value, . . . we might gain some advantage from letting the Russians feel that they were admitted to discussion at an early stage. We want to work with them not only during the war but after.'[154]

Ultimately it was agreed that the three Foreign Ministers should meet in conference as a preliminary to a meeting at heads of government level. But as the balance of the Allied war effort shifted away from Britain, Eden became peculiarly sensitive to the protocol of such an occasion:

> Roosevelt has had his way again and agreed to Moscow for the Foreign Secretaries' Conference with alacrity. His determination not to agree to a London meeting . . . is almost insulting considering the number of times we have been to Washington. I am most anxious for good relations with U.S. but I don't like subservience to them and I am sure that this only lays up trouble for us in the future. We are giving the impression . . . that militarily all the achievements are theirs and W[inston], by prolonging his stay in Washington, strengthens that impression.[155]

When Roosevelt relented, it was Stalin who now baulked at the idea of a conference in London, but Eden still tended to blame the Americans for their failure to back the British proposal from the beginning.[156]

In many ways the Moscow Conference of Foreign Ministers of October 1943 proved the most businesslike of all wartime Allied meetings.[157] Considering that his bargaining position was the weakest of the three leading delegates, Eden had some cause for satisfaction. The Russians accepted a revised date for the Second Front;

there was full agreement on the question of Arctic convoys; and it was arranged that a Soviet mission would work with the Yugoslav Partisans. Eden took particular pleasure in the decision to establish a European Advisory Commission in London as a forum for the discussion of European problems. But the body which met for the first time on 14 January 1944 was less wide-ranging than he had hoped for – a reflection of his failure to secure wholehearted American backing. Eden had wanted a body which could consider European political problems as they arose, both before and after the armistice. But the fact that the Commission was to meet in London was enough to arouse American suspicions that it was a device to facilitate undue British influence on the post-war settlement. In practice its deliberations were restricted to German issues and considerable delays ensued before any agreements were reached, of which the most important concerned the division of Germany into zones of occupation.

In general Eden was struck by the way the conference failed to deliver those advantages which Churchill confidently expected to flow from the priority he consistently gave to Anglo-American relations. Eden needed Hull's support to carry the day on key issues, especially relating to Eastern Europe. It was not forthcoming. The Americans seemed, if anything, keen to undermine the British position. Eden recorded them 'making claims to a share in the bomber offensive which is by no means justified but further dims our glory'.[158] In his proposal for a Declaration on Joint Allied Responsibility Eden sought American and Soviet endorsement for the principle that none was looking to set up spheres of influence in any part of Europe. He wanted an equal right with Russia for the United States and Britain to exert influence on the settlement in Eastern Europe. The American response was cool.[159] Particular disagreements were apparent over the future of Poland. Hull, anxious like his President to delay difficult questions until the peace conference, especially as a presidential election was due in a year's time, would not allow Eden to raise the matter with Molotov.[160] As a result, 'the Soviet Union emerged from the conference with no restrictions on its freedom of action in Eastern Europe'.[161]

The Moscow Conference began the most hectic period of Allied diplomacy in the entire war. Eden returned briefly to London to report to the Commons on the outcome of his talks. On 23 November he and Cadogan left England again, arriving in Cairo the following day to join Churchill for talks with Roosevelt and Chiang Kai-shek. Eden and Churchill than flew to Tehran early on the 27th for the first of the wartime meetings of the 'Big Three'. On 2 December he was on his way back to Cairo for further discussions. Compared with Moscow, Eden's own role at this series of international gatherings was peripheral. In general his diplomatic travels confirmed for him the worrying aspects of the Foreign Minister's Conference without that meeting's compensating features. At Cairo even Churchill admitted the difficulties of working with Roosevelt: 'FDR was a "charming country gentleman", but business methods were almost non-existent, so W. had to play the role of courtier and seize opportunities as and when they arose. I am amazed at patience with which he does this.'[162] Eden found the Americans 'inclined to run out' of the European Advisory Commission and used 'some pretty plain language' to keep the idea in being.[163] He resented the Americans' abrasive behaviour over the conference communiqué and, still unconvinced by the United States' efforts to inflate China into the status of a great power, foresaw future difficulties over South-East Asia.[164] Despite strenuous efforts, Anglo-American military differences seemed 'no better'.[165] As Eden later told the War Cabinet, Roosevelt made promises to the Chinese which had not been endorsed by Churchill, and the Americans and British had had to go to Tehran 'without having had an opportunity of reaching a decision upon our combined plans for Europe in 1944'.[166]

At Tehran Roosevelt made it clear that his priority was now to secure a lasting relationship with the Soviet Union. From an American point of view Britain's moment had passed. The concept of a 'Big Three' now concealed the reality of the Grand Alliance. Though the future of Poland was firmly on the agenda for discussion, Eden failed to secure the backing he needed from the United States. 'A difficulty', he noted, 'is that the Americans are terrified of the subject which Harry [Hopkins] called "political dynamite" for their election.'[167] France occasioned renewed problems, but Eden had to confront not only a sceptical United States but also a hostile Soviet Union, with Molotov inclined to treat the entire French population as a nation of collaborators. To his credit Eden refused to abandon this central aim of his foreign policy. 'We have just got to build up a France somehow,' he insisted to Churchill as 1943 drew to a close. 'Though the Bear's manners are steadily improving I have still no ambition to share the cage alone with him.'[168]

*

For the rest of the war Eden fought a losing battle to retain something like parity in the Anglo-American relationship. His tone became increasingly bitter. He observed that the Americans had 'a much exaggerated conception of the military contribution they are making in this war. They lie freely about this.'[169] Harvey noted that Eden was getting fed up at having to refer everything to the United States. 'Can't we really have a foreign policy of our own?' Anything referred to the Americans, Harvey added, was at once blocked by Hull or Roosevelt, afraid of things being done by anyone except themselves.[170] The Americans seldom reciprocated. In June 1944, for example, they withdrew their ambassador from the Argentine without first consulting London. But it was the Far Eastern theatre which really showed the growing weakness of the British position. Eden felt acutely the political need of having the British Empire reconquered mainly by British troops. 'If we are merely dragged along at the tail of the Americans in the Pacific', he warned, 'we shall get no credit whatever for our share in the joint operations.' He had 'a feeling that the Americans are not anxious that we should play any notable part in the Pacific war. We want to make it plain to the world that we have played our part in regaining our Far-Eastern Empire.'[171] As far as it was possible, Eden wanted to recreate the disposition of world power which had existed before the outbreak of war. But his quest for this impossible goal was tempered by a dose of realism. Though the Foreign Office sought to resist concessions to the Americans over oil which would result in the introduction into the Middle East of a 'foreign power to rival our influence', Eden wisely cautioned against unrestricted commercial competition at the present time.[172]

Europe offered more scope for British initiatives. 'I have little doubt myself', Eden noted in May, 'that FDR does NOT want us to take the lead in Europe and equally no doubt that we should.'[173] He sought to persuade Churchill to take the management of the Polish–Soviet dispute into British hands. 'Our position . . . is very different [from America's]', he argued, 'because of our alliances with Poland and the Soviet Union.'[174] His conviction grew that only Britain could win any concessions for the Poles. Roosevelt 'will not be embarrassed by them hereafter, any more than by the specific undertakings he has given to restore the French Empire'.[175] Ironically, it was over the issue of France, upon which Eden had hitherto found most difficulty in achieving Anglo-American co-operation, that he claimed his greatest measure of success in the closing stages of the war. The compromise over recognition reached at Quebec had not worked out in practice. Little seemed to have changed, as an entry in

Eden's diary for March 1944 makes clear: 'W. wanted to send message to Halifax agreeing with President's attitude to French which I don't like at all. President's absurd and petty dislike of de Gaulle blinds him. It would be folly to follow him in this.'[176] But as the Allied invasion of Europe drew near it was obvious that decisions would soon have to be taken. 'Surely we cannot make over the decision as to our relations with France to an American general,' Eden complained in April.[177] He still found Churchill reluctant to ruffle American feathers over this issue. The Prime Minister insisted that, whatever Britain decided, it would be necessary to inform Roosevelt and 'obtain his agreement beforehand'. He still failed to see in de Gaulle the future leader of France whom it was necessary to court in Britain's own interests. On the contrary, 'we have a very hostile man without any forces worth speaking of, trying to thrust himself into the centre of vast and deadly affairs and calling himself "France"'.[178] Despite persistent rebuffs Eden plugged away with that combination of patient argument and persuasive charm, peppered with flashes of passion, which had characterized his tactics throughout. Between early March and mid-June he sent Churchill forty-one minutes on the single subject of France.[179] If Eden's temper sometimes snapped, this was scarcely surprising. 'It's a girl's school,' judged Cadogan. 'Roosevelt, P.M. and – it must be admitted de G – all behave like girls approaching the age of puberty.'[180]

On the eve of D-Day Eden wrote a paper for the cabinet on the French situation and its possible effect on Anglo-American relations. It was a rarity for him to find time for this sort of activity and the result pleased him.[181] Failure to reach agreement with the French committee, he warned, would have serious repercussions. De Gaulle had been allowed to wrong-foot the Allies because of the obstinacy of the American President, but if Roosevelt 'renounces the paternity of France, surely he must allow us to do the schooling in our own way'.[182] On D-Day itself Churchill and Eden were still quarrelling over the same issues:

> I was accused of trying to break up the government, of stirring up the press on the issue. He said that nothing would induce him to give way, that de Gaulle must go. F.D.R. and he would fight the world. I didn't lose my temper and I think that I gave as good as I got. Anyway I didn't budge an inch.[183]

After an 'acid not to say bitter' cabinet meeting on 7 June,[184] Eden turned his attention to de Gaulle, persuading him to open discussions with the British, without American participation. 'AE tired but triumphant this morning', noted a private secretary, 'having brought de Gaulle along considerably at dinner last night and then carried [Churchill], though it took him till 3 a.m.'[185]

Churchill's attitude had not in fact changed. At a meeting with de Gaulle on the eve of the Normandy landings he revealed his true feelings. 'Each time we must choose between Europe and the open sea, we shall always choose the open sea. Each time I must choose between you and Roosevelt, I shall always choose Roosevelt.' Significantly, de Gaulle noted Eden shaking his head, seemingly 'quite unconvinced' by these remarks.[186] The Foreign Secretary kept up his pressure in the following weeks, while Churchill stood firm on the priority of American wishes.[187] In practice the course of events in France ultimately decided matters in Eden's favour. As he later put it: 'As France was liberated, its administration fell without question into the hands of the Resistance which acknowledged General de Gaulle as its chief and this chapter of our difficulties, the cause of so many hard feelings, naturally resolved itself.'[188] The United States had little alternative but to recognize the French Committee of National Liberation as the *de facto* authority in July and formal recognition as the Provisional Government was accorded on 23 October 1944. 'At last,' sighed Cadogan with evident relief. 'What a fuss about nothing!'[189]

Though the issue of de Gaulle gradually resolved itself to his satisfaction, Eden remained wary of America's – and particularly Roosevelt's – capacity to impede British national interests in post-war Europe. As Churchill travelled to Quebec in early September for a further meeting with the American President, the Foreign Secretary warned against the idea of retaining the post of Supreme Commander, Allied Expeditionary Force in Germany, after that country's occupation. In view of Eden's wish to build up a Western European security group there would be obvious complications if British forces were to 'continue for an indefinite time under the command of an American general'.[190] When Eden joined Churchill in Quebec he was alarmed to learn of the problems which the Prime Minister's amateur diplomacy had already created. Churchill had put his name to the Morgenthau Plan for the pastoralization of post-war Germany. Not surprisingly Eden 'flew into a rage' when he learnt of this agreement and in a heated exchange reminded Churchill that the Foreign Office, with the backing of Molotov and Roosevelt himself, had been at work for several months on a plan to come into force once Germany surrendered.[191] At one point he shouted 'you can't do this' to an evidently startled Prime Minister.[192] Eden 'felt somehow irritated by this German Jew's bitter hatred of his own land and also wasn't sure that scheme was all that good from our point of view. Anyway this is no way to discuss it, so was rather peevish.'[193] But Churchill, having pledged himself to Roosevelt, would not be moved. It was the only time, Eden recalled, that Churchill showed his displeasure with him in front of the President.[194] Once again, the course of events ultimately undermined this ill-considered product of the Churchill-Roosevelt relationship.

It was some advantage to Eden that, by early 1945, Churchill had dropped his earlier arguments and embraced the Foreign Secretary's desire for a strong bloc of Western European states including a rejuvenated France as the linchpin of Britain's post-war strategy.[195] As a result the two men presented more of a united front at the Yalta Conference in February than had often been the case hitherto. But Eden still approached this conference with considerable misgivings, not least because of mounting apprehension about the post-war ambitions of the Soviet Union. Indeed, his anti-American tendencies were becoming tempered by his belief that Britain and the United States must combine to withstand possible communist expansion, though he still doubted Roosevelt's willingness to co-operate. The President's inclination towards diplomatic procrastination only intensified under mounting burdens of tiredness and ill-health. While Eden understood the attraction of postponing difficult decisions, he also saw the dangers inherent in this approach. 'Europe will take shape or break up while [Roosevelt] stands by and it will be too late afterwards to complain of frontiers or the nature of conflicts, e.g. Poland and Greece.'[196] As a result Eden feared that the Yalta discussion would be 'chaotic [with] nothing worthwhile settled'. Stalin was 'the only one of the three who has a clear view of what he wants and is a tough negotiator. P.M. is all emotion in these matters. F.D.R. vague and jealous of others.'[197] With this in mind he pressed Churchill for a preliminary meeting of the Foreign Secretaries in Cairo before the heads of government came together at Yalta.[198] In the event Eden met the American delegation in Malta immediately prior to the opening of the full conference at Yalta, and there impressed upon Hopkins and the new Secretary of State, Edward Stettinius, the need for a firm Anglo-American stand on Poland and the dangers of making undue concessions. He found them both in 'complete agreement' with him, but had little time to work on Roosevelt before departing for Yalta.[199]

At the main conference Eden continued his long battle on behalf of France, supported now by Churchill. As Hopkins recorded, 'Winston and Anthony fought like tigers for France'.[200] Indeed Eden had earlier tried to secure French participation in the political discussions of the Big Three. As he explained:

> We must plan for the future and I find it difficult to contemplate a future in which France will not be a factor of considerable importance. She must be interested in almost every European question. If we do not have her cooperation, she will be able – not at once perhaps – to make difficult the application of any solution which does not suit her.[201]

It was largely because of British persistence that France emerged from Yalta reinstated with at least the trappings of great-power status. Roosevelt ultimately relented and Stalin, anxious for concessions in other spheres, fell into line. France secured not only a zone of occupation in post-war Germany but also a seat on the Control Commission to supervise that occupation.

Elsewhere Yalta gave less cause for British satisfaction. The handling of Far Eastern matters confirmed Eden's worst suspicions about American intentions in that part of the world. At Malta he had suggested that it would not be necessary to bribe the Soviets to enter the Japanese war. Their self-interest would dictate such a course.[202] At Yalta, however, he was presented with a *fait accompli*. 'F.D.R. simply told W. the agreement had been reached and he and Stalin proposed to sign it.' Despite objections from Eden and Cadogan, Churchill insisted on signing the agreement. 'I don't think he could bear the idea of the other two signing anything he wasn't in on!'[203] The elastic settlement on the future of Poland has occasioned an enormous historical controversy in its own right, which is largely outside the scope of the present work. Though the West's 'surrender' probably owed more to the strategic realities of February 1945 than to the physical and mental deterioration of the American President, it is none the less interesting to note Eden's contemporary reaction to Roosevelt's state of health. The President gave 'the impression of fading powers. . . . Impossible even to get near business.'[204] When Roosevelt suggested that differences with the Soviet Union were largely a matter of the use of words, Eden concluded that he was deluding himself.[205]

For the remainder of the war Eden seemed torn between a growing recognition that the world conflict had greatly diminished Britain's capacity to determine the course of international events and a continuing desire to make the best of a bad job and stick to the foreign policy objectives which he had held throughout. In a wide-ranging survey of the effects of Britain's deteriorating external position circulated in March, he spoke of areas of the globe where 'we might have wished to assert ourselves had it been possible'.[206] To the American Secretary for the Navy he suggested that there was an analogy between the position of Austria and England after 1815 and the position of England and America in 1945. 'The United States has taken the place of England and England has taken the place of Austria.'[207] This implied a readiness to accept American involvement in European affairs to a degree he had once eschewed.

At the same time Eden insisted that 'we couldn't allow them to dictate our foreign policy and if they were wrong we would have to show independence'.[208] He was still worried that the British war effort was being discounted by American opinion. 'We should use every effort', he advised, 'to bring home to the Americans what we have done and are doing.'[209] The death of Roosevelt on 12 April 1945 was obviously a watershed in Anglo-American relations, particularly from Churchill's point of view. Eden formed a favourable first impression of his successor, Harry S. Truman, though he was sorry to see Stettinius replaced as Secretary of State by J. F. Byrnes, possibly because he had found it relatively easy to dominate the former.[210] Under the new presidency Eden remained concerned at signs of America taking the lead in areas, especially in the Far East, where Britain retained the primary interest.[211] Before the atomic bomb brought a sudden end to the Japanese war he was determined that Britain should take part in that country's military occupation and have a full share in whatever system of political and economic control was set up. 'Unless we open discussions with Washington in the near future', he warned Churchill, 'there seems to be a real risk that American views may crystallise before we have time to influence them.'[212] And, as VE Day brought the war against Germany to a close, Eden enthusiastically endorsed Orme Sargent's assessment of Britain's position within the future constellation of great powers. Sargent, soon to become Permanent Under-Secretary at the Foreign Office, had written of the perception

in the minds of our two big partners, especially in that of the United States . . . that Great Britain is now a secondary Power and can be treated as such, and that in the long run all will be well if they – the

United States and the Soviet Union – as the two supreme World Powers of the future, understand one another. It is this misconception which it must be our policy to combat.[213]

*

As the war ended, *The Economist* protested against Churchill's 'policy of deference' to the United States and the way it had hindered efforts to evolve a distinctively British approach to world affairs.[214] The trouble, Eden once explained, was that Churchill was half American and regarded Roosevelt with almost religious awe. Eden never shared these feelings, seeing Roosevelt as an astute politician, but also a man of great personal vanity and obstinacy.[215] He found the President's reluctance to confront difficult issues particularly irksome. 'It was no excuse for us to be without a foreign policy in Europe', he later recorded, 'simply because we were waiting for the United States.'[216] The danger also existed that Churchill would seek to emulate in Britain the sort of control which Roosevelt enjoyed in America. 'He sees himself in Roosevelt's position as the sole director of war. It is not what country wants', concluded Eden in 1942, 'nor does it produce good results.'[217]

Over the question of France, in particular, Eden displayed a surer grasp of British national interests than did Churchill. Too often the latter allowed his emotions to overcome his reason. A personal animosity towards de Gaulle was exacerbated by a sometimes unquestioning subservience to Roosevelt. Thus the triangular relationship between Britain, the United States and France emerged, as Eden once described it, as the 'principal [diplomatic] problem of the war'.[218] Cadogan identified the reality of the situation: 'P.M. ended with a tirade against de G. "He has been battening on us and is capable of turning round and fighting with the Axis against us." That is just untrue. Tiresome he may be, but sound on essentials.'[219] Eden showed more understanding than Churchill of France's plight after the catastrophe of 1940. He once conceded that it was only the 'hazard of geography' which had preserved Britain from the same fate.[220] He was as aware as anyone else of how exasperating de Gaulle could be. But de Gaulle was 'not a permanency'. France, 'we hope, will be and we want to build her up'.[221] Having identified this central goal of British diplomacy, Eden stuck to it tenaciously, even when it meant defying the American President.

In their differing estimations of the United States Eden and Churchill shared a delusion about the extent of British power in the post-war world. Both Eden's idea of an independent British foreign policy and Churchill's concept of an Anglo-American partnership of something like equals were based on a failure to anticipate the speed with which Britain would descend from the top rank on the international stage. Both notions, writes Elisabeth Barker, 'ended in the dustbin of history'.[222] But it was a delusion shared by the overwhelming majority of the political generation to which the two men belonged, not least by the senior members of the Labour cabinet which took power in July 1945. That said, Eden's attempt to save what he could of Britain's freedom of action for the post-war world was not without its merits. Most of his calculations were based on the expectation that America would not offer a physical presence on the continent to provide the necessary stability for the Europe which emerged from the Nazi yoke – that she would, as after Versailles, retreat into isolation. Certainly Eden took a more balanced view of the Anglo-American relationship than did his master. Nearly two years after Pearl Harbor, Churchill summed up his feelings:

> It was . . . a blessing that Japan attacked the United States and thus brought America wholeheartedly into the war. Greater good fortune has rarely happened to the British Empire than this event which has revealed our friends and foes in their true light, and may lead, through the merciless crushing of Japan, to a new relationship of immense benefit to the English-speaking countries and to the whole world.[223]

There was truth in this. But the alliance was, as Eden saw, a more complicated relationship than Churchill wished to accept. His experience of working with the United States during the Second World War would have lasting implications for the rest of Eden's political career.

7

Eden and the Russians

In that era, we at least knew with whom we were dealing and the man [Stalin], for all his ruthlessness, had something of greatness too.[1]

'WE shall work this war together.' So, according to Eden, said Winston Churchill in October 1940.[2] Throughout the conflict, indeed, Eden held high office. He was for five years a member of the Defence Committee and for four and a half Foreign Secretary. The latter post entailed membership of the War Cabinet. From the end of 1942 he was also Leader of the Commons. But it was clearly the Foreign Secretaryship which enabled him to emerge in the popular mind – and in reality – as the second most important figure in the government. Yet it can be argued that Eden's appointment to the Foreign Office in December 1940 took him, at least in the short term, away from rather than towards the centre of the war directorate. Churchill seemed to sense this when he discussed with Eden four months earlier the possibility of a move from his existing post at the War Office. According to the Prime Minister, whose main concern was to bring Eden into the War Cabinet, War Office work was expanding while that of the Foreign Office contracted. He seemed relieved that Eden preferred to stay where he was.[3] By early October Churchill's tone had changed. He suggested now that at the Foreign Office Eden could 'help much with U.S.A' – an argument which failed to take into consideration the proprietorial attitude which Churchill held to keeping the ultimate control of Anglo-American relations firmly in his own hands.[4]

By the end of the year the deed was done and, though Eden 'came home sad at thought of leaving soldiers', a month later the new Foreign Secretary was prepared to concede – at least to Churchill – that he was pleased with the change.[5] Yet at the time of Eden's move the scope for diplomatic initiatives was distinctly limited. Britain's central task was that of military survival rather than the niceties of diplomatic negotiation and post-war planning. Indeed, in the eyes of some it was the vacancy at the Washington embassy – which coincidentally enabled Eden to return to the Foreign Office – which was the most important post of the moment. Washington, suggested Leo Amery, was the 'major part of foreign affairs' with Halifax the outstanding candidate and Eden a 'possible but by no means sufficient alternative'.[6] Strictly speaking, Britain was never fighting alone in the year between the fall of France and the Nazi attack on the Soviet Union. The dominions were an important factor in the nation's war effort, while the courting of the United States as an ever more sympathetic non-belligerent took up much necessary time and effort. But the former were the departmental responsibility of the Dominions Secretary and the latter a jealously guarded preserve of Churchill himself. By contrast with the traditional range of diplomatic activity in peacetime, the Foreign Office faced a somewhat anomalous situation. Eden thrived on work and the international climate which confronted him was not best suited to his temperament. He perhaps sensed this himself in his determination to retain a seat on the Defence Committee after leaving the War Office.

Such factors may also help to explain the somewhat uneasy relationship which developed between the Foreign Secretary and his senior officials during 1941. With Eden just three months into his new job, 'Chips' Channon noted that the Foreign Office, having longed for his return, was now both bored and disappointed with him. 'Apparently he is nervous, exigent, fretful and the F.O. boys are pining for the Halifax days again.'[7] Hugh Dalton heard that Eden was 'writing many "peevish Minutes", but is not really a strong man. He goes about saying, "There is no control here. I can't imagine what Edward [Halifax] did with himself all day."'[8] Of particular interest was the reaction of Alexander Cadogan, the Permanent Under-Secretary who had worked closely with Eden during the 1930s both at Geneva and in London. 'A. in rather a flap,' he recorded within a week of Eden's appointment. 'When he was at W.O. he seemed admirable, but I fear that here he is getting as jumpy as ever.'[9]

By April Cadogan had grasped both the difficulties which confronted his new master and the character weakness which made him susceptible to this situation:

We are in that awful period when everything is going wrong and those in authority feel they have got to *do* something. . . . I got A.

alone and said 'Please don't do anything for the sake of *doing* something! Don't throw in small packets here and there to get chewed up. I know it's disappointing and humiliating to look forward to another year of the defensive, but *don't* squander the little you've got.' He professed to agree. I begged him to believe that diplomacy could only be prepared in our munition factories. But will o' the wisps have a fatal attraction for him and Winston.[10]

Cadogan judged that the urge to 'do something' diplomatically made little sense in view of the country's military weakness. In fact, 'diplomacy is completely hamstrung'.[11] In the circumstances he wondered whether Eden might find it easier to work with a new Permanent Under-Secretary.[12] For his part Eden was 'appalled' by the state of affairs bequeathed by his predecessor and regarded all the Under-Secretaries as 'completely negative'.[13] He found the Foreign Office 'dead, there was never a constructive suggestion from below, only a bureaucratic machine, gently ticking over'.[14] As late as October 1941 Harvey recorded Eden 'groan[ing] at the inactivity, at the lack of results of work at the F.O.' He 'longs again for Army or the W.O.'[15]

The Middle East and the Balkans provided some scope for Eden's restless energy in the first months of his new appointment. With Germany looking to make good recent Italian setbacks in the area, the War Cabinet decided to dispatch Eden and General Sir John Dill, the Chief of the Imperial General Staff, to assess the situation at first hand. This first of the many overseas visits of his wartime Foreign Secretaryship did little to enhance Eden's reputation. British policy was to offer help to Greece, to coax Turkey into the conflict on her side and to dissuade Yugoslavia from succumbing to Nazi blandishments. All this was to be achieved without compromising General Wavell's recent military successes in North Africa. From the outset, however, many observers were sceptical about Eden's journey. Vansittart, still for the time being the government's Chief Diplomatic Adviser, regarded it as a great mistake, while Cadogan also questioned its wisdom. The Foreign Secretary's 'stunt trip' was 'a diplomatic and strategic blunder of the first order', he recorded in March.[16] None the less Eden committed himself wholeheartedly to aiding the defence of Greece in the hope thereby of stiffening the will of Turkey and Yugoslavia. He regarded this as agreed government policy which it was merely his task to facilitate.

Even when Churchill and the government – notwithstanding the guarantee which Britain had given in April 1939 – showed increasing doubts about a British military commitment to Greece, Eden remained undaunted. He was perhaps unduly swayed by his rapturous reception in Athens. 'How in such surroundings', wondered Oliver Stanley, 'could he keep his judgement clear.'[17] Eden's unequivocal recommendation was considered by the War Cabinet on 7 March:

> Collapse of Greece without further effort on our part to save her by intervention on land, after the Libyan victories had . . . made forces available, would be the greatest calamity. Yugoslavia would then certainly be lost; nor can we feel confident that even Turkey would have the strength to remain steadfast if the Germans and Italians were established in Greece without effort on our part to resist them.[18]

The fact that Eden's views were endorsed by Dill and Wavell left the cabinet with little room for manoeuvre. He had 'committed us up to the hilt,' noted Cadogan, while remarking also that Kingsley Wood, John Anderson and A. V. Alexander were now 'evidently out for A's blood'.[19] In fact Eden was placing himself in an increasingly vulnerable position. As Cadogan explained:

> Actually the Cabinet's doubts were resolved for them by Anthony signing an agreement with the Greeks! When this was revealed to them, it gave rise to mixed emotions in some of the members − annoyance that they should have been rushed in this way, secret satisfaction that if the thing went really wrong there was a good Scapegoat handy![20]

Suddenly everything did go wrong. On 6 April 1941 the Germans invaded Yugoslavia, forcing her surrender in less than a fortnight. Greece was the next Nazi target and fared no better. The balance of advantage in the Middle Eastern theatre had been dramatically reversed and the whole British position in the eastern Mediterranean appeared endangered. Not surprisingly, Eden incurred considerable criticism on his return home. The trip, he insisted, had been worthwhile, though 'it was the most harrowing experience I have ever been through'.[21] His position was scarcely helped by a poor speech in the Commons on 6 May but, with Churchill's backing, he survived.

In later years the Greek disaster was an episode to which Eden's critics repeatedly referred and on which he remained particularly sensitive.[22] He could, of course, point out that his own assessment had been backed up by that of the government's competent military advisers. Granted that Eden had estimated German military strength in the Balkans at twenty-three divisions – a figure which Britain was in no position to match – any hope of success depended upon the by no means certain intervention of Turkey and Yugoslavia.[23] Yet Eden privately admitted at the time that there was only a small chance of 'pulling off anything really big' – by which he meant a Balkan front.[24] In his defence Eden always maintained that the Greeks had already decided to resist Germany and were not pushed into accepting offers of British help, and that Germany's commitment to the Balkans had at least delayed the Nazi attack on the Soviet Union by several weeks. Such points are difficult to verify. But it is interesting to note the testimony of General Francis de Guingand, then Wavell's Chief of Staff, who had been asked to assess the possibilities of successful action in Greece in the event of a German attack:

> All the military evidence . . . was absolutely against the campaign. It was clearly going to be a disaster from the start. But Eden had obviously made up his mind that we had got to do it, and . . . he was quite unscrupulous and tried to make [me] fudge the figures, counting rifles as guns and things like that . . . In fact the campaign had not been more of a disaster only because of very good luck and some extremely good generalship.[25]

*

It is against this unpromising background that Eden's wartime diplomacy towards the Soviet Union must be assessed. After a period when the scope for diplomatic initiative had been limited, after the setback of his two-month odyssey to the Middle East and in the context of an administration in which Churchill would always reserve for himself some of the main components of British foreign policy, relations with the Soviet Union came to assume a central importance for Eden, particularly after the Nazi invasion of June 1941. Diplomacy towards a country which was certain to emerge – assuming an Allied victory – as one of the great powers of the post-war world became a key issue for Eden – *his* preserve in a way that Anglo-American relations could never be.

Though Eden believed that there were right-wing sections of the Conservative Party which regarded him as 'more than half a bolshevick' [sic], it was not ideological affinity which led to his emergence as one of the Soviet Union's more consistent friends.[26] But his experiences as a minister in the 1930s made it easier for him than for many of his colleagues to do business with Stalin's regime. His sentiments were based more on traditional suspicions of Russian foreign policy than upon an ideologically motivated anti-Bolshevism. Stalin, he noted, was the 'true lineal descendant of Chingiz Khan'.[27] Having been the first British minister to visit the Soviet dictator in 1935, he had offered his services again in 1939 as a negotiator when the question of a Russian alliance had finally forced itself on to the cabinet's agenda. It was an alliance of which Eden had become a convinced advocate. Even after the conclusion of the Nazi–Soviet Pact he still hoped that ongoing trade negotiations with the Soviet Union could form the starting-point for a more comprehensive political agreement.[28] In 1940 and early 1941, however, when the USSR behaved more like an ally of the Nazis than a potential friend of Britain, Eden's attitude became distinctly cool. Early in 1940 he even seemed ready to contemplate a British declaration of war on Russia in response to her invasion of Finland.[29] By the time that he became Foreign Secretary, Eden's opinion had changed again and he now proclaimed to Sir Stafford Cripps, Britain's ambassador in Moscow, that the improvement of relations was one of his first priorities.[30] Even so, when Eden embarked upon his ill-fated tour of the Middle East, he was wary of the idea that he should include Moscow on his itinerary. A letter to Cripps in mid-January was the very model of Foreign Office realism and caution: 'I feel that we can only possess our souls in patience until such time as we have sufficient success in our military operations to inspire in the Soviet Government some of the fear and respect which they now feel for the Germans.'[31]

According to his memoirs, Eden returned from the Middle East in April 1941 with 'Britain's relations with the Soviet Union . . . high on my list of priorities', impressed that the recent conclusion of a Soviet–Yugoslav Pact indicated that the Russians would consider helping countries threatened by a Nazi attack and convinced that the time had come 'for a smoothing-out of relations between our two countries'.[32] Almost certainly this account gives an exaggeratedly prescient impression of his assessment of Anglo-Soviet diplomacy early in 1941. His contemporary stance was more cautious. Though keen to do nothing to increase Soviet suspicions, and willing even to concede de facto recognition of the Soviet take-over of the Baltic republics, Eden was reluctant to open discussions until the Soviets had given tangible evidence of their readiness to cease co-operation with Germany.[33] 'I do not want to indulge in useless gestures', he stressed, but 'I do want to disabuse the Soviet Government of some of the suspicions which, however unjustifiably, they do entertain of us.'[34] But a meeting with the Soviet ambassador, Maisky, on 5 June left him thinking that the Russians would continue to give in to Nazi pressure 'unless their skin is asked of them'.[35]

None the less, even before the Nazi attack on 22 June, Eden had emerged as the leading advocate inside the War Cabinet of aid to the Soviet Union. His attitude was hard-headed and unsentimental. In the event of a German invasion, he explained to Cordell Hull only four days before the blow fell, Britain would not become the ally of the Soviet Union, but 'we should have a common enemy and a common interest'.[36] Soviet hypocrisy sometimes appalled him. In September Maisky made a 'nauseating speech' about his country's consistent support for victims of aggression. Eden 'contrived not to vomit'.[37] But the Foreign Secretary, like his Prime Minister, immediately recognized that the escalation of the war might transform Britain's strategic outlook at a time when the United States remained a non-belligerent, albeit a benevolent one. And, while Churchill seemed content to respond to the new situation with gestures comparable to 'favourable references to the devil in the House of Commons', Eden saw that the Soviets would want more tangible sustenance, including diversionary military operations. To Churchill he wrote:

> It is surely at least possible that a month from now the Russians will still be in the field and fighting back, though in all probability very hard pressed. I fear that if even by then we are still unable to stage any land operation to relieve the pressure on Russia, the effect upon our position internationally will be bad.[38]

Churchill's refusal to respond as positively as Eden wanted provoked some of the Foreign Secretary's more acid comments on the premier's shortcomings, at least to those in Eden's immediate circle if not, characteristically, to Churchill himself.

But of equal interest to any differences of policy between the two men are the clear signs of Eden's determination to keep this area of diplomacy firmly within his grasp. He was particularly irritated when Churchill seemed ready to begin the same sort of personal correspondence with Stalin which already characterized his relationship with Roosevelt. To the Prime Minister Eden's words were circumspect – 'I think that we must keep [direct communications] for occasions of capital importance and I do not want you to become involved in the day to day details of diplomacy'[39] – but the truth was that he found 'W[inston]'s dictator moods irritating'.[40] Eden

183

privately compared Churchill's initiative with Chamberlain's personal dealings with Mussolini in the winter of 1937–8 and confessed to being 'very fed up' with his 'monopolistic tendencies'.[41] It was the time in the war when Eden was most critical of Churchill's leadership. The latter exercised a 'devastating' effect on planning, his views on military matters were 'dated' and Eden was convinced of the need to appoint an independent Minister of Defence.[42] Churchill's existing methods were 'wearing out' the Chiefs of Staff and to no purpose. Eden seemed to win his point over communications with Stalin when the Prime Minister apologized for being 'so tiresome' over his personal telegrams, but Harvey soon noted that Churchill was again developing a 'regular correspondence' by private telegraph, while from Moscow Cripps warned that the florid tone of these messages was having the worst possible effect on Stalin.[43]

In the last resort Eden had to accept that there would be moments when Churchill would continue to take the lead in whichever aspect of British foreign policy attracted the attention of his fertile, but not always predictable, brain. Having accepted that the Prime Minister was essential in embodying the national will to continue the struggle against Hitler, Eden knew that this was a price he had to pay.[44] But it made him more than ever intent on resisting the incursions of anyone else into his departmental responsibilities. This was a sensitive point for Eden throughout the war and particularly during his first twelve months as Foreign Secretary.[45] During 1941 Stafford Cripps in Moscow, Hugh Dalton at the Ministry of Economic Warfare and the government of India were all accused of stepping on his ministerial toes.[46] The following year Harold Macmillan and even the ever-loyal Richard Law were added to this list of transgressors. 'Am I Foreign Secretary or am I not?' he once asked Cadogan.[47] Though his rhetorical question required no answer, the boundaries of diplomatic activity within the overall administration were bound to become blurred in a time of total war.[48] Not surprisingly, critics interpreted Eden's attitude as little more than petulance. He was 'like the little boy trying to clutch all the toys,' judged Dalton. 'Edward [Halifax] used to give his toys away to other children.'[49]

As the Soviet military position deteriorated in the second half of 1941, Stalin's demands became ever more insistent, yet harder to fulfil. Eden clearly wished to do everything in Britain's power to relieve the hard-pressed Russian army and people. He urged upon Churchill the speeding up of a mission to Moscow to discuss supplies, which had been agreed with Roosevelt. In what became one of his recurrent themes, he warned that delay would only 'arouse suspicions of our sincerity'.[50] He found Churchill's response to Stalin's request for supplies wholly inadequate and, after a prime-ministerial explosion, succeeded in bringing him round to his point of view.[51] Eden even pressed for a diversionary landing in France, while recognizing that the military case against it was 'very strong'.[52] In October he put to the War Cabinet the arguments in favour of acceding to a Soviet request that Britain should declare war on Finland, Romania and Hungary. Though there were obvious practical objections to this move, the Soviets attached considerable importance to it and 'we naturally are anxious not to rebuff or discourage them. . . . A refusal might arouse their latent suspicions as to our motives.'[53] Progress was difficult, especially when Church-ill's latent anti-Bolshevism began to reassert itself. 'After his first enthusiasm', noted Harvey, the premier was 'now getting bitter as the Russians become a liability. . . . No one stands up to him but A. E.'[54] Not surprisingly, the autumn saw Eden's credentials as a convinced friend of the Soviet Union soar. According to Maisky he was the one British statesman who enjoyed his trust, while Beaverbrook, soon to lead the 'Second Front Now' campaign through the pages of his newspapers, found himself 'seeing eye to eye with that man'.[55]

By early November Stalin was pressing for agreement on war aims and for an extension of the Anglo-Russian military alliance. Eden informed Cripps that Britain would be prepared to continue collaboration with the Soviet government after the war in order to work out the details of the new settlement of Europe, but seemed a little hesitant at the idea of a post-war alliance.[56] Angered by the tone of a message from Stalin on 8 November, Churchill was reluctant to make further concessions. The Soviet leader's need of Britain was, he told Eden, 'greater than our need for him'.[57] But Eden, urging Maisky to build bridges, helped to secure a Russian apology, while the Foreign Office concluded that Stalin's wish for an understanding on war aims and plans for post-war organization probably derived from a suspicion that Britain was seeking an Anglo-American peace from which the Soviets would be largely excluded.[58] The decision was now taken to send Eden to Moscow. As he explained to the exiled Polish premier, General Sikorski, Britain 'had such evidence of the deep suspicion which had implanted itself in the Soviet Government's mind that there was no alternative but for him to go'.[59] Here, it seemed, was the opportunity for real diplomacy for which Eden had been waiting.

The Foreign Secretary clearly felt himself better equipped than Churchill to deal with the Soviets. 'If it were only method', he noted, 'there is not much difference between us, but there are two problems: Winston's instinctive hatred of Red Russia and his deep reluctance to consider postwar problems at all.'[60] Eden also prided himself on his professionalism and insisted that he was reluctant to make the journey unless the political ground was prepared in advance.[61] In fact he was entering a potential diplomatic minefield. Precisely what he could achieve remained unclear. Having insisted upon significant territorial concessions, including the annexation of the Baltic States, at a time when the Soviet Union was in grave danger of military defeat, Stalin was hardly likely to be more accommodating now that a major counter-offensive was in preparation around Moscow. His demands were likely to be unacceptable not only to Britain but also to American opinion. Hopkins and Winant had already warned against any British commitments to definite frontiers for any country before the final peace treaty. The United States did not want to enter the war 'to find after the event that we had all kinds of engagements of which they had never been told'.[62] On 4 December, in an attempt to allay American anxiety, the State Department was informed that Eden would try to assuage Soviet suspicion and resentment without entering into binding commitments. But the American government already believed that Britain would look indulgently on Soviet domination of post-war Eastern Europe and continued to watch Eden's visit with considerable misgivings, insisting on being kept informed of the progress of his negotiations and warning against any agreement which infringed the Atlantic Charter.[63]

In the days before Eden set off on what was, in winter and in wartime, a potentially hazardous journey, the limitations on his freedom of manoeuvre and his relatively weak bargaining position became apparent. 'An unsatisfactory discussion about Russia', he recorded on 27 November. 'Winston most reluctant that I should commit myself at all. Wanted me to be negative about post-war etc. Argued impossibility of this. Max supported me, but we made no headway.'[64] His formal memorandum for the War Cabinet insisted that the limits of his discussions with Stalin should be defined in advance and their objectives fixed, but his own main goal still seemed to be to dissipate Soviet suspicions.[65] The cabinet meeting on 4 December went less favourably for Eden than he implied in his memoirs.[66] In his diary he wrote of a 'poor cabinet' at which, in

Beaverbrook's absence, he received little support – 'all maintaining that I needed nothing, and ignoring that Stalin had been told I was bringing armies!'[67] 'Eden does not wish to go to Moscow without something in his pocket', noted Bruce Lockhart, 'i.e. declaration of war on Finland etc., as Stalin requested.'[68] But the cabinet specifically urged him to avoid being drawn into precise discussions on territorial changes and boundaries.[69] The two-hour meeting concentrated on whether Eden should tell Stalin that Britain could not let him have troops for his front line but would help out with more tanks and aeroplanes. The discussion, thought one of those present, could have been concluded in fifteen minutes, but Eden was 'very sticky about facing Stalin with what he still thinks will be a disappointment'.[70] Even so, Eden's announcement that he would tell Stalin that the second front already existed – in Libya – revealed scant awareness of Soviet sensitivity over this issue.

Inside the Foreign Office doubts grew about the wisdom of the mission:

> Altogether Eden's Russian visit seems foolish. He cannot achieve very much. . . . On other hand, disadvantages are great. . . . Not at all good for Foreign Secretaries to go wandering round in wartime, and this is second time in less than a year. . . . It is a big risk.[71]

Meanwhile Cripps was worried that Eden's propensity for 'nice phrases and friendly gestures' would prove positively harmful since they would be interpreted by Stalin as being 'intentionally used as substitutes for action of some kind'.[72] A contrast was emerging between what was actually in Eden's 'hamper for Russia' and the impression given to Stalin in Churchill's formal announcement of the visit:

> He would be accompanied by high military and other experts, and will be able to discuss every question relating to the war, including the sending of troops. . . . I notice that you wish also to discuss the post-war organization of peace. . . . The Foreign Secretary will be able to discuss the whole of this field with you.[73]

After a gloomy meeting with Beaverbrook, with the latter contemplating resignation over the government's Soviet policy – 'War Office and Air Ministry were not playing the game . . . impossible to talk to Winston any more about Russia' – Eden set off on the morning of 7 December.[74] Richard Law, his newly appointed Under-Secretary, made a last-minute effort to dissuade him from making the trip in view of possible developments in other theatres of the war and the likelihood of a domestic political crisis, but to no avail.[75] Eden still hoped for success. It was obviously his most important diplomatic mission since returning to the Foreign Office. Events, however, continued to conspire against him. Reaching Invergordon by the following morning, Eden was diagnosed as suffering from gastric influenza. By then news of the Japanese attack on Pearl Harbor and America's entry into the war had been confirmed. Though transforming Britain's longer-term prospects, these events did little to ease Eden's immediate diplomatic difficulties. He was 'staggered' to learn that Churchill now proposed an immediate visit to Washington to meet Roosevelt – an event which would remove the limelight from his own journey, while simultaneously rekindling anxieties as to what commitments the Prime Minister might enter into if his own penchant for diplomatic initiatives were given too free a rein.[76] In addition, Britain now faced the need to divert men and supplies to the Far East, making it even harder to sustain a supply of aid to the Soviets. On 12 December Churchill told the War Cabinet that Eden's offer of RAF squadrons to aid the Soviet war effort – in effect the one ace left in his pack – had had to be withdrawn. Somewhat unrealistically, Eden was now to suggest that there would be a great advantage if the Russians declared war on Japan, once they felt strong enough to do so without imperilling their European front.[77] Even in terms of the direct Anglo-Soviet balance, Eden's hand was visibly weakening. The German advance had been checked and pushed back from Moscow. The immediate threat to the Soviet capital had been lifted.

In these inauspicious circumstances, Eden's party arrived at Murmansk on 12 December. The town was shrouded in a thick fog, and the objectives behind the British mission were little clearer. Despite the activity involved in its preparation, the Foreign Secretary's diplomatic and military portfolio looked distinctly slender. His meetings with Stalin quickly revealed the weakness of Britain's negotiating position. By contrast, the Soviet leader seemed thoroughly in control of the situation. Less susceptible than many others to Eden's charm, he was 'a quiet dictator in his manner. No shouting, no gesticulation, so that it is impossible to guess his meaning, or even the subject of which he is speaking, until the translation is given.'[78] During the talks Stalin produced a detailed set of proposals. These included the need to clarify British and Soviet war aims, plans for the organization of the peace, agreements on aid, the issue of declarations of war on Finland, Hungary and Romania and the need for British military support in the Soviet Union itself if a second front in France remained impossible. Stalin wanted a treaty of alliance for the duration of hostilities and a second treaty for the post-war era which would commit Britain to recognizing the Soviet frontiers as they existed on 21 June 1941. This meant accepting the gains made since the signing of the Nazi–Soviet Pact, including Russia's annexation of the Baltic states and territory seized from Finland, Romania and, most significantly, Poland. When Eden explained that he could not give any commitment on frontiers at the present time, Stalin calmly replied that in that case he would prefer to have no agreement at all.[79] Granted his instructions from the cabinet and in view of the ever watchful eye of the United States, the Foreign Secretary's reticence was reasonable enough. There was, however, an element of unreality in the British position, as Harvey quickly recognized:

> We cannot commit ourselves now, without much further thought and consultation, to definite frontiers for E. Europe, but if at the end of the war Russia is in occupation of the Baltic States no one is going to turn her out. The Baltic States clearly must go back to Russia but their fate cannot be signed away by us without any further thought.[80]

Overall, Eden seems to have been surprised by the extent and clarity of the Soviet demands. Attempts to satisfy the Russian delegation with anodyne drafts which omitted any precise reference to frontiers got nowhere. Eden had largely ignored Cripps's exhortations to draw up 'something definite and concrete' with which to meet Stalin's fixed aspirations.[81] The best that he could do after two days of near deadlock was to persuade Stalin that America would have to be consulted before any treaty was signed. This limited success was achieved by leaving Stalin with the impression that the obstacles to an Anglo-Soviet treaty embodying his territorial claims could be quickly removed. This seems to have been something which Eden genuinely – if mistakenly – believed. As Harvey recorded: 'If Stalin could sign our proposed agreements, it would be fairly easy for A.E. to move Winston to next stage of getting the Cabinet to agree to get Roosevelt to tackle the whole question.'[82] It therefore proved possible to put together a communiqué to satisfy the niceties of diplomatic protocol, but neither Eden's assurance to Churchill that his mission had been a success nor his description of the talks to the press as 'full, frank and sincere' reflected reality.[83]

Aware that the stance he had tried to maintain in Moscow had proved untenable, Eden returned from Russia convinced that the Soviet case would have to be conceded. On the question of frontiers he now emerged, at least in his own mind, as a hard-headed and far-sighted realist whose task was to educate Churchill in the real facts of life, however distasteful these might be.[84] Yet neither Churchill nor the War Cabinet proved as responsive as he expected. At the same time Eden, by his attitude towards Soviet claims, was laying himself open to those charges of cynical *realpolitik* and even appeasement to which he was always particularly sensitive. With Churchill still in Washington, Eden argued the case for the immediate recognition of Stalin's demands, while asking for the matter to be taken up with Roosevelt. His central theme was still the need to dispel Soviet suspicions. 'I am clear', he insisted, 'that this question is for Stalin acid test of our sincerity and unless we can meet him on it his suspicions of ourselves and United States Government will persist.'[85] Churchill was unmoved, though 'greatly surprised with Anthony'.[86] He reminded Eden that Britain had never recognized the 1941 frontiers except *de facto*. They had been acquired by acts of aggression 'in shameful collusion with Hitler'. Rather than respond to Stalin's concerns about the sincerity of the West, Churchill preferred to focus on the 'sincerity involved in the maintenance of the principles of the Atlantic Charter'.[87] Not for the last time, Eden faced considerable problems in winning round his master to his own point of view. 'We're going to have great difficulty over this,' noted Harvey. 'AE will have a lone fight as no one else in the Cabinet will speak up to the P.M.'[88]

Eden was motivated by a number of considerations. In the first place he doubted Britain's capacity, even at the end of a victorious war, to influence the political configuration of Eastern Europe. 'It must in any case be borne in mind', he wrote in January 1942, 'that we shall not be able to affect the issue at the end of the war by anything we do or say or refuse to say.'[89] Just as importantly, Eden now argued vigorously that Stalin was a man with whom it was possible to do diplomatic business and whose policy objectives need not be interpreted in strictly ideological terms. The Soviet dictator was 'much more the heir of Peter the Great than of Lenin, quite indifferent to ideological issues but very determined on getting the territory and the strategic frontier he wanted'.[90] He developed his thoughts in a letter to Halifax:

It is very difficult to know what [Stalin's] real motives are but I believe him to be genuinely concerned for the future security of his country. . . . I do not believe that at the moment he is greatly interested in international Communism. Of course, it is true that he would have concealed the fact from me if he were. On the other hand, international Socialists and Communists have a certain mental outlook with which we are familiar and which it is not so easy to conceal.[91]

This was perhaps a reasonable calculation, but in making it Eden seemed more concerned to dispel Soviet suspicions of Britain than to allay the equally understandable fears entertained by many of his colleagues in relation to ultimate Soviet intentions. Ironically, it was leading Labour figures such as Clement Attlee who were most convinced that Russia was set on post-war expansion.[92] Churchill's famous comment, made in 1939, that the Soviet Union was 'a riddle wrapped in a mystery inside an enigma' suggested that he too lacked Eden's certainty concerning Soviet motivation.

Eden's attitude towards Stalin's territorial aspirations also needs to be set in the context of the increasing thought given by the Foreign Office from the beginning of 1942 to the problems Britain would face in the post-war world. Such forward planning made sense once the American entry into the war created a Grand Coalition against which Hitler and his allies were unlikely to prevail. Despite his more immediate wartime preoccupations, Eden was more active in such planning than his critics have sometimes allowed, though not perhaps as active as the fertile brains of some Foreign Office officials would have liked. At the end of January 1942 Eden warned that, with a defeated Germany and a temporarily weakened France, there would be nothing to stop Soviet domination of Europe at the end of hostilities. The defeat of Hitler would inevitably leave Soviet forces entrenched further westwards than they had been in 1941 and a victorious Soviet Union was bound to claim more generous frontiers than she had held before the Nazi attack. Such circumstances compelled Britain to seek immediate political arrangements with the Soviet government. At the same time he urged his colleagues to avoid policies which would imply that Britain's aim was an Anglo-American peace. Indeed the tone of Eden's memorandum implied that such vital British interests were at stake that an American veto on discussion of the question of frontiers would not be acceptable.[93] Such reasoning was not shared by Cadogan: 'We shall make a mistake if we press the Americans to depart from principles, and a howler if we do it without them.'[94] Eden's concern with the post-war balance of power extended beyond Europe. He believed that Soviet collaboration offered a means to maintain Britain's position in the Middle East. As early as June 1941 he had suggested that the Russians should promise not to interfere in this area, particularly in Iran and Iraq, as a first step towards an understanding with Britain.[95]

By the beginning of February 1942 clear divisions had emerged within the War Cabinet and the Foreign Office on the question of recognizing the Soviet Union's 1941 frontiers. Among senior ministers Eden had the backing of Beaverbrook, Bevin and Morrison, but found Churchill and Attlee opposed.[96] The cabinet finally agreed to place the matter before the American government in a way which presented both the arguments for accepting Stalin's demands and the objections to doing so.[97] Eden envisaged the possibility of a German collapse in the course of the year, which would lead to an 'exclusively Soviet victory with all that implies'. This situation called for an urgent agreement with Stalin. 'This may not prevent him from double-crossing us, but it will at least lessen pretexts.'[98] Whether the position was sufficiently grave to justify a breach with the Americans remained, however, the crucial issue. Cadogan took a cynical view of Eden's attitude: 'How funny A. is! Because it fits in with his trip, he is quite prepared to throw to the winds all principles (Atlantic Charter) which he has not drafted. This amoral, *realpolitik* line was never his.'[99] Ironically, it was the worsening situation in the Far East which helped Eden win the argument. To Churchill he implied that some sort of bargain might be struck whereby Britain should recognize the Soviet Union's territorial claims in return for Russian action against Japan. 'The Far East connection', thought Eden, 'should appeal to Roosevelt.'[100] On the following day, 7 March, Churchill told Roosevelt that the increasing gravity of the war had led him to conclude that the Atlantic Charter should not be so construed as to deny the Soviet Union its frontiers of June 1941. 'I hope therefore that you will be able to give us a free hand to sign the treaty which Stalin desires as soon as possible.'[101] From the Washington embassy Halifax viewed the premier's volte-face with understandable surprise: 'Having called Anthony every name from a dog to a pig for suggesting composition with Stalin, he now goes all out for it himself in a message to the President.'[102]

Fortified by this success – though securing Churchill's support was not the same as gaining Roosevelt's – Eden continued to urge all possible help for the Soviet Union. He regarded the forthcoming German offensive against Russia as potentially decisive for the whole war. If it failed to inflict decisive defeats upon the Red Army, Germany's fate would be sealed. If, on the other hand, it succeeded in breaking up effective Soviet resistance, then the war might be prolonged for another four years.[103] Britain's interests as much as Russia's therefore demanded that all possible supplies should be sent to the Soviet Union.[104] Above all, Britain must not give the impression that she was ready to fight the war to the last drop of Russian blood. Concessions in the political sphere might at least convince Stalin of British good faith and rule out a compromise peace between Germany and the Soviet Union. For this reason, and faced by continuing American opposition, Eden came to favour telling Roosevelt that Britain would go ahead with a treaty on the lines that Stalin wanted in the hope that the United States would simply acquiesce.[105] 'Anglo-Russian relations cannot be left in suspense at this critical moment in the war,' he warned.[106] Not for the last time, Eden was prepared to risk ruffling the Anglo-American relationship in the interests of what he saw as compelling British interests. Others such as Cadogan were less convinced:

A., determined to go ahead, gallops gaily over the ground, which will give way under him, one of these days. We're selling the Poles down the river, and everyone will suspect we're going to do the same to them, and we're annoying and disgusting Roosevelt. And to what purpose?[107]

In this situation of still divided counsels detailed discussions of an Anglo-Soviet treaty got under way. In an effort to assuage American anxieties Eden included in the British draft a provision to allow emigration from the Baltic States. He also took up the idea that the Soviets should be asked for a declaration to accompany the formal treaty promising a measure of autonomy for the republics.[108] Eden set out his thinking in a frank letter to Halifax:

the fundamentals . . . seem to me to be these. We, Russia and the United States have got to work together if there is to be a sane world after the war. Stalin regards these Baltic States as strategically part of Russia. He will never feel safe unless he has them and he will look with suspicion on anyone who he thinks doesn't want him to have them. In a sense he regards our attitude to this claim of his as a touchstone of our sincerity when we say that we wish for permanently good relations with Russia. His attitude may be unreasonable, but nothing we can say will shake him out of it.[109]

Though it was a 'disagreeable business' to consign the peoples of Estonia, Latvia and Lithuania to Soviet overlordship, perhaps for all time, Eden seemed not to anticipate how vulnerable he now was to the charge of appeasement.[110] Domestic opposition to the proposed treaty took him by surprise. 'It's curious', commented Cadogan, 'that A, of all people, should have hopes of "appeasement".'[111] When the backbench Foreign Affairs Committee appeared ready to draw uncomfortable parallels, the Foreign Secretary tried to nip such thoughts in the bud. 'Munich', he insisted, 'was a collapse before a foe. . . . It has, of course, no resemblance to the present project.'[112] None the less Bruce Lockhart found him 'a little worried. . . . Conservatives said he was selling Europe to Bolshevism.'[113] Among ministers opposition came from Duff Cooper, now Chancellor of the Duchy of Lancaster, who had honourably resigned from Chamberlain's cabinet over the Munich settlement. According to him, Germany's actions against Czechoslovakia and Poland were more easily excusable than the Soviet occupation of the Baltic republics.[114]

Eden was less inclined to accept criticism from the Lord Chancellor, Lord Simon. The latter sent Churchill a long letter couched in high moral tones protesting against the course proposed by the Foreign Secretary and stating that he could not defend it in the House of Lords. Eden found it difficult to countenance such a stance by a man whom he regarded as among the most culpable of the appeasers of the 1930s.[115] But Simon did have a point in suggesting that Stalin's aim might be to cause a rift in Anglo-American relations which would only create problems for the settlement of the post-war world. 'Is it too much to say', he asked, 'that our guiding principle should be that we sign no agreements with anybody that either have not the approval of the United States or to which the United States cannot be a consenting party?'[116] These were words which Eden might one day have cause to ponder. For the time being the misgivings of ministers such as Cooper and Simon at least strengthened the case for caution. Churchill warned that he had no wish to face 'a bunch of resignations'.[117]

The course of the Anglo-Soviet negotiations suggested that there was something to be said for the warnings of Eden's critics. Preliminary meetings between Eden and Maisky were thoroughly discouraging. The ambassador emphasized the bitterness felt by the Russian people at the inadequacy of British support for their superhuman efforts against the common enemy – an argument which did scant justice to Britain's own military predicament. More worryingly, Maisky seemed to raise his demands at each meeting. 'It is a tough job to make progress with these people,' noted Eden. 'They always ask for more.'[118] It was a remark which could have been made by those who had embarked upon the equally unrewarding task of negotiating with Nazi Germany a few years earlier. In the circumstances it was scarcely surprising that Eden made sure that he had a fall-back position if the negotiations ended in deadlock. While presenting the cabinet with a new draft treaty which omitted the provision for emigration from the Baltic States, he also took up Cadogan's idea for an alternative treaty making no mention of frontiers at all.[119] Cadogan took comfort that Eden was not prepared 'to throw the Poles down the drain', but remained wary that he might give way under the influence of Oliver Harvey.[120] Churchill and Eden began discussions with Molotov in London on 21 May. These revealed that the Soviets would insist that the Soviet–Polish frontier should be settled directly between the two countries. This gave Eden the opportunity to put forward Cadogan's alternative draft of a twenty-year pact of mutual assistance leaving the tangled question of frontiers to one side. By this stage and facing continuing opposition from the United States and from parliament, Eden was 'longing to get out of his promise about frontiers'.[121] Somewhat to his surprise, Molotov eased his difficulties by agreeing to accept the new proposals and on 26 May 1942 the treaty was signed.

Eden thus seemed to have brought the negotiations to a most successful conclusion. He was pleased at the almost universal support which the treaty enjoyed. Beaverbrook was among those who offered congratulations, urging Eden to 'take full credit for it'.[122] It was, Eden suggested, a great relief that events had fallen out as they had and a considerable step forward that Britain and the Soviet Union had finally put their relationship on a sounder footing. But his assertion that he had been concerned lest the treaty should lose the greater part of its value 'owing to the offence which it caused to our other allies' scarcely reflected his attitude over the preceding months.[123] In many ways Eden had been lucky that it had proved possible to sign a treaty without doing considerable damage to Anglo-American relations and to Eden's domestic political position. Though the question of Soviet frontiers would not go away, immediate difficulties had been overcome, leaving him to savour his most striking achievement in the war so far. It was a good moment to reassert his sovereign control over British diplomacy. To Churchill he complained that ministers were making speeches on foreign affairs without consulting Cadogan or himself:

> I fully understand that the procedure may well be regarded as tedious by my colleagues, but now that conjecture on the terms of the postwar settlement and such like problems is becoming rife it is more than ever necessary that the Foreign Office should know what is being said.[124]

*

The conclusion of the Anglo-Soviet treaty left Eden optimistic about the future – but not blindly so, though his public manner could sometimes give the impression that he wanted to 'sheer away from difficulties'.[125] He had been impressed by the Soviet readiness to take a 'broad European view of our problems' and, to his surprise, had even found Molotov co-operative. He did not, however, forget that 'we have yet to win the war and the peace'.[126] Even so, Eden's memoirs give too clear an impression of his mounting apprehension about Soviet intentions. As Churchill also found, the Cold War generated an irresistible temptation to lay claim in hindsight to a degree of prescience largely lacking at the time. Though Eden was from time to time beset by doubts, he still did not reach any consistently hostile interpretation of Soviet policy before the very last months of the war.

For the time being Anglo-Soviet relations could be relegated to a less prominent place among Eden's concerns, though the issue of the Second Front could always complicate inter-Allied diplomacy, especially when Molotov returned from Washington having apparently secured an American commitment to the invasion of Europe before the end of 1942. As the war entered its second half the Foreign Office gave increasing attention to the possibility of negotiating peace terms with anti-fascist elements in Italy and Germany. For his uncompromising attitude on this matter Eden has incurred much criticism.[127] But the impact of such negotiations on the ever suspicious Soviets was certainly a factor in his calculations. Eden's scepticism about the German resistance movement, in which he was influenced by his friend and future designated biographer, John Wheeler-Bennett, was probably well justified. But a stronger argument for a possible 'missed opportunity' exists in relation to Italian peace initiatives. Even as Mussolini's regime moved towards collapse, Eden continued to justify his refusal to contemplate abandoning the policy of unconditional surrender. In February 1943 he wrote:

> If we want to go further and get some group in Italy to cooperate with us on that basis, I realise that we shall have to hold out at least some hope in regard to the future of Italy, in order to secure their cooperation. But there is nothing very definite we could promise the Italians. We can give them no comfort about their overseas possessions. We cannot guarantee the territorial restoration of metropolitan Italy, owing to the pledge we have given to the Yugoslavs to espouse their claim to Istria after the war.[128]

Almost certainly Eden's encounters with Mussolini during his first spell at the Foreign Office had, consciously or otherwise, left their mark in the form of lasting Italophobia. Even Cadogan, finding Eden reluctant to have anything to do with Italian peace-feelers, judged his attitude 'silly! I'd start talking to them at once.' 'It's the old complex,' he concluded, 'the old ideology – that *terrifies* him and makes him see ghosts.'[129] When it became clear that Harold Macmillan, Minister Resident in North Africa, was among those seeking an early peace with Italy, Eden no doubt felt additional cause to stand his ground. Macmillan had joined the list of those who, in Eden's mind at least, were attempting to intrude upon the Foreign Secretary's prerogatives.[130]

It was also at this time that Eden began to examine the possibility of a new international organization to guarantee the future peace. As a former League of Nations man he was always more sympathetic to such ideas than was Churchill. Eden submitted his first paper on the subject to the cabinet in January 1943.[131] It argued for political and economic world councils and stressed the importance of the great powers accepting the notion of the indivisibility of peace and being prepared to act together to maintain it whenever and wherever it was threatened. The unpalatable alternative was a world 'in precarious balance, with the great powers, each with its circle of client states, facing each other in a rivalry which will merge imperceptibly into hostility'.[132] Churchill's attitude remained, however, a substantial impediment to serious consideration of such matters by the cabinet. Eden became irritated by the Prime Minister's 'regular joke' that 'I suppose it is thought that the war is now going so well that we needn't trouble about it any more, and can amuse ourselves this afternoon'.[133] Eden thus faced the dual task not only of ensuring that Britain and her wartime partners approached the peace with a common world policy, but also of elevating Churchill's eyes beyond the immediate and still formidable job of defeating the enemy. Yet from November 1942 Eden had significantly less time to devote to these all-important tasks of diplomacy, foreign and domestic. In addition to his existing responsibilities, he now took over from Stafford Cripps the not inconsiderable burden of leading the House of Commons. Thereafter, Cadogan's diary is liberally sprinkled with the complaints of an exasperated civil servant, unable to secure the attention of his political master to the pressing daily concerns of the Foreign Office, let alone to longer-term planning.[134] It was, judged Cranborne, 'too great a burden for anyone'.[135] Even Harvey admitted that under the new arrangements foreign affairs were 'bound to suffer'. They were left 'to the fag end of the day or to hasty moments snatched between parliamentary business. He cannot give them first attention or adequate reflection.'[136]

*

Throughout 1943 Eden strove to make a success of the Grand Alliance which now existed between Britain, the United States and the Soviet Union. It was no easy task and he sometimes despaired of success. 'Russian policies all dark and uncertain,' he confided to his diary during one moment of doubt. But he also understood that, as peace drew nearer, national rivalries were bound to grow. 'A town whose fate don't [sic] matter much when victory is remote suddenly becomes significant when its fate is to be decided.'[137] His attitude was encapsulated in the answer he gave to Roosevelt in Washington in March, when the President asked whether he believed that the Soviets had a grand design to establish communist regimes across Europe. Eden confessed that he could not tell for certain, but insisted that, even if this fear proved well founded, nothing would be lost by attempting to co-operate with the Soviets.[138] During these same discussions the issue of the Baltic States resurfaced. Roosevelt now seemed ready to take a more pragmatic attitude towards the Russian claim, but hoped to use Anglo-American agreement to it as a bargaining counter to win concessions in other areas. As regards Poland, Roosevelt envisaged fewer difficulties in the way of an acceptable settlement than did the Foreign Secretary. Waving aside Eden's warning that its people had fixed ambitions to emerge from the conflict with their pre-war boundaries intact, the President suggested that Poland would be content to receive East Prussia in the west in compensation for agreeing to the Curzon Line as her eastern frontier.[139]

In the case of Poland, and in line with his own 'realistic' attitude towards Soviet aspirations in the Baltic, Eden had already reached a conclusion not dissimilar from Roosevelt's. But he was under no illusion as to the difficulties of selling such an extensive territorial revision to the Poles themselves. Britain's position was inevitably more delicate than that of the United States, notwithstanding the substantial number of American voters who claimed Polish ancestry. Roosevelt felt no strong moral commitment either to the London-based Polish government-in-exile, or to Poland's inter-war frontiers. Britain, by contrast, had gone to war in defence of Poland – albeit that country's independence rather than its territorial integrity – and had continued to recognize Sikorski's exiled administration. In late September 1943, after reading a memorandum from his staff, Eden confessed that he felt 'taken aback' by the extent of Britain's commitments.[140] The quest for a Russo-Polish settlement was therefore high on Eden's agenda, though he was fully aware that its achievement would be a task of Herculean proportions.

The Polish–Soviet dispute began to re-emerge in the first months of 1943 with an exchange of public statements between the two governments concerning their disputed frontier. Such chances of a satisfactory resolution as did exist were dealt a severe blow when, on 12 April 1943, the German news agency Transocean announced the discovery of the mass graves of about 10 000 murdered Polish officers in the Katyn Forest near Smolensk. Eden's reaction reflected his somewhat over-optimistic desire that this development should not damage the prospects for good Anglo-Soviet and Polish-Soviet relations. He told the cabinet that he was urging the Poles not to let 'this incident' divert them from more pressing matters. The whole affair, he argued, was designed to sow discord between the Allies – which was probably true – and it would be best to ignore it as far as possible.[141] The problem was that the available evidence pointed overwhelmingly, as the Germans claimed, to Soviet guilt. When the Polish government invited the Red Cross to conduct an inquiry, the Germans announced that they too had asked the Red Cross to intervene. The Soviets argued that these simultaneous approaches indicated German–Polish collusion and used this excuse to break off relations with the London Poles.[142] Stalin's attitude dealt a severe blow to Eden's hopes of building upon the Anglo-Soviet relationship and of working out an acceptable settlement for Eastern Europe. Cranborne found Eden 'low and depressed . . . very much discouraged by the Polish-Russian row'. It was, concluded Cranborne, an ominous foretaste of great-power co-operation in the post-war world.[143] As for the 'guilt of Katyn', Eden seemed not to wish to recognize the truth since that truth conflicted with his overwhelming desire to make the Soviet alliance work. To this goal the interests of Poland would always take a poor second place. As late as February 1944 – and in contrast to what was then widely acknowledged in Whitehall – he argued that evidence about Katyn was 'conflicting and whatever we may suspect, we shall probably never know'.[144]

At the end of July Eden, wary of Stalin's reaction to a further bilateral meeting between Roosevelt and Churchill, suggested that he should visit Moscow straight after the Quadrant Conference scheduled for August in Quebec. Indeed, as he left for Quebec Eden took with him three Foreign Office papers all urging the need to get round a table with the Russians.[145] Inside the Foreign Office the continuing quest for good Anglo-Soviet relations was becoming entwined with a perceived need to place some limitations upon post-war Soviet domination of Central and Eastern Europe. Yet government thinking about British objectives in such areas remained rudimentary, a fact which Harvey attributed to the strain imposed on Eden by 'that wretched H of C'. 'Brilliant empiricism and improvisation, which has served him so well in the turmoil of the war, cannot here replace solid work and thought.'[146] In discussion with Maisky in late August Eden said that Britain would prefer to accept the right of both countries to exercise an interest in all parts of the continent rather than divide Europe into spheres of influence.[147] Precisely what this would mean in practice remained, however, unclear.

Though a meeting of Foreign Ministers in Moscow was arranged for October, the problem – as Eden appreciated – was that the Soviets were unlikely to make any concessions if all Britain had to offer was confirmation that there would be no Second Front that year and that no progress had been made in Anglo-American thinking on the question of Russia's frontier claims.[148] Fortified by Hopkins's private assurance in Quebec that Roosevelt's mind was 'almost exactly the same as my own' on the frontier issue, Eden tackled Churchill once more.[149] The extension of Allied operations on the European mainland, he argued, would necessitate consultations with the Soviet Union on an ever growing number of questions affecting the post-war settlement:

I believe the Russian attitude in these discussions would be likely to be affected for the better if they knew our views about their western frontiers. Stalin said to me . . . in December 1941 that he regarded the question of the USSR's western frontiers as 'the main question for us in the war'. The obstacles hitherto have been the difficulty of getting American agreement and our obligations to the Polish Government. But we have had indications that President Roosevelt's attitude may have changed somewhat, and, as regards the Poles, I doubt whether any improvement can be expected in Polish-Soviet relations while the Soviet Government are still uncertain whether we endorse full Polish claim to their pre-1939 frontier with Russia.[150]

When, however, Roosevelt insisted that any Soviet query about frontiers at the Foreign Ministers' Conference should be met by a non-committal response from the US representative, there was little more that Eden could do, especially as Churchill was now 'getting dangerously anti-Russian' again.[151] The Prime Minister was becoming worried that Britain might have no alternative but to fight the Soviet Union once Germany was defeated. Eden recorded Churchill's thoughts:

G[ermany] and J[apan] had been the great restraints upon R[ussia]. We were committed to destroy both. R. would then be immensely powerful . . . it might be that I should still see many years of war, perhaps all my life. I admitted that all this might be true but argued that only possible basis for a policy was to try to get on terms with Russia.[152]

The War Cabinet met in early October to consider Eden's brief in advance of the conference. Though his December 1941 meeting had been characterized by the precision of Soviet demands, Eden clearly hoped that something less definite would satisfy the Russians on this occasion. The American attitude precluded any formal agreement on frontiers, but Eden favoured private discussions containing a clear hint for Molotov that, when the appropriate time came, Britain would not obstruct Russian claims to their 1941 frontiers. He had in fact already made a statement, along these lines, to Maisky.[153] Though Poland would have to be treated separately, Eden again favoured exploring the situation in Moscow, before, as he hoped, a more substantive decision at the meeting of the Big Three at Tehran.[154] What was worrying from Eden's point of view was the evidence of Churchill's mounting hostility towards the Soviet Union. The Prime Minister talked 'great nonsense' to the effect that Germany should not be weakened too much as 'we may need her against Russia'.[155]

It was thus in circumstances hardly more promising than those existing at the time of his last visit in 1941 that Eden prepared to journey again to Moscow. The main difference, perhaps, was a lower level of confidence on his part about the outcome of his mission. With far greater attention attaching to the heads of government meeting in Tehran a month later, the prospect existed that 'poor A. E. would have only stonewalling with Molotov'.[156] It was probably these limited expectations which accounted for Eden's undoubted satisfaction at the ultimate outcome of the conference.[157] It represented, he later recorded, 'the high tide, if not of good, at least of tolerable, relations between us'.[158] Churchill agreed and pronounced its achievements 'prodigious'.[159] Eden certainly prepared himself well and the conference was widely seen as a personal triumph for him. Hastings Ismay, Churchill's Chief Staff Officer, who accompanied Eden to Moscow, later penned a fulsome tribute. Eden's handling of his interview with Stalin was masterly,

> as indeed was his performance throughout the Conference. I had always liked and admired him, but I had hitherto been inclined to think that he was one of fortune's darlings . . . and that his meteoric success had been primarily due to charm of manner and a lucky flair for diplomacy. I now saw how wrong I had been. His hours of work were phenomenal. . . . Nothing was too much trouble and he never went to a meeting without making sure that he had every aspect of the problem at his finger tips.[160]

Yet, notwithstanding a number of undoubted achievements and an impressive individual performance on Eden's part, Moscow was no more than a partial success, especially as regards Anglo-Soviet relations.

The conference, which lasted from 19 to 30 October, proved remarkably harmonious, especially after Molotov had been reassured that the cross-Channel invasion would definitely take place in the spring or early summer of 1944. Only when Churchill, in correspondence with Roosevelt, seemed to question the Allied commitment to Operation Overlord did Eden become alarmed, even contemplating the possibility of resignation.[161] Eden must have been pleasantly surprised that the Soviets made no great fuss about the delay in launching a Second Front – 'there was no recrimination'[162] – and that they seemed much less concerned about future frontiers than during his last visit to Moscow. But whatever his relief at avoiding difficult issues and unpleasant confrontations, it is hard to escape the conclusion that in key areas either no progress was made or that Eden's hopes suffered a significant setback. He did urge the Soviets to reopen diplomatic relations with the exiled Polish government, but received no backing on this issue from Cordell Hull, and Molotov quickly dismissed the matter from further consideration.[163] The failure to work out an agreed policy on Poland was particularly worrying in view of the continuing successes of the Red Army which left the Soviet position in Eastern Europe ever more commanding. Eden's best hope was that the Poles would benefit from the improved relations between the three great powers which he believed to be the conference's principal achievement.[164] But it is at least as plausible to argue that this was the moment when Britain effectively conceded that post-war Poland would fall within the Soviet sphere of influence.[165]

In other areas Eden had to abandon policies which the Foreign Office had been nurturing for the previous two years or more. One was the idea that the states of Eastern Europe should be encouraged to federate with their neighbours to produce larger and stronger entities. If this avoided the fragmentation of Eastern Europe which followed the First World War, Eden hoped that these larger units could fill the power vacuum which would inevitably exist after the defeat of Germany, while serving as an effective barrier against her early revival. During the final negotiations for the Anglo-Soviet treaty in 1942 Eden had assured Molotov that such federations would never be directed against the Soviet Union, but it would be naïve to imagine that their potential value in this respect had not been appreciated as doubts about the ultimate reliability of post-war Russia grew.[166] Though in 1942 the Soviets had not opposed the principle of federations, the military situation a year later allowed them to adopt a more assertive line and Eden's proposal received short shrift. A similar fate befell his suggestion that the three powers should endorse the notion of joint rather than separate spheres of influence in Eastern Europe and also the so-called 'self-denying ordinance', the idea that they should not make treaties with smaller powers except after consultation and agreement with one another. As he strove desperately to save at least some of these ideas, Eden passed a note to Hull to explain his purpose. 'I am sorry to take your time', it read, 'but behind all this is a big question: two camps in Europe or one.'[167] But in the absence of American support Eden gave up the struggle, content perhaps that over other issues such as the setting up of a European Advisory Commission he was making more progress.

Put together, these setbacks represented a considerable diplomatic defeat for Eden, marking the end, in the opinion of one historian, 'of British attempts to limit or constrain a free hand for the USSR in Eastern Europe'.[168] Yet Eden's telegram to Churchill exuded a curious sense of satisfaction: 'There may of course be snags later, but for the moment we are in unexpectedly smooth waters.'[169] Providing the Soviets were treated as equals, they were now 'in the current to move with us in all matters'.[170] Eden felt less pleasure at the outcome of the Tehran Conference in November, not least perhaps because his own role was inevitably reduced by Churchill's presence. Cadogan noted him 'in despair about this hazy conference'.[171] He had hoped for a deal on Poland, including the Curzon Line with compensation in the west, and agreement on the return to Warsaw of the exiled Polish government as soon as the military situation permitted.[172] In the event little was achieved at the political level. 'If we could get on to the business soon', recorded a frustrated Eden, 'we might be able to hammer something out', but American caution stood in his way.[173] Though Eden, Molotov and Hopkins were supposed to confer on the detailed questions raised by the Big Three, they made only slow progress. Tehran was most notable for its agreement on the cross-Channel invasion the following spring.

In finalizing the arrangements for Overlord, Tehran at last removed one particular bone of contention in Anglo-Soviet relations. Yet this only helped focus those relations on the question of the post-war settlement within which the future of Poland remained the dominant issue. Returning from Tehran, Eden argued that the Curzon Line was 'defensible' and that the Poles would be well advised to accept a western frontier along the River Oder, 'extreme' though such a solution was.[174] Those who heard his report to the War Cabinet probably got an exaggerated impression of what had been achieved at the two conferences.[175] 'Assuredly', Eden minuted in December, 'I do not want to throw the poor Poles to the Russian wolves. That was never in my thought, even though my power to help them may be limited.'[176] As his final words implied, however, Eden was conscious that Britain had few cards left to play. When, as the Red Army crossed Poland's 1939 frontier, the Poles asserted their 'indestructible right to independence' and called upon the Russians to respect that right, Eden noted that the overwhelming Soviet victories were giving rise to public impatience with the Poles.[177]

They would, he thought, be 'mad' if they failed to come to terms. 'Unhappily signs are they are mad.'[178] The popular mood seemed to affect even Churchill who showed how far his own thinking had changed since 1942:

> The tremendous victories of the Russian armies, the deep-seated changes which have taken place in the character of the Russian state and government, the new confidence which has grown in our hearts towards Stalin – these have all had their effect. Most of all is the fact that the Russians may very soon be in physical possession of these territories, and it is absolutely certain that we should never attempt to turn them out.

As the Allies sought agreement on Poland's eastern frontier, Churchill was inevitably aware that 'the Baltic States, and the questions of Bukovina and Bessarabia, have very largely settled themselves through the victories of the Russian armies'.[179] In reply Eden detailed Soviet aspirations as regards their borders with Finland, Romania and the Baltic States and reiterated that 'we should agree to all these claims'. In the case of the Baltic States it might be wise to avoid a public statement appearing to violate the Atlantic Charter. It could be explained privately to Stalin that Britain had to take an evasive line in public but that 'we had no intention of disputing the Soviet claims'.[180]

Poland, however, remained the litmus test of Soviet good faith and indeed of the overall prospects for post-war co-operation between the British and Russian governments. Eden's own attitude tended to oscillate between alternating moods of doubt and hope. His memoirs, perhaps inevitably, give a rather too consistent picture of mounting anxiety about the chances of continuing collaboration. Eden notes a minute he wrote in early March – 'Is Soviet regime one which will ever co-operate with the West?' – and another written a month later in response to a Foreign Office memorandum on the Polish question and at a time when the Russians had unilaterally declared that they would open diplomatic relations with the Italian government of Marshal Badoglio: 'I confess to growing apprehension that Russia has vast aims and that these may include the domination of Eastern Europe and even the Mediterranean and the "communising" of much that remains.'[181] But at much the same time Eden insisted that Churchill could still make a good case to the Commons for the continuing reality of great power collaboration, stepped in to prevent the publication of an anti-Soviet tract produced for the army educational service, advised against an early showdown with Stalin and argued that, exasperating as recent Soviet behaviour had been, 'I feel we ought not to jump to the conclusion that they have decided to go back on the policy of co-operation'.[182] Furthermore, if Soviet territorial claims seemed incompatible with the Atlantic Charter, Eden no longer regarded this declaration as sacrosanct. Britain would soon need to insist that the Charter did not necessarily apply to enemy states, so as to assert her right to detach pieces of territory from Germany at the Peace Conference.[183]

Eden might well be criticized for pursuing a policy based on little more than hope and one which flew in the face of his underlying expectations. On occasions his reluctance to admit the possibility of a post-war Soviet menace seemed almost wilful. When Duff Cooper argued for the effective organization of the states of Western Europe to confront this danger, he responded:

> It is above all important that any proposals for closer association between ourselves and the Western European allies, or even with the States of Western Europe, should be for the sole purpose of preventing a renewal of German aggression. It would be fatal, as I see it, to let it be understood that there is any other purpose in such an association.[184]

But the rationale for Eden's approach was best summed up in a minute written in early July:

> It is all very difficult, but at least we are convinced that we are trying to operate the right policy. The Russians may make it impossible. If we fail it should not be through our fault, nor through an undue display of weakness on our part towards Russia.[185]

With the rejection at the Moscow Conference of his notion of shared responsibility in Eastern Europe, Eden's mind turned increasingly during 1944 to a possible agreement with the Soviets on designated spheres of influence. However sympathetic he may have been to Russia's territorial claims, it had never been his intention to accept that the whole of Central and South-Eastern Europe should fall under exclusive Soviet domination.[186] In its attitude towards the Balkans, in particular, the Foreign Office always had one eye on the likely configuration of forces which would emerge after Germany's defeat.[187] In Yugoslavia the main problem was the existence of two rival resistance movements, the communist Partisans and the anti-communist Cetniks, which gave every impression of being more interested in fighting each other than in defeating the Germans. Over this question Eden found himself constantly at odds with Churchill, the Special Operations Executive (SOE) and the Ministry of Economic Warfare under Lord Selborne – 'this chip off the old blockhead and his disorderly crew'.[188] The issues in dispute had as much to do with Eden's right to keep control of British diplomacy in his own hands as with genuine policy differences. None the less Eden became a doughty supporter of the anti-communist Cetnik leader, General Mihailovic, from his emergence to prominence in the autumn of 1941. A year later, when Mihailovic's star had fallen, Eden still argued the case for continuing to support him 'in order to prevent anarchy and communist chaos after the war'.[189] Churchill, by contrast, was increasingly attracted by the Partisan leader, Tito, seeing in him, something of 'a Bonnie Prince Charlie', and ultimately prevailed against Eden's better judgement.[190] Both Prime Minister and Foreign Secretary placed too much faith in the youthful King Peter, who proved to be a slender straw upon which to base any British policy.

Similar machinations in Greece saw the communist-inspired EAM emerge as the dominant element in the resistance movement. By the spring of 1944, therefore, Britain's Balkan policy seemed in some disarray, for which Eden was keen to disclaim responsibility:

> It is we ourselves who have built up Tito and his Partisan communist organisation. And the same applies to Greece where it was SOE who originally launched and supported EAM. In both cases we backed them for immediate operational reasons in spite of their being communist movements. . . . The time has come for us to consider from the long-term view what is going to be the after-war effect of these developments instead of confining ourselves as hitherto to the short-term view of what will give the best dividends during the war.[191]

Securing the British position in Greece – which meant removing the communist influence – now emerged as a policy priority:

> If, as is possible, strong Russian armies were in Roumania and Hungary and if Bulgaria were pro-Russian, with Tito in charge in Yugoslavia, the whole of the Balkans would pass under Soviet influence. Apart from the direct damage to our strategic position this would have a lamentable effect in Turkey and in the Near East generally.[192]

The obvious way forward, as Eden saw, was to establish some form of *modus vivendi* with the Soviets over the Balkans. This was the basis of his talks with Gusev, the new Soviet ambassador, at the beginning of May. From these Eden believed that he had the makings of an agreement to enable Britain to take the lead in Greece in exchange for Russian predominance in Romania.[193]

When, therefore, Eden came to review Soviet policy in the Balkans at the time of the Normandy landings, he warned against assuming that conflict with Britain was inevitable. 'We should not hesitate to make our special interest in the Eastern Mediterranean and therefore Greece and Turkey . . . clear to the Russians; but in any steps which we take to build up our influence, we must be most careful to avoid giving the impression of a direct challenge.'[194] A second cabinet paper in August suggested similar tactics for Central Europe. Britain should 'avoid any direct challenge to Russian influence in . . . countries adjacent to the Soviet Union', but avail herself 'of every opportunity to spread British influence'. Overall, Eden seemed confident that a new equilibrium could emerge in Europe, founded on the existing Anglo-Soviet alliance, which would not be inimical to British interests. It was important that the Soviet Union should not think that Britain and America aimed to build up a combination of European states, including a revived Germany, to keep her in check. Indeed, it should be Britain's constant aim to 'strengthen the hands of the collaborationists [in Russia] by paying regard to the Soviet Government's reasonable demands and views'. Eden believed that the Soviet Union would respect legitimate British interests in Eastern Europe, concentrate on reconstruction and economic development and pursue a policy of co-operation.[195] Such thinking inevitably brought him into conflict with the Chiefs of Staff who were beginning to stress the possible dangers from the Soviet Union after Germany had been defeated. 'Apparently', noted Alan Brooke, the CIGS, 'the Foreign Office could not admit that Russia might some day become unfriendly.'[196] Eden produced a memorandum to enlighten his military colleagues. Suggestions of a Western European security bloc involving German participation should be avoided at all costs as this would quickly destroy all hope of preserving the Anglo-Soviet alliance.[197] Not surprisingly, a meeting on 4 October failed to produce agreement. There had been a growing tendency in certain quarters, Eden argued, to focus too sharply on the possibility of future difficulties with the Soviet Union. The immediate problem, he insisted, was to guard against renewed German aggression and not an unproven Soviet threat.[198]

Against this background the famous percentages agreement reached between Churchill and Stalin in Moscow in October appears as something more than a casual late-night improvisation by the two men. As he left for Moscow Eden still hoped for an understanding with the Soviets on South-East Europe without being over-optimistic of success: 'I have a feeling that our difficulties are multiplying and that we shall "muck it up" beyond a doubt.'[199] Like many others, he had had his confidence shaken by Russia's failure to support the Warsaw Rising in August and September. It seemed that the advancing Soviet forces had deliberately held back, allowing the Germans to crush the rebellion. Eden, however, was at least prepared to believe that there might be military reasons for the Soviet delay and was in any case increasingly irritated by the behaviour of the Poles themselves. 'I must say that I think that Stalin comes pretty well out of the Polish record,' he minuted in September.[200] Though he seemed uneasy with the mathematical precision of the percentages agreement, Britain had little bargaining power. 'I fear we must simply accept the realities of the situation, however disagreeable,' he telegraphed to London.[201] With eerie exactitude Churchill and Stalin divided up the Balkans into spheres of influence: Romania 90 per cent to Russia, Greece 90 per cent to Britain, Yugoslavia and Hungary 50/50 and Bulgaria 75 per cent to Russia (changed by Stalin to 90 per cent). In subsequent haggling with Molotov, Eden held out for a 50/50 division in Yugoslavia, but accepted 80/20 for the Soviets in Hungary.

Eden's own account of the Moscow Conference stresses the Polish question rather than the Balkan agreement. The successes of the Red Army meant that the Curzon Line was no longer really at issue. 'It was what happened in Poland that mattered.'[202] He favoured raising the question straight away, telling Stalin that this was something which could ruin Anglo-Soviet relations and asserting the right of the Poles to settle their own affairs.[203] Lengthy discussions involving both the Soviet-backed Lublin Committee and Mikolajczyk, premier of the Polish government-in-exile, produced few results, though Mikolajczyk did agree to return to London to try to persuade his colleagues that the Curzon Line had to be accepted as part of a general settlement. Overall Eden found Molotov and Stalin more amenable than expected. The latter talked 'very good sense' and Eden 'agreed with him more often than with W[inston] who is emotional in his approach to foreign affairs'.[204] Not for the first time, a period of personal contact with the Russian leadership served, for the moment at least, to dispel doubts and anxieties.

This was a recurring phenomenon of the war years, helping to explain the fluctuation of Eden's mood between genuine hope and near despair. Eden thus left Moscow, as did Churchill, enthused by the general atmosphere of goodwill. Sharing a drink with Harold Nicolson just before Christmas, he admitted to 'a real liking for Stalin. He says that Stalin has never broken his word once given.'[205] The problem was that, 'as almost always happens with the Russians, once the personal contacts are broken the difficulties come surging up again'.[206]

As war in Europe moved inexorably towards its close, Eden's relations with Churchill became, if anything, more difficult, even though they now regularly spoke with one voice about a possible Soviet threat in the post-war world. 'As the purely military problems simplify themselves', noted Harvey, 'the old boy's tireless energy leads to ever closer attention to foreign affairs.' At the same time Eden's own combination of offices offered no respite. 'He never has time to read our papers now', Harvey recorded in November, 'and so his handling has become very superficial. A mixture of "intuition" (ominous word!) and past experience.'[207] Over Greece Eden found Churchill particularly tiresome. For more than a year Eden had tried to persuade him that the only practicable way of establishing a provisional government was to induce the exiled king to give way to a regent so as to encourage the guerrilla resistance leaders to join the administration. To Eden's dismay Churchill decided that they should both go out to Athens, where they arrived on Christmas Day 1944. Surprisingly, this mission proved a success in so far as Churchill finally withdrew his opposition to a regency under Archbishop Damaskinos. 'One might well ask, was this journey really necessary,' recorded Pierson Dixon. 'It was, if only because . . . the P.M. would not agree to follow Anthony's ideas unless he went out and saw things for himself.'[208] Paradoxically, where prime ministerial attention really was required, Churchill could be as obstinate and short-sighted as ever. When in January 1945 Eden sent him a memorandum on the probable German reaction to defeat, Churchill urged him to circulate it to the cabinet. But 'I will read it when I can. We have not defeated them yet.'[209]

By this time attention was beginning to focus on the forthcoming meeting of the Big Three at Yalta in the Crimea. Eden saw that decisions on Poland could not be delayed much longer. 'Unless we can get a free and independent Poland', he warned, 'our future co-operation with [Stalin], whether we will or not, is bound to be affected.'[210] If Eden was not exactly looking for a showdown, he was certainly moving in that direction. He set out his goals in advance of the conference. The Soviets might not 'be averse' to strengthening the Lublin provisional government with other representative Poles providing such figures favoured collaboration with Russia and these additions might include 'one or two' from the London-based government. In addition, he attached great importance to free elections. Such a package might seem a tall order, 'but something of the kind we must get from the Russians if Poland is to be really independent and if Anglo-American-Russian relations are to be possible upon that basis of cordiality necessary for the future peace of Europe'.[211]

At Yalta Eden's initial reaction was unfavourable. Stalin's attitude towards small countries struck him as 'grim not to say sinister'. There were too many delegates and 'no steady flow and brisk exchanges as at Teheran'.[212] He became increasingly impatient with Churchill who had not prepared himself to cope with the agenda.[213] Initial conversations on Poland produced a 'very dusty answer' from Stalin, but Eden was determined to return to the fray and secured the backing of Churchill and Roosevelt to do so.[214] When the matter was raised again on 9 February, the Soviets seemed unprepared to discuss Eden's paper, 'so I fairly let 'em have it, told 'em something of British opinion, said I would far rather go back without a text than be a party to the sort of thing they wanted etc.'[215] In the circumstances the agreement secured, promising some enlargement of the Lublin Committee, was perhaps the best that could be hoped for. Though couched in ambiguous and imprecise language, the agreement on Poland seemed to meet one of the key objectives which Eden had outlined before the conference opened. Granted that a truly independent Poland was never a practical proposition, he was probably more content with this outcome than his official biographer implies.[216] 'The PM and Anthony are well satisfied – if not more', reported Cadogan, 'and I think they are right.'[217] Eden considered that the subsequent complaints of several Conservative backbenchers were unjustified.[218]

One decision finalized at Yalta caused considerable retrospective controversy in the last year of Eden's life. This was the agreement he signed on behalf of the British government for the repatriation of all Soviet citizens who had served with the German forces. The cases for and against Eden on this matter have been rehearsed elsewhere and need not be repeated here.[219] But it is important to note that the crucial decisions were taken in the summer and early autumn of

1944, when Eden was still relatively hopeful of post-war Soviet co-operation. It was in Moscow in October that he had promised Molotov and Stalin that Britain would return all their subjects – in full knowledge of their likely ultimate fate. His notes and minutes – 'we cannot afford to be sentimental about this' – suggest that Eden was less susceptible to humanitarian arguments than some of his colleagues, especially Selborne and, at one time, Churchill.[220] In general Eden was keen to avoid adding to the growing list of impediments in the path of Anglo-Soviet co-operation. By the time of Yalta he was also understandably concerned about the implications of any policy reversal for British prisoners liberated by the advancing Red Army. Overall, 'it was not a nice decision to have to make'.[221]

The problem with the main Yalta accords was less the agreements reached than Soviet reluctance to carry them out over the following weeks. As he left Yalta Eden was probably as unsure about ultimate Russian intentions as he ever had been. 'Our great problem today', he told Toby Low, 'is the Bear. Can he be brought to co-operate and work with us? I do not yet know this and on this point much, if not all, depends.'[222] But Soviet behaviour towards Turkey and an unhelpful attitude towards the forthcoming conference in San Francisco on the United Nations Organization gave further cause for concern. As had happened before, Eden's mood rapidly hardened once the Yalta delegates had gone their separate ways. 'I take the gloomiest view of Russian behaviour everywhere,' he confessed in late March. 'Altogether our foreign policy seems a sad wreck and we may have to cast about afresh.'[223] With the Soviets behaving 'so abominably in every respect', Eden urged Churchill to reduce his personal messages to Stalin to a minimum. He doubted whether there was any longer much point in going to San Francisco. 'How can we lay the foundations of any new world order when Anglo-American relations with Russia are so completely lacking in confidence?'[224] But his trip did give him the opportunity to meet the new American President, Harry Truman, to whom he argued that, if in the past there had been concern lest Britain and America should be seen to gang up against Russia, in the present climate a display of Anglo-American solidarity might produce better results in Moscow. He was pleased to note that Truman proposed to use a forthcoming meeting with Molotov to explain 'in words of one syllable' the importance of making progress on the Polish issue.[225] Eden himself favoured postponing the opening of the San Francisco Conference until the Soviet Union showed that it would proceed on the basis of the Yalta decisions.[226] In the event Eden took the opportunity of the conference for a frank talk with Molotov: 'I told him that we did not understand what had happened since Yalta. There all had been agreement and friendly understanding. Since then only difficulties have been encountered.' Eden got the impression that Molotov was looking for some sort of deal on the basis of allowing the Soviet Union a free hand in Poland in return for continuing Russian non-intervention in British spheres of interest, but he gave no encouragement to this idea.[227] Though Eden was the success of the conference, enjoying the Geneva-like atmosphere and basking in media attention, his underlying mood remained pessimistic.[228] 'I sometimes feel', he confessed, 'that we are entering period like that of Second Balkan War transferred on to world stage.'[229]

Illness removed Eden from the centre of affairs for a month during the early summer. By the time that he returned to the Foreign Office, 'feeling very good though a little shaky on my legs', preparations were in train for 'Terminal', the last of the wartime meetings of the Big Three, to be held at Potsdam.[230] Eden found the world outlook 'gloomy' with 'signs of Russian penetration everywhere'.[231] On 10 July he sent Churchill a note enumerating the cards which Britain still held 'for a general negotiation with the Russians, in the shape of things which the Russians want from us and which it is in our power to give or withhold'. British bargaining counters included the German navy, credits, portions of German industry and the Soviet desire for Italian ships and for access to the eastern Mediterranean. There was little to induce Stalin to release his grip over Central and Eastern Europe, but Eden urged Churchill not to surrender such advantages as Britain held without corresponding Soviet concessions.[232]

Not surprisingly, Potsdam was the least successful of the Second World War summit conferences. The atmosphere of compromise permeating earlier meetings had disappeared. With Churchill's health 'so far deteriorated that he has no energy left to seize his opportunities', much depended on Eden.[233] 'Prepare for the afternoon meeting', noted Cadogan on 22 July, 'which means giving Anthony sufficient material for him to be able to explain to Winston what it's all about.'[234] Eden found Churchill unhelpful, fearing that he had again fallen under Stalin's spell. He had largely lost his earlier confidence in Truman, and in the conference sessions he clashed bitterly and frequently with Molotov, that 'most able but ruthless automaton'.[235] In an attempt to strengthen Churchill's will to resist Soviet demands, Eden sent him what was probably the clearest warning he had yet issued of Russian aggrandisement:

> The truth is that on any and every point, Russia tries to seize all that she can and she uses these meetings to grab as much as she can get. . . . I am deeply concerned at the pattern of Russian policy, which becomes clearer as they become more brazen every day.

What made the problem worse was that Soviet designs were not confined to Europe: 'Is their interest in the Lebanon the first stage to an interest in Egypt, which is quite the last place where we want them, particularly since that country with its rich Pashas and impoverished Fellahin would be a ready prey to communism?'[236]

The outlook seemed so hopeless that Eden succumbed to the private thought that the British general election might free him from his responsibilities. 'If it were not for the immediate European situation', he confessed, 'I am sure that it would be better thus, but that is a big "if", I admit.'[237] The British electorate duly obliged. With little progress made, the Potsdam Conference was suspended to allow the British delegation to leave on 25 July to learn the results of their election. Against the expectations of at least the Soviet leaders, it was Attlee and Bevin and not Churchill and Eden who returned.

*

Eden's wartime policy towards the Soviet Union has inevitably been judged in the light of what followed – four decades of armed antagonism between East and West which history has dubbed the Cold War. In such a context his continuing quest for Soviet co-operation and friendship, often in adverse circumstances, risked appearing, as he himself came to realize, at best as naïvety, at worst appeasement. Those who had to deal with the Russians as wartime allies would naturally have preferred in later years to be attributed with prescience about the coming Cold War rather than blamed for excessive optimism. But even Churchill was not as clear-sighted about the future Soviet menace as he later claimed. Until 1943 he harboured few anxieties, believing that the Russians would emerge considerably weakened by the war and in need of Western support and goodwill.[238] Eden wrote of Churchill's attitude:

> It has to be remembered that his record was written, tho' not his minutes, after Soviet policy had revealed itself in all its stark brutality.

W's attitude to Stalin in conference and conversation was not so consistently stern or firm. He fell at intervals under S's spell.[239]

But such words were as true of himself as they were of Churchill.

Yet this is scarcely a matter for reproach. Wise men may argue that the only satisfactory outcome of the war from Britain's point of view would have been the elimination of both Nazi Germany and the Soviet Union, and one or two said this at the time. In March 1940 Robert Vansittart was already 'convinced that we have got to aim at the destruction of both Nazism and Communism in this war, and that we shall not have won it or peace if we don't succeed'.[240] But such a goal never entered the parameters of practical politics. When the Soviet Union joined the conflict in June 1941, Britain's military fortunes were at a low ebb, victory more a long-term aspiration than an immediate objective. Thereafter the Red Army played the leading role in destroying the Nazi war machine as Soviet casualty statistics – and indeed those of Germany – eloquently testify. In winning over the Russians to at least a tolerable rapport with their new British allies of convenience, Eden played a major part. In November 1943, when relations were probably at their best, Bruce Lockhart recalled the situation before Eden's return to the Foreign Office:

> It was a period when most of Whitehall was anti-Russian; a period when prejudice seemed bent on driving us to self-slaughter. Certainly, at this time, policy played with the maddest fires. . . . If you hadn't come back to the Foreign Office, I should not like to envisage what Anglo-Russian relations would be today. I think anything might have happened.[241]

Lockhart later testified that he had seen no one handle the Russians more tactfully or more successfully. 'Perhaps it is because he has a dash of Slav blood in his veins.'[242]

It was of course only too evident that British and Soviet objectives in Europe were in many ways incompatible. Eden's aim was to strike a balance between satisfying Soviet ambitions in the interest of both wartime and, it was hoped, post-war co-operation, and doing his best to preserve the independence and strength of the states of Eastern Europe. It was inevitable, granted the disposition of available power and the way in which the war developed, that he would be more successful in the first objective than the second. Two views of the Soviet Union coexisted within the British war directorate as a whole and within Eden's own mind and, if he sought first to find evidence of Soviet goodwill, his was a less one-sided assessment than has sometimes been suggested.[243] If he usually took a more optimistic view of Russian intentions than did the military hierarchy and the Post-Hostilities Planning staff, Eden always recognized that he might be wrong. His underlying desire to think the best of Russia led him to adopt an equivocal and inconsistent attitude towards the future of post-war Germany. One historian has written of 'a long period of tacking backwards and forwards on the issue of German partition, which left both Allies and colleagues in doubt as to his real views'.[244] One half of him reflected the traditional Foreign Office view that Germany 'should eventually find a place again in the family of nations',[245] but, as has been seen, he remained wary of German resurgence and opposed to the inclusion of Germany in any anti-Soviet alignment.

If Eden's minutes and memoranda, his letters and diary entries convey no settled conviction about ultimate Russian intentions, he never actually concluded that the Soviet Union, as the leader of an ideological crusade, was an impossible partner for Britain.[246] Given the time for leisured reflection he might have reached a different, or at least more consistent, conclusion. Even before he became Foreign Secretary, Amery complained that Eden had 'very little gift of looking ahead'.[247] This complaint became the stock in trade of later critics.[248] But if the Foreign Secretary tended to develop policy from day to day, this was above all a function of the crushing burdens which he carried for most of the war, burdens which finally reacted on his health in the summer of 1945.[249] As Harvey recorded as the war came to an end: 'A. E. had latterly become quite exhausted. [He] could no longer look at the problems properly or read the papers about them. It had become mere improvisation.'[250] The formation of the Grand Alliance made it probable, if not certain, that Britain would achieve her military objectives; yet those in the realm of foreign policy became, paradoxically, more elusive. The basic shape of post-war Europe was decided on the battlefield and not in the conference chamber. Granted the probably insoluble dilemma which surrounded Britain's wartime diplomacy, it would seem harsh to judge Eden on the basis of relatively meagre results. Overall, his performance as Foreign Secretary between 1940 and 1945 probably marked the peak of his career.

8

The succession

The long era as crown prince was established, a position not necessarily enviable in politics.[1]

FOR a total of about seventeen years, punctuated by a short interval around the outbreak of the Second World War, Anthony Eden was widely regarded as Britain's 'next Prime Minister'. With such a lengthy record as heir apparent, first to Chamberlain and then to Churchill, he was understandably often described as a natural 'number two'. Such an evaluation is inevitably reinforced by the failings of his comparatively brief premiership and its tragic denouement. Yet it is unlikely that Eden himself would have endorsed such a judgement. For him the trials of waiting for the top job in British politics imposed a strain which at times he found almost impossible to bear. Foreign Secretary before the age of forty, Eden was in his late fifties and, many have argued, in a precarious state of health, by the time that he finally claimed his inheritance. The question of the succession was seldom far from his mind – nor that of his immediate entourage – for most of the fifteen-year period in which Churchill led the Conservative Party. This fact inevitably added complexities to the relationship between the two men to which neither gave public expression. At the same time the years of waiting revealed serious flaws in the character of Eden, the political animal. He was certainly ambitious enough to want to be Prime Minister. Yet, to the exasperation of his friends – and in a sense of himself – there were strict limits on what he would do to achieve this goal.

Aspirations towards the premiership were not particularly relevant when Eden re-entered the government as Dominions Secretary at the outbreak of the Second World War. His inevitable disappointment at failing to secure a leading post in the War Cabinet was no doubt underlined by an uneasy recognition that, in the public mind,

Churchill's star now outshone his own. As a former Foreign Secretary, Eden later described his new position as 'humiliating'.[2] This was an exaggeration. He stood at the top of the second rung of ministerial appointments, but his position effectively ruled him out of contention should Chamberlain fail to stay the course and a successor be needed. He was a 'constant attender' at the War Cabinet rather than a full member, but inevitably felt constrained when matters outside the immediate concern of the Dominions Office were discussed. Loyal Chamberlainites still saw in Eden a source of potential danger. His 'screen star attitude and moustaches, his mannerisms and faultless costume ensure[d] united enthusiasm among the flapper vote',[3] and it is certainly true that his loyal band of backbench adherents continued to meet in his absence. Indeed the Eden Group became a more effective watchdog of the government without its head than it had been with his presence. But others believed that Eden's moment had now passed. 'It was difficult to realise', noted Chamberlain's PPS after one unimpressive parliamentary performance in December, that 'one was listening to a speech by a man who had been Foreign Secretary.'[4]

His supporters were understandably angry and disappointed that the removal of Hore-Belisha from the War Office in January 1940 was not used to bring Eden into the War Cabinet.[5] While Chamberlain and Churchill formed a more effective partnership in government than has sometimes been recognized, the latter's relationship with Eden stayed relatively cool. On the one hand Eden remained conscious of the 'loose cannon' element in Churchill's make-up, while on the other Churchill was reported in February as insisting that he would rather have Chamberlain than Eden as Prime Minister 'by 8 to 1'.[6] The two men certainly clashed in this period over whether Britain could accept Eire's declaration of neutrality. In the spring, as Chamberlain's popularity began to decline, Channon judged that Eden was 'on the fringe of the plot and watching and waiting for his chance'.[7] But according to another observer he had no independent point of view and no intention of upsetting the political *status quo*. Caught up in the routine of government, he was in fact 'a very small straw in the current of events, with no ambition to be anything else'.[8] At all events there is no evidence that he played any significant role in Chamberlain's replacement by Churchill in May. From this crisis Eden emerged as Secretary of State for War, an office in which, as he recognized, it would 'not be easy to maintain harmony with W[inston]'.[9] This represented promotion, though he still lacked a formal seat at the War Cabinet table.

Eden's performance as War Secretary met with mixed assessments. On the whole he at least enhanced his popular position and moved nearer the centre of the government's war machine. It was at this time that he became, perhaps for the first time, a true follower of Churchill, convinced by his performance during the summer of 1940 that he was the only conceivable national leader. Even so, his admiration was never unqualified. Churchill's hours of work were not to Eden's liking and Eden could still voice doubts about Churchill's judgement and even express his view that 'Winston would be a better P.M. if he did not try to argue the details of war himself'.[10] By August, with Chamberlain ailing and Churchill concerned about the lack of support he was receiving from the War Cabinet, the Prime Minister frequently discussed with Eden a governmental reconstruction. The possibility was raised that Eden might replace Halifax as Foreign Secretary, though not in terms likely to tempt Eden to make the move:

> Would I prefer to go there? To him it seemed F.O. was contracting job and W.O. expanding. I replied I would prefer to stay where I was. Winston seemed relieved. 'I know where I can find another Foreign Secretary, he is here in the room with me, but where am I to find another S of S for War. Can anyone tell me that?'[11]

Chamberlain's advanced cancer finally compelled his resignation from the government at the end of September, necessitating a ministerial reshuffle and the election of a new party leader, for Chamberlain had retained this position after being ousted as Prime Minister. Eden expressed a readiness to do whatever Churchill wished. Part of him was happy to stay where he was, but he was also attracted by the idea of succeeding Chamberlain as Lord President of the Council with effective charge of the home front. 'Said Prime Minister wanted someone to hold his hand and he was clearly the one to do it and he could more or less stipulate to be on the Defence Committee and have a hand in settling strategy.'[12] Churchill, possibly influenced by the opinions of senior Conservatives, ultimately decided to appoint Sir John Anderson as Chamberlain's successor, leaving Eden where he was.[13] Other changes were kept to a minimum. The view of the dying Chamberlain that the supersession of Halifax by Eden would create trouble within the party was probably sufficiently valid to inhibit Churchill from a more drastic reconstruction.[14] As Eden noted, Churchill 'lamented that he could not give me F.O., and thus bring me into War Cabinet, and seemed distressed at this. "It is not what I want" he repeated many times.'[15] Churchill duly and inevitably succeeded Chamberlain as party leader, but the possibility that Eden might be nominated as his deputy was quickly nipped in the bud.[16] Yet if Eden had still to be fully accepted back into the party fold – 'I am beginning to make my peace with the Tory Party', he had noted in August[17] – he must also have realized that there was now a distinct dearth of rival contenders for the leadership. Churchill, moreover, was already sixty-six and pushing himself at a tremendous pace. 'If he is not careful Winston will be in his grave or a lunatic asylum,' warned one senior Tory a few months later.[18] At the time of Chamberlain's final resignation Churchill himself reminded Eden that he was now an old man, that he would not make Lloyd George's mistake of carrying on after the war and 'that the succession must be mine'.[19]

There seems little doubt that Churchill was determined to remove Halifax from the Foreign Office at the earliest possible opportunity. Halifax's readiness to contemplate a compromise peace in the dark days of the early summer had probably determined his fate in the long term. Over a range of issues, moreover, he had emerged as Churchill's leading antagonist inside the War Cabinet.[20] The sudden death in December of Britain's ambassador in Washington gave Churchill his chance. After briefly considering the appointment of the aged Lloyd George, the premier's mind turned to his Foreign Secretary. Halifax, after a vain and somewhat desperate attempt to persuade Eden that he was a more suitable candidate for the vacant embassy, was obliged to accept.[21] Eden now returned to the Foreign Office as a full member of the War Cabinet. He was not simply a Conservative appointment, but was liked and respected by most leading Labour figures, partners in Churchill's coalition since its formation.

Churchill was relieved by the change. Though, as has been seen, there had seldom been a total identity of views between Eden and himself, he was certainly more at ease with Eden than with Halifax. The former had at least managed to distance himself from the worst features of Chamberlain's appeasement policy. Two weeks into the new regime Churchill asserted, 'I think Anthony is putting his hand on the Foreign Office. I see a different touch in the telegrams.'[22] Indeed there were those who interpreted Eden's appointment as a sign that Churchill wanted to rid himself of all restraint in the control of the government. It had been Halifax's contention about his own position that 'what little use he might be here lay in restraining W. C.'[23] Harry Hopkins, Roosevelt's personal envoy, was among those who took a jaundiced view of Eden's promotion. 'I am sure the man has no deeply rooted moral stamina. . . . I fancy Churchill gives him high office because he neither thinks, acts – much less say[s] – anything of importance.'[24] Maurice Hankey also judged the governmental reconstruction as the moment when Churchill succeeded in surrounding himself with yes-men.[25]

This was certainly unfair to Eden who was never a mere cipher. He reacted strongly when the Australian premier suggested as much. In a moment of exasperation after a cabinet meeting in April 1941, Menzies asked, 'Has no one in the Cabinet a mind of his own?' Eden replied 'with doubtful modesty that I hoped I had!'[26] Having a surer grasp of many of the key diplomatic issues of the war than his master, Eden's view ultimately prevailed on a surprising number of occasions. He knew when to give way to Churchill, when to stand his ground and when to wage a subtle war of attrition to achieve his goal. Many years later Eden insisted that the war years were the hardest and proudest of his life. 'My friendship with Sir Winston was the closest political association that I have known.'[27] This was probably true enough, but the opening of public and private archives has shown that Eden's was at best a partial assessment. As has been argued above, there were real tensions about the man-agement of the war effort and serious differences over key areas of policy. Churchill always regarded himself as having ultimate control over Britain's wartime foreign policy and not just in its broad out-lines. This was a fact which caused Eden immense irritation but which he had ultimately to accept.

Eden tried to protect his position as far as possible. The decision that the exiled Halifax should retain a seat at the War Cabinet whenever he was back in London prompted him to remind Churchill that responsibility for advising the War Cabinet on the conduct of foreign policy rested with the Foreign Secretary alone. Alluding to his own unhappy experiences in the thirties Eden insisted that 'we can none of us wish to re-enter a period of divided responsibility'.[28] But the sharing of authority with Churchill was something Eden could do little about. What was often a difficult relationship could always fall back on a bedrock of mutual respect and affection. A little of Churchill's emotion and charm usually smoothed Eden's ruffled feathers. 'I regard you as my son,' he explained after one intimate late-night dinner. 'I do not get in your way nor you in mine.'[29]

*

What is striking is the speed with which the question of Eden's succession to the premiership re-emerged as a live topic of speculation once he had returned to the Foreign Office. This was the case both among those who welcomed the prospect and among those who did not. Some believed that Eden himself was obsessed by his own future prospects. His weakness, noted Edward Spears, was that he was always thinking about his career. 'Therefore he is not so bold as he should be, and too much inclined to conciliate those in authority.'[30] Menzies judged Eden not 'of sufficient tonnage' for the premiership, while according to Walter Monckton he was too conventional a thinker to make a great leader.[31] Wendell Willkie, Roosevelt's defeated opponent in the recent presidential election, considered that Britain would have to think up 'somebody better than that' to succeed Churchill.[32] Such a prospect, insisted R. A. Butler, was in any case remote since 'the Party will never follow him'.[33] But Eden loyalists were already preparing for the day when their man would enter 10 Downing Street. As early as February 1941 Cranborne was trying to enlist support to build Eden up for the succession, while a few months later Harvey stressed the importance of getting Eden's own people into key posts within the parliamentary party.[34] It was very much a repetition of the 1930s debate between those for whom Eden represented the hopes of the coming generation and those who found his qualities limited and insubstantial.

For much of 1941, however, consideration of Eden's future prospects was largely academic. This was essentially a function of the progress – or rather the lack of it – of the war effort. If Churchill's popularity wavered in the face of military setbacks, the same was yet more true of Eden. His extended trip to the Balkans and the Middle East, designed to rally the Allied forces, was largely a failure. While the Greek expedition began as Churchill's brainchild, responsibility for its collapse was increasingly attributed to Eden. Though it is probably an exaggeration to suggest that Churchill was ready to drop his Foreign Secretary to save himself, by April speculation was rife that Eden might have to leave the Foreign Office.[35] He scarcely improved his position by his performances in the Commons, especially that in May after the evacuation of British troops from Greece.

'He opened the Debate with an appallingly bad speech,' recorded Channon:

> no cheers greeted him and he gave a dim account of his travels and failures. He sat down amidst complete silence. I have never heard an important speech so badly delivered: and Duff Cooper said afterwards that the most damaging speech against the Government was Anthony's. . . . Winston looked uncomfortable and aware of Anthony's shortcomings.[36]

By the summer Churchill was even ready to mention Oliver Lyttelton, Minister of State in Egypt, as a serious competitor for the succession.[37]

The entry of the Soviet Union into the war in June provided a new and more promising focus of attention. The crisis for Eden soon passed. If Churchill's support had wavered, by the end of the year he was again making repeated assurances that the Foreign Secretary's claims on the succession were secure.[38] Churchill may have used this device as a means of limiting Eden's opposition on specific issues of policy.[39] 'It will not be long before you are in control,' he told Eden in November. 'Then you can do as you like about relations with the Soviets.'[40] Such tactics served their purpose from Churchill's point of view, especially as they had the effect of tying Eden's destiny firmly to his own. In the last resort Eden was readier to await his formal annunciation as Churchill's chosen successor rather than try to seize the crown for himself. But he was also undoubtedly encouraged by the Prime Minister's remarks to believe that the handover would come sooner rather than later. Indeed the collapse of Britain's position in the Far East, and in particular the fall of Singapore in mid-February 1942, left Churchill's wartime fortunes at their lowest ebb. A major political crisis seemed a distinct possibility. Harvey insisted that Churchill would have to reform his cabinet 'or he must be got to go'.[41]

No doubt encouraged by a Gallup Poll which suggested that half of those expressing an opinion believed that he would make the best successor to Churchill, and informed on a confidential basis of concern over the condition of the premier's heart, Eden even began to plan the details of his own cabinet. 'Long talk with A.E. today about the future,' recorded Harvey as the Far Eastern situation worsened.

> I tell him he must be prepared to take over. I think he is. He is remarkably calm about it. He said he would make Bobbety Foreign

Secretary and have a separate Minister of Defence. . . . He would get rid of the Beaver if he could. . . . He feels he is on good terms with Bevin who would work readily with him.[42]

Eden's own regard for Churchill's war leadership had certainly waned since the summer of 1940. The latter's refusal to create a separate Ministry of Defence was a source of particular frustration. 'So far as the conduct of the war itself goes', he informed Harvey, Churchill had become 'a nuisance'.[43] He even seems to have expressed his readiness to serve as Minister of Defence in a government headed by the socialist Stafford Cripps.[44] 'A.E. oscillates between wanting to be P.M. and wanting to stay where he is,' noted Harvey.[45]

If Eden was never prepared to engage in plots to bring about Churchill's political demise, he was fully involved in attempts early in 1942 to persuade him to mend his ways. These efforts were motivated by a desire to bring about the most effective possible conduct of the war effort. But the furtherance of Eden's own ambitions was an important secondary consideration. Thus he had doubts about the Labour leader, Clement Attlee, being given the title of Deputy Prime Minister as 'giving some blessing to idea of him as successor' and only seemed to develop an attraction for the post of Leader of the Commons after it had already been offered to Cripps, since it appeared to represent 'a stepping stone to being P.M. later' and he did not want Cripps 'to groom himself for Prime Minister'.[46] When Cripps accepted, Eden was left 'rather sore'. 'He doesn't want to lead H of C now, but he doesn't want either to prejudice his chances of the succession.'[47]

Churchill's cabinet reshuffle in February 1942 had little practical effect on the way government business was managed. What were largely cosmetic changes – 'the truth is that Winston does not want to change. He likes to move the pieces on the board'[48] – were played out in the growing expectation of a major political upheaval. Harvey sketched out a possible scenario:

The crisis may come in two stages, first an uneasy reconstruction with either a vice minister of defence or possibly a separate minister with Winston remaining, and then a later one involving Winston's departure. I don't know whether it is better to have two stages or one. . . . What I am wondering is whether you should become Minister of Defence, if it were offered to you, in the first stage. . . . I am afraid it would break down in the end . . . but it might be necessary to give it a trial.[49]

What was no longer as clear as it had appeared a few months earlier was whether Eden would be the beneficiary of such a crisis. The star of Stafford Cripps was now in the ascendant as briefly, but surely, this ascetic socialist emerged as a serious candidate for the premiership. Though Gallup still found in May 1942 that Eden was popularly regarded as the most suitable leader should anything happen to Churchill, Cripps now followed only three percentage points behind. According to Beaverbrook, if there were to be a change 'it would now be Cripps. Everything else was out of the picture.'[50] Similarly Channon noted that Eden's shares were 'very down just now' and that he had joined an ever growing band of ex-future Prime Ministers.[51]

At all events Eden was no more satisfied with the machinery of government after Churchill's changes than before and there is evidence that he gave momentary thought to resignation.[52] There was

> no day-to-day direction of war except by Chiefs of Staff and Winston.
> I would not object to this, if it gave results, but it doesn't. Chiefs of
> Staff only too readily compromise where issues should be decided
> and Winston's unchecked judgement is by no means infallible. When
> Cripps gets back [from India] he and I and Oliver will have to have a
> heart to heart and then tackle Winston.[53]

But Eden was always unlikely to take his own advice. He invariably found it easier to share his grievances with his Foreign Office subordinates than to bring them to Churchill's attention. Oliver Harvey, Richard Law, J. P. L. Thomas and others were fully aware of their master's exasperation, but the picture inside 10 Downing Street was probably rather different. According to Churchill's private secretary, Eden would frequently arrive to protest 'eyes ablaze and at the end of his tether' but, falling victim to Churchill's personal charm, would return to the Foreign Office 'soothed and relaxed'.[54] Almost certainly this placid demeanour lasted no longer than Eden's first encounter back at his ministry with an appropriate sounding board upon whom to vent his renewed wrath.

Two events of the early summer helped to ease Eden's disgruntlement. The first was his successful negotiation of the Anglo-Soviet treaty, for which he received widespread applause. Interestingly, his entourage encouraged him to view even this success in terms of his future career: 'I told A.E. that all this reinforces the need for him to go to U.S. in the autumn, as he is thinking of doing. He must not get into the position of being the Red Eden – to work the future he must stand equally well with Russia and with America.'[55] The second development was his formal anointment as Churchill's heir. On 16 June 1942 the Prime Minister informed the King that, in the event of his own sudden death, he should call upon Eden to succeed him. Paradoxically this made it even more difficult for Eden to be numbered publicly among Churchill's critics, of whom there was still a sizeable number, especially after the fall of Tobruk that same month. If anything, Eden may now have felt it necessary to align himself more fully with Churchill and by July Harvey was complaining that Eden should see more of his governmental colleagues and avoid the risk of becoming as isolated as he had been at the time of his 1938 resignation.[56] Eden explained his position to Harvey. He would

> do nothing against Winston, now or ever, and he would be ready to serve under any of his colleagues. I took him to task rather for this passive attitude, but he said he didn't wish to appear in any way as intriguing. He thought his position was pretty secure in actual fact in the country and also owing to the P.M's famous testament . . . recommending that the King should send for him if anything happened to himself. . . . He feels himself that he hasn't any serious rival.[57]

Cripps's ascendancy passed almost as rapidly as it had arisen. By the second half of 1942 his power was evidently weakening and his challenge to Churchill effectively dissolved after Montgomery's victory at El Alamein in November. Thereafter the premier's position was never again under serious threat for the remainder of the war. Despite Eden's earlier apprehension, he and Cripps had worked well together. As the latter recognized, their joint resignations might have led to Churchill's own downfall, but Eden 'won't think of it'.[58] Cripps himself left the War Cabinet in October, exchanging the leadership of the House for the Ministry of Aircraft Production, an important but less politically sensitive position. The vacant Commons leadership was taken by Eden. The combination of this new post – at which he proved an undoubted and popular success – with the Foreign Secretaryship, membership of the War Cabinet and his growing role as Churchill's midnight confidant left him unquestionably the government's number two figure. Whenever it should come, the succession was indubitably his.[59] Even Beaverbrook, of whom Eden remained warily suspicious, now subscribed to this view.[60] Harvey noted in August that Eden's interest in the premiership was 'increasing all the time'. Six months before he had been one of several alternatives; now 'he only could succeed Winston'.[61] But the accumulation of offices which he had acquired by the end of the year, though no doubt to his political advantage, was scarcely conducive to the effective conduct of government business, of which Eden himself had been so critical only a few months earlier:

> He must spend all day in the H of C three days a week, and mostly on the bench itself. He has no time to see Ambassadors and others, which it is part of his work to do, and paper work duly piles up all day. . . . It may be and probably is good for him politically to gain this further experience. . . . But it is bad for the Office and for foreign affairs which do not get that care and reflection they require.[62]

Not surprisingly, Eden soon became 'more tired than I have seen him for a long time'. The obvious solution was to delegate some Foreign Office work, especially as Richard Law, his junior minister, was devoted to Eden personally and represented no political threat. But Eden was characteristically reluctant to give up any authority because 'he hates to let the controls go out of his hands even for a minute'.[63]

Tiredness, in fact, provides the most convincing explanation for Eden's equivocal response to the suggestion early in 1943 that he should take over the Indian Viceroyalty from the outgoing Lord Linlithgow. His failure to reject Churchill's offer out of hand seems in retrospect as surprising as the premier's readiness to make it, especially as Eden had just returned from a successful visit to the United States where he had 'put himself across everywhere . . . as the successor to the P.M.'[64] By 1943 Churchill's position was much more secure than it had been and he probably felt that he needed Eden less than at any time over the previous two years. He may also have concluded that life would be more agreeable with a more pliant Foreign Secretary than the present incumbent. Eden himself believed that Churchill's readiness to see him become Viceroy was partly due to his growing desire to take the management of foreign policy into his own hands.[65] As so often in his career when confronted by a difficult personal decision, Eden was 'pathetically anxious to canvass opinion'.[66] Had the advice received been of one mind this peculiar episode might not have dragged on through the first months of 1943. The reaction among Eden's circle in London was a combination of alarm and disappointment. Not for the first time in his career his supporters seemed to hold higher hopes for Eden and to have a clearer picture of how these could best be realized than he did himself. As Harvey put it: 'I hope no one will take this seriously. For A.E. to go would be to lose all he has now gained – to miss the P.M.ship, to miss the vital peace-making years, to confound his friends and confirm his critics.'[67] But he found Eden attracted by Churchill's supplementary offer of the direction of the war in the Far East and understandably influenced by the Prime Minister's assertion that he now hoped to 'go on some years yet'.[68] Cranborne, Orme Sargent and Lord Baldwin all counselled against Eden's acceptance.[69] But the decisive intervention appears to have been that of the King who told Churchill that Eden's qualities were too important for the war effort, however valuable might be his contribution in India.[70] Thereafter the idea gradually faded and the succession to Linlithgow finally passed to General Wavell.

With this particular danger evaded, Eden's political circle expended much effort in the last months of 1943 to persuade him to reduce the burden of his ministerial responsibilities. 'You are a prodigious worker', conceded Thomas, 'but even you cannot face this position and we must recognise the fact and deal with it now.'[71] According to Law the pressures upon Eden's time meant that many important issues were going by default. Leadership of the Commons was an indulgence which the country could not afford now that it was 'seriously threatened with an outbreak of peace'. Omissions on Eden's part at this stage might mean 'sowing dragon's teeth which will be harvested five or ten years from now'.[72] But Eden was temperamentally disinclined to delegate responsibility and the problem remained unresolved.[73] By early 1944, however, concerned at the growing unpopularity of the Foreign Office, Eden seemed ready to relinquish this post altogether to Cranborne to concentrate on the Commons. 'He wants to get out before the criticism of the F.O. affects his chances of the succession,' noted Bruce Lockhart.[74] To make it clear that the change should not be seen as demotion, Eden hoped to take the title of First Lord of the Treasury from the Prime Minister and thus 'become the obvious coadjutor and successor'.[75] He may also have been influenced by renewed doubts over Churchill's health following a serious bout of pneumonia at the end of 1943. Eden dined with Cranborne on 6 March when the latter pronounced that the only hope for the future of the Conservative Party lay in Eden's leading it and that if he went on to the home front for a brief spell

this would 'make the transition all the easier'.[76] Churchill, however, was not attracted by the prospect of working closely with
Cranborne. That effectively brought this particular scheme to an
end, for it was not the sort of issue upon which Eden would
challenge Churchill's authority. 'W muttered to me as we joined
the others: "You will have to go on just as you are for a few months
longer".'[77]

Before long, however, Eden had swung around full circle and was
suggesting that he might give up the leadership of the House, either
to R. A. Butler or to John Anderson, and concentrate upon the
Foreign Office. Exhaustion was a factor in his vacillation. 'I am so
weary', he confessed, 'I hardly care so that I am released from one.'[78]
Observers were understandably sceptical. 'I shan't believe it till I see
it,' recorded Harvey, while, according to Brendan Bracken, Eden
'never knows his own mind, gets up with a new idea every day
and abandons it as soon as he has to risk trouble with the P.M.'[79] His
inability to bring the matter to a head appeared somewhat pathetic.
'I feel – as I have always felt', noted Lockhart early in 1945, 'that he
will go on talking about this problem and be able to do nothing – at
least as long as Winston remains.'[80] This prediction proved accurate.
Eden retained his almost unendurable combination of offices and
functions for the remainder of the war.

He was sustained by growing confidence that nothing could now
stand in the way of the fulfilment of his own ambitions – and
probably without too long a delay. He still, of course, had his critics
and detractors. Beaverbrook had turned hostile once more while
right-wing Conservatives were pressing the claims of John
Anderson, the Chancellor of the Exchequer.[81] Randolph Churchill,
who probably saw Eden as having usurped his own rightful place at
his father's side, favoured Harold Macmillan. 'Anything is better
than Anthony,' he assured Diana Cooper.[82] According to the *Sunday
Express* many Conservatives were not sure that Eden was the man
they wanted to see succeed Churchill. They still felt that he was too
remote from the organized life of the party. His lack of interest in
home affairs was a weakness waiting to be trundled out, together
with the feeling that he was not quite 'the immaculate Tory' they
were looking for in their next leader. R. A. Butler, Oliver Lyttelton,
Richard Law and R. S. Hudson were all mentioned as possible
alternatives.[83] In practice, however, such speculation – the stuff of
politics – is more significant as an indication that the succession to
Churchill was a live issue than as evidence that Eden's claims were
under serious challenge. Any attempt to divert what was now an

almost pre-ordained pattern of events would have come up against tremendous odds, not least a huge groundswell of popular opinion. In a poll taken in April 1945 as many as 84 per cent of those questioned named Eden among the six most successful members of the wartime government. A year earlier Gallup found that 55 per cent of the population wanted Eden to succeed once Churchill left office. Cripps, once a serious rival, still came in second place but with a mere 5 per cent support. By October 1944 the *News Chronicle* reported that Eden enjoyed the backing of 21 per cent of the electorate for the leadership of the post-war government with 24 per cent still supporting Churchill. Attlee, in third place, commanded just 7 per cent. Before the end of the year Eden had overtaken the Prime Minister as the popular choice to lead the country once the fighting was over. In January he opened up a lead of 11 percentage points over Churchill with no other candidate in sight. The future Conservative leadership was not, of course, at the disposition of the electorate, but such statistics were difficult to ignore.

*

The outcome of the 1945 general election had obvious implications for Eden's position within the Conservative Party. He had faced the prospect of victory without real enthusiasm, not least because such an outcome would probably have produced no changes in the highest reaches of the government. 'Am beginning seriously to doubt whether I can take on F.O. work again,' he confided in July. 'It is not work itself which I could handle, but racket with Winston at all hours. He has to be headed off so many follies.'[84] Harvey believed that Churchill's interfering ways would have killed Eden before long had the Conservatives stayed in power.[85] But a Labour victory, with an overall majority of nearly 150 in the new Commons and every prospect of a full five-year term for the new government, was an altogether different proposition. After an emotional final meeting of the Churchill cabinet the Prime Minister spent half an hour alone with Eden. When Churchill declared that he, unlike Eden, would never sit in the cabinet room again, he seemed to be stating no more than the obvious. Churchill was now nearly seventy-one years old. By all logical calculations it was unreasonable to suppose that he would hold office again. The British electorate's verdict, if it seemed to many to lack gratitude, had surely brought his long political career to a close. As Eden put it, 'while there is much gratitude to W as war leader, there is not the same enthusiasm for him as P.M. of the peace. And who is to say that the British people were wrong in this?'[86]

Eden knew that Churchill would find it difficult to relinquish his hold on the reins of power. But what he and others failed to appreciate was the old man's burning desire to win a general election in his own right. After 1945 the Churchill–Eden relationship entered a new phase. During the war, whatever the strain and notwithstanding moments of real doubt about Churchill's management of the government, Eden had in the last resort accepted that Churchill was the right man in the right place at the right time. With the end of hostilities and the Conservatives' loss of office, such reasoning no longer applied. Questioned many years later about the restrained account in his memoirs of his difficulties with Churchill after 1945, Eden conceded that Churchill's decision to stay on after the war had been 'so far as one can be objective' a mistake.[87]

To begin with, Eden was torn between elation at his newly found freedom and a desire to get hold of the Conservative Party and mould it in his own image. He took comfort that no one could hold him responsible for the disastrous tone of the party's election campaign, from which he had been largely absent through illness, and from evidence that his own reputation remained high. According to Lockhart, everyone under fifty now regarded Eden as the Conservatives' only hope.[88] There was the fear that Churchill would 'stay on and get everything wrong', but few imagined he would have the stomach for five years of opposition politics.[89] 'Don't trouble yourself about that,' Eden assured a sceptical Thomas Barman, anxious about the 'curious group' surrounding Churchill, 'the old man won't be there much longer. The people you have in mind will go with him.'[90] But old politicians are less adept than their military contemporaries at simply fading away. 'He ought to resign now and write', argued Eden at the end of August, 'but he won't.' The ex-premier's moment had passed and he was not likely to evolve into an effective opposition leader:

> Winston is not suited to the times; does not know or understand the public; does not even know the House of Commons intimately. Comes in for the big occasions and is good. But never attends to the routine work; does not know individual members.[91]

Convincing Churchill of these facts was, however, a different matter. The danger existed that, if the older man remained in place for too long, Churchill's would not be the only reputation to suffer. Eden's might be dragged down by association.

Towards the end of 1945 Churchill embarked upon an extended visit to the United States, obliging him to hand over to Eden the day-to-day conduct of the opposition in the Commons. Churchill hinted that, if Eden performed well in his absence, he might formally give up the leadership on his return. Such uncertainty was scarcely to Eden's liking. He looked to the party to 'bell the Winston cat', in other words to tell him that it was time to go, but would not act himself. 'Anthony says Winston and he have been through so much together that he himself will not do it.'[92] In this unsatisfactory position Eden became excited by the prospect of becoming the first Secretary-General of the United Nations. 'He was almost like a schoolboy in his exuberance.'[93] It was very much a re-run of the Indian Viceroyalty episode of 1943, especially in his 'pathetic remark' that after three or five years with the UN he could come back to lead the party.[94] A new factor, however, was the possible impact of Eden's decision upon his now ailing marriage. J. P. L. Thomas explained:

> The whole trouble . . . was Beatrice. She was in love with this fellow [C.D.] Jackson and wanted to marry him. She had a bargain with Anthony. If Anthony remained in politics and had a chance of going to No 10, she would not have a divorce or make a scandal. If he went to UNO, that would be a different matter. Naturally she preferred UNO.[95]

Once more Eden's circle sought to strengthen his resolve. Baldwin concluded that, if Eden went to the UN, Butler would become party leader.[96] Cranborne, privately avowing that if Eden did this he was 'through with him', tried flattery. It would, he said, be an absolute catastrophe if Eden left British politics at the present time:

> You, and you alone, can hold that portion of the centre vote on which we shall have to depend absolutely if we are to win the next election. . . . Moreover, I am greatly afraid that your departure might well mean a new split in the Conservative Party. People like Dick [Law] and many on the backbenches and even more in the country, faced by an undiluted dictatorship by Winston, backed by James [Stuart] and Ralph [Assheton], would simply refuse to play. . . . I quite see the odious position in which the present situation puts you. But personally I do not believe that it will last long. In my view, Winston's day

is over. . . . He now belongs to the past, and even he is bound to find this out.[97]

In the event the decision was taken out of Eden's hands. Despite considerable lobbying on his behalf there was never much chance that he would secure even the nomination of Attlee's government, let alone the Secretary-Generalship itself.[98] The post went to the Norwegian politician, Trygve Lie.

With this issue resolved, Eden seemed 'quite keen' again on the idea of the party leadership.[99] Everything depended, however, on what Churchill decided on his return from America. His famous anti-Soviet speech at Fulton, Missouri, in March 1946, which clearly took Eden by surprise, scarcely suggested an early withdrawal from the political limelight. Eden at least took comfort that this latest initiative might indicate that Churchill was less interested in leading a parliamentary opposition than an anti-Russian crusade independent of the Conservative Party.[100] The two men dined together on Churchill's return. Churchill made it clear that, while he might be ready to relinquish control in the Commons, he intended to retain the overall party leadership and to lead the Conservatives at the next general election. Plagued by the recurrence of a duodenal ulcer, Eden was understandably depressed. The effect upon his nervous system was becoming apparent. 'Several times', noted Lockhart, 'when we were discussing what should be done and what he wanted to do, he said, rather testily: "Oh God, I do wish the old man would go".'[101]

By early April Churchill − possibly at the prompting of Beaverbrook − even appeared to withdraw his earlier offer of dividing the leader's functions. An exchange of letters with Eden showed how completely he still dominated their relationship. Churchill said that he had concluded that it was best not to make a formal change at the present moment. His health was much improved, he had more strength and he knew that he could count on Eden's help 'in an ever increasing measure'. As so often over the previous five years the carrot of the succession was dangled tantalizingly before Eden's eyes. Churchill was 'most anxious' that matters should be so handled as to make 'the formal transference, when it occurs, smooth and effectual'. Though no date was mentioned, the tone suggested an early handover. Neither man could have believed that a further nine years would elapse before Eden took charge.[102] By contrast, Eden's reply sounded formal and stiff:

> Thank you so much for your letter in which you tell me that you have reconsidered the position since our talk and would now like to carry on with the leadership of the Opposition in the House. As I told you the other day this is entirely agreeable to me and I am delighted at the doctors' good report.

But there was, he ended, no room for the post of a deputy opposition leader in the Commons.[103] Churchill could now make Eden's reluctance to deputize for him whenever he was away appear an act of selfish petulance.[104] As usual, Eden gave way, agreeing to step in when Churchill was abroad and to continue to preside over meetings of the chairmen of the party's parliamentary committees.[105]

Eden could only soldier on. There was always a chance – 'more of a fervent hope than a likely expectation' – that Churchill might use his party conference speech to announce his resignation.[106] Eden could hardly bring himself to believe that Churchill would still be leader by the time of the next election, but the whole business was 'so exhausting that one is seriously tempted to accept pleasant city tasks that would allow one a leisured life'.[107] By the autumn, and with few signs of Conservative revival, he had determined on a new approach. Through a series of speeches he would seek to outline his version of Conservatism. Besides giving himself an opportunity to develop his interest in domestic politics, this would enable the party to decide what – and whom – it wanted. 'WSC has no policy except to change the name of the Party.'[108]

By 1947 Churchill had become little more than a part-time and semi-detached leader of the opposition. His main efforts were concentrated upon his war memoirs and even during the parliamentary session he seldom spent more than three days a week in London. From Eden's point of view this was most unsatisfactory:

> WSC does not come near the House for several days, does not attend the Shadow Cabinet, does not know what line of conduct has been decided and then arrives to say that he is going to speak. Anthony says that WSC has aged very much and is at times almost 'gaga'.[109]

Towards the end of the year several shadow cabinet members, significantly not including Eden, made concerted moves to persuade Churchill to stand down. The Chief Whip, James Stuart, was given the unenviable task of suggesting to the now seventy-three-year-old leader that it was time to step aside for a younger man. But little could be done in the absence of any formal procedure to change the leadership and in the face of Churchill's stubborn insistence that he would stay on.[110] Similarly, Macmillan's attempt early in 1948 to convince Churchill of the need to adopt a more business-like approach to the tasks of opposition met with a brisk rebuff: 'I do not agree with what you propose, and I do not think our colleagues would either. I propose to continue the present system as long as I am in charge. I do not think things are going so badly.'[111]

The crises of 1947, domestic and international, encouraged the belief that the Labour government, despite its overwhelming parliamentary majority, might not last its full term. There was serious talk of a coalition, in which Eden's claims to the premiership would be strong, granted the cross-party support he had long enjoyed.[112] Surveys made at Conservative Central Office showed that Eden was widely admired by Labour voters.[113] The government, however, survived and Eden was left to take a somewhat macabre interest in the apparently continuing decline in Churchill's health. By the end of 1947, his physician later recorded, Churchill was 'sliding, almost imperceptibly, into old age'.[114] Eden was sent a fuller picture: 'There is a change in him physically and beneath the forced Herculean efforts to work and the ubiquitous gaiety there are signs of acute strain and of a general break up.'[115] In the summer of 1949 Churchill suffered a stroke. A few months earlier J. P. L. Thomas reported that his powers were decaying rapidly and that he would not last until the general election. Eden 'accepts this testimony as the best way out of a horrible impasse'.[116] Yet to be sustained by expectations of Churchill's physical collapse was far from ideal and there were moments when Eden again seriously considered abandoning politics altogether.[117] The longer an intolerable situation continued, the greater was the temptation to leave public life to secure his own financial position in the City. By any logical calculation, however, the longer Eden could endure his ordeal the sooner Churchill's retirement was bound to be. Having come so far and waited so long, part of Eden always recognized the sense of hanging on that little while longer.

In the event Churchill was still at the helm when Attlee called an election for February 1950. From Eden's point of view the outcome was the worst of all possible worlds. A clear Labour victory would have convinced Churchill that he must now retire, his goal of forming a peacetime administration finally frustrated. A Conservative victory, satisfying his ambition, would probably have enabled him to leave office after a renewed but brief taste of power. But the narrow return of the Labour government not only persuaded

Churchill to stay at the helm for yet another, if presumably brief, parliament, but encouraged him to adopt political tactics of which Eden strongly disapproved. As Lockhart explained:

> He disapproves very strongly of Winston's parliamentary tactics (sleepless nights and get the government out at any cost). . . . With old age and declining power this haste on [Churchill's] part is for him a personal race. He knows that, unless he can get the government out quickly, time will prevent him ever being P.M. again.[118]

After six years of Labour government, therefore, Eden still stood, where he had been a decade earlier, as Churchill's natural successor. Just before the general election of 1951 Gallup revealed that even among Conservative voters a majority would prefer to see Eden as party leader. Yet this meant little when set against Churchill's still iron determination. To the outside world Eden maintained the façade of the loyal lieutenant, anxious to do his master's bidding. 'About Winston', recorded Malcolm Muggeridge in April 1951, Eden 'said gravely that his health was not as good as people thought, and that he could speak of this with an easy conscience because, as far as his own private inclinations were concerned, he'd vastly rather be Foreign Secretary than Prime Minister.'[119] But those closer to him recognized the extent to which waiting for the succession was telling on Eden's nerves.[120]

At least his position as Conservative number two was never seriously threatened during this period. Harold Macmillan's standing certainly rose during the years of opposition and there were those who came to believe that he possessed a sharper intellect than Eden.[121] Eden himself remained wary of Macmillan, as he had been during the war, fearing perhaps a potential rival, and he seems to have helped dissuade Churchill from appointing him as party chairman in succession to Ralph Assheton in 1946.[122] That post went to Lord Woolton and in later years Eden heard that he too had once fancied his chances to become leader.[123] In a sense the new intellectual climate of Conservatism in the late 1940s, epitomized by Butler and the Research Department, worked to Eden's disadvantage, notwithstanding his own attempts to broaden his competence in domestic political issues. It was 'the age of the don, the economist, the man of outstanding intellectual ability'.[124] And there were always those on the party's right wing who would never welcome Eden's accession to the leadership,[125] though their influence was probably balanced by his ability to appeal beyond the confines of his own party. As has been seen, even Eden's most loyal supporters were sometimes exasperated by his reluctance to bring matters to a head. 'If he wants to be regarded as a leader', snapped Cranborne, 'he must act like one.'[126] His longstanding ally, Arthur Mann, of the *Yorkshire Post*, even warned that, by ignoring Eden's speeches, the BBC were unwitting partners in a plot to keep him from his just deserts.[127] For the most part, however, the general perception was that Eden had done little to damage his own claims. Lord Derby, who had watched more Conservative leaders than most, probably expressed the majority opinion: 'I have never had great faith in him but I do see now that he does make himself into a really possible successor to Winston.'[128] But when Winston would need a successor was entirely another matter.

*

Following the October 1951 general election a Conservative government was installed at Westminster with Churchill as Prime Minister and Eden, inevitably, as his Foreign Secretary. Churchill's intentions on re-entering Downing Street can only be guessed. It must have galled Eden in later years to be told by Christopher Soames, Churchill's PPS throughout his final premiership, that the old man did not enjoy his job and was 'often bored':

> Christopher told me how sometimes, when over in H of C after questions, he would telephone Christopher to come to his room, tell the Private Secretaries he was not to be disturbed, then produce two packs of cards and play whatever was his favourite game of the moment with C.[129]

It seems unlikely that Churchill expected to remain in office for three and a half years until he was in his eighty-first year. According to John Colville, he had it in mind to serve as Prime Minister for one year only before handing over to his loyal lieutenant.[130] This seems also to have been Eden's expectation, and not necessarily one which much enthused him, since by then the shine of electoral triumph would have vanished, leaving only the practical problems of running the country.[131] But Churchill's conviction of his own indispensability grew rather than diminished with age and there was always, in his own mind at least, good reason to postpone his departure. There is evidence to suggest that nagging doubts about Eden's suitability for the succession may have been a factor,[132] though Churchill would probably have felt the same about any nominated successor, and any hesitations never reached the point where he seriously considered another candidate. In retirement Butler, in particular, liked to play up the idea of Churchill's doubts, implying that the premier had begun to consider him a more suitable replacement, with the succession remaining an open question until just before Churchill's retirement.[133] This was an exaggeration and caused Eden considerable irritation:

> Up to the last meeting which took place, according to Butler, between Churchill, Butler and myself and which was held to decide that Churchill must go, the issue of a successor was still unsettled: 'As we went into the room, neither of us had any idea which would finally be chosen.' I need hardly say that no such meeting ever took place.[134]

Eden, however, was certainly nervous about his position as heir apparent and sought from Churchill the reassuring title of Deputy Prime Minister.[135] He did see Butler as a possible rival, especially when the latter's stock rose through his generally successful stewardship of the national finances. Favourable publicity for the Chancellor did not necessarily meet with the Foreign Secretary's approval and Eden and his second wife, Clarissa, even seemed ready to use the Foreign Office to conduct a propaganda battle against Butler.[136] It was in the minds of other senior Tories, rather than of Churchill himself, that the strongest doubts were felt about an Eden succession, doubts which intensified the longer Churchill hung on. 'Eden would make a very terrible Prime Minister,' judged Beaverbrook.[137] Macmillan's assessment was more measured: 'If he had the first place, he might easily rise above all these faults. It may really be that he has been Prince of Wales too long.'[138] Objectively, such verdicts suggested less that Eden might yet be thwarted of his ambition than that, when his time finally came, he would still need to prove his fitness for the highest office.

With Churchill and Eden occupying the same positions as between 1940 and 1945 there was a strong element of *déjà vu* about the new government. Indeed, both in his cabinet appointments and in his attitude to key policy issues such as great-power relations, it appears that the Prime Minister was rather pathetically trying to recreate the atmosphere of the war years. From Eden's point of view, however, there was no possibility of going back to the relationship of that period. By the early 1950s, foreign policy was once again an entity in its own right and, according to Eden, the preserve of the responsible minister. He was now much less tolerant of the Prime

Minister's intrusions than he had once been, especially as Churchill's judgement seemed altogether less sound and his capacity to grasp the essential issues substantially reduced. 'It was very galling to A.E.', noted Evelyn Shuckburgh, now Eden's private secretary, in November 1952,

> when he returned to London and reported to the Prime Minister that all Winston had to say was 'What have you done about getting the Duke of Windsor invited to the Embassy in Paris?' This sort of comment, which is constant, and the irrational interference in all sorts of small matters . . . is making A.E.'s position most difficult and irritating for him.[139]

Anthony Seldon, in his pioneering study of Churchill's last government, broadly concluded that the Prime Minister was equal to the demands of his post, at least until his stroke in June 1953 and possibly until as late as October 1954.[140] Yet it is doubtful whether Eden – or indeed other leading cabinet members – would have shared this assessment. Harry Crookshank, the Leader of the Commons, described Churchill some time before his serious illness as 'terribly drooling . . . fast losing his grip'.[141] Eden in fact no longer believed that Churchill was physically or mentally up to his job.[142] After Churchill's stroke that conviction was doubly confirmed.[143]

All this made for an increasingly difficult partnership between the government's two leading figures. Those subjected to one of Eden's periodic outbursts of pent-up emotion no doubt got an excessively jaundiced view. 'Chips' Channon was one. In a 'sudden burst of nerves and temper' and 'almost hysterical', Eden poured forth on his problems. 'I get all the knocks; I don't think I can stand it much longer.'[144] Such displays had always been part of Eden's make-up, but now became significantly more frequent. At best the Churchill–Eden relationship was one of 'affection tempered with tension'.[145] At worst the affection was difficult to discern. Yet still, despite his advancing years, Churchill seemed to hold the whip hand. He would 'sometimes tease him, sometimes pet him, occasionally thwart him'.[146] Dean Acheson, the American Secretary of State, recalled with amusement an occasion when Churchill ribbed Eden over the former's plans to transform the view from the garden of 10 Downing Street.[147] With hindsight the whole saga of the succession appears in the same light, with Eden dangling helplessly at the end of a string, Churchill still the master puppeteer, entirely in control of events. Be that as it may, Eden's obsession with the succession was scarcely

conducive to the effective management of his departmental responsibilities. Not only did he and Churchill clash over key policy issues,
but the conviction – or perhaps just the hope – that his own tenure
at the Foreign Office would be relatively brief encouraged that short-
term, *ad hoc* approach to diplomatic problems, neglecting a more
considered strategical analysis, for which he was widely criticized at
the time and since. Matters were not helped by the suspicion, in
Eden's mind and that of his closest circle, that figures near to the
Prime Minister, if not Churchill himself, helped to orchestrate a
press campaign against the Foreign Secretary.[148]

Churchill seems to have had some misgivings about his capacity to
carry on as early as February 1952, but judged that the unpopularity
of austerity measures made it a bad time for Eden to take over.[149]
Two months later concern was being expressed about his lack of grip
and some believed that Eden's presence was needed on the home
front to improve the government's performance.[150] Churchill, however, now expressed the hope of being able to work with Eisenhower
after November, when the General was expected to succeed Truman
as American President.[151] By the autumn, with the Prime Minister
apparently impervious to the promptings of the Chief Whip that he
should now retire, Eden was 'very impatient' to take control.[152]
Churchill assured his niece Clarissa – who was also Eden's new
wife – that he was merely looking for an opportunity to go, providing only that he was allowed one last visit to the United States as
head of the government.[153] Salisbury, who had grown understandably suspicious of Churchill's manoeuvres, urged Eden to press the
old man about his intentions. If this did not work, senior cabinet
ministers, excluding Eden, would need to discuss a formal
approach.[154] In the event Eden could only elicit a vague declaration
that, *when the time came*, Churchill would 'hand over his powers and
authority with the utmost smoothness and surety'.[155] The premier
had become 'an increasing liability', which it was singularly difficult
to remove.[156] With Churchill openly criticizing his policy towards
Egypt, Eden even considered retiring from politics at the time of the
coronation.[157]

*

Eden was often described as a lucky politician. If so, his good
fortune ran out early in 1953. He had been feeling unwell for
some time and X-rays revealed gallstones. There is no need to add
to the account of Eden's condition and the three operations he

endured in the course of the year, as chronicled by his official biographer,[158] though his own words offer a convenient summary:

> What Brendan used to call 'your tough north country constitution' could not make up for the accident of the knife which slipped. . . . It necessitated a second operation which was almost fatal and during which I was saved by the junior surgeon, Blackburn, while the senior kept repeating: 'Blood, I have never seen so much blood.' It was . . . [the American surgeon] Cattell who . . . saved my life [with a third operation in Boston]. . . . Moreover, this operation must be a long one and difficult. (C[larissa] says I was eight hours in the operating theatre.)[159]

The result was that Eden would have been in no position to become Prime Minister for most of 1953, even had Churchill been minded to retire. As it was, the old man's fortunes seemed, astonishingly, to be reviving just as Eden's own health collapsed. 'For some time he has been hanging on to power while his strength was failing,' noted Moran in February. 'And now, without any apparent reason, he seems to have taken a new lease of life.'[160] In Eden's absence Churchill himself took charge of foreign affairs and, to the alarm of the majority of his cabinet, called on 11 May for a summit conference of the great powers. On the evening of 23 June, however, Churchill suffered a severe stroke which, for a man of his age, seemed certain to end his career.

Removed from the centre stage by his own illness, Eden relied on others to keep him informed. Colville wrote confidentially on 26 June in terms which would have made it very difficult for Eden to act, even had he been in a physical condition to do so:

> He intends to remain P.M., devolving his duties on others, for one reason only – namely that Mr. Eden should have the time to recover fully without in any way hurrying the process, and to return to take over from him. This he has discussed with me on several occasions and said as much at luncheon today to Bobbety and to Rab – both of whom expressed agreement and approval. Should he deteriorate, and have to resign, I think (with pretty good reason) that the Queen will ask Bobbety to form a Caretaker Government for six months.[161]

In the circumstances Eden could ask no more than this. The extraordinary device of an interim premiership held by his closest political ally would keep the Prime Minister's seat warm until he was ready to take it over himself. Butler reported an instinct for 'definite survival' on Churchill's part but confirmed that Eden's interests were safe. There was no need to 'rush you out of your well earned convalescence'.[162] J. P. L. Thomas agreed that, while Churchill might live on for weeks, months or even years, active life was over 'unless a miracle happens'.[163] Yet happen it did. By early July Colville was writing of a 'fantastic improvement'. Hitherto the best that could be hoped for seemed to be 'partial, but crippling paralysis'. Now it seemed that Churchill could remain in office without undue strain until parliament reassembled after the summer recess.[164] Though Salisbury still thought it unlikely that the premier could return to active political life, Churchill himself had decided against taking any decision until September.[165] Colville concluded that there was no certainty that he intended to retire. 'I surmise that he still hopes to bring off some final triumph.'[166]

Eden's staff knew how unwelcome such news would be. Shuckburgh took it upon himself to warn the Foreign Secretary of the marked change in the tone of private comments emanating from Chartwell. With Colville suggesting that it would be unwise for Eden to broach the subject of the prime ministerial handover when he returned from his convalescence, Shuckburgh concluded that the situation might 'still want some careful handling'.[167] Salisbury agreed, knowing, it seemed, which way a clash between the two men on this most delicate issue was likely to go: 'I have been through all this a hundred times. The fact is that the P.M. is much tougher than Anthony. He very soon brings Anthony to the point beyond which he knows he will not go and then he has won the day.'[168] When Prime Minister and Foreign Secretary met, a 'fragile' Eden followed this advice and avoided mentioning the succession.[169] It was clear that the old man would be 'difficult to shift'.[170] Though Churchill seemed to contemplate carrying on until the Queen returned from a tour of Australasia the following spring, Eden hoped either that the party conference would reveal his incapacity or that he would be unable to cope with the day-to-day demands of government once parliament reassembled in October. 'I would bet any money', insisted Thomas reassuringly, 'that the public appearances just can't be done without something awful happening.'[171]

As all turned on estimates of Churchill's staying power, Shuckburgh suggested that Eden might arrange to call at Beaverbrook's villa in the South of France, where the Prime Minister was holidaying, to form his own assessment.[172] Unhappy about the political situation generally, Salisbury tried to precipitate matters by offering his resignation on the grounds that he had reached the age of sixty and that it was important to reduce the average age of the cabinet. Churchill, nearly twenty years his senior, was unimpressed and now spoke of carrying on with much the present team of ministers until six months before the next general election, before handing over to Eden.[173] With the latter still not fully fit, there was renewed discussion that he might exchange the Foreign Office for the less onerous leadership of the House. Robert Carr urged the advantages of this move in allowing Eden to consolidate his position in the party while regaining his strength.[174] But Eden feared that to refuse the Foreign Office might, by indicating continuing physical weakness, furnish Churchill with further grounds for procrastination. After meeting the premier on 1 October, Eden held private discussions with Butler, Salisbury and Monckton. None could see an effective way forward. 'In short, they are all stymied by Winston as usual.'[175] When Churchill astonished almost everyone with the success of his party conference speech a few days later, it was clear that talk of his imminent political demise was premature.

By the end of 1953 it seemed that, short of an unforeseeable crisis, Churchill would only leave office at a time of his own choosing. The unedifying charade over the succession continued to be played out over the next fifteen months doing little credit to either man. Having been the beneficiary of Churchill's attachment to power during the summer of 1953, when he himself had been *hors de combat*, Eden was now in a weak position to force the pace. In due course it was always possible that, with the backing of senior party figures, Eden might bring matters to a head. 'Anthony might strike,' noted Moran. But 'somehow it is not a word we associate with Eden'.[176] In practice the 'Greek tragedy' would 'simply go on and on, until it pleases God to end it'.[177]

Almost certainly Churchill was no longer capable, by any objective criteria, of running the government effectively. At times he seemed perfectly normal, but at others he was obviously very feeble and his mind tended to wander, usually to irrelevant wartime analogies. Growing deafness and an unwillingness to use a hearing aid only exacerbated the problem. At international conferences he was little short of an embarrassment as Eden found at Bermuda in December 1953 – 'amiable but completely vague'.[178] Yet Churchill would only say that he would retire before the next election. As that did not need to be held before the autumn of 1956, this was scarcely a firm commitment. Eden professed to have reconciled himself to a situation he could do little to change. He had, he told Selwyn Lloyd, ceased to worry about the future, content to leave events to develop without his interference.[179] This, however, was a façade. Eden had spoken of retirement from politics before, but early in 1954, encouraged by his wife, he seemed merely to be looking for the most opportune moment to go. In April he confessed that he was at the end of his tether and could not stand it much longer.[180] Dragged to Washington by Churchill in June on what he judged a futile mission, Eden claimed to be counting the hours to liberation. 'I cannot go on like this with this old man. I must escape somehow.'[181] Robert Carr tried to offer comfort. Gallup in February showed that 54 per cent of the electorate and even 38 per cent of Conservative voters believed that Churchill ought now to retire.[182] But such statistics counted for little in the face of the premier's obstinacy.

In March Churchill indicated that he would retire in May or at the latest by the end of the summer.[183] Optimistically Eden began to enquire about the constitutional propriety of a new government being formed at the beginning of a summer recess and not meeting parliament before the start of the autumn session.[184] Nutting, however, preferred to keep his fingers crossed 'until I see the furniture vans arrive!'[185] In the event, when Eden wrote in early June, suggesting a transfer of power at the end of that month, Churchill argued that he could not now leave in view of talks with Eisenhower about a possible summit. He managed to make Eden's impatience appear selfish in contrast to his own disinterested statesmanship: 'I am increasingly impressed by the crisis and tension which is developing in world affairs and I should be failing in my duty if I cast away my trust at such a juncture or failed to use the influence which I possess in the causes we both have at heart.'[186] It was a sign of Eden's desperation that, on their return journey from Washington, he effectively struck a bargain with Churchill that in return for agreeing to a meeting with Malenkov – a policy goal of which he was personally most sceptical – the Prime Minister would resign at the end of September.[187] Colville was struck by the difficulty Eden had in choosing the right moment to raise the matter. 'I thought, and said, how strange it was that two men who knew each other so well should be hampered by shyness on this score.'[188] But Churchill could always backtrack on such an agreement by reference to the international mission which he alone could perform. He had had good medical reports and did 'not feel unequal to [his] burdens'. By late August he declared that he had no intention of abandoning his post in the present international climate. He still had a role to play in the interests of peace and hoped he could count on Eden's 'loyalty and friendship' during this important period.[189] The offer of the deputy premiership coupled with the leadership of the House in place of the Foreign Office scarcely suggested an imminent hand-over.[190] But, as the *Manchester Guardian* argued, 'to enter into yet another probationary period of no fixed length as a domestic min-ister would be to expose himself to dangerous political hazards, with no compensatory gain'.[191]

Shortly before Christmas Churchill chaired a meeting of senior ministers to discuss the date of the next election. In practice the real issue was who would be leading the Conservatives at that time.[192] The premier now talked of carrying on until June or July 1955, a prospect which, Eden argued, would deny a new government the chance to establish itself before going to the country. In such a situation he would prefer to 'complete my immediate work at the F.O.' and then 'retire into private life'.[193] Not until the New Year did Churchill finally agree to leave office in the spring. Even then, as late as March, he showed signs of wanting to backtrack, leading, accord-ing to Macmillan, to an angry and emotional outburst from Eden at the cabinet table.[194] A proposed visit by Eisenhower to Europe for the tenth anniversary of VE Day momentarily offered Churchill one last escape route from his impending retirement. As the formal transfer of power approached he began 'to form a cold hatred of Eden' for having thwarted him, though such emotions – undeserved as they were – did not last long and could never fully overcome the genuine affection he had developed over the years for his chosen successor.[195]

Finally, on 6 April 1955, almost a decade and a half since Churchill's first private intimation that the succession would be his, Eden entered 10 Downing Street as Prime Minister and First Lord of the Treasury. 'No two men have ever changed guard more smoothly,' he subsequently wrote.[196] Except in the sense that his succession was not challenged by any other candidate, this was a laughable assess-ment. Yet it was significant itself that, even in his memoirs, Eden had no wish to write anything which might hurt Churchill's feelings or damage his standing in history. Churchill, convinced that retire-ment was synonymous with death, had failed to understand why Eden was so impatient to wear his crown. He himself had been in his sixty-sixth year when the royal summons had arrived in 1940. But the two men's careers had followed very different patterns and objectively there is a strong argument that Churchill would have been wise to retire from the front line after the 1945 general election. Instead, Eden had to spend a further decade as crown prince, with one eye fixed anxiously on the succession to the detriment of his nerves, his health and, to some extent, his performance as a politician. These years of waiting saw Eden at his best and at his worst. Overriding loyalty to Churchill is visible throughout. He would not stoop to the intrigues which many others placed in his position would have seized upon as the only escape from an otherwise impossible situation. Almost certainly, Eden could have forced Churchill's resignation on several occasions in the ten years after the end of the war. Yet there was an obverse side to this admirable trait. Churchill's capacity to defy not only age and infirmity but also his chosen successor for so long revealed a flaw in Eden's political make-up which did not augur well for one who now took over the supreme post in British politics. It seems likely that the premiership came just in time, before the anxiety of protracted expectation finally sapped his will to carry on. It seems also reasonable to conjecture that he now felt all the more need to prove his capacity for the highest office of state.

9

The domestic politician

Few talk today of what should be the fundamental aims of the Tory Party. How particularly I wished that you could have taken over the leadership several years earlier.[1]

I T has become almost a truism for those who chronicle Eden's career to remind their readers that he had no great feel for, nor understanding of, the great issues of domestic politics. In so far as this judgement is true, it must be more a statement of fact than of criticism. As an American commentator put it shortly before Eden succeeded Churchill in 1955, 'more than nine-tenths of his world and life is what he happens to have chosen as his craft – foreign affairs'.[2] If his unpaid appointments as PPS are included, Eden's ministerial career began in February 1925. Just over thirty years later he became Prime Minister with, as Lord Home once put it, ultimate responsibility for 'everything from domestic drains to peace and war'.[3] In the intervening period Eden served at the Foreign Office, in one capacity or another, for about seventeen and a half years, at the Dominions Office for eight months, the War Office for seven months and, in his first appointment of all, as PPS, at the Home Office for just under a year and a half.[4] His time at the Foreign Office encompassed the rise of the dictators, most of the Second World War and an important period of the Cold War. In the circumstances Eden can be forgiven if his major contribution to British politics was not delivered on the home front. Yet the imbalance of Eden's pre-premiership career has been cited by his detractors as much more than a statement of factual record. It has been seen as evidence of his unsuitability, through the narrowness of his ministerial training, for the highest office, as one explanation, in other words, for the 'failure of the Eden government'.

It seems probable that Eden entered politics with no fixed political philosophy nor indeed any clear picture of his own position within the spectrum of the Conservative Party. His campaign speeches in the 1920s offered few signs of originality, nor indication of a distinctive approach to the key political issues of the day. As an early biographer put it,

> It wasn't brilliant stuff. To say 'our primary need is for more markets,' when he meant that we would have more money if we sold more goods was not only to state the obvious but to do so in a tedious way. But this verbal orthodoxy was backed, at least, by youth, charm and manifest sincerity.[5]

His approach was entirely pragmatic. Many Tories of this era engaged in the debate between free trade and protection with dogmatic conviction. Not so Eden. 'It seems to me', he confessed, 'that the only useful test which can be applied in these fiscal controversies . . . is the result which is actually achieved.'[6] Eden's domestic vision was determined above all else by his experience in the First World War, engendering an ill-focused but by no means insincere desire to improve the lot of ordinary British people, with whom his time in the trenches had given him a lifelong empathy. David Carlton perhaps makes too much of the readiness of a young and ambitious PPS to deliver speeches supportive of his political master – in this case Austen Chamberlain – to create the impression that Eden's earliest attitudes bordered on the reactionary.[7] Eden's main preoccupation was always with the wider world – as his decision to embark upon an extended tour of the Empire in July 1925, only five months after his appointment to the Home Office, surely confirms. In so far as he gave thought to the key domestic issues of the day, the experience of the 1926 General Strike was no doubt the most important factor.

This, and the way the Prime Minister handled both it and its aftermath, seems to have attracted Eden to the sort of consensual politics which Stanley Baldwin represented. The early stages of Eden's career show that 'his first regard was to seek conciliation and to avoid controversy'.[8] Baldwin ignored the shrill advice of many of his colleagues and refused to use the government's victory in the General Strike as the justification for a vindictive attack on the trade union movement. Increasingly, the premier gave voice to the honourable, if conventional, aspirations of that generation of young Conservatives whose lives had been so deeply scarred by the war. When, particularly after 1929, his position came under threat, there was never any doubt about Eden's loyalty. 'I gave public support to

S.B. in his difficulties, from principle not expediency,' Eden confided to his diary. 'I thought at one time I had backed the wrong horse.'[9] For so long as Baldwin remained Conservative leader, Eden publicly declared in 1930, 'so long will its "right" wing be unable to dominate the Party's counsels and narrow its purposes – of this the Trade Disputes Bill was sufficient example; so long also will confidence persist that the Conservative Party can remain truly national both in the sources of its strength and in the objectives of its policy'.[10] A decade later, when stocks in Baldwin's reputation were at a low ebb, Eden had not changed his mind: 'I like him very much and feel most sorry for him now. We owe him a debt for what he did to unite England. History will do him justice for that.'[11]

The Conservatives' loss of office in 1929 gave Eden somewhat more scope to define his position within the political spectrum. He joined a group of young MPs which included several future cabinet ministers such as Oliver Stanley, William Ormsby-Gore, Walter Elliot and W. S. Morrison. But it was a lesser-known member, Noel Skelton, who coined the phrase 'a property-owning democracy' to which image and goal Eden would cling tenaciously for the rest of his political career and beyond. The group shared a left-of-centre attitude on social questions which set them apart from many of their older colleagues in the Commons – 'you have no idea how awful they were,' Eden later recalled[12] – but which wedded them firmly to Baldwin's leadership. In a speech at the end of 1929 to the Unionist Canvassing Corps, Eden talked of the idea of co-partnership in industry and proclaimed his wish to see 'every worker a capitalist'. Such phrases made a good impression on the public platform. They seemed to portend a radical shift in the distribution of national wealth. But it can scarcely be claimed that Eden had given much thought to what their practical application involved or to the obstacles which might lie in his path. In any case, his primary interest remained the overseas arena and by September 1931 he was back in government as Under-Secretary at the Foreign Office. Not surprisingly, Harold Laski, writing in the *Daily Herald* in 1933, complained that Eden had done little to define an attitude in domestic politics. 'Young Tories in a vague and inchoate way would like to do the decent thing if they only knew what is the decent thing to do.'[13]

Whatever other failings have been laid at his door, few have accused Anthony Eden of neglecting his political duties. The opposite was in fact the case. Eden drove himself remorselessly, often to the detriment of his health. Six and a half years at the Foreign Office before his resignation in 1938 kept him fully occupied with the worsening diplomatic climate. Even as a cabinet minister he tended to keep his own counsel when matters clearly removed from his departmental responsibilities were discussed. 'Eden has unfortunately no experience of how a great Department is run other than the F.O.,' lamented Thomas Jones in 1936.[14] But as he was still under forty years of age there seemed plenty of time to remedy such a deficiency in his political education. It was not in fact until he left the government that there is much further evidence of Eden's attention being drawn to domestic issues. Then it was largely at the prompting of his closest advisers, acting perhaps on the assumption that his time might soon come to head the government himself.[15]

Yet the war years – notwithstanding the intense strain of his ministerial workload, compounded by membership of the War Cabinet and, latterly, responsibilities as Leader of the Commons – did focus Eden's mind on the domestic political scene more sharply than hitherto. The reasons were twofold. On the one hand Eden was sure, from an early date, that the war would 'bring about changes which may be fundamental and revolutionary in the economic and social life of the country'.[16] Unlike Churchill he did not believe that such developments could be completely ignored, however compelling the primary task of defeating Hitler. Churchill's claims to greatness as a war leader are beyond question, but he was always 'pretty bored with anything except the actual war'.[17] On several occasions Eden found himself put firmly in his place when he tried to raise the premier's horizons beyond the plane of grand strategy. 'As you know', Churchill minuted in May 1941, 'I am very doubtful about the utility of attempts to plan the peace before we have won the war.'[18]

In the second place Eden, who never throughout his long career claimed a monopoly of rectitude for his own side in politics, clearly found the experience of coalition government after May 1940 congenial. His confession to Hugh Dalton in January 1941 that he had been agreeably surprised how easy his new colleagues were to work with denoted considerably more than Eden's habitual charm.[19] To Oliver Harvey he confirmed that he found both Clement Attlee and Ernest Bevin 'very decent people' and he wondered whether he could not work with them after the war.[20] Such hankering after political realignment was encouraged by uncertainty as to the future evolution of the Conservative Party and his own place within it. For some time it was by no means clear that the change from

Chamberlain to Churchill had been matched by corresponding changes within the balance of the party as a whole. It was not insignificant that Chamberlain retained the party leadership until illness forced his withdrawal from public life in the autumn of 1940. The following year Eden found himself subjected to a hostile press campaign, for which he held Beaverbrook responsible, 'as a preliminary to getting the appeasers back again'. The experience made him more than ever 'conscious of his incompatibility with the Right'.[21] At the same time the backbench 1922 Committee complained that Eden only promoted left-wingers within the Foreign Office. Small wonder that, in a moment of exasperation, he confessed to hating 'the old Tories'. He 'would rather join the Labour Party if they remained dominant'.[22] Even when these particular battles had been won, it remained unclear whether the Tory Party at large and Eden, as Churchill's heir apparent, were entirely of the same ilk. 'There were no orthodox Tories left in the War Cabinet,' noted R. A. Butler early in 1942 of a body of which the Foreign Secretary was a prominent member. 'Eden was not liked.'[23] As late as 1943 Eden expressed continuing distaste for the Conservative Party. 'That is not where my supporters come from.' Harvey confirmed that the real basis of his strength lay among the general, largely non-party, mass of opinion in the country.[24] Eden thus came to believe that he stood, as in the 1930s, for the aspirations of a younger generation determined to achieve concrete gains from the flames of war.

Eden gave serious, if somewhat unstructured, thought to the possible creation of a centre party of left-wing Tories, right-wing Labour and Liberals as the best vehicle for the construction of a new and better Britain. The nucleus for such a grouping would come from younger Conservatives such as Butler, Oliver Lyttelton and Oliver Stanley. These men could work easily enough with Labour leaders like Bevin, while leaving parliamentary opposition to the far left. Labour was 'so conservative' and Eden 'so progressive' that they could never function as a two-party system.[25] Eden toyed with such ideas for at least two years. Despite the lessons of history, he remained, as late as the summer of 1944, inclined towards a centre coalition to carry out post-war reforms – always assuming that Churchill retired in time.[26] Conversations with Bevin led him to believe that the wartime coalition could be carried into the peace in return for the retention of some limited state control of industry.[27] Such a prospect seemed more attractive than the leadership of a Conservative Party still encumbered by a substantial right wing for whom the very concept of post-war controls was anathema. 'You do not imagine that I shall ever consent to lead a party of Southbys and Waterhouses?' recorded Harold Nicolson of a conversation with Eden in July 1944: 'I will do no such thing. I will not represent the moneyed interests. There are young men who think differently from the old men; young men who see clearer than we do the prospects of a new Toryism. It is them whom I wish to lead.'[28]

The fact that Eden was more than twenty years Churchill's junior inevitably gave him a greater incentive to pay attention to the problems of the peace. If Churchill was the man to see Britain through its present crisis, Eden's entourage – and some beyond it – encouraged him to believe that his task would be to take the country into the years of peace and reconstruction.[29] It was important that the initiative should not be conceded to the Americans in the international context and the Labour Party in the domestic. Each of Eden's speeches

> must be marked by a further clarification of our war–peace aims so that he thus builds up his doctrine for the peace settlement and identifies himself with those who are thinking of the future. He must come to be regarded as their spokesman and leader.[30]

Despite a lack of enthusiasm from Churchill – 'I do not think anyone knows what will happen when we get to the end of the war, and personally I am holding myself uncommitted'[31] – Eden used a speech at the Mansion House in May 1941 to give publicity to the ideas of the Cambridge economist John Maynard Keynes. 'It is continually being impressed upon me', noted Eden, 'how desirable it is that we should say something to make plain why Hitler's New Order cannot succeed and give an indication of our ability to help to meet the material needs of Europe which he is unable to supply.'[32] The speech anticipated some of Labour's post-war appeal while recalling the internationalism upon which Eden had built his own reputation over the previous decade:

> Social security must be the first object of our policy after the war, and social security will be our policy abroad no less than at home . . . the free nations of America, the Dominions and ourselves . . . have the will and the intention to evolve a post-war order which seeks no selfish national advantage.[33]

Eden's incursions into the unfamiliar field of social and economic policy were understandably hesitant. At times he doubted whether his grasp would ever be sufficient for a prospective Prime Minister. 'I do not really feel confident in myself as No. 1 at home', he confessed in 1942, 'and it looks as if, *faute de mieux*, I might drift that way.' In such moods, the attractions of a new career outside politics to provide financial security for his family and his own old age could appear compelling, especially as he was overspending at the yearly rate of £1000 or more from his private income. 'Trouble is I feel too tired to tackle these post-war problems. I am desperately in need of a change and I do not know enough of economics.'[34] His lack of self-confidence was very apparent to those around him. A speech to his constituents in September 1942 on the future world order 'went well'. But

> how cautious and timid he is when speaking of postwar, as hesitant as he is bold and certain when speaking of foreign affairs. We push and push and each time get him a little way but the time has long come for a big bold lead and damn the consequences. I want him to take the wind out of the sails of F.D.R. . . . but he still gives the impression of lagging behind.[35]

Others were less charitable. One Conservative backbencher deemed his 'purely a press-made reputation'. A short spell at a domestic ministry would exhibit Eden as a complete failure. Only at the Foreign Office would 'mere platitudes and amiable generalities' suffice.[36] Leo Amery, who, though a contemporary of Churchill, gave plenty of thought to the problems of the peace, feared that Eden was a 'sentimental internationalist with no ideas on Empire or economic questions'.[37]

Eden's capacity to fulfil the post-war expectations of his followers and to defy the gloomy prognostications of his critics was clearly linked to the debate which arose periodically throughout the war as to whether he should leave the Foreign Office to broaden his ministerial experience. Malcolm MacDonald, who had been well disposed towards Eden throughout the 1930s, considered it a disaster that he had been reappointed Foreign Secretary in 1940. As a future Conservative leader Eden needed wider domestic experience.[38] Within the Foreign Office J. P. L. Thomas and Richard Law, who always regarded their duties as Eden's political advisers as of equal importance to their diplomatic functions, were of the same mind. They warned that those Tories who were hostile to Eden wanted to keep him where he was so that 'when the time comes, they will be able to say that you're a wonderful fellow but a specialist, and after all you know your stuff and it would be a pity to take you away from it'. Law continued:

> You must leave familiar pastures and toughen your digestion. . . . You may have something to contribute and . . . you will be able to unite and lead the country in the very difficult times that will lie ahead when this silly war is over. At any rate I know that large numbers of people like myself would not tolerate some of the alternatives and would fight to the death to bring them down.[39]

By November 1942 Eden had accepted the post of Leader of the Commons with its wide-ranging functions and responsibilities. But as he added his new duties to those of the Foreign Office the effect was not what his supporters had hoped. The double burden was almost insupportable and the diaries of Alexander Cadogan, who was well placed to observe Eden's working day, contain several disparaging comments on the effect on the conduct of Foreign Office business. 'He leaves London lunch-time Fri.,' Cadogan noted soon after Eden's new appointment, 'gets back lunch-time Mon., and will now have to spend Tue., Wed. and Thurs. in the House. Quite impossible!'[40] From time to time Eden reached the point of giving up one or other of his demanding posts. Cranborne was an obvious potential successor at the Foreign Office, but his uneasy personal relations with Churchill, combined with Eden's innate reluctance, despite near exhaustion, to relinquish any of his powers, stood in the way of a satisfactory resolution. Yet from the point of view of Eden's post-war role the problem remained. Early in 1944 Amery advised a reshuffle which would leave Cranborne Foreign Secretary, restore Attlee to the Dominions Office and enable Eden to take on the Lord Presidency with its extensive committee work. If Eden was going to be party leader in the future, 'it was important that he should develop contact with every kind of problem rather than continue with specialised experience'.[41] In the event this good advice was ignored and Eden struggled on with his crippling workload for the duration.

The nearer the defeat of Hitler appeared, the more minds turned to questions of post-war reconstruction. In British politics the critical moment was the publication of the Beveridge Report in December 1942. When the report was debated early the following year, Butler urged Eden to give a lead to Conservative MPs and show the country that it was the party's intention to carry out an unprecedented policy of social reform.[42] In the event the chief government speakers were John Anderson and Kingsley Wood. Their patent lack of enthusiasm for Beveridge's proposals provoked a backbench revolt among Labour and Liberal MPs who nominally took the government whip. Preoccupied by the intractable conundrums of Allied diplomacy, Eden, notwithstanding a few speeches on domestic issues,[43] never had the opportunity to emerge as the spearhead of a new Conservatism. It was increasingly young backbenchers of the Tory Reform Committee, Peter Thorneycroft, Quintin Hogg and Lord Hinchingbrooke, who stole the initiative.

At the defeat of Germany Churchill gave each party leader within the coalition the option of remaining in the government until the end of the Japanese war or else withdrawing immediately so that an election could be held in July 1945. Attlee expressed a preference for an autumn election, an option Churchill had not offered. Accordingly, the country went to the polls in July. For a combination of reasons Eden approached the election without enthusiasm. The first was his genuine distaste for this partisan side of politics. In a striking diary entry in September 1941 Eden noted: 'I thought as I listened to Max [Beaverbrook] and Winston revelling at every move in these old games, and even Winston, for all his greatness, so regarding it all, that I truly hate the "game" of politics.'[44] Even more revealing was a scribbled remark on the top of a batch of electoral literature presented for his inspection as the contest drew near: 'Please keep all this muck together for me somewhere out of sight until I have to splash into election manure.'[45] Innate reactions were compounded by extreme tiredness, mounting ill-health and personal tragedy. In June 1945 Eden was diagnosed as suffering from a duodenal ulcer. His mother died that month and, during the campaign itself, his elder son was reported missing in action. On 22 July it was confirmed that Simon Eden was dead. This tragedy proved one of the last straws for a marriage which, for several years, had been at best unconventional. Eden's illness, which largely kept him out of the campaign, probably came as something of a relief. But others regretted the absence of the one man who combined a high standing in the country with the ability to endow the Conservatives with a progressive and forward-looking image. Arthur Mann of the *Yorkshire Post* wanted the Tories to engage a battleground which Labour threatened to monopolize:

I believe a great many people will pay particular attention to what you may say in relation to domestic problems. The great thing is to convince the people, especially the younger people who have never voted before, that there is going to be a change in the economic set-up that will check the exploitation of scarcities by selfish interests; that public control of industry need not and must not mean public management of industry and that the controls we have in mind while checking abuses will not hamper any legitimate private enterprise that sets the true service of the community as its primary aim.[46]

Eden's misfortunes no doubt contributed to a mood of uncertainty about his place in post-war politics. With election fever mounting he recorded:

Read my Gibbon and thought about politics a good deal. I am afraid I am not really much use as a party man. I dislike our extreme right more than somewhat and I seem for ever to be seeing the other fellow's point of view. In other words I am not a political warrior like Winston but only a civil servant. All the same I am not sure that that's fair for I do care deeply for our people. . . . I would like to feel that these men were working with me, that they liked me and that together we could build a better England. But how to do this through the sordid medium of party politics and how else to do it?

This entry in Eden's diary concluded with the almost plaintive remark, 'Perhaps I could make something of the C[onservative] party if I had it'.[47] That privilege, however, still rested with the seventy-year-old Churchill and it was he who set the tone of the Tories' election campaign, brushing aside Eden's criticisms with the dismissive, 'You, my dear Anthony, know nothing about home affairs'. The campaign was the 'cheapest and dirtiest' Eden could remember. He found it hard to accept that Churchill could so easily rekindle the partisan side of party politics at the expense of men with whom he had collaborated so closely and so recently in the business of government. In Eden's view Churchill had gravely compromised his prestige while destroying national unity and unnecessarily reviving class hatred.[48] But Eden's one significant contribution – a radio broadcast in late June – could do little to redress the balance of the Conservative campaign. The broadcast was the most impressive single element on the Tory side. Eden sought to differentiate Conservatism from Labour's proposals without descending to the level of partisan abuse.[49] From the Foreign Office Robin Hankey wrote enthusiastically, hoping that Eden had struck exactly the right note to turn the tide against Labour. 'Provided you get people to realise that the Conservatives can and will be progressive, I think the government will get a large majority. All this nationalisation talk is very theoretical and puts most people off.'[50]

*

Hindsight, however, suggests that the 1945 election was one which the Conservative Party could not win no matter how it managed its campaign. Labour had succeeded, probably quite early on in the war, in capturing the popular imagination and convincing the electorate that it was the party which could and would implement that blueprint of reconstruction and social improvement whose appeal extended far beyond Labour's traditional voters. The outcome was the election for the first time in the country's history of a majority Labour government enjoying a massive parliamentary margin of 146 seats over all other parties combined. Eden had sometimes expressed the hope that Labour would win but seems to have expected a Conservative victory.[51] He accepted the verdict philosophically enough. His subsequent analysis lacked the sophistication of the professional psephologist but seems accurate none the less: 'the country wanted a Government of the Left and got it! Tactics might have saved us a few seats but would not . . . have altered the general verdict.'[52] In many ways the result came as an evident relief. He could 'never have stood another Government as No. 2 to Winston and Leader of the House plus the F.O.'[53]

Precisely what the Conservative Party achieved in terms of putting its own house in order during the six years of Labour government remains a matter of some historical debate. On the one hand any notion of a complete transformation of outlook and philosophy does less than justice to the liberal strand of inter-war Conservatism which had clearly existed in the party of Baldwin and Neville Chamberlain.[54] On the other, whatever success the Conservatives did have after 1945 in improving their electoral image was slow to translate itself into success at the ballot box. The Tories recorded no single by-election gain at Labour's expense during the Attlee government and returned to power in 1951 still commanding a smaller proportion of the popular vote than Labour.[55] In many ways, moreover, the conduct of opposition politics during this period left much to be desired. 'What a deplorable opposition we have been,' concluded Richard Law in 1950.[56] Perhaps the most advantageous change occurred with the least effort on the Conservatives' part. A beneficial effect of electoral defeat in 1945, the loss of 173 seats and the retirement of a significant number of older MPs, was to clear out a large amount of parliamentary dead wood at a single stroke. The average age of the parliamentary party in 1945 was only forty-one and Eden himself expressed satisfaction that the 'new material' was 'excellent'.[57] Yet for all these qualifications it seems undeniable that the scale of Labour's success in 1945 did compel the Conservatives to integrate some of Labour's thinking into their own programme more clearly than they had yet done. By 1950 the party presented a different image, clearly committed to the welfare state and the mixed economy, and able to project itself as a party of moderation, while trying to portray its Labour opponents as class warriors.

What then was Eden's role in this process? Can it be said that he did more than strike 'sensitive poses in domestic politics'?[58] There is reason to believe that Eden resented the amount of credit tradition-ally given in this respect to R. A. Butler and the backroom staff of the Conservative Research Department.[59] But any assessment of Eden's contribution must start with an understanding of his position in the 1945 parliament. In most essentials he was the day-to-day Leader of the Opposition. The performance of the official leader was generally deplorable. Probably no figure other than Churchill could have got away with it. Churchill was frequently absent from parlia-ment, and indeed Britain, concentrated upon his other interests, especially the writing of his memoirs, took little interest in anything of a routine nature, and regularly left Eden to chair the shadow cabinet and preside over the weekly meetings called to discuss parliamentary business.[60]

Granted his wide-ranging responsibilities, Eden was probably well advised to stick to the broad brush rather than involve himself in working out detailed policy proposals. Freed from the constraints of office, the former Foreign Secretary engaged in a wider range of policy issues than at any earlier stage of his career. The scope for making a decisive contribution on foreign policy questions was in any case limited, granted the difficulty Eden would have had in significantly distancing himself from the views of Labour's Foreign Secretary, Ernest Bevin. At a time, moreover, when the notion of shadow ministerial portfolios was less clearly defined than it has since become, Eden often found himself answering the government from the opposition front bench on questions far removed from his particular area of expertise. In 1946 he led for the Conservatives in opposing the government's Coal Nationalization Bill and that to repeal Baldwin's Trade Disputes Act of 1927. Harold Macmillan later paid tribute to Eden's performance:

The House listened to him attentively. It was the fashion to say that he only shone in debates on foreign affairs. In fact he often made a far more profound contribution to the solution of our many economic problems than appeared from his modest and debonair approach.[61]

Eden would never have claimed to be an intellectual. The ideas which he popularized during the years of opposition were scarcely original. He did not escape in this period of his public life, any more than in any other, from his tendency to litter his public speeches with platitudes and clichés. 'We base ourselves upon the individual', he told the party conference in 1946, 'upon the need to develop the individual personality; we recognise that the individual can develop only through membership of a living united community. But unity is not mere uniformity.'[62] Such has been the stock in trade of many politicians on the public platform, but some have had a greater capacity to swathe the banal in a cloak of profundity. Unlike, for example, Macmillan, Eden seemed unable to enliven a commonplace idea with a strikingly original phrase. Orme Sargent, who had worked closely with him in the Foreign Office, noted how his former master tended to lean on others for his opinions and Reginald Maudling, then a promising young recruit to the Conservative ranks, later recalled that one of his early tasks was to help Eden with his speeches on financial and economic matters:

> He was, of course, the supreme expert in Foreign Affairs, but his knowledge of domestic subjects was more limited and, therefore, the scope for helping him was more considerable. . . . Happily for me, the sort of ideas which I was developing coincided with his own.[63]

But this was not an easy time for even the most original Tory thinker to make his mark. The intellectual climate of the late 1940s belonged indubitably to the left and the most Conservatives could hope to do was to offer a distinctive modification of those ideas which had swept Labour to power, the ideas of what came to be called the post-war consensus or settlement.

In the circumstances Eden did not acquit himself badly. He focused on two themes. One was a revival of that notion of a property-owning democracy to which he had been attracted a decade and a half before. As the government was effecting a substantial shift in the balance between the public and private sectors of the British economy, this was good politics. It showed that capitalism was not necessarily synonymous with the exploitation of the working class but could be compatible with a more prosperous, yet still individualistic, society. The second was the basic relationship between freedom and order, the place of the individual in a society from which the strict concept of *laissez-faire* seemed to have been permanently excluded. *Freedom and Order* was the title used by Eden for a collected edition of his speeches published in 1947.[64] Yet there was an unmistakable reliance on others even here. If the phrase a 'property-owning democracy', had been coined by Noel Skelton, 'freedom and order' was taken from Maudling's fertile brain.

Eden repeated the words 'property-owning democracy' so often as to create a cliché of his own. Robert Carr has testified that it was at the core of his political creed. 'Never a year', he wrote in 1976, 'and scarcely ever a month has gone by over these last twenty years without Anthony writing to me or telephoning me urging the Conservative party to awaken from its apparent disinterest in this basic theme.'[65] Eden clearly hoped that the government of his old Chief Whip, Edward Heath, would revive the idea in the 1970s.[66] But cliché or not the idea was not without substance. For Eden it was the only alternative to a steady trend after 1945 towards a more socialistic society – what later commentators have called the ratchet effect.[67] He tried to interest Churchill soon after the electoral defeat of 1945:

> As it seems to me, we want to build up the party on the broadest possible basis and our appeal should be to the people as a whole. We should champion a way of life which is the antithesis of socialism, but which is not just a way of life for a few sections of society and protection for property. . . . we want a modern version of Tory Democracy. . . . We can seek to broaden the basis of our appeal so that the people may have a constructive alternative to turn to when socialism wearies or revolts them as, in time, it most certainly will.[68]

For the most part, however, the rethinking of Conservative philosophy went on with at best the unenthusiastic toleration of the former Prime Minister.

Eden did not envisage a return to unbridled free enterprise. The demands of post-war reconstruction effectively ruled that out. Close co-operation between government and private industry could bring valuable financial and economic benefits to the country, he told an audience in Walthamstow in August 1946.[69] By then his thoughts, assisted no doubt by others,[70] were beginning to crystallize. He was determined to put forward a new formulation of progressive

Toryism based on the idea of everyone having a maximum stake in the country. It would include co-partnership in industry and assistance for tenant farmers to buy their own farms.[71] As the party conference drew near, pressure mounted for a clear statement of the Conservative position in domestic politics. It fell to Eden to answer this call. In a well-received speech, which also reaffirmed the Tories' commitment to full employment and national insurance, Eden outlined his thinking:

> The objective of Socialism is state ownership of all the means of production, distribution and exchange. Our object is a nation-wide, property-owning democracy. These objectives are fundamentally opposed. Whereas the Socialist purpose is the concentration of ownership in the hands of the State, ours is the distribution of ownership over the widest practicable number of individuals.

It was important, Eden argued, that the ownership of property should not be regarded as a crime or a sin, but as a reward, a right and a responsibility to be shared as equitably as possible among all citizens. The condition for achieving a wider distribution of ownership was not redistribution of wealth but an increase in its production and, in particular, in the productivity of industry.[72]

Changes in the practices of British industry were central to Eden's vision. As early as 1946 he took part in a series of small dinners at the Savoy at which leading bankers and industrialists made personal contact with Conservative figures including Butler, Woolton and Stanley. These contacts continued throughout the years of opposition and into the first years of Conservative government in the 1950s.[73] Eden believed that the necessary sense of industrial partnership could best be achieved through the promotion of schemes of share ownership by employees, employee participation and profit sharing. People needed to have a direct interest in the prosperity of the firm and industry in which they worked. 'All this', recalled Lord Carr, 'he saw as an integral and central element in the development of his wider concept of a property-owning democracy.'[74]

Much of Eden's rhetoric would not have been out of place in the Conservatism of the 1980s. He doubted the capacity of government to run British industry:

> Initiative, adaptability, a capacity to keep abreast or ahead of our competitors, these qualities are all indispensable to our national survival. Can you imagine a Government department displaying these

qualities? In other words, there is no country whose industrial future is less suited to government ownership and control.

He poured scorn on the notion that so-called 'public ownership' was the same thing as ownership by the people. 'I know that Socialists believe that once industries are nationalised the electors will be convinced that the industries belong to them.' But, while it was the professed socialist objective that ownership should be ever more closely vested in the state, the Conservative purpose was to vest it in the people. There was every difference between owning something oneself and having it owned on one's behalf.[75] Workers needed incentives if production was to be increased. Extra reward should therefore always follow extra effort and initiative. For this purpose reductions in direct taxation were desirable, but would run the risk of inflation until government reduced its own expenditure. Overall the burden should be shifted from direct to indirect taxation.[76] There was little here to which a later Thatcherite could take exception.

Some greater coherence was given to Conservative industrial policy by the publication in the spring of 1947 of the party's Industrial Charter. This was the work of the Industrial Policy Committee, set up the previous autumn, building upon the earlier efforts of the Research Department. Eden played no significant part in this process other than by helping to secure Churchill's tacit approval. But his was probably the leading voice in popularizing the Charter in speeches over the following six months. Eden's efforts at amplification remained couched in the sort of vague generalization of which he had become a master. At Cardiff in May he declared:

> We believe in planning, but in planning for freedom. We believe not in equality of misery imposed from the centre, but in equality of opportunity achieved by common effort. We believe that the benefits both of Government leadership and of individual initiative can be secured.[77]

But, as John Ramsden has argued, the primary importance of the Industrial Charter was political rather than industrial or economic. Though many of its detailed recommendations were never implemented, it played a significant part in re-establishing the Conservatives' credentials as a political force in the middle ground.[78]

Eden, however, remained concerned that the party was making insufficient headway against the incumbent government. Churchill was still a major problem. Eden feared that a snap election would find the Tories unprepared, but with Churchill, his adrenalin pumped up by the renewal of electoral battle, returning from the political periphery to 'run the show'.[79] Uncertain himself whether or not to speak out, his colleagues urged him to seize the initiative at the 1947 party conference. 'We shall get nothing concrete from the old man', warned Thomas, 'so you must just take the law into your own hands and give the hungry bread.'[80] Eden duly obliged. His speech to the party faithful reads well after nearly half a century:

> We do not want a Britain where a soulless State presides over and determines every single movement of our lives. Nor do we want a Britain of extremes of wealth and poverty. We want the farmer to own his farm, the working man his house, the artisan to have an interest in the work he does, with liberty and justice and opportunity for all. . . . 'Laissez-faire' remembered freedom and denied order. Socialism remembers order and has to deny freedom. The message of Conservatism to a generation growing up in a regimented age is this. Our faith is freedom within the law for all men. Come, help us reconcile this with the order which the modern world must have.[81]

Most political analysts agree that it is the performance of governments rather than oppositions which determines the electoral fortunes of both. So it no doubt was in the late 1940s.[82] But the temptation for opposition politicians is inevitably to assume that salvation lies in their own hands. By-election setbacks, particularly that in South Hammersmith in February 1949, led Eden to doubt whether the Conservative Party could secure a large enough majority at the next general election to enable it to do its job – 'and a small majority would be a calamity'.[83] Butler was equally despondent. In his view the party's propaganda organization had never followed up the Industrial, and subsequent Agricultural, Charters which were now 'almost dead'. 'Our "hungry sheep" who speak and work "look up and are not fed".'[84] Churchill faced a critical meeting of the 1922 Committee at which several members demanded a detailed policy statement and in April the party's Policy Committee, of which Eden was chairman, decided that such a document should be issued. *The Right Road for Britain*, upon a revised version of which the Conservatives would fight the next general election, appeared in July. It was largely the work of Quintin Hogg, amended by the Conservative Research Department, and was the most detailed policy statement produced during the years of opposition. Eden helped to launch it with a speech on the BBC.

Throughout this period Eden's principal importance lay not in the formulation of specific policies but in his capacity to symbolize the progressive face of Conservatism. As in relation to his internationalism in the 1930s, the image was as significant as any reality behind it. It was probably beyond Churchill's capacity, as he advanced through his eighth decade, to convince the electorate that the Conservatives had accepted the general progressive thrust of the post-war settlement and would not tolerate a return to the conditions and attitudes of the 1930s. This was Eden's role. His official biographer, himself a Tory politician of a later generation, has written that he 'always found [Eden's] brand of humane, liberal and progressive Conservatism, born in the trenches on the Western Front in 1916, the only version that appealed to me'.[85] Robert Carr, Eden's PPS in the early 1950s, developed the same point:

> Certainly as a young prospective candidate between 1946 and 1950 who had at that stage never met him, I felt he gave me more assurance than any other of our leaders (including Rab) that the Tory Party was of a kind to which I could commit myself. At that stage I had never met him and so my impression was entirely uninfluenced by any personal connection. I think my view is probably shared by large numbers of people and I think this illustrates how Anthony's influence at most times of his life may have been greater outside the political world than it was inside the top circle of politics.[86]

This importance as a symbol of the general direction in which the future of Conservatism lay should not be underestimated. Eden's elegance may have been on a higher plane than his eloquence, but he gave expression to 'what the average person was thinking quite competently'.[87] The Conservative Party's difficulties were clearly revealed by the general election of February 1950. It was widely assumed that the Conservative and Labour parties polled at something like their full strength in this contest.[88] The outcome was a Labour Commons majority of five over all other parties combined. Yet between them the Tories and the Liberals captured about 1.85 million more votes than the victorious Labour government. The conclusion was not difficult to draw. The problem was how best to get those who had voted Liberal on this occasion to vote Conservative at the next – and probably not long delayed – election. Eden had recognized the importance of Liberal votes some time earlier. He had opposed the attempts of Conservatives such as Peter Thorneycroft to do a deal with the Liberals, even though he had been led to believe that the latter would support an electoral pact with himself but not with Churchill.[89] In view of the size of Labour's 1945 majority it was reasonable to argue that the Liberals needed to be extinguished if the Tories were to secure a majority of the popular vote. 'I am still confident that we can get Liberals', he confided to Thomas in 1947, 'but hand will have to be played with skill and policy and tactics dictated by one man not six.'[90] There was good reason to believe that traditional Liberal voters were more likely to rally to a party headed by Eden than anyone else. He accepted the received wisdom of his old mentor, Baldwin, that elections were won or lost north of the Trent and in this area was confident of picking up former Liberal support.[91]

As the Conservatives do seem to have been returned to power in 1951 – when Labour secured more votes than at any time in the party's history – on the basis of their ability to take the votes of a by then almost moribund Liberal Party, Eden's importance cannot be ignored. The floating vote was so critical in the electoral arithmetic of 1950–51 that Eden's image as a man of the middle way may have been decisive. At least one observer judged that the Conservatives owed their return more to Eden's standing than to any other Conservative agency.[92]

*

If there is a disappointing side to Eden's performance as a domestic politician it lies less in the concepts he espoused than in the extent of their implementation. For the rest of his life he felt regrets that his vision of a property-owning democracy was not carried into greater effect by the Conservative governments of the 1950s and early 1960s. Eden's party-political broadcast during the 1951 election campaign spoke optimistically of encouraging a wider spread of ownership among all sections of the British people. His remark that too much wealth had been concentrated 'in the clutches of the state' and that it was necessary to 'distribute it more fairly' might have been taken to presage a major programme of denationalization, but the Conservatives were committed, as Eden stated, only to take iron and steel out of public ownership.[93] The idea of returning the great utilities to private hands would have been dismissed, in the intellectual climate of the day, as the impractical brainchild of a few eccentric free-marketeers. But existing Conservative policies were unlikely to achieve the transformation of which Eden spoke. While the Churchill government's drive to build a record number of new houses led to an increase in the proportion of owner-occupiers, other initiatives were less fruitful.[94]

The question of industrial profit-sharing provides a good illustration. During the years 1953–55 a modest crop of company employee shareholding schemes was introduced by such well-known firms as ICI, AEI, Courtaulds, Rolls-Royce and Rugby Portland Cement. Thereafter the idea rather went out of fashion, though in late 1958 it was briefly rumoured that Macmillan, Eden's successor as Prime Minister, was beginning to take an interest in the question of wider share ownership.[95] Workers understandably proved keener to share profits than to risk the possibility of losses. Later Eden attributed these disappointments to Macmillan's opportunistic attitude to politics. 'I could never get Harold even remotely interested in a property owning democracy', he recalled, 'and he was quite determined that there should be no profit-sharing in the publishing business! It would be good if Ted [Heath] could be a little more warm-blooded. He has much ability and courage.'[96] Not until the privatization schemes of the 1980s was there, against the background of a rather different industrial philosophy, a significant broadening of popular capitalism in the sense Eden had envisaged.

With his return to government in October 1951 Eden lost that freedom to range across the spectrum of domestic politics which he had enjoyed during the years of opposition. As Foreign Secretary his workload was undoubtedly heavier than in the 1930s. The best he could now hope for as the unquestioned number two in the government was to exert some influence over the general trend of domestic politics, though he appointed Robert Carr as his PPS partly in the hope that Carr's first-hand industrial knowledge would expand his own understanding of the home front. It was Carr, for instance, who inspired a suggestion from Eden to the Minister of Labour, Walter Monckton, that the government should not allow Labour to seize the initiative on the question of industrial welfare, but should itself bring forward new Factory Acts and an overhauled Factory Inspectorate.[97] From time to time, as during the war, it was suggested that a move from the Foreign Office would be in Eden's best interests to broaden his experience of domestic politics before taking over the premiership from Churchill. In the spring of 1952 the Cabinet Secretary advised Churchill that Eden should give up the Foreign Office and assume the chairmanship of the government's Home Affairs Committee from Lord Woolton.[98] For Eden the proposal created something of a dilemma. If his touch deserted him on his transfer to domestic politics, he might endanger the succession; if, on the other hand, the change worked well, it might give the ageing premier one more excuse to delay retirement.[99] In the event Eden stayed where he was. As Churchill entered the last year of his premiership, his physician recorded him wondering how well Eden would 'get on with all the Home stuff'.[100] In many minds it was still an open question.

Despite his primary commitment to foreign policy, Eden's intervention could still, on occasion, prove decisive on the domestic front. This was certainly the case in relation to the Treasury's plan in February 1952, known as Plan Robot, to float the pound. While the pound found its own level, all sterling balances held outside the dollar area would be blocked. The aim was to take pressure off the external balance of payments and place it on the internal economy. As the value of sterling fell, imports and exports should come back into equilibrium. In practice, however, Robot would have meant abandoning the commitment to full employment and a managed economy to which both Labour and the Tories had pledged themselves at the election. The majority of the cabinet had more or less accepted that Robot was necessary by the time that a Treasury official, Eric Berthoud, was sent out to the NATO conference in Lisbon to acquaint Eden with its details. Eden sent a cautious response urging that more careful thought should be given to the whole idea. On his return he attacked the plan vehemently on social grounds, particularly in view of the serious impact it would have on unemployment. Any other policy, including a sharp cut in the rearmament programme, was to be preferred.[101] Eden took much credit for his role in killing the plan which, whatever its merits, was probably not practical politics in the climate of the time.[102] His opposition to Robot helps define his position on economic policy within Conservative thinking, since it was the nearest the party came to accepting a market philosophy before the 1970s. The episode led to a considerable worsening of his relations with the Chancellor. Even so, Butler's subsequent mild budget seemed to indicate that the gravity of the crisis which had prompted Robot in the first place had been exaggerated.

With hindsight it became popular to disparage Britain's economic management in the 1950s on grounds of complacency. Fundamental issues were ducked with serious consequences for later decades. But politics in a democracy remains, as Butler famously stressed, the art of the possible and, after the years of wartime sacrifice and post-war austerity, the growing affluence of the new decade, fostered by favourable international conditions, scarcely provided the appropriate environment for a fundamental challenge to the structural bases of the British economy which later commentators deemed necessary. To all outward appearances the government did possess the tools to manage the economy successfully. Churchill was content to allow the British people to enjoy the rewards for which they craved after a decade of deprivation. The fact that his retirement was always assumed to be on the imminent horizon hardly encouraged long-term planning and it is difficult to escape some feeling of drift as far as the government's domestic politics were concerned. By the end of 1953 Macmillan felt that the administration was in need of 'a new theme or a new faith'.[103] All the same it is not easy to believe that any available alternative leadership, including Eden's, would have made that much difference.

Contemporary opinion judged Butler's stewardship of the Exchequer under Churchill to be a success. The Chancellor himself, however, was periodically struck by doubts about underlying weaknesses in the economy. In February 1954 he complained to Eden that the cabinet was not aware of the very serious situation that had arisen over government expenditure: 'The burden of defence, the runaway expense of the new agricultural policy, the housing programme etc., combined with the continuing liabilities in Egypt and the forecast of increased liabilities in Germany are almost more than the Exchequer can bear.'[104] These anxieties evidently transmitted themselves to Eden and the twin questions of controlling inflation and reducing government expenditure were at the heart of his domestic priorities by the time he formed his own government in April 1955.

*

In following as dominant a figure as Churchill into 10 Downing Street Eden understandably felt the need to stamp his own personality on the new administration and to secure a renewed electoral mandate. As pressing a reason as any was the lingering doubt felt by many senior colleagues as to his capacity for the top job, particularly in relation to domestic politics. Shortly before the prime ministerial change-over, Butler characteristically declared that Eden's limitations were not a matter for concern. He, Butler, would 'manage him from behind'.[105] Eden tried hard in his acceptance speech after his elevation to the party leadership to remind his audience of his position in Conservative politics. The alternative to socialism, he proclaimed, was to build on the work that had already been done and 'create a truly national property-owning democracy' – a phrase of which less had been heard since the Tories' return to power in 1951.

Yet it has often been suggested that Eden missed an important opportunity to make a distinctive mark at the outset through his failure to effect a radical reshuffle of the cabinet he inherited from Churchill. His options, however, were limited. Butler and Macmillan were clearly the leading figures for whom senior positions needed to be found. Eden resisted the temptation to move his old ally, Lord Salisbury, to the post which he himself had vacated at the Foreign Office, largely because of possible constitutional difficulties about appointing a peer. The premier seems to have believed that it was time to move Butler from the Treasury after nearly four years in that office. He felt constrained, however, by the fact that Butler's wife had recently died from cancer. In the circumstances moving Macmillan to the Foreign Office and leaving Butler as Chancellor pro tem was probably the best available solution. But the impression was of a hesitant start. More culpably, perhaps, Eden also avoided a wide-ranging reconstruction after his election victory in May. As Robert Carr later recalled:

> I felt at the time and have gone on feeling it even more strongly since, that if only Anthony had restructured his Government in May 1955, he would have put his own stamp on affairs in a way which he never succeeded in doing, particularly on our domestic politics.[106]

Over the calling of the general election Eden appeared, to the outside world at least, rather more decisive. The conclusive factor may have been Treasury warnings about the economic outlook.[107] But it was a matter to which Eden had given considerable thought before succeeding Churchill. A meeting with the Prime Minister designate on the day of Churchill's resignation left Woolton complaining of Eden's indecision – 'which isn't a very hopeful sign'.[108] Even so, in going to the country as early as May, Eden ran the risk of going down in history as the shortest-serving Prime Minister this century. Salisbury emphasized the importance of 'putting across your own philosophy, which I believe differs materially from Winston's'.[109] In the event the Conservative campaign stressed the party's achievements in the diplomatic sphere since 1951 – for which Eden could personally take much of the credit – and a desire to unite the country. Anticipating the appeal which Mrs Thatcher made in the late 1970s, Eden saw the need to attract the votes of the skilled industrial working-class 'who could be expected to benefit most from the kind of society we wanted to create'.[110] Yet the Conservative manifesto, *United for Peace and Progress*, made fewer specific promises than any post-war manifesto.

Eden was certainly the most prominent figure in the campaign. He appeared forward-looking in contrast to the Labour leader, Clement Attlee, whose distinguished career was obviously nearing its end. Eden travelled throughout the country and spoke in targeted marginal seats, mostly in short extempore addresses to the crowds which gathered to meet him. He refused to indulge in partisan attacks on his opponents who were always referred to as 'our socialist friends'. 'He said nothing memorable', concluded David Butler, 'but said it very well. He stirred up no excitement but he made no enemies.'[111] If it was not a thrilling campaign, the outcome left Eden with considerable cause for satisfaction. The Conservatives emerged with an overall majority of 58 seats. For the first time this century a party in power had improved on its performance at the previous election. It was a 'remarkable victory in view of . . . the disarray and loss of morale within the Party in the latter days of Winston's leadership and the view expressed by Lord Woolton [the party chairman] only nine months earlier about the danger of an Election disaster'.[112]

By the time of the election concern was being voiced inside the government over the growing number of strikes in British industry. Churchill had appointed Walter Monckton to the Ministry of Labour in 1951 with a specific brief to avoid confrontation on the industrial front. He had also intervened personally in the 1951 campaign to emphasize that there would be no legislation affecting trade unions in the lifetime of the next parliament.[113] Over a period of time, however, the result had been to encourage unions to believe that they could always secure what they wanted if they were prepared to create enough trouble. Eden and Churchill had clashed over a dispute on the railways during the last months of the latter's government.[114] On 3 June 1955 the cabinet considered the problem of industrial relations. Proposals discussed included legislation to introduce secret ballots, improve arbitration procedures and impose a compulsory cooling-off period before strike action could be taken. It is clear that Eden was lobbied by Lord Nuffield, founder of the Morris Motor Company, who, on the basis of his long experience of the car industry, argued that secret ballots would have a dramatic effect in improving the industrial climate.[115] In the face, however, of opposition from both the unions and the majority of leading employers, the government drew back from legislation. This reticence has been widely criticized with the suggestion that a great opportunity for trade union reform was thereby missed. Such a failure, according to conventional wisdom, contributed to the process by which Britain began to lag behind her industrial competitors. Yet, while the union leaders of the mid-1950s were undoubtedly more moderate men than those who thwarted the industrial relations legislation of the Wilson and Heath governments nearly two decades later, it remains doubtful whether the climate of the time – full employment and mounting prosperity – was right for the sort of plans which the government discussed. Eden himself seemed more alive to the kind of problem legislation might create than was Edward Heath after 1970: 'Legislation to make all strikes illegal unless preceded by a secret ballot would be practically impossible to enforce. You could not fine or imprison large numbers of workers for coming out on strike without having voted to do so.'[116] In any case, much of the difficulty resulted from unofficial strikes called by militant shop stewards, often in breach of their own unions' agreed rules and procedures.[117]

Eden remained, however, convinced that the improvement of industrial relations was an urgent priority. The starting-point had to be to establish a greater sense of partnership in industry as an integral element in the evolution of a property-owning democracy. He used his speech to the 1955 party conference to reiterate this theme – 'the core of my faith in domestic politics' – and called on every firm in the country to consider urgently introducing measures to create this sense of partnership.[118] With industrial relations legislation ruled out, Eden concentrated on the need to determine the causes of industrial discontent and unrest. He concluded that inflation was the first among them.[119] This was the central domestic issue for the remainder of his premiership. 'I am sure that our task must be to try to get the cost of living down', he wrote in August, 'and to this many things must be subordinated.'[120] This diagnosis was certainly correct, but the question remains as to whether his remedies were adequate.

Eden believed in the need to take money out of the economy by means other than taxation. He probably sympathized with Lord Chandos, the former Oliver Lyttelton, who wrote to him shortly after the election to warn against 'the new doctrine which is inherently Socialist that taxation is not to raise revenue – even primarily

– but is also and perhaps principally a weapon with which to regulate the economy'.[121] Reducing government expenditure was a priority and, despite Eden's understandable preoccupation with Britain's overseas obligations, the defence budget was far from sacrosanct. Projected figures suggested that the cost of defence would rise from £1527 million to £1929 million by 1959. 'We had to call a halt in defence expenditure and hold it over a period of years.'[122] The Defence Minister, Selwyn Lloyd, made some progress, despite opposition from the services, but the drive for economies tended to languish under his successor, Monckton, after December 1955.[123] Though often accused of a distaste for strategic planning, Eden initiated a thorough re-examination of Britain's long-term needs in the early summer of 1956, for 'it is only in the context of our long-term needs that we can make plans which . . . will give our economy a better chance of survival'.[124]

Eden was also concerned that inflationary pressure was coming from the unprecedented expansion of the building industry into which, of course, Churchill's government had put considerable resources. In August 1955 he asked Butler to consider reimposing some of the restrictions which the Conservatives had removed. It would be 'unpleasant' to do this, but better than having to endure a serious crisis in the balance of payments.[125] Also on his agenda were the freezing of dividends and incentives to encourage savings, so as to mop up excess spending power in the economy. The unions, he believed, would respond to the argument that rising costs would inevitably price Britain out of world markets with their members the first to suffer.[126] Overall, it is clear that Eden was not, as one critic has argued, obsessed with economic growth.[127] Indeed, no sooner had he replaced Butler by Macmillan at the Exchequer in December than he advised the new Chancellor that 'our excessive spending as a nation is due . . . to the fact that both Government and industrialists have been captivated by the idea of an "expanding economy"'. It had been right to expand, but now the country was 'bursting at the seams'.[128]

Eden's proposals needed time to bear fruit and were not perhaps wide-ranging enough to cope with the wage-push inflation which was the most obvious manifestation of economic malaise. Butler's monetary policy was inadequate to deal with the inflationary effects of the tax cuts he had introduced before the election. Demand sucked in imports, leading to a balance of payments deficit, while pressure on sterling mounted. An emergency budget in October reversed many of the tax cuts introduced in the spring. The belated replacement of Butler by Macmillan seemed to represent a repudiation of the government's economic policy so far, while leaving the new Chancellor disgruntled at his rapid upheaval from the Foreign Office. By the end of the year the Prime Minister's honeymoon period was clearly over. 'I was disappointed by Eden's lack of grip on home affairs,' recalled his press secretary. 'He seemed to know even less about economics than I did.'[129] A former Labour Chancellor concurred: 'I do not think he really has a clue about how to deal with inflation . . . but he has been warned by the more inflationary minded people of the dangers of deflation.'[130] City analysts called for a showdown with the unions together with swingeing cuts in public expenditure.[131] But at least one right-wing voice still sang Eden's praises. According to Brendan Bracken he was 'a fine man', prepared to take 'almost any risk' to prevent further increases in the cost of living. With some prescience Bracken predicted that Eden would soon clash with his new Chancellor.[132]

Before embarking on a visit to America at the beginning of 1956, Eden spoke in Bradford. He seemed, in his relatively simple way, to anticipate the breakdown of that Keynesian consensus which had dominated economic thinking since the war. The problem, he said, was how to maintain full employment but combine it with stable prices. He still favoured expansion, but it must be 'balanced and restrained'.[133] A White Paper on the Economic Implications of Full Employment was issued in March 1956, immediately before the budget. It stressed the imperative need to reduce the annual increase in costs which was outstripping the rise in the value of industrial output. In the meantime, however, Eden faced a crisis in his relations with Macmillan.

Returning from the United States, Eden was confronted by a memorandum from the Chancellor which congratulated him on the 'wonderful job' he had done in the diplomatic arena, but which went on to warn of a 'lot of trouble' on the home front.[134] Macmillan followed this up three days later with a letter pointing out how serious he considered the inflationary position and the general economic situation to be and asking for Eden's support. He wanted to cut subsidies on bread and milk to help stem the continuing flight from sterling. Several ministers warned that Eden, fearful of the immediate effect of the cuts on the cost of living, would be opposed, but Macmillan somewhat theatrically implied that it might be a resignation matter if his will did not prevail. Eden tried unsuccessfully to assuage the Chancellor by proposing a Capital Gains Tax. In the end, after seven cabinet meetings, a compromise was reached by which part of the subsidy was removed immediately and the rest later in the year.[135] Macmillan's diary comment that he had very little personal contact with Eden during the crisis perhaps indicated that the two men were already not working as harmoniously together as was desirable.[136] In a trial of strength between them it was the Chancellor who had prevailed. Some of those around Eden felt their nagging doubts being confirmed. 'Does he really have views', wondered William Clark, 'or is he a hollow man? Is he really opposed to Macmillan's rather strict economic policies? Does he really put "one nation" above all else? I don't know and I realise how vague his impact is on me. . . . Is there really no one behind the door, where "the Consul is busy"?'[137] The possibility that Macmillan might replace Eden in 10 Downing Street began to be whispered.

In a move which assumed additional importance after the onset of the Suez Crisis, Macmillan warned Eden in early April to expect an economic crisis and probably a devaluation, some time between August and November.[138] In response Eden stressed the need for wage restraint and, perhaps recalling his lost battle over food subsidies, emphasized the government's role in maintaining price stability where that lay within its power.[139] By early July he seemed reasonably optimistic that the economic strategy was showing results. Conversations with the leaders of both sides of industry were helping to stabilize prices, while restrictions on home consumption were getting rid of 'over-full employment'.[140] When, however, the crisis over Nasser's seizure of the Suez Canal broke later that month, consideration of the short-term problems of the British economy gave way, in Eden's mind at least, before the overwhelming conviction of the long-term threat to British interests posed by the Egyptian leader's actions.

What, however, is striking is the speed with which Eden reverted to economic problems once the dust of the diplomatic and military emergency had begun to settle. With Antony Head, the newly appointed Minister of Defence, Eden raised the disturbing warnings received from the Governor of the Bank of England: 'I know how many your difficulties are. I am, however, sure that we have all of us got to contribute to cut expenditure all round for this coming financial year or we shall not survive to do better later on.'[141] A few days after Christmas, and still hopeful of carrying on as head of the government, Eden wrote to Macmillan in terms which scarcely indicate a Prime Minister who found it difficult to hold his own in discussions of economic policy. His letter suggests more feel for the country's underlying problems than Eden has usually been credited with and merits extensive quotation:

> I agree with the Governor in thinking that the measures you had taken were gradually getting our economy healthier, and this improvement would have continued if we had not been at the mercy of the 'confidence' factor. I feel that the Governor is right in saying that both the credit squeeze and the increase in Bank Rate have gone far enough. Capital investment has already been quite sufficiently discouraged. We also know that the problem of cutting public expenditure raises acute political difficulties, as we have seen in successive examinations of social services and other similar expenditure, quite apart from defence. All of this prompts me to suggest that we should ask the governor whether there is any means of reducing our vulnerability to the 'confidence factor'. This inevitably springs from our position as bankers for the Sterling Area. As such we no doubt derive benefits. But do these offset the damage our reserves suffer when sterling is under pressure? The gains we painfully win from improved trading returns vanish almost overnight. Is there any way of meeting this? Why must sterling continue always to bear all of these burdens?[142]

If Eden was working towards a fundamental reappraisal of the role of sterling, he did not, of course, have the time to take this process very far. Early in 1957 he prepared a wide-ranging review of the lessons of the Suez adventure. Most of this concerned the country's international position. As regards the domestic front, Eden noted that the events of the last few months would create a measure of deflation which 'may not be entirely unhealthy'. He drew attention to the 'alarming increase' in the cost of the welfare state, not all of which was 'directly related to our struggle for existence', and commented upon the deterrent effect of existing rates of personal taxation.[143] These documents from the last weeks of his premiership offer a tantalizing glimpse of the possible domestic agenda of an Eden government allowed to run its full course. On 9 January 1957, however, the Prime Minister resigned.

*

At the meeting called formally to elect Eden's successor as Conservative leader, Lord Salisbury, his closest colleague in the outgoing cabinet, understandably paid tribute to the retiring premier's long record of achievement in the international arena. But he also reminded the audience that Eden was 'equally concerned with home politics and, in particular, with problems of industrial peace and better industrial relations'.[144] It was a necessary, if slightly exaggerated, corrective to the image created by a career self-evidently slanted towards the overseas rather than the home front. This is not to suggest that Eden made any profound or particularly original contribution to the intellectual development of Conservative thought. In the 1990s it is perhaps easier than it once was to appreciate how few front-line politicians make such a mark. Even so, until the Second World War Eden's analysis of domestic issues was rudimentary in the extreme. Even after 1945 he found it all too easy to revert to the banal and unimaginative. Particularly on the public platform Eden often came near to expressing faith in the political equivalents of motherhood and apple pie. In general he drifted along with the prevailing intellectual climate – a climate which, from relatively early in the war until the end of his career and beyond, belonged indubitably to the political left. Only in the sense of mounting frustration with rationing and bureaucratic controls had the tide begun moving back towards the Conservatives by the early 1950s. As with the Labour Party after 1979, even the most imaginative of post-war Tories found it difficult to carve out a distinctive and viable political creed. The Conservative role was to respond as constructively as possible within parameters set by their opponents. Within this context Eden performed a not insignificant role.

Throughout the post-war era, and particularly during the Labour government, Eden symbolized better than anyone else did or could that the Conservatives had become a forward-looking party capable of responding to the social and economic aspirations of the mass electorate. In his own memoirs Eden wrote next to nothing about his period out of power, leaving 'six years untilled soil', as he told his prospective biographer.[145] But his role in popularizing the new Conservatism may have been crucial granted the narrow margin of victory in 1951. Back in government his domestic role was inevitably curtailed by the constraints of his office and his absence through illness for much of 1953. If at this time his understanding of economic issues was limited, it could be argued that this weakness was shared by most of the contemporary political élite. In 1952 Robert Hall, head of the cabinet's economic section, grudgingly conceded that he was 'a bit better' than Churchill.[146] Faint praise perhaps, but the latter had at least served as Chancellor of the Exchequer for four years in the 1920s. Churchill's post-war administration has recently been styled a 'government of Tory "wets", for whom social harmony was a higher priority than economic efficiency'.[147] This is accurate enough, though the slightly condemnatory tone derives as much from retrospective wisdom as from any contemporary perceptions.

At the head of his own government, Eden was readier than his predecessor to examine some of the fundamental issues of domestic politics. He was an interventionist Prime Minister, who often irritated his colleagues by interfering in their departmental affairs. 'Fussy' is an adjective frequently ascribed to his conduct of business.[148] 'Can you explain to me', he enquired of the Minister of Agriculture, 'why if we have such a large surplus of milk, we cannot make more butter and cheese, instead of pouring milk down the gutter?'[149] In the field of education Eden's intervention proved more fruitful. He took a noticeable interest in the proposals of his minister, David Eccles, to advance technological education. The White Paper in March 1956 proved to be a landmark in the evolution of the polytechnics and the new universities of the 1970s. It was perhaps the most lasting domestic achievement of the Eden government.[150] For all his activity, however, to talk of the Eden government in terms of 'missed opportunities' in respect of industrial relations legislation or immigration controls – another area where the cabinet considered, and then rejected, legislation – is to take too broad a view of the boundaries of political possibility. British history since 1945 suggests that major changes in domestic policy take place when an intellectual climate exists to sustain such initiatives. The success of innovative legislation after 1945 and 1979 seems to confirm this analysis. Few believed that the state of industrial relations in the mid-1950s, or the level of migration from the new Commonwealth, merited such an approach. Similarly, Eden may have spoken of the fundamental importance of property ownership in his vision of Conservatism, but the policy of the right to buy council housing, which could have translated his aspiration into reality, may not have been practical politics during his premiership. Even so, there are tantalizing hints that in some areas of domestic policy Eden was moving, perhaps not entirely purposefully, towards a more 'Thatcherite' analysis than his traditional credentials within the one-nation school of Toryism might suggest. In particular it seems likely that the fight against inflation would have occupied a more prominent place in the late 1950s in an extended Eden government than it did under Macmillan. In the last resort, however, Eden might have been constrained from taking difficult decisions by the strong vein of humanity which always characterized his political outlook. 'We must not appear like the hard-faced men of 1918,' he noted when contemplating the worsening economic situation in the summer of 1955.[151] Such a trait made him more attractive as a human being, if less effective as a politician.

10

Eden and Europe

As you know, my feeling about the Common Market is that the
whole subject should be treated as a business negotiation and
not as a religion.[1]

A MONG the varied assaults launched against Eden's historical
reputation after his retirement from public life, that relating to
Suez inevitably caused him most concern. It was, however, at least
an attack which occasioned little surprise. Suez had been an exceptionally controversial episode at the time; there was a widespread
belief that much damaging information had been concealed from
public knowledge; and, in his heart of hearts, Eden must have
known that his own account in *Full Circle*, upon which he consistently refused to elaborate,[2] was unlikely to remain unchallenged.
Subsequent disclosures, revelations, reinterpretations and, inevitably, criticisms, though unwelcome to Eden, were at least not unexpected. After Suez, however, the issue which proved most damaging
to his historical standing was one which he had largely failed to
anticipate. It concerned his role in guiding Britain's response to
those steps towards European integration which culminated in the
signing of the Treaty of Rome in March 1957.

Eden offered his version of this story in *Full Circle*, though without giving the impression that European integration was a question
where he needed to defend his own actions. Strikingly, though, he
made only one passing reference to the 1955 Messina Conference
which came to be seen as a crucial stage on the road to the setting up
of the European Economic Community or Common Market as it was
then widely known. This omission reflected no desire on Eden's part
to conceal, but merely his estimation of the conference's relative
unimportance from a British point of view. Almost as soon as *Full
Circle* was published, however, the climate of opinion began to
change and accepted wisdom swung to the view not only that

Britain's 'destiny lay in Europe', but also that Britain had lost a unique chance to 'seize the leadership' of the European movement in the first dozen years after the Second World War. In the search for a scapegoat, Eden's centre-stage position made him an obvious target. If, as was widely believed, the post-war Labour government had missed opportunities, the failure of the supposedly Europhile Churchill to change the direction of British policy after the Conservatives' return to power could be largely attributed to the baleful influence of his Foreign Secretary and ultimate successor. Within this scenario it was conveniently symbolic that the European movement reached an important landmark with the signing of the Treaty of Rome just weeks after Eden's guiding hand was finally removed from British diplomacy.

The first attacks on Eden's handling of the European question came from former colleagues within his own party. As the possibility of British membership of the Common Market became a live issue for the first time, Eden was stunned by an attack launched upon him in the House of Lords by the maverick Tory peer, Robert Boothby. Speaking on 2 August 1962 Boothby claimed that the former premier was 'not only the architect but the creator of the Six' because of his refusal to participate in the concept of a European army in the early 1950s.[3] Boothby, who repeated his charges in print, had never ranked among Eden's intimates,[4] but similar criticisms were voiced in the memoirs of his former cabinet colleagues, Lord Kilmuir (David Maxwell-Fyfe) and Harold Macmillan.[5] Eden was not only hurt but genuinely surprised by these attacks. In the case of Kilmuir he actively contemplated legal action. A note written to a research assistant two months after Boothby's speech in the Lords captures his feeling of bewilderment:

> What rather bothers me is the number of vague speeches about the Conservative Government having missed the earlier chance to go in the Common Market and I have not the slightest idea of what they are referring to unless it be E.D.C. [European Defence Community]. We know the answer to that but do they mean some later occasion?[6]

None the less, by 1964, when Miriam Camps published her pioneering and still valuable account of Britain's early relations with the EEC, she could write confidently of the post-war decade as one of missed opportunities in which the leadership of Europe had been there for the taking.[7]

Such views became part of the accepted historiography of post-war Britain, casting a cloud over Eden's pre-Suez stewardship of British diplomacy. Critics saw in his European policy evidence of Eden's underlying shortcomings as a statesman:

> He did not see the importance of Europe, either in economic or political terms, I think largely because Eden was essentially a tactician in politics, in foreign policy. He was not a strategist. He did not see the broad picture . . . of Western Europe with Germany as a great industrial and economic giant as it is today. He could not see this.[8]

Five years after Eden's death as well disposed a Conservative MP as Nigel Fisher argued that he had been guilty of a 'grievous mistake' which had 'cost his country dear'. Eden had 'underestimated from the outset the political will of the six and for a quarter of a century Britain has paid and is still paying for his refusal to join the negotiations which led to the Treaty of Rome'.[9] Not even Eden's official biographer would take up this particular cause on his behalf, concluding sadly that Eden simply 'did not consider Britain to be a European nation'.[10] Writing in 1986, and with the authority attaching to a former President of the European Commission, Roy Jenkins summed up the received wisdom on Eden's culpability. He 'more than anyone else was responsible for allowing Europe to be made without us in the crucial years from 1951 to 1956'.[11]

*

Few of those who had been privy to Eden's thinking in the first days of the Second World War would have anticipated this indictment. Seeking to raise the sights of Britain's war aims, at a time when few had had much opportunity or inclination to think beyond the immediate goal of thwarting Hitler's aggression, the then Dominions Secretary concluded that the 'only possible solution' was 'on the lines of some form of European federation'. This would comprise a European defence scheme, a customs union and a common currency. Eden recognized that his suggestion might seem 'somewhat wild', but argued that it merely extended the proposals launched by the French Foreign Minister, Aristide Briand, a decade earlier.[12] Such visionary moments seem not to have lasted, especially after Eden's reappointment as Foreign Secretary in December 1940. As a Foreign Office man he soon adopted that cautious scepticism on the European issue which characterized the attitude of his ministry – and, it must be said, of the British political establishment generally – for the remainder of his political life. It was not often that Churchill, obsessed with the immediate task of winning the war, encouraged – let alone led – speculation on the post-war world. In November 1942, however, the premier voiced enthusiasm for a pan-European scheme, only to be met by the caution of his Foreign Secretary. Eden was 'dubious and did not see how anything European could work unless we and Russia and also to some extent America took a continuous part in it'.[13]

It was no doubt optimism engendered by Montgomery's recent victory in the North African desert which prompted Churchill to this rare incursion into post-war planning. The mood was still with him in the New Year when he penned his celebrated 'Morning Thoughts', calling for the establishment of a Council of Europe, and announced that he had agreed with the American president on the creation of an 'instrument of European government'.[14] This visionary phase in Churchill's thinking culminated in a broadcast on 21 March 1943 in which he repeated his call for a Council of Europe and, in due course, 'the largest common measure of the integrated life of Europe that is possible, without destroying the individual characteristics and traditions of its many ancient and historic races'. Churchill was always vague on the question of Britain's precise relationship with any European organization, but his views caused some alarm within the Foreign Office. It was left to Eden to pour cold water on his speculative zeal. Visiting Washington in March, Eden reported to Churchill that, according to Roosevelt's confidant Harry Hopkins, there should be no attempt to set up a European council, since this would 'give free ammunition to the [American] isolationists who would jump at the chance of sitting back in a similar regional council for the American continent'.[15]

It was not that Eden and the Foreign Office had set their faces firmly against any form of European grouping for the post-war era. Their objections were based on the priority accorded at this stage of the war to securing American – and if possible Soviet – participation in a world organization. Eden had no thought that Britain could turn its back on the continent once the war had been won. His concern to secure the restoration of France as a great power, and his readiness to stand up when necessary to both Churchill and Roosevelt on this issue, confirm this fact, but there was always an element of ambiguity in Eden's attitude. This was captured in a paper placed before the War Cabinet in November 1942:

> We have to maintain our position as an Empire and a Commonwealth. If we fail to do so we cannot exist as a World Power. And we have to

accept our full share of responsibility for the future of Europe. If we fail to do that, we shall have fought this war to no purpose, and the mastery of Europe which we have refused to Germany by force of arms will pass to her by natural succession as soon as the control of our arms is removed.[16]

At this stage Eden was probably ready to subordinate planning about post-war Europe to the necessity of securing American involvement in a wider world organization. As the war progressed and his doubts about the intentions and reliability of the United States increased, these priorities were modified, but the dilemma was always present, as when the question of possible post-war European integration stirred again in the course of 1944.[17] This time the initiative came from the continent itself. In July 1943 Eden had anticipated that some of the smaller countries of Western Europe would want to enter into 'very close relations' with Britain.[18] He cannot, therefore, have been surprised the following March when the Belgian Foreign Minister, Paul-Henri Spaak, asked whether Britain would take the lead in forming a 'Western bloc'. At much the same time the Foreign Ministers of Holland and Norway were also insisting that they had learnt the bitter lessons of 1940 and were determined to collaborate closely with Britain and build upon wartime arrangements for the training and equipment of their reconstituted national forces. Once again Eden's response was cautious.

Keen to avoid commitments which Britain might lack the capacity to fulfil, Eden advised that the sort of bloc envisaged would have the disadvantage of appearing to be directed against the Soviet Union, with whom Britain had a formal treaty of alliance.[19] In July he reiterated the point, previously used in response to Churchill, that a European bloc might help American isolationists argue that Europe had no need of US support for its post-war defence.[20] But Eden was also coming under pressure from another direction. In late May Alfred Duff Cooper, the British representative with the French National Committee in Algiers, had put forward a plan for a closely integrated 'Western Union' to be formed by Britain and the Western European Allies. Significantly, Cooper's proposals, though stressing the need to counter post-war Soviet expansion, were envisaged primarily in economic terms. Eden, clinging still to the Anglo-Soviet alliance which he had done so much to bring about, replied formally two months later that Cooper's proposals would only increase the dangers from Russia which they were designed to diminish.[21]

The Foreign Secretary's opposition was not, however, as total as Cooper no doubt concluded. It was the latter's overt anti-Sovietism which worried Eden. He was by no means dismissive of the concept of a Western bloc *per se*, but it would have to fit within the wider framework of great-power co-operation for which Eden was actively working as plans materialized for a United Nations Organization. Inside the Foreign Office Gladwyn Jebb had taken trouble to analyse Spaak's initiative in depth and Eden himself, before sending his discouraging response to Cooper, had invited the Chiefs of Staff to consider the question of a regional security system to cover Western Europe as part of a wider scheme of European security. The Chiefs replied that from a military angle there was considerable merit in a proposal which offered the sort of depth which was becoming increasingly necessary to Britain's defence. Significantly, they drew attention to a possible future clash between Britain and the Soviet Union in which a Germany integrated in the Western system would play an important role.[22]

By November Eden found himself engaged in a renewed debate with Churchill on the question of some form of European integration, but with the roles essentially reversed from those of 1942–3. The premier could now see little merit in the creation of a Western bloc. Until a strong French army had been recreated – and that might take ten years – there was nothing for Britain in the countries of Western Europe 'but hopeless weakness':

> That England should undertake to defend these countries, together with any help they may afford, before the French have the second army in Europe, seems to me contrary to all wisdom and even common prudence. . . . I cannot imagine that even if we go on with taxes at the present rate, which I am sure would be ruinous for economic revival, we could maintain an Expeditionary Army of 50 or 60 divisions, which is the least required to play in the Continental war game.[23]

Eden sought to answer Churchill point by point. He agreed that it would be absurd to enter into any defensive commitments except in conjunction with the French and as part of a general plan for containing Germany evolved under the aegis of a world organization. It was also true that the countries of Western Europe had behaved 'very foolishly' between the wars, but Britain's own record had not been 'entirely praiseworthy'. The lesson of 1940 was the need to build up a common defence association in Western Europe to thwart another Hitler *'whencesoever he may come'*.[24] Contrary to Churchill's fears, a properly organized Western Europe could provide Britain with defence in depth and large resources of manpower which would greatly ease Britain's burden, enabling her to avoid a huge standing army which might cripple her economy. Even so, Britain would have to reconcile herself to a rather larger land commitment than 'the famous two divisions which was all we had to offer last time'. Eden argued that the Soviet Union would not oppose an organization directed specifically against German resurgence and took rather optimistic comfort in an assurance given by Stalin three years earlier that he had no objection to Britain taking the lead in a regional security pact.[25]

In practice little further progress was made on this issue before the end of the war, not least because de Gaulle made it difficult to conclude a bilateral Anglo-French treaty – upon which any proposed wider grouping would have to be based – by proceeding to sign his own pact with the Soviet Union. It is, however, possible to draw certain tentative conclusions about Eden's position on the European question at the time he gave up responsibility for British foreign policy in July 1945. He was clearly by no means opposed to the idea of some form of Western European union, provided this could be accommodated within the framework of a world organization. The Foreign Office's internal discussions reveal that the need for defence against a revived Germany was seen as paramount.[26] Shortly before the Labour government took over, Eden stressed that any grouping should be limited to countries under direct German threat, since otherwise Soviet hostility and suspicion were bound to be aroused. This meant the exclusion of states such as Spain, Portugal and Italy.[27] Even so, in his more pessimistic moments Eden was not blind to the difficulties of maintaining great-power co-operation with the Soviet Union once the cohesive force of Hitler's menace was removed. There is little evidence that questions of sovereignty, supranationalism and federation, which would ultimately dominate discussions of Britain's relationship with continental Europe, were seriously considered. When Eden wrote of a European federation in September 1939, he did so without apparently attaching particular significance to that concept. It is, however, clear that discussions of a Western bloc in the later stages of the war were, notwithstanding Duff Cooper's contribution, essentially confined to the military sphere. Gladwyn Jebb, head of the Foreign Office's Economic and Reconstruction Department, who believed that a logical corollary was the formation of a customs union, recognized that his was the view of a tiny minority.[28]

*

Over the following six years, when the stewardship of Britain's external relations passed into other hands, crucial developments took place setting Western Europe on the path towards integration. For these Eden can bear no direct responsibility. But his role and attitude must be considered if his own and his party's conduct after 1951 are to be understood and the extent of his divergence from supposed Conservative Euro-enthusiasts evaluated. It is now clear that Labour's Foreign Secretary, Ernest Bevin, was not as unsympathetic to the idea of European integration – at least until 1948 – as was once thought. During his first three years in office Bevin explored several ideas for closer co-operation including, in 1946–47, a European customs union, and might have made more progress but for the inherent caution of the Treasury and the Board of Trade.[29] Thereafter, however, Bevin became increasingly hostile to the growing continental view that integration should proceed on supranational lines. His distaste for such a development helped encourage the view that a more positive attitude towards European union would have to await the re-election of a Conservative government.

During these years of opposition Churchill, operating as little more than a part-time party leader, was more than ever inclined in his public speeches towards the sweeping and the visionary. His declarations in Zurich in September 1946 and the Hague in March 1948, calling for, but scarcely defining, a United States of Europe, provoked a response which was emotional rather than intellectual, but none the less struck a chord with a growing mood in Europe, and in some sections of British political opinion. In January 1947 he formed a cross-party British United Europe movement, but the decision taken at the Hague to establish a permanent European Assembly failed to meet the aspirations of the increasing band of European federalists. The Council of Europe which met for the first time in Strasbourg in August 1949 was a purely consultative body, soon dismissed by many as little more than a talking shop.

Developments which then appeared to be of altogether greater significance were also taking place in the defence field, culminating in the formation of NATO in 1949. Eden applauded these, fully aware of their constitutional implications. As he told the Commons in May 1948, 'there must always be some merger of sovereignty in any common central organisation for any international purpose, whether it is to maintain peace or even to fight a war'.[30] Despite Britain's deteriorating relationship with the Soviet Union, his thinking on defence was still partly conditioned by the fear of a German revival. As Bruce Lockhart recorded:

> He still thinks that in the long run Germany will benefit from the wrangles of the Allies and will regain her power and strength. He still believes in the danger of a German menace and thinks that we have forgotten this aspect of the European problem in our dread of Russia.[31]

These were, Eden believed, practical problems requiring practical solutions. He was altogether less committed to the schemes to which Churchill put his distinguished name.

Eden attended the celebrated Hague Congress and subsequently wrote an article expressing support for the nations of Western Europe in their will to unite. Significantly, however, he went on to define the distinct position of the United Kingdom which was bound to distance the country from the powers of the continent:

> The fact that we are not only a European power but the heart and centre of a great empire is generally recognised and indeed welcomed. Europe has not forgotten that for a year this country, with its partners in the . . . Empire overseas, stood alone. This event, more than any other, has brought European opinion to understand how essential it is that no step should be taken which could in any way weaken the contribution that the British Commonwealth can collectively make. This is fortunate, because there is not for us any European advantage which could justly be weighed in the balance against this free association with our kinsmen overseas.[32]

Eden did not take part in the United Europe movement nor join the Conservative delegation to the Council of Europe. It would be misleading to suggest that there were no differences among leading Tories over Europe in the late 1940s, but these tended to be of emphasis rather than fundamental principle and were never as clear-cut nor of such serious consequence as Eden's critics later suggested.[33] A letter which Boothby sent to Eden after the inaugural meeting of the Council of Europe helps clarify the position.

According to Boothby Britain faced a 'choice of fearful consequence'. The country must 'go right in with the United States' in an attempt to build an English-speaking bloc, or 'take the lead in creating a genuine Western European Union'. The idea of an

English-speaking union had always, Boothby stressed, held attractions for him and deserved 'most careful consideration' since it would guarantee Britain's political security and her standard of living. On balance, however, he favoured the European option. But Boothby's own vision of what could be achieved merits attention:

> What I do feel most strongly . . . is that we have no right to act as a perpetual brake on effective European unity of any kind. If we bring what is left of the sterling area with us, they will accept a comparatively loose form of federation, under our leadership; because cut off from the agrarian countries of Eastern Europe, they desperately need the complementary economy of the sterling area. If we continue, indefinitely, to shiver on the brink, it would be better to let them go ahead and form a tighter federation, with our moral support – but nothing more.[34]

Though there was much here with which Eden would come to agree, events showed that Britain's capacity to determine the shape of European unity was more limited than Boothby supposed.

Increasingly at this time Eden drew attention to the priority which Britain had to give to her Commonwealth connections in any consideration of policy towards Europe. 'For us', he told an Australian audience in February 1949, 'Empire must always come first.'[35] Similarly, in his adoption speech for the 1950 general election, he stressed that, while Conservatives welcomed the idea of closer collaboration with Western Europe, Britain's European neighbours fully understood that 'for us the Commonwealth and Empire must always be the first consideration'.[36] With the inestimable benefit of hindsight it has become usual to point to the rapid evaporation of Britain's imperial heritage which, within a generation, removed the Commonwealth as a factor of particular significance in Britain's calculations. Thus Eden stands accused of a lack of foresight for which the country ultimately paid a heavy price. If it be part of a statesman's required credentials to predict the future, such a charge has its merits. But the priority of the Commonwealth was not a point which Eden needed to defend against the arguments of Conservative colleagues and Labour opponents. It was an article of faith which scarcely merited further discussion. His old friend, Oliver Harvey, now installed as ambassador in Paris, spent his time 'telling people that we cannot separate ourselves from Europe'. But 'at the same time, we are part of the Commonwealth', and 'without the Commonwealth Britain would be a doubtful asset to Western Europe'.[37]

The Schuman Plan, the most significant of all post-war schemes for European integration, was presented in May 1950. From it emerged a year later the European Coal and Steel Community, consisting of the six countries which in March 1957 became the original signatories of the Treaty of Rome. The aim of Jean Monnet, its author, was to forge the coal- and steel-producing capacity of France and Germany into a single unit so that war between these countries would become an impossibility. The plan held few attractions for Britain's Labour government, especially as the intention was to subordinate the coal and steel industries of participating states to a supranational High Authority. The Conservatives, arguing that the goal of Franco-German *rapprochement* was so important that this initiative must not be allowed to fail, nor to succeed without British participation, set out to censure Bevin's handling of the matter. It fell to Eden to move a critical Commons motion on 26 June. He would later refer to this speech as evidence that he was in no sense a 'tepid European'.[38] But according to his critics the speech provides grounds for disappointment and surprise that the Conservatives failed significantly to move on from Labour's policy when they formed a government less than eighteen months later and proof that, at this time at least, Eden did not regard the loss of sovereignty entailed in the High Authority as an insuperable obstacle.[39] As has been seen, however, the issue of absolute sovereignty was not one upon which Eden wished to make a stand. More important was his stipulation that Britain should join the discussions of the plan on the same terms as the Dutch, which meant that Britain would not necessarily be bound by any conclusion reached.[40]

The significance of Eden's reservations became clear in August 1950 when Harold Macmillan and David Eccles, on behalf of the British Conservative delegation at Strasbourg, proposed a non-supranational alternative to the ECSC to try to make it more acceptable to British opinion. This was interpreted in France as a thinly veiled attempt to wreck the whole scheme and was rejected out of hand.[41] Indeed Eden always argued that it was the fact that the Schuman Plan was firmly in place, complete 'with its federal implications', by the time of the Conservatives' return to office which determined the new government's attitude.[42] Even Macmillan ruled out the possibility that the British people would 'hand over to any supranational Authority the right to close down our pits or our steelworks'.[43]

Yet some confusion in Conservative policy towards Europe did persist. On 11 August Churchill addressed the Council of Europe on the theme of a European army. Eden later suggested that this speech had been 'misunderstood at the time', raising hopes that a Conservative Britain would be willing to merge its forces with those of its European neighbours 'to an extent which Mr. Churchill certainly did not contemplate'.[44] At this stage Churchill probably had no clearly defined plan in mind. 'His purpose', argued Macmillan, 'was to throw out general ideas and give an impetus towards movements already at work.'[45] But Churchill said enough to convince many that Britain under the Tories would play a leading part in a European defence organization, rather than simply applauding from the sidelines while another important stage in Franco-German *rapprochement* was passed. Lord Salisbury's alarm at Churchill's proposals probably reflected Eden's thinking:

> Then, there is Winston's Plan for an European Army and an European Commander-in-Chief and Minister of Defence. . . . I see Winston wants his plan discussed in the House of Lords as well as the House of Commons. . . . I shall find it very difficult not to speak against it, and I expect you will be in the same position. I wish he wouldn't rush ahead so fast.[46]

When in fact the Conservatives came to debate proposals for a European army in February 1951 Eden was careful to pick up a theme, implicit in Churchill's Strasbourg speech, that British participation should be based within an Atlantic framework designed to secure an American commitment.[47]

*

On 28 October 1951 Eden resumed responsibility for foreign policy after a gap of more than six years. The surviving documentation provides little evidence that he believed that the incoming government had much scope to change the direction of policy towards Europe, or that he faced serious opposition to his policy from within the cabinet. Eden later reflected on the situation as he saw it:

> By the time we took over, the European Defence Community had been adopted by leading ministers in six European countries and there was much fervour in support of the plan which was essentially a federal one, for the scheme involved a European Minister of Defence and a European Parliament to which he was to be responsible. Nobody at that time thought that we could have joined it and to have attempted to substitute a non-federal plan . . . would have created confusion

and resentment. The only course to follow was to help the EDC from without as the previous government had proposed.[48]

One of Eden's first acts on returning to office was to reaffirm the principles laid down by his Labour predecessor in the Washington Declaration of September 1951 that Britain, along with America and France, would aim for the inclusion of a democratic Germany on a basis of equality in a continental European Community which should itself form part of a constantly developing Atlantic Community. Here was a clear signal that Eden had accepted the fundamental premises of the Labour government's policy.[49] It was essentially the policy of the Foreign Office as set out in a briefing paper prepared by Eden's former private secretary, Pierson Dixon, on 31 October. Britain would play an active part in all plans for European integration drawn up on an inter-governmental basis, but considerations of national defence, the Commonwealth and the sterling area precluded any subordination of British policy to a supranational authority. Britain would, however, encourage any countries wishing to proceed on such lines to do so and would consider 'methods of associating' the United Kingdom with any resulting supranational bodies or schemes for integration. Closer European union offered the only way to resolve the age-old antagonism between France and Germany – 'an aim to which we attach the highest importance'. Eden accepted this summary of Britain's position, while emphasizing his personal commitment to the idea of a European army.[50] Above all he was concerned that the Conservative government must not appear 'less forthcoming than our predecessors', as he put it to Butler when urging the Chancellor to make a personal appearance at the consultative assembly of the Council of Europe.[51]

On 22 November the cabinet met, in Eden's absence, to authorize a statement to be made by the Home Secretary, David Maxwell-Fyfe, at the consultative assembly concerning the creation of a European Coal and Steel Community and a European Defence Community. According to Macmillan's later account the precise wording of this statement was a compromise reflecting the 'temperamental divergences of opinion' within the cabinet.[52] Characteristically, the anodyne cabinet minutes provide no support for this interpretation.[53] It is, however, clear that Churchill's attitude was less ambiguous in government than in opposition. In a paper on 'United Europe' which he circulated a week later, he made it clear that he was not opposed to a European federation provided it came about naturally and gradually.

But 'I never thought that Britain or the British Commonwealth should, either individually or collectively, become an integral part of a European Federation, and have never given the slightest support to the idea.'[54]

The suggestion that Churchill had been heavily influenced by Eden and even that he was anxious not to embarrass his designated successor by taking Britain further down the integrationist road than would be welcome to Eden is difficult to sustain.[55] Churchill was seldom reluctant, even in his final premiership, to take on his Foreign Secretary when, as happened not infrequently, their views diverged. The problem was that too much attention had been paid to the grand flourishes of Churchill's earlier pronouncements and too little to the small print. Untrammelled by the constraints of office, Churchill, through his indefinite vision of the late 1940s, had created expectations which he never tried to fulfil on returning to government. A Conservative backbencher voiced the hopes which the new premier's oratory had aroused: 'Churchill's return has, certainly, created the psychological background for us to take up the leadership of Europe, and the present state of the Schuman and Pleven Plans seems to offer a practical opportunity to take the initiative.'[56] But as Eden in later years took pleasure in recalling, Churchill regarded the European Defence Community which emerged as no more than a 'sludgy amalgam'.[57] His attitude towards European integration was more consistent than has often been claimed. Macmillan's lament that there had been 'almost a betrayal' is certainly an exaggeration.[58]

According to John Colville, reinstated somewhat reluctantly in his wartime post as Churchill's private secretary, the question of European unity did not figure high on the new Prime Minister's list of priorities. An altogether more important task was to recreate the Special Relationship with the United States which, he believed, had been allowed to tarnish under the Labour government.[59] Britain should consider herself 'part of Europe but not comprised in Europe'.[60] Such thinking almost certainly coincided with Eden's own and with that of the majority of those who subsequently laid claim to the credentials of ardent Europeanism. Evelyn Shuckburgh, now installed as Eden's private secretary, saw him as aiming to make the most of Britain's influence in the world. In this context 'we must see ourselves as an active and enlightened European nation with a world role and not as a limb of Europe'.[61] These thoughts conditioned Eden's attitude to the EDC. A system of Western defence would help to reduce tension in the wider world and Britain should encourage the European states to come together to achieve this goal, even though she did not wish to join herself. 'If the European army plan fails', Eden confided a few days after returning to office, 'then will be the time for Britain to bring forward an alternative plan.'[62]

It thus seems probable that there was less division inside the government over its European policy than has sometimes been claimed. None the less the circumstances surrounding Maxwell-Fyfe's pronouncement in Strasbourg on 28 November and Eden's comments at a press conference in Rome a few hours later gave rise to the canard that Eden had, almost single-handedly, destroyed any lingering hope that Britain could still seize the leadership of the European movement.[63] Maxwell-Fyfe's statement welcomed European initiatives for closer integration, while signalling Britain's reluctance to move beyond the realm of inter-governmental co-operation. The Home Secretary's words, which Eden had now approved, were in line with the cabinet's decision of 22 November, though in his memoirs Maxwell-Fyfe amalgamated what he said at the assembly with what he said to the press shortly afterwards.[64] Eden's declaration that Britain would never join a supranational Defence Community, and that the most she could contemplate was association with such a body, struck a more negative note than had his cabinet colleague's statement, but there was no difference of substance between the messages conveyed by the two men. Shuckburgh, who was also in Rome, sensed no disposition on Eden's part to sabotage a pro-European decision of the cabinet – 'that is a great distortion by Maxwell-Fyfe'.[65] Yet, argued one critic, the Home Secretary's cautiously phrased speech came as 'no disappointment' to his audience. By contrast when the news arrived of Eden's press conference 'the effect on the mercurial atmosphere [at Strasbourg] was nothing short of shattering'.[66] The impact, confirmed Boothby, was 'catastrophic', 'a shattering blow'. Every delegate at the Council of Europe 'thought we had betrayed Europe, as indeed we had', while Maxwell-Fyfe was 'on the brink of resignation'.[67] The Belgian, Paul-Henri Spaak, actually did resign as President of the Council of Europe.

When Maxwell-Fyfe voiced these charges in his memoirs, published in 1964, Eden sought legal redress. It brought to an effective end his relationship with a colleague who had been among his most loyal lieutenants during the Suez Crisis. When Macmillan repeated the charges five years later, Eden was genuinely taken aback and made every effort to persuade him to withdraw his accusations, largely without success. The episode left an unpleasant atmosphere in relations between the two former Prime Ministers. Eden set down his case for the benefit of posterity:

> There were no differences about EDC in Cab. Telephoned . . . Harold . . . and told him I would send letter and some notes. He said whole business was Churchill's fault, but he don't [*sic*] say that in book. He says Cabinet split and W over-persuaded. I also sent him evidence that Fyfe–Eden contrast was nonsense. He argued that his book made plain my Rome press conference was not damaging. That is hardly line taken by first edition I saw. Oliver C[handos] . . . remarked Harold should have been a Cardinal in Middle Ages, when he could have intrigued all week, confessed and been absolved on Sunday and start whole business again on Monday.[68]

Contemporary evidence seems to support Eden's interpretation of events, that the supposed difference between himself and Maxwell-Fyfe was 'a complete fabrication'.[69] Eden's words may have added some clarity to those of his colleague, but were scarcely at variance with them. As the longer and more authoritative pronouncement, the Home Secretary's words inevitably attracted most attention at the time. *The Times* attached no particular significance to Eden's remarks in Rome, but noted that Maxwell-Fyfe's statement caused 'deep disappointment' to 'two thirds of the Assembly of the Council of Europe'.[70] The former French premier, Paul Reynaud, a convinced federalist, spoke immediately after the Home Secretary but before Eden's press conference. His remarks, 'tinged with sarcasm and occasional malice, were strongly critical of our attitude'.[71] Reynaud described Maxwell-Fyfe's statement as a 'direct repudiation' of the motion which had been adopted by the Consultative Assembly in August 1950. A memorandum by Mr Gallagher of the Foreign Office began from the premise that the recent change of government had been expected to produce a dramatic change in Britain's attitude. Maxwell-Fyfe's speech had therefore come as a severe shock to European federalists. Minuting what he judged a 'brilliant account', Eden's first inclination was to circulate this paper to the cabinet. 'But would it offend the Home Secretary? Might it be wiser to send it under cover to P.M. and certain selected colleagues.'[72] Even Anthony Nutting, whose later writings anticipated the Boothby–

Kilmuir–Macmillan thesis of Eden's culpability, struck a different note in an internal Foreign Office memorandum written only a month after the Strasbourg and Rome statements. The French 'passion for federation', he argued, arose largely from the 'utter bankruptcy' of French political leadership. If further attempts were made 'to coerce us into joining a European Federation', Britain should withdraw from Strasbourg altogether.[73]

Traduced, as he saw it, by former colleagues, Eden took comfort in his retirement from the testimony of continental observers. René Massigli, a former French ambassador in London, assured him that his own forthcoming memoirs would contain 'multiples preuves que personne chez vous n'a souhaité entrer dans l'EDC et que les Six s'accommodaient très bien de cette situation'.[74] Even Spaak was equally emphatic: '[Maxwell-Fyfe] ne laissait planer aucun doute à cet égard. . . . La participation britannique à la formation de l'Europe était ainsi indiquée. Elle était dérisoire.'[75] According to Spaak, the Strasbourg delegates now recognized that, if progress towards European integration were to be made, all hope of British participation had to be abandoned.

Yet not all the criticism of Eden was the product of later invention as he was sometimes prone to imply. As early as 3 December 1951 seven members of the Conservative delegation to the Council of Europe, including Boothby and Julian Amery, wrote to Churchill to emphasize that Maxwell-Fyfe's speech, 'while not coming up to the expectations which had been aroused in some quarters, held out considerable hopes'. By contrast, Eden's statement had come as a 'shattering blow to most members of the Assembly'.[76] This interpretation is difficult to square with much of the evidence presented above. In all probability the Conservative delegates read into Maxwell-Fyfe's statement what they wanted to find in the knowledge that the Home Secretary had been a prominent member of Churchill's entourage in establishing the European movement in the late 1940s. Eden, by contrast, had never attempted to hide his scepticism. 'I know that you have always taken a dim view of the Council of Europe,' Boothby had written a few days earlier.[77] At all events Eden easily dismissed these protests, at least to the satisfaction of Churchill and himself. Perhaps more significantly, Boothby revealed a rather different emphasis in a private letter to Eden later in December. He now suggested that Spaak's resignation was 'quite a good thing'. 'By the end . . . he had lost his touch, his head and his temper.' Furthermore, Boothby, supposedly an ardent Europeanist, confessed that he was 'frightened of the Germans' and warned that a European army without British *and American* participation 'must sooner or later be directed by a revived German General Staff'. 'Pretty cool', noted Eden, 'from the man who has always led in advocating German rearmament.'[78]

Over the following weeks Eden sought to clarify the British government's position, though it would be wrong to imply that policy towards continental Europe was ever the Foreign Office's central preoccupation. Eden saw no value in fudging the issue and pretending that certain courses of action were viable when this was evidently not the case. It was an attitude he maintained to the end of his life on this question. This meant being clear about where schemes for European integration were heading. The purpose behind the Pleven Plan for a Defence Community, he reminded Churchill, was, at least in French and Italian minds, to 'pave the way for federation'.[79] Yet a European army could make a vital contribution to Western defence and would avoid the need to admit an independent Germany into NATO.[80] The government would therefore continue to play an active part in plans for uniting national efforts on an inter-governmental basis. But 'it is only when plans for uniting Europe take a federal form that we cannot ourselves take part, because we cannot subordinate ourselves or the control of British policy to federal authorities'.[81] This strategy seemed to satisfy the United States. General Eisenhower, then Supreme Commander of NATO's European forces, stressed the great benefit which derived from Britain's continuing to carry out her worldwide responsibilities, but urged Eden to emphasize Britain's approval of projects such as the Pleven Plan and her readiness to associate with them on a co-operative basis so as to advance the development of collective security.[82]

By mid-December an important policy statement on Europe emerged from the Permanent Under-Secretary's Committee inside the Foreign Office. This concluded that

the United Kingdom cannot seriously contemplate joining in European integration. Apart from geographical and strategic considerations, Commonwealth ties and the special position of the United Kingdom at the centre of the sterling area, we cannot consider submitting our political and economic system to supranational institutions. . . . But while it is neither practicable nor desirable for the United Kingdom to join the integration movement, there would seem to be advantage in encouraging the movement without taking part in it. This is, in fact, the policy which the United Kingdom is now following.[83]

In a widely circulated message to British representatives abroad a few days later, Eden underlined the critical difference between co-operation in Europe and the federation of Europe. For him this distinction had become fundamental.[84] Then, visiting the United States in January 1952, Eden made his most categorical and, it must be said, most negative public statement on this whole issue. Addressing an audience at Columbia University, he confronted the challenge of federalism head on. 'If you drive a nation to adopt procedures which run counter to its instincts', he warned, 'you weaken and may destroy the motive force of its action.' The idea of Britain joining a European federation was 'something which we know, in our bones, we cannot do'.[85]

In later years Eden tended to claim that there had been no significant differences of opinion inside the Conservative cabinet over European policy. After seeing an advance copy of Macmillan's memoirs in 1969 he complained:

> This is the first I have heard of this and I confess that I resent this new habit, started by Kilmuir, of colleagues who did not speak up in opposition in the Cabinet at the time, still less resign on the issue, ventilating their criticism in their biographies fifteen years after the event.[86]

At least in Macmillan's case such criticism is unfair. Over a period of several weeks from the end of 1951 the Minister of Housing waged a characteristically languid campaign to modify government policy, of which Eden was fully aware at the time. But Eden, during his final term as Foreign Secretary, was more than ever intolerant of his colleagues' incursions into his own area of specialization, and impatiently brushed Macmillan's efforts aside, while excluding them from his future recollection. Furthermore, the debate between the two men was never as clearly defined along pro- and anti-European lines as Macmillan, claiming a vital principle to be at stake, later implied. After Eden and Churchill had clarified Britain's position during a visit to Paris in mid-December 1951, Macmillan wrote to Eden. His tone could scarcely be described as critical: 'May I congratulate you on your success, and the change of climate that has followed your return to the Foreign Office. It is a triumph for you.'

On specifically European policy Macmillan seemed ready to concede
to Eden's case:

> I have felt that our case had not been understood. Now it is so. If we
> had been 'in' on the Schuman Plan or European Army Plan two years
> ago, we might have moulded both to our liking. But it would have
> been impertinence, as well as folly, to butt in at the last stage of the
> negotiations. It could even have been called a wrecking policy.[87]

None the less, in response to the policy statement of the Permanent Under-Secretary's Committee, Macmillan submitted a set of
alternative proposals to Eden in mid-January. He did so in terms
which scarcely conveyed the importance he supposedly attached to
the issue. Macmillan had 'amused [him]self' in the intervals of his
work in composing what was 'at any rate a point of view' which the
committee 'might find it worth looking at'. He left it to Eden to
decide whether he would like anyone else to see it.[88] In the circumstances it was scarcely surprising that one of Eden's staff minuted:
'The S of S knows of this letter and the enclosures, but does not
want to read them at the moment.'[89] Macmillan claimed that it was
still possible for the United Kingdom to divert the movement of
European integration away from its present course – something he
had appeared to doubt in his private letter to Eden a month earlier.
Conceding that Britain could not, 'of course', join a federation or be
party to any arrangement which limited her membership of the
Commonwealth or her association with the United States, he
reminded Eden that federation was not the only form of constitutional association between nation states. Reverting to the sort of
scheme he had unsuccessfully put forward as an alternative to the
Schuman Plan in 1950, Macmillan wondered whether it might be
possible to offer the continent the alternatives of a European con-
federation with Britain or a continental federation without her.[90]
What Macmillan seemed to envisage was the organization of Europe
along lines comparable to those of the British Commonwealth.

Eden, however, had an altogether clearer picture of the realities of
the European situation. The movement for unity, he argued, was
proceeding along two tracks. One was the Atlantic Community, a
wide association of states which was achieving a considerable unity
of purpose in the creation of Western defence without a formal
surrender of sovereignty. The other, a European Community, consisted of a small group of states moving towards political federation
through the progressive establishment of supranational bodies in

specified areas.[91] Macmillan wanted Britain to be ready with a non-federal replacement scheme in case the existing EDC proposals collapsed.[92] This was very much what Eden would devise in the crisis of 1954. For the time being, however, he did not believe that the advantages of putting forward such a fall-back position would outweigh the inevitable charge that Britain was seeking to sabotage the existing programme of the continental powers. As Dixon pointed out, Macmillan's proposals overlooked the very strong commitment on the continent to proceed along federal lines, with the long-term goal of a European political authority. There was no indication that the continental states would accept an inter-governmental approach as a satisfactory alternative, even with British participation.[93]

When, therefore, the issue came before the cabinet in mid-March it was no surprise that Eden's views carried the day. He argued that Macmillan's plan for the creation of a European union on the pattern of the Commonwealth was unlikely to satisfy the aspirations of the continental states.[94] The Minister of Housing registered his disappointment in a further memorandum to Churchill, recalling again the hopes raised by the latter's pronouncements in opposition, but without effect. According to his own later testimony Macmillan now seriously considered resignation.[95] But Eden, and for that matter Churchill, could be forgiven for thinking that only a minor difference over tactics had been at issue. The Prime Minister submitted Macmillan's final protest to the Foreign Secretary, but doubted whether 'we can do any more than [Eden] is now planning'.[96] Macmillan himself seemed ready to toy with other ideas, telling Colville in May that the development of the British Empire into an economic unit as powerful as America and the Soviet Union was 'the only possibility'.[97]

With the limits of British policy established to his satisfaction, Eden was better placed to pursue the more positive aspects of his European strategy. He aimed to build bridges between the supranational communities of the Six and those countries, including Britain, which preferred inter-governmental co-operation. In March 1952 he presented the outline of the so-called Eden Plan to the Ministerial Council of the Council of Europe. While Britain could not become part of Europe, she was determined not to be isolated as a result of the supranational initiatives of the Six. Eden convinced his audience that Britain's aim was not to put a break on the march to federation, but to establish genuine organic links with the Coal and Steel Community and the EDC.[98] 'We have established a formal and special relationship between the United Kingdom and the European Defence Community,' he told the Commons in April.[99] Two months later Eden summed up British policy in terms which can scarcely be described as anti-European:

> Our stock in Strasbourg now stands high. This is due to our initiative in launching proposals which have been accepted as proof of our willingness to establish close political, as well as technical, links with the European Communities. If, as I hope, the proposals are translated into action, we shall have rounded off an awkward angle and have brought our relations with the Council of Europe into line with our general policy of encouraging the movement towards federation in Europe and associating the United Kingdom as closely as possible with it short of a merger.[100]

Eden doggedly pursued his policy of 'association' with Europe over the next two years. As the continental powers considered the idea of a European Political Authority, the Foreign Office stuck to its chosen course. 'The best line for us to take', argued Dixon, 'seems to be not to discourage the European countries from undertaking this venture, but to try to ensure that the European Political Authority develops in a spirit friendly to us and in a form which permits close association with us.'[101] The Eden Plan was certainly modest in scope. It envisaged such unexceptionable links between the Six and the Council of Europe as joint meetings and shared facilities. In April 1954 a treaty of association was negotiated with the EDC allowing for the inclusion of a British army division in an EDC corps. Within the cabinet Eden had been cautious about any assurance to France that Britain would retain troops in Germany at a predetermined level since it was certain that once the EDC was ratified the Americans would respond by reducing the number of their troops in Germany.[102] The Foreign Office hoped for closer ties with the Coal and Steel Community. Eden told Monnet at the end of August 1952 that Britain's task was to lay the foundations for an intimate and enduring association with the Community.[103] At the beginning of 1954 he even seemed ready for Britain to join a common market in steel. Opposition inside Whitehall and the steel industry itself led to the dilution of this proposal and the treaty of association actually signed in December was more limited than Eden might have wished.[104]

Much of the Whitehall establishment was far more sceptical than Eden of the developments taking place in Europe. The British representative on the Organization for European Economic Co-operation wrote of a deep malaise in France, Germany and Italy which had led to the search for supranational solutions to difficulties which were intractable on a national scale. The continental governments were groping reluctantly towards a federal solution while seeking at every turn to leave doors open for Britain to associate herself with any new developments. Thus the 'tendency to partial European federation' was based upon uncertainty and fear rather than 'an urge to European brotherhood on a healthy basis'.[105] The junior Foreign Office minister, Anthony Nutting, for all his later Euro-enthusiasm, struck a similar note. The Six no longer believed in national sovereignty, 'as we do', because it had let them down in two world wars. In desperation, therefore, they sought safety in a unified state.[106]

With all the benefits of hindsight it is clear that such sentiments did less than justice to the positive forces animating many in Europe which would come to fruition in the signing of the Treaty of Rome as early as March 1957. But such thinking was by no means uncommon in the British official mind of the early 1950s and it helps to set Eden's more constructive policies in context. If the Foreign Secretary is to be blamed for a lack of vision as to what the forces of European supranationalism could achieve, his guilt must be shared, with very few exceptions, by a whole political generation. An enormous leap of imagination would have been required to lift Britain from the non-European tradition in which her history had placed her. When the Conservative government came to office in 1951 Britain's foreign trade and industrial output were still as large as those of France and Germany combined.[107] All gave priority to the Atlantic dimension of Britain's external policy which would preclude entry into Europe except in association with the United States. Britain's status as an independent world power was a reality, however short a term history would set on this role. Few anticipated how quickly the Commonwealth would decline in Britain's economic and commercial life over the next two decades. As late as 1955 the Commonwealth accounted for almost half of Britain's imports, more than half her exports and two-thirds of her overseas investments.[108] When at a cabinet meeting in March 1952 Salisbury stressed that Britain was not a continental nation 'but an island power with a Colonial Empire and unique relations with the independent members of the Commonwealth', he was stating a fact to which all his colleagues would have subscribed rather than opening up a controversial idea for debate.[109] Shuckburgh recalled Eden saying that in the postbag of a typical English village 90 per cent of letters coming from abroad would be from beyond Europe, from countries where British troops had been stationed or where Britons had their relatives living.[110]

Above all Eden never believed that the European option on offer – with its commitment to federalism – was a viable one for Britain. In later years it has become common to suggest that Britain, through a combination of negligence and wilfulness, missed the 'European bus' waiting to transport her to her rightful destiny. But Eden would have wanted to know the precise destination of such a vehicle. As a research assistant put it when reviewing Eden's role in the 1950s, it was surely wiser, and indeed more honest and honourable, to admit from the outset that Britain could not support a policy whose ultimate aims were unacceptable.[111] Those ultimate aims were less obscure as the Defence Community and the Coal and Steel Community came into being than has sometimes been suggested. The head of the British delegation to the European Coal and Steel High Authority had no doubts:

> There could be no sense in creating an ECSC as an end in itself. It would be natural, as leading figures in the Community have indicated in conversations and speeches, for the principles which lie at the root of that Treaty to be extended progressively throughout the economic field and to result in a single market with its accompanying measures of a common currency or linked currencies and a unified transport system. From there to a common foreign policy and a common budget for all but local requirements would be almost inevitable developments.[112]

Accepting this interpretation, and convinced that this was not an acceptable course in the prevailing climate of British opinion, Eden had little room for manoeuvre. 'I am quite certain in my own mind', he concluded many years later, 'that had we pursued any other course than that which we followed, the result could have been disastrous for Europe.'[113]

*

If the federal implications of the EDC made it unacceptable for Britain, rather different considerations caused hesitations in France and, though the Defence Community treaty had been signed in May 1952, it awaited ratification by the French National Assembly. It was after all less than a decade since France and Germany had been at war and a background of conflict since 1870 was not easily forgotten.

In December 1952 Eden discussed the situation with the French President, Vincent Auriol. The latter conceded that the EDC was necessary to prevent independent German rearmament, but that without Britain Europe would be only a sort of Holy Alliance in which Germany predominated. At this stage Eden remained convinced that creating a European army would be a thousand times better than admitting an independent German state into NATO.[114] One consideration, he told Churchill, who also seemed interested in a looser arrangement than the EDC involving a German national army, was the need to support the West German Chancellor. Dr Konrad Adenauer's 'whole foreign policy' was based upon Franco-German reconciliation and co-operation. The breakdown of the Defence Community would be a serious blow to Adenauer's position and might well lead to his defeat in the next German elections.[115] It was, though, inevitable that the longer EDC ratification was delayed the more Eden and the Foreign Office turned their minds to alternatives in the event of its complete collapse. This had been part of his policy from the outset, though unlike Macmillan he had not wished to put forward alternative schemes publicly so as to give the EDC every chance of success. By late 1953, however, Eden was wondering whether it would not be possible to contain Germany within NATO.[116] This might involve a permanent commitment to maintaining troops on the continent.[117] But he remained anxious to retain as much British freedom of action as possible.[118]

When, therefore, on 30 August 1954 the French Assembly failed, on a procedural motion, to ratify the EDC treaty, the British government was better prepared to respond than has sometimes been implied. Three days earlier Eden had warned the cabinet that there was every likelihood that the treaty would soon be rejected, though he remained insistent that, until the critical vote, no public encouragement should be given to any idea of an alternative to the treaty. Subject to safeguards, 'he could, however, envisage no better alternative' than to bring Germany into NATO in agreement with France and America.[119] Eden's later suggestion that his solution was a bath-time inspiration is less than convincing.[120] The idea of developing the 1948 Brussels Pact, of which NATO was essentially an enlargement, had been under consideration for some time and Foreign Office figures such as Frank Roberts and Anthony Nutting have since laid claim to its parentage.[121] But this is not to deny to Eden the overwhelming credit for resolving the crisis which the French decision created. As Jebb, now Britain's ambassador in Paris, put it, 'instant and far-reaching decisions will have to be taken'.[122]

Eden rose to the occasion. His aim was less to assert Britain's belated leadership of Europe than to resolve the vexed but specific question of German rearmament.[123] With a conference of the European members of NATO together with the United States and Canada called for 28 September in London, Eden wisely decided on a preliminary tour of the six European capitals.[124] His difficulties came from two quarters. In Paris premier Pierre Mendès-France was initially obstructive. Eden warned that a negative policy would drive Germany into the arms of the Russians and encourage the United States to retreat into 'fortress America'.[125] More important was an unwanted intervention by the American Secretary of State, John Foster Dulles, who made a sudden flying visit to Adenauer on 16 September. In later years Eden attributed Dulles's interference to little more than personal malice:

> He flew to Bonn while I was in Rome and pressed Adenauer against it. When I met him in London, having been to Paris and got M.F.'s agreement, he was against the London meeting and made every pretext. In truth he was against anything in which U.K. and perhaps above all self could take a lead. A meanly jealous man.[126]

Dulles, however, was worried that an enlarged Brussels Pact would be less acceptable to the American Congress than the EDC, since the former was clearly inter-governmental in conception and could not be presented as the precursor of European federation.[127] But Dulles's inability to propose any alternative course of action allowed Eden to proceed.

As chairman of the London Conference Eden was in his element. To the patient diplomatic skills for which he was renowned he added, when Mendès-France again proved obstructive, a timely display of the temper he usually reserved for more private occasions. It was, recorded one observer, 'more violent than anything I have ever experienced in international meetings'.[128] To clinch his deal Eden offered an apparently dramatic commitment to maintain four British divisions on the European mainland, together with a tactical air force, for as long as necessary. France seemed finally to have extracted from Perfidious Albion the sort of cast-iron guarantee she had craved for fifty years. Eden made the concession after overcoming unexpected resistance from Churchill at a crucial Downing Street meeting two days earlier. This particular aspect of Eden's solution to the EDC crisis probably was a last-minute expedient. Both Nutting and Jebb recorded not only their surprise at the British commitment but also Eden's steadfast opposition to such a step only days before. To the suggestion of the need to make a permanent guarantee of troops, Nutting recalled, Eden had reacted like a 'kicking mule'.[129] Similarly, less than a month before Eden's historic declaration, Jebb had been instructed to protest to Mendès-France for having reverted to this 'shop-soiled' idea.[130] Eden himself recorded a discussion with Nutting and Jebb at the Paris embassy on 15 September 1954: 'Some rather fierce talk about what we should offer French in addition. I was against anything at this stage and was rather unpleasant about it goaded by my fatigue and Tony's assurance that Chiefs of Staff were willing.'[131] Yet it remains possible, as Nutting suspected, that Eden, playing his cards close to his chest, had deliberately kept Mendès-France waiting until the last moment before offering his carrot to secure the deal.[132]

Eden's scheme allowed for the extension of the Brussels Pact to include West Germany and Italy. This expanded organization became the Western European Union in 1955. Britain and 'the Six' were thus joined in a non-supranational body which Britain might logically expect to lead. The Allied occupation of West Germany would now end, allowing that country to become a full member of NATO. Yet the price Britain paid for all this was perhaps less than it appeared. Her 'unprecedented commitment' was 'merely a more definite and succinct statement of [her] intention than had been contemplated under the terms of Britain's association with the EDC' in April.[133]

At all events Eden's achievement won widespread praise. Selwyn Lloyd, his Minister of State, sent 'hearty congratulations' from New York: 'Everyone here is delighted . . . and your ears must be tingling at the things which are being said about you.'[134] Even Dulles applauded Eden's success. In later years Spaak paid tribute to Eden's diplomatic expertise and wrote of his having 'saved the Atlantic Alliance'.[135] There followed the Wateler Peace Prize from the Carnegie Foundation together with the Order of the Garter from the Queen. The London Conference was a major element in the most successful single year of his Foreign Secretaryship. Eden had pursued a consistent policy since 1951 of supporting European integration from the outside and only coming forward with an alternative and less ambitious scheme once the EDC itself had collapsed. That policy appeared to have been crowned with success. The Western

European Union, Eden's inter-governmental alternative, also removed the gap between the Foreign Secretary and his Conservative critics, who had never wanted to pursue the federal option but had believed that Eden should have launched a WEU-type scheme at an earlier stage. Boothby even took credit for having 'helped to kill EDC'. Few things in his public life, he told the Commons, had given him more satisfaction.[136] Another Tory critic now wrote to Eden in revealing terms: 'I always hoped for a solution on these lines rather than for EDC itself. As things have turned out, however, I feel bound to admit that your previous support of EDC has greatly strengthened our position for carrying out the new policy.'[137]

What Eden failed to realize was the capacity of the European federalists to recover from their setback of 1954, to build upon the Coal and Steel Community – which, of course, had survived the EDC débâcle – and to create the European Economic Community in a remarkably short space of time. In the spring of 1955, by which time Eden had become Prime Minister, the representatives of the Six, meeting at Messina, were ready to consider extending the principles of ECSC co-operation into other sectors of the economy or even the setting up of a full-scale Common Market. Yet Eden's later critics showed no greater prescience. Writing to Eden in July 1953 Boothby had clearly underestimated Europe's capacity to go its own way without British participation:

> I am sure that there is no head of steam behind the attempt of Spaak and Monnet to create a 'little federation'; and that, if it is persisted in, it will collapse for lack of mass appeal or popular demand. This I believe to be the main reason for the continuing refusal of the French to ratify the EDC Treaty. They feel that it might push them into a rigid political federation from which we should be excluded; and that they don't want, and won't take.[138]

As late as July 1956 Boothby was still telling the Commons that the EEC would fail without Britain and that its members would be happy to change their course and submit to British leadership on whatever terms Britain chose to impose.[139]

From the Messina Conference there emerged the Spaak Committee which engaged in detailed discussions of the twin issues of a Common Market and collaboration on nuclear energy. To this the Eden government agreed to send a representative, the Board of Trade official, Russell Bretherton. The broad policy of Eden the Prime Minister was little different from that of Eden the Foreign Secretary

In Tasmania, 1925

2 Lobby Journalists' Luncheon, 23 July 1935

3 On top of the Empire State Building with former Govern
of New York State Alfred E. Smith, December 19

Freedom of Durham, 1945. Eden with troops of the 50th Division, Durham Light Infantry

At the Acropolis, March 1941

6 Eden and Sir Alexander Cadogan leave the Capitol after President Truman delivered his first address to a joint session of Congress and the House of Representatives, 16 April 1945

Arriving at Vancouver, 1949

With Robert Carr, Rochester, New York, August 1951

12 In Garter robes

left:
11 Arriving for a meeting of the Cabinet for the first time since his illness,
2 October 1953

13 The Prime Minister, May 1956

right:
15 Commonwealth prime ministers meet at 10 Downing Street, July 1956

below:
14 With John Foster Dulles

16 The Prime Minister leaves 10 Downing Street for the House of Commons, 12 September 1956

17 Sir Anthony and Lady Eden at London Airport, just before departure for Jamaica, 23 November 1956

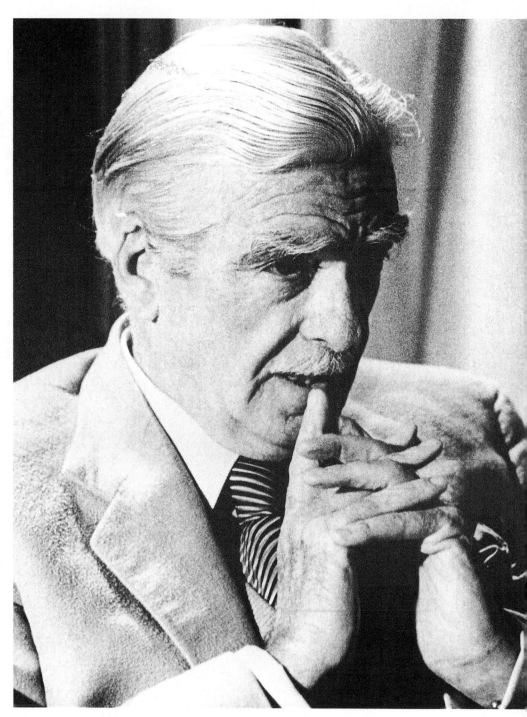

18 Interviewed for Thames Television, April 1972

– benevolence towards the initiatives of Britain's European neighbours while avoiding federalism – though representation at the
Spaak Committee did suggest a readiness to try to steer the outcome
of these initiatives in an acceptable direction. It remained the prevailing Whitehall assumption that Britain could always associate
herself with whatever emerged from Europe, if only because the
continuing mutual suspicions of France and Germany made a British
presence desirable to both parties. The promptings of Eisenhower
and Dulles that Britain should consider joining the proposed European Community probably caused Eden to question the reliability
of America as Britain's leading ally more than it prompted a recon-
sideration of his European policy. The supposedly pro-European
Macmillan was now installed at the Foreign Office, but failed to
change the course of British policy, less because he was constrained
by Eden than because of his involvement in other issues such
as Cyprus and East–West relations. No more than Boothby did
Macmillan see the revival of European federalism as a challenge
demanding an urgent British response. The 'Monnet concept', he
told Tory MPs in November 1955, 'was doomed to failure'.[140] In fact
the Treasury and the Board of Trade, rather than the Foreign Office,
were the most important government departments in the determination of European policy at this time, as Bretherton's appointment to
the Spaak discussions confirmed. This emphasized that the issues
were seen primarily in economic and not political terms. Even had
the situation been otherwise, British policy would have been unlikely to follow a different course. The Foreign Office, including such
later Euro-enthusiasts as Gladwyn Jebb, was overwhelmingly
opposed to British membership of a European Common Market.[141]

But as the Spaak Committee proceeded the difficulties of the
British position became apparent. Bretherton was concerned that
the more Britain got involved in the committee's deliberations the
more she would appear committed to the ultimate outcome. If, on
the other hand, Britain stayed aloof, she would limit her capacity to
exclude distasteful features from the final Spaak report.[142] The head
of the British delegation to the Organization for European Economic
Co-operation summed up the problem:

> If it is not our intention to become members of the Common Market
> or be associated . . . it is far better to be outspoken at the outset. . . .
> I would regard it as a fatal mistake to abandon our present status in

order to become a member of a purely European grouping which by implication would require some abandonment of sovereignty.[143]

Such considerations led the British to distance themselves from the Spaak Committee – Bretherton was withdrawn in early November – and produced a distinct change in policy by the end of 1955. The government now tried much more actively than before to steer the Six away from supranationalism by proposing a British alternative to the Common Market in the form of a wider free-trade area, based on the OEEC. Tactically the move was reminiscent of the wrecking policy which Macmillan and Boothby had proposed in the first days of the Conservative government. The British ambassador in Washington assumed that Macmillan's 'real object' was 'to hasten the end of the Common Market'.[144] But the singular failure of the new approach adds a measure of justification to the policy upon which Eden had insisted after 1951. The British initiative, widely interpreted as an exercise in sabotage, merely served to concentrate the minds of the European federalists more purposefully on their own endeavours.

Spaak wrote to Eden in early February 1956 expressing his disquiet at the new British attitude and reaffirming his personal commitment to supranationalism, but received only a non-committal reply.[145] During 1956 the British government in general and Eden in particular had much else besides Europe with which to concern themselves. Even so, the slow manoeuvring of Whitehall departments scarcely seemed in tune with the momentous events taking place on the continent. Eden later remembered 'occasional discussions about such topics as the liberalisation of trade in the OEEC'. But there was 'never any suggestion from the Treasury or Board of Trade that there was a bus to be missed'.[146] Finally on 12 July Harold Macmillan, now Chancellor of the Exchequer, and Peter Thorneycroft, President of the Board of Trade, presented a joint memorandum to the cabinet suggesting how the pressure for tariff reduction might be handled at the forthcoming meeting of the OEEC. Considering that, following a meeting of the Foreign Ministers of the Six in Venice in May, the European powers were now working on a formal treaty, the cabinet still showed no sense of urgency. As Eden recalled, it was proposed by Macmillan and Thorneycroft

that the Departments concerned should examine what would be involved for us in possible arrangement which would result in ourselves and the Messina group progressively removing tariffs over a large part of the trade between us, excluding agricultural and horticultural products. At the same time we ought, they contended, to try to protect the position of the Commonwealth in our market for agricultural products and, our colleagues suggested, safeguard the advantages we enjoyed through Imperial Preference in the Commonwealth markets.[147]

Plan 'G', as the British scheme for a free trade area became known, was soon overshadowed by Nasser's nationalization of the Suez Canal on 26 July. Ironically that crisis momentarily brought Britain and France closer together than at any time since 1940. But the circumstances of its ending – a unilateral British decision merely communicated to the French – helped determine the latter to proceed with all speed towards their European option, despite earlier misgivings over the Common Market proposals. The cabinet discussed Plan 'G' at two meetings in mid-September.[148] But not until 26 November, with the Middle Eastern crisis past its peak, did Macmillan announce to parliament the intention to begin negotiations for an *industrial* free-trade area within the OEEC. By this stage, however, the Messina powers were well advanced in drafting what became the Treaty of Rome. This contained the germ of an agricultural policy which would exclude British participation unless Commonwealth interests were set to one side. The debate in the Commons, just six weeks before Eden's final resignation, revealed no significant differences of opinion. Macmillan referred to the great markets in the United States and the Soviet Union and to the desirability of founding a single market for *manufactured* goods in Europe. But significantly he added:

> I believe that we all agree that it is quite impossible for the United Kingdom to join such a Customs Union. . . . I do not believe that this House would ever agree to our entering arrangements which, as a matter of principle, would prevent our treating the great range of imports from the Commonwealth at least as favourably as those from the European countries.[149]

As Eden later stressed, it was not a case of 'tepid Europeans . . . propounding policy and holding back colleagues who would have entered a customs union'.[150]

*

The ignominious failure of the Suez venture provided an obvious occasion to reassess Britain's place in the world. In early January

1957, just before Eden left for Sandringham to inform the Queen of his decision to resign, Selwyn Lloyd, the Foreign Secretary, proposed to the cabinet that the time was now ripe for Britain to pool its resources with 'our European allies so that Western Europe as a whole might become a third nuclear power comparable with the U.S. and the USSR'.[151] As Lloyd recalled, he did not get much support from his colleagues, most of whom felt that the first priority was to mend Britain's badly damaged fences with America.[152] But a recently declassified document suggests that Eden's mind was moving on the same lines as Lloyd's. At the end of a brief assessment of the lessons of Suez the Prime Minister concluded:

> While the consequences of this examination [of Britain's world position and domestic capacity] may be to determine us to work more closely with Europe, carrying with us, we hope, our closest friends in the Commonwealth in such development, here too we must be under no illusion. Europe will not welcome us simply because at the moment it may appear to suit us to look at them. The timing and the conviction of our approach may be decisive in their influence on those with whom we plan to work.[153]

Too much, however, should not be read into these words. It would be easy to suggest that, after a cathartic tragedy of classical Greek proportions, Eden had finally achieved a self-knowledge and understanding which led him inexorably towards locating Britain's destiny inside Europe.[154] In fact Eden's apparent change of heart, as indeed Lloyd's, reflected above all else a man who considered himself to have been grievously let down by his American allies.

The authentic Eden retained into his long years of retirement a consistently cautious, yet intellectually honourable, attitude towards the question of European integration.[155] Europe was one of the few issues which could persuade a man by then under strict doctors' orders to limit his involvement in public affairs to return to the political arena. When in the early 1960s British entry into the EEC became a distinct possibility, Eden was as concerned as ever that the full implications of this step should be considered. He felt that the government of his successor, Harold Macmillan, failed to come clean with the British public about what membership of the Common Market entailed. In particular Macmillan was 'wrong to funk the federation issue'.[156] It had always seemed to Eden a logical and almost inevitable development of the Common Market that those countries which joined it should ultimately federate:

In these circumstances it does not seem to me unreasonable, or offensive, to any other country to ask that H.M. Government should make its position clear before it embarks upon negotiations which may lead to the acceptance of conditions intended by others to promote federation. This seems to me indispensable, if we have no intention of travelling the full way ourselves. . . . What we must not do . . . is to join any organisations without a full understanding of its implications and then find ourselves being swept further than we intended. . . . Better clarity now than recrimination a few years hence.[157]

*

It is a truism that each generation rewrites its history in the light of contemporary perceptions and preoccupations. In the mid-1990s, in the wake of the Maastricht Treaty and with a British government arguing – not for the first time – that Europe is now ready to move in a direction more congenial to this country, Eden's attitude towards European integration appears less culpable than it did thirty years ago. In the early 1960s, with intellectual opinion moving strongly in favour of British membership of the EEC, it became usual to criticize those who had missed earlier, supposedly golden, opportunities for Britain not only to participate in the European movement but to dominate it. 'The Common Market enthusiasts', wrote the former Foreign Office official William Strang in 1962, 'have evolved a kind of myth about the British attitude to Europe which has gained a currency which it does not deserve.'[158] In so far as such opportunities ever existed, personalized attacks on Eden distort the reality of British politics in the 1950s. As a recent biographer put it, 'Eden was not some politician from a vanished era struggling in vain against an integrationist flood tide in Western Europe, let alone Britain'.[159] When in 1961–2 British membership of the Community became a practical proposition, many voices in both major parties were raised against the Macmillan government's proposals. 'In the mid-1950s the political consensus wholly precluded such a venture', as was apparent in the Commons debate on trade policy in November 1956.[160] Richard Crossman accurately captured the reality of the situation in June 1961:

The fact that Britain must now consider entering the Market is a demonstration that the attempt to maintain ourselves as an independent Great Power, which all parties have been making since the war, has come unstuck. Even three years ago no one was dreaming of proposing such a thing.[161]

The consensus of Eden's political generation was that Britain must be regarded as a world power and not merely 'as a unit in a federated Europe'.[162] Evelyn Shuckburgh, by no means an unqualified admirer of Eden, did believe that opportunities had been missed, but refused to place the blame on Eden's head:

> When everyone has had his say, history will not blame any individual (with one exception [de Gaulle]) for the pathetic stumblings and tumblings of the European nations in their progress towards unification. History (assuming that it will be written in the long run by genuine historians searching for truth, and not dictated by dogmatists) will surely show that the trouble lay in the unfortunate nonsynchronisation of the realisation, in the different European countries, of their need for unity and of the inadequacy of their national status.[163]

Eden was in no sense narrowly anti-European. But he did stress the established props of Britain's great-power status and feared that these were incompatible with full participation in an insular Europe.

None of this, of course, stopped others from scrambling on to the European bandwagon a few years later, claiming a clarity in their earlier vision which had never really existed and heaping scorn on Eden for his shortsightedness. 'That is called the benefit of hindsight, I believe' was Eden's acidic comment on such charges.[164] His critics could always point to Churchill's prophetic speeches in the late 1940s as proof that Eden must have diverted the Conservative Party from its natural course over the following decade. Yet even Spaak came to realize that he had read too much into Churchill's words.[165] Looking back in 1966 Boothby was sure that the distinction between Eden and himself had been clear-cut. 'You never wanted us to go into Europe. I did. That was a perfectly legitimate difference of opinion.'[166] A decade earlier, however, Boothby had conceded that he and Eden had never seriously disagreed on any major aspect of policy, at home or abroad, during a political association over thirty-five years.[167] Certainly Boothby had never favoured a federal solution. In 1953 he declared his opposition to all idea of a six-power European federation, advocating a wider, looser and more flexible European association.[168] Macmillan's charge against Eden was more subtle. Giving Eden grudging credit for his search for solutions to short-term problems, Macmillan left no doubt that he was operating on a higher plane. 'My eyes, rightly or wrongly, were fixed upon a more distant future – the organisation of Europe in the second half of the century, and the place which Britain and the Commonwealth should hold in a great design.'[169] Macmillan's most recent biographer stresses tactical factors in terms of the internal hierarchy of the Conservative Party in his decision to take up the European issue.[170] Be that as it may, Boothby and Macmillan singularly failed to translate into practical terms their idea of a middle course between Eden's notion of inter-governmental co-operation and the federalism of Spaak and Monnet, which they themselves rejected.

Yet, at a time when the European Community is no longer widely proclaimed as the panacea for Britain's ills, Eden's role may be viewed in a more constructive light than simply as part of the collective myopia of a political generation. When the uneasy relationship of Britain and continental Europe over the last half-century is viewed as a whole, perhaps the most striking feature to emerge is British reluctance to confront the reality of the European movement head on. At no time have the full implications of a potentially federal community been openly discussed. After Eden's retirement the wider significance of the Treaty of Rome was pushed to one side as politicians sought to persuade a sceptical public of the advantages of British membership, largely in terms of selfish economic benefit. The debate over the last thirty years has been dominated by what are in practice issues of detail – Commonwealth preference, farm subsidies and budget contributions – with discussions of funda-mental political questions either ignored or postponed to a suppo-sedly more appropriate date. To his credit Eden regarded such tactics as essentially dishonest. His concern was to address those issues which really mattered. A letter written as Britain's first application to join the Common Market reached its climax sums up his central beliefs:

> The authors of EEC always intended it to lead to a closer political union, and if it is to succeed it must do so. . . . Would we opt for such an arrangement now? May be the majority would, but all I am concerned about is that the British people should know where they are going before they wake up and find themselves where they do not want to be.[171]

Three decades on, and with European integration again at the fore-front of the British political agenda, Eden's words seem no less relevant or important.

11

Eden and the Cold War

It is a challenge to our whole way of life from the Elbe to San Francisco, from the Balkans to China. We must never forget this in anything we do.[1]

THOSE charged with the direction of British foreign policy after the Second World War could not fail to recognize that they were operating in a vastly different environment from that occupied by their predecessors. The war had not only destroyed the aspirations of Nazi Germany to world domination but also the primacy of Europe in international affairs. Britain had been victorious on the battlefield but at an enormous cost to her status as a great power. In effect the country was bankrupt. She came out of the conflict 'the greatest debtor in the world'.[2] Churchill discussed the situation with President Truman at the Potsdam Conference:

> I spoke of the melancholy position of Great Britain, who had spent more than half her foreign investments for the common cause when we were all alone, and now emerged from the war with a great external debt of three thousand million pounds. . . . We should have to ask for help to become a going concern again, and until we got our wheels turning properly we could be of little use to world security or any of the high purposes of San Francisco.[3]

But Britain had not fought the war in order to relapse into the status of a second-class power. She still stood at the head of a vast Empire and Commonwealth whose independent members had, with the exception of Eire, rallied to Britain's aid in 1939. And Britain retained her position in the Grand Alliance, even though her standing was, as one official put it, like that of 'Lepidus in the triumvirate with Mark Anthony and Augustus'.[4] Across the political spectrum the role of the world's third greatest power, symbolized by Britain's presence at Yalta and Potsdam, was cherished almost as an end in itself. To this extent the decision of the electorate in July 1945 made little difference to the overall thrust of British policy. Indeed foreign policy played only a negligible part in the election campaign and Churchill, in taking Attlee with him to Potsdam, had explained that the two party leaders thought alike on foreign affairs.

Yet the fact remained that the post-war world would be dominated by the two superpowers which had emerged from the war, the United States and the Soviet Union. The wartime partnership of these two countries was rapidly transformed into the state of latent antagonism which history has dubbed the Cold War. Britain's position within this divide was never in doubt, notwithstanding Eden's hopes that the Anglo-Soviet alliance could be maintained into the years of peace. The Cold War dominated the remainder of Eden's political life. Nearly all the diplomatic issues of the next decade contained at least a Cold War dimension. Yet it was a feature of this conflict that contact between the two sides rapidly disappeared. In practice, therefore, the fundamental questions for Britain as a significant but second-rank player on the Cold War stage concerned more her relations with her leading ally, the United States (and secondarily with friends in the Commonwealth and Western Europe) than those with her newly designated enemy, the Soviet Union. This was particularly the case by the time that Eden re-emerged as Foreign Secretary in 1951. Indeed, circumstances beyond his control drove him into a more restricted partnership with America than his wartime diplomacy suggests he would have preferred. For six years, however, Eden was no more than a spectator on the sidelines while the crucial decisions of post-war British foreign policy were taken by his domestic political opponents.

The leading figures in the new Labour government had of course served in Churchill's War Cabinet and to that body the premier had explained in April 1945 that the dominating facts in the world situation now emerging were the positions of the United States and the Soviet Union. In this new environment the British Commonwealth could only hold its own through unity, together with superior statecraft and experience.[5] Yet it soon became clear that Britain's pretensions to great-power status depended on something more – and that was American money. Such dependence was itself paradoxical since it scarcely implied that degree of freedom of action upon which great-power status ultimately rested. But when in December 1945 the Labour government accepted a 3.75 billion dollar loan at 2 per cent interest, negotiated by Lord Keynes, it did so to help restore Britain's standing in the world arena rather than as a means to cushion a process of inevitable decline. Shortly before leaving office Eden had enthusiastically endorsed as 'excellent' a memorandum by Sir Orme Sargent entitled 'Stocktaking after VE-Day'. Whatever Britain's troubles, this paper had scarcely envisaged that it was now time to accept a diminished role in world affairs. On the contrary,

> we shall have to take risks, and even live beyond our political means at times. We must not, for instance, hesitate to intervene diplomatically in the internal affairs of other countries if they are in danger of losing their liberal institutions or their political independence. In the immediate future we must take the offensive in challenging Communist penetration in as many of the Eastern countries of Europe as possible. . . . We must not deviate from this course or be discouraged even if the United States give us no help or even if they adopt a policy of appeasement towards Russian domination, as well they may.[6]

It was perhaps the post-war Labour government's main achievement to ensure that the United States played a more positive and co-operative role in the emerging Cold War than Sargent feared.

*

Eden was relieved that his successor as Foreign Secretary turned out to be Ernest Bevin rather than Hugh Dalton as many had anticipated.[7] His relations with Dalton had generally been cool, especially when, in Eden's eyes at least, Dalton had used the Ministry of Economic Warfare to challenge the Foreign Office's sovereign control over Britain's external relations. By contrast, Eden had got on well with Bevin, a fellow member of the War Cabinet, and regarded him as the 'best man they had available'.[8] For his part Bevin had told Smuts in 1944 that he would, on certain conditions, be willing to serve under Eden in a post-war coalition.[9] Their friendship was an unlikely one. Bevin was a self-made man who made few efforts to disguise his working-class origins. His blunt turn of phrase contrasted with Eden's exquisite good manners and charm. But both were capable of displays of temper and they shared a tendency towards vanity.

Knowing that the Foreign Office was in safe hands encouraged Eden to take a relaxed view of the country's problems, at least for the time being. 'I don't change my view that God takes care of England, and that, even in this, it may later mysteriously so appear.'[10] Certainly there is no evidence that losing office came to

Eden as anything like the shattering blow it did to Churchill. Those
who saw him in his first weeks of opposition found him relaxed and
enjoying his newly found liberty. It was, thought Lockhart, a perfect
opportunity to restore his health and free himself from Churchill's
'clamping influence'.[11] After a weekend with Eden in late August
1945, Lockhart concluded that he was 'blooming with health', looking years younger and full of energy. A hectic afternoon's gardening
was followed by three vigorous sets of tennis.[12] The former American Secretary of State, Edward Stettinius, also found Eden very
relaxed and not in his usual highly strung frame of mind – in fact
'a normal healthy person for the first time since I have known
him'.[13] Eden admitted that he was enjoying his leisure and that he
liked Bevin and had no wish to embarrass him.[14] He watched the
resumed Potsdam Conference with interest, particularly the performance of Britain's new representatives. It was perhaps inevitable
that he should be privately critical, feeling that they had made 'little
fight' over Poland's western frontiers. 'I am sure that it is folly.'[15] Yet
it is unlikely that Eden and Churchill would have fared more
successfully on behalf of the Poles. Certainly if Bevin failed it was
not on account of any lack of firmness. His American opposite
number, J. F. Byrnes, was rather taken aback by Bevin's forceful
manner and later concluded that 'Britain's stand on the issues before
the conference was not altered in the slightest, so far as we could
discern, by the replacement of Mr. Churchill and Mr. Eden by Mr.
Attlee and Mr. Bevin'.[16]

By the time that parliament debated the King's Speech on 20
August Eden was at his most conciliatory. 'What [Bevin] has said',
he declared, 'represents a foreign policy on behalf of which he can
speak for all parties in this country.' This was scarcely surprising.
As he thought back to foreign policy discussions inside the War
Cabinet, Eden could not 'recall one single occasion when there was a
difference between us'.[17] It was always Eden's contention that who-
ever guided British diplomacy should strive to pursue bipartisan
policies. As he later wrote, 'Foreign Secretaries ought to be as far as
possible above the battle; I always had it in mind that one day I
might have to go down to the House of Commons and tell the nation
that it was at war.'[18] The continuity in foreign policy and accompanying popular quips to the effect that the bulky Bevin was
effectively his double – 'Hasn't Anthony Eden grown fat?' – greatly
pleased him.[19] That continuity was cemented by Bevin's decision,
against the expectation of some, to keep the senior officials of the

Foreign Office at their posts. Even Halifax, a long-time Conservative politician, held on to the Washington embassy until 1946.[20] Bevin and Eden maintained regular contact, especially through the intermediary of Pierson Dixon, now Bevin's private secretary, a post he had also filled under Eden.[21] During the London Foreign Ministers' Conference in September 1945, for example, Eden seems to have advised Bevin to pursue a policy of 'absolute firmness' towards the Soviets.[22] Even Prime Minister Attlee took the trouble to enquire whether Eden had any objections to his proposed visit to Washington in the autumn to discuss atomic energy with Truman and the Canadian premier, MacKenzie King.[23]

At this time Eden had not abandoned all hope of co-operation with the Soviets in shaping the peace. Indeed it would be misleading to suggest that, except in the minds of a few zealots, the lines of the Cold War had yet been firmly drawn. Eden's comparative optimism was evident when he spoke in the Commons in November. Indeed, he struck a more idealistic note than at any time since the early 1930s. He could not, he said, see a way to make the world safe for atomic energy 'save that we all abate our present ideas of sovereignty'. He wanted a world in which the relations between nations could be transformed in a given period of time as the relations between England, Scotland and Wales had been transformed. As a positive move in the right direction he suggested that the UN Charter, drawn up in a pre-atomic age, should be revised to remove the 'anachronism' of the veto. This speech was made in the wake of the failure of the London Conference to reach agreement. Eden did not try to hide the difficulties in the way of a settlement, but he suggested that they were not insuperable, appealing to Bevin to persevere in efforts to arrange another conference. Soviet moves in Eastern Europe, Eden insisted, were directed against the possibility of German resurgence and not at the Western democracies.[24] The decision to try again for agreement in Moscow secured his approval.[25] Almost certainly Eden's renewed bout of internationalist thinking was conditioned by his own aspiration to become UN Secretary-General, returning, he hoped, after three or five years to lead the Conservative Party.[26] A diary entry recording a meeting with the American ambassador on Christmas Day 1945 is revealing:

He is much troubled about prospects for UNO. . . . He thought I should make it plain I was available. He appeared to hope I would do the job and repeated that if UNO was not made to work by someone with statesmanship . . . it must fail and result would be another war and atom bomb. He may well be right.[27]

Tory MPs tended to be less keen than Eden on the pursuit of a bipartisan foreign policy. Trouble arose over the terms of the American loan. Though more generous than anything offered to any other nation, many believed that the United States should have made an outright gift. Churchill and Eden would have been prepared to vote for the measure in the Commons, but the attitude of a number of backbenchers obliged them to adopt an official party stance of abstention. 'The loan was extremely unpopular here', explained Eden, 'and the object of our inaction was to try to prevent the Tory Party collectively voting against it.'[28] A further difficulty was the mounting deadlock in relations between the wartime allies. By early 1946 Eden was becoming concerned at the steady deterioration of Britain's relations with Russia and even with the United States, but he remained anxious that the Conservatives should not be committed to an anti-Soviet policy.[29] The motive force in Soviet foreign policy, he told the Commons on 21 February, was still the experience of German invasion, and Russia remained sincere in its desire to co-operate with Britain and the United States.[30]

Such a stance may or may not have been realistic granted Soviet behaviour in Eastern Europe, but the loose cannon which Eden could not control was his own party leader. At this stage of his career Eden took comfort in the thought that the septuagenarian Churchill was a figure from the past, unlikely to shape the course of a future Conservative government, but this was not how he was universally perceived either in Britain or abroad. The great war leader enjoyed a unique prestige. His pronouncements on world affairs still carried a weight which not even Labour's leading cabinet members, let alone Eden, could emulate. All the evidence suggests that Churchill took no steps to consult Eden before making his famous speech of March 1946 in Fulton, Missouri, and that its contents came as an unwelcome surprise. In the presence of Truman, Churchill declared – with rhetorical flourish, if geographical inexactitude – that 'from Stettin in the Baltic to Trieste, in the Adriatic, an iron curtain has descended across the continent'. He proposed a special relationship between the British Commonwealth and the United States to counteract the communist menace, which had already dominated the ancient states of Eastern Europe and whose secret agencies were striving remorselessly to extend Soviet power far beyond Russia's frontiers. Churchill had exploded the myth of great-power unity. But in the eyes of many, particularly on the Labour benches, he was accentuating and accelerating, rather than merely describing, the division of the world into geopolitical blocs.

It is striking that Eden had no advance notice of his leader's words. Questioned in parliament on what Churchill would say only hours before the speech was delivered, Eden replied:

> The hon. Gentleman said that . . . Mr. Churchill was going to make a speech this evening, in which he was going to put Russia on the spot. I do not know from where he got that information. I certainly have not heard anything of the kind from my right hon. Friend, and, may I add, I do not believe it for one single moment.[31]

Both Cranborne and Eden took Churchill's speech as a sign that he might be ready to give up the Conservative leadership in order to head an anti-Soviet crusade. But in the short term his intemperate language embarrassed Eden and made it no easier for him to offer support to Bevin for a policy of moderation. Not surprisingly Eden spoke to his leader on the telephone to discourage him 'from further polemics with Stalin'.[32] Unfortunately from Eden's point of view, the worsening international situation merely convinced Churchill that he still had a role to play in national and international politics. The gap between the two men was not closed. Indeed, the government's determination to grant Indian independence at the earliest possible moment set them further at odds.[33] 'Anthony had the conviction', recorded Lockhart, 'that WSC almost wants a war in order to stage a comeback.'[34] Eden judged that Churchill's handling of the Russian issue had 'put us all in the wrong', while his 'laments on India and Burma . . . do us no good'.[35] When Eden spoke at Watford of the necessity for a new approach to the Soviet Union, Churchill denounced his thinking as 'Wallacy'.[36] This speech showed Eden's continuing reluctance to accept that a conflict between East and West was inevitable. 'If our Soviet Allies are building their foreign policy on any other premise than our friendship', he declared, 'then they are building on a false reading of the facts.' But his assertion that the two ideologies could live together in peace if both would accept not to 'back their fancies in every other land' involved a curiously optimistic, perhaps naïve, understanding of the mainsprings of Soviet policy.[37]

As has been argued above, Eden's concerns were not restricted to foreign affairs during the years of Labour government. This was the time, more than any other, when he attempted to broaden his interest in and understanding of the key issues of domestic politics; in Churchill's absence he had to fulfil many of the routine functions of opposition leadership; and he himself took advantage of his freedom from office to travel widely abroad. In addition, Eden suffered from several periods of ill-health, including three weeks' convalescence following the removal of his appendix. But diplomacy remained, as always, the field where he felt most at home. As in the war years, so in the immediate post-war period, Eden was torn between his desire to think the best of the Soviets and the mounting evidence that his trust was misplaced. By 1947 there were signs of a hardening of his attitude. He expressed concern at developments in Hungary and welcomed growing American involvement in the affairs of Europe after the issuing of the Truman Doctrine, which seemed to commit the United States to an open-ended defence of democracy against communist subversion. A meeting with Bevin in June at which he urged the latter to respond positively to the Harvard speech of Secretary of State, George Marshall, which had indicated an American willingness to offer Europe economic aid, may have influenced subsequent British policy. Bevin now enthusiastically grasped at what became the Marshall Plan, to the enormous economic benefit of Western Europe.[38]

By 1948, as Czecholosvakia slipped under Russian control, Eden began to show more concern with the ideological content of Soviet foreign policy, speaking in October of the 'Trotskyite principle of world communism'.[39] The crisis over Berlin left him 'worried by the international situation' and 'really anxious about the future'.[40] His speech to the party conference referred to the idea of an 'iron curtain' and showed that he had now caught up with Churchill in the latter's perception of world politics. Like many at this time Eden was forced to abandon lingering hopes of Soviet co-operation, gradually assuming the outlook of a conventional Cold Warrior. The constant objective of Soviet foreign policy, he concluded in October 1949, was the expansion of Soviet influence.[41] In speeches and articles Eden emphasized themes which became recurrent refrains. On the one hand he criticized the government for relying exclusively on American help: 'We cannot become the permanent pensioners of the United States. We have a role of our own to play as the heart and centre of a great Empire.'[42] On the other, when dealing with Russia, 'no advantage is to be gained merely by a policy of appeasement . . . by yielding upon some issue of principle to which we are pledged, merely in the hope that by so doing we shall placate the Soviet Union.'[43]

Not surprisingly, Eden found it difficult to adopt a partisan stance during the election campaign of February 1950. Indeed, when the new Labour government with its much diminished majority met the Commons, Eden was ready with words of reassurance. Parliamentary arithmetic would not change his basic line and the bipartisan approach of the previous parliament would be maintained.[44] The outbreak of the Korean War in June saw this policy put into practice. He welcomed Truman's forthright response to communist aggression and the endorsement this received from the Labour government.[45] British troops entered the war under the authority of the UNO with the full backing of the Conservative opposition. Though differences between the parties began to emerge over the question of European integration and the correct response to the Iranian nationalization of the Anglo-Iranian Oil Company's installations at Abadan in May 1951, an essentially bipartisan line was maintained to the very end of the Labour government. This was still the case after Herbert Morrison replaced the ailing Bevin as Foreign Secretary in March, though personal relations with the new minister were never as close as with his predecessor.[46] Eden believed that the first six months of 1951 might well prove 'the most critical internationally in this post-war world'. Such circumstances demanded a united home front: 'the whole is greater than the part and class hatred the favoured best agent of communism'.[47]

*

While Conservative spokesmen made much of the continuing signs of a siege economy, Britain showed considerable evidence of recovery after six years of Labour government. As Alec Cairncross has concluded, Attlee's administration was 'successful in achieving a fast growth in exports, eliminating . . . the external deficit and then the dollar deficit and sustaining a high level of industrial investment in spite of the virtual cessation of personal savings'.[48] While no power could yet seriously challenge Britain's third place in the international league table, however, the gap between her and the two superpowers had only widened since 1945. But as the prospect grew of his return to office Eden began to show signs of that illusion of British greatness which the growing prosperity of the 1950s, the new Elizabethan age, would soon confirm. Speeches made during an American tour in the summer of 1951 suggest that, like Churchill, he hoped to re-establish the Special Relationship on that basis of equality which the diplomacy of the later war years should have convinced him was unrealistic. In Denver he argued:

The boundaries of British responsibility and influence may seem to have shrunk in recent years, but are they so changed after all? It does not seem so when we look, even now, at the dispersal of British forces in Korea, Hong Kong, and Malaya, in the Middle East, in Austria and Trieste, and nearer home, in Germany. In all this I do not suggest for a moment that the British effort is at its peak. But it is a salutary reminder, none the less, to those who underestimate our present endeavour.

In Chicago a fortnight later he emphasized the importance of the Anglo-American alliance:

It is upon us – the United States and the British Commonwealth – that the chief responsibility for leadership must fall. We have a common language, not only in our words, but in our hearts which beat as one. If we falter or fall apart, all will be lost. If we stand and work together, we can surmount the dangers of the next few years and open a new vista of hope to the world.[49]

The emphasis on the Commonwealth is striking. Like Churchill, Eden had concluded that the unity of the white dominions was a key factor in Britain's ongoing status as a world power.[50]

Though the American contribution in Korea greatly exceeded Britain's own, Eden was keen that the United States should appreciate Britain's part in the joint crusade against communism in other parts of the world, notably Malaya. Korea, Indo-China and Malaya were all parts of what he described as a 'world front'.[51] There is much to be said for David Carlton's judgement that Eden's unwillingness to accept the role of junior partner in the Anglo-American alliance was to be the most significant feature of the remainder of his political career.[52] As tension mounted between Morrison and his American opposite number, Dean Acheson, Eden warned that 'much work will have to be done if we are to keep Anglo-American relations running smoothly'.[53] These were prophetic words from a man who, five years later, would leave those relations at their lowest post-war ebb.

On 28 October 1951 Eden returned to the Foreign Office where he would remain until April 1955. Assessments of his performance as Foreign Secretary in this period have undergone something of a metamorphosis. It was once possible for even a generally hostile writer to suggest that, by April 1955, 'no man could claim a prouder record of international settlements than Anthony Eden'.[54] Certainly he could point to a string of diplomatic successes, especially in 1954, his *annus mirabilis*, which seemed not only to enhance his own and his country's standing, but also to leave the world a safer place than he had found it. Many recent writers, however, have been less complimentary. In the context of an increasing obsession with Britain's long-term decline, the 1950s have come to occupy a prominent place as a decade of complacency during which seeds of later decay were planted by sins of commission and omission. Britain, it has been said, spent too long in an unequal struggle to maintain the illusion of world power, while harder-headed competitors renounced the priorities of a bygone age and invested in the long-term strength of their domestic economies. Part of this critique is of strictly economic policy, but part relates also to an alleged failure to reduce Britain's overseas commitments to a level commensurate with her status as a power of the second rank. 'Only abdication as a great power', judges one critic, 'would have released sufficient resources to arrest economic decline.'[55] Eden's unique position in the management of British diplomacy between 1951 and 1957 makes him an obvious target. Vision was needed and he failed to provide it. Thus, as one writer concludes,

> Framing realistic objectives which adequately register changes in the international environment and Britain's own shrinking material base constituted the ultimate test of British statesmanship in the 1950s. Eden failed the test and the result of this failure was that the gulf between national resources and the multitude of political problems whose outcome the UK desired to influence steadily widened.[56]

In sum Eden 'lacked the attributes of realism, breadth of vision and the capacity to think ahead which mark the true statesman'.[57]

There is some truth in this analysis, but it is by no means a complete picture. Two broad points help explain the denigration of Eden's performance. In the first place many of those who have looked at his whole career have found difficulty in reconciling the diplomatic catastrophe of his premiership with the success once widely attributed to his earlier years. If Eden's conduct at Suez is not easy to rehabilitate, then the quest for flaws in the pre-Suez era has obvious attractions in the construction of a more coherent whole. Secondly, much of the criticism of his last Foreign Secretaryship relates to the supposed missed opportunity to take the lead in European integration which has been discussed above.[58]

As Foreign Secretary in the 1950s, Eden probably did fail to give sufficient attention to longer-term questions. It was a feature of his three periods at the Foreign Office that he spent too much time embroiled in diplomatic minutiae at the expense of broader issues. This was the obverse side of his undoubted virtues of diligence and application, compounded by a chronic inability to delegate. As the diplomat Frank Roberts recalled, Eden was 'a great stickler for details'. He 'took the closest personal interest and at times nearly drove us to distraction by insisting on commas being adjusted and adjectives changed, as we thought, unnecessarily'.[59] Eden's preferred method was to deal with diplomatic problems one by one through patient negotiation. In an early speech to the UN General Assembly he called for others to join him in grasping 'definite and limited problems and working for their practical solution'. This message was repeated to the Commons in mid-November 1951.[60] If this reflected an 'intellectual disability',[61] it was at least one which appealed to an intellectual on the Labour benches. Richard Crossman recorded: 'Eden opened the . . . debate with an extremely adroit performance – not the appalling tour of the horizon to which Ernest Bevin and Herbert Morrison subjected us, but a well-composed, well-delivered speech on a few outstanding topics.'[62]

Eden was also the victim, like ministers before and since, of the sheer pressure of departmental work. He was immediately struck by the way the volume of Foreign Office activity had increased even since 1945, especially through the setting up of new agencies such as the OEEC, NATO and the Council of Europe. Compared with pre-war days, he told Morrison, the workload had doubled.[63] He was soon complaining of being dragged off to so many international conferences that he was left with little time to think. During his first five weeks in office he spent only eight days 'fairly stationary' in London.[64] A junior minister later described Eden's typical working day: 'Working from before breakfast to the early hours of the next morning his day was a non-stop series of Cabinet meetings, conferences and interviews, with questions or debates in Parliament to answer at least once a week.'[65]

Dealing with Churchill had become no easier than during the war. If anything, the old man's governmental techniques were even less business-like. He tended to treat the cabinet as a 'deliberative body', recalled Macmillan, who was well placed to judge. 'I felt that Eden, who was often overwhelmed by anxious and difficult problems requiring immediate decisions, suffered under this deliberate and sometimes rambling procedure.'[66] On top of his other difficulties, Eden's health was never robust during this period. His complete collapse in 1953 is well documented, but there is evidence that his underlying medical problems were already apparent on his return to office. Evelyn Shuckburgh, who now became his private secretary, was soon aware of the situation and of Eden's increasing dependence on medication. 'It is discouraging that his health should be so bad,' he remarked as early as November 1951.[67] A year later, as he travelled to New York, Eden himself noted his 'wretched old pain' starting up, necessitating his giving himself two injections. These afforded some relief, though he found it 'hell' giving press, radio and television interviews on arrival.[68]

Despite these burdens, it would be unfair to Eden's reputation to suggest that his final Foreign Secretaryship was simply an exercise in crisis management, displaying an 'inability to think and plan strategically'.[69] His languid manner sometimes concealed the depth of his powers of thought and analysis. It was a style which Americans in particular found difficult to accept: '[David] Bruce feels Eden has been a disappointment as foreign minister. He is too slipshod. He has an easy-going, nineteenth-century manner and asks in an off-hand fashion, "well, Bob [Dixon], what do we think about this?" That "doesn't go", David says.'[70] Nor was Eden helped by a certain weakness in expressing himself clearly in speech and print. Too often he sounded bland and superficial. His delivery in the Commons tended to be dull though he sometimes showed a capacity to respond effectively when winding up a debate.[71] One old weakness was never eradicated. Eden 'was inclined to leave no stone unturned and no avenue unexplored in his search for the missing cliché'.[72] The blurred edge and studied ambiguity, which sometimes secured a compromise in diplomatic negotiations, did not always convey the impression of an incisive mind. But others perceived more substance behind the diplomatic façade:

> He had something much more valuable. That was a genuine flair for foreign affairs that let him see them in the broad sweep of strategy across the world. He was attracted by the complexities of a problem, and would – partly intellectually, partly intuitively – work out different ways forward. At the same time he could sit back and speak lightly and colloquially, often most amusingly, about the grind of day-to-day business and the ponderous people he had to meet. It was at such times that one saw the paradox in him most clearly. The gallant Sir Galahad of the public image frequently spoke heavily

and often prosaically when on the platform whereas the hard-working professional spoke lightly, pithily, and with engaging modesty in private. . . . He would constantly be testing ideas on himself or on others.[73]

Eden did appreciate that Britain's power and influence had declined, though he found it difficult to reconcile himself to the logical consequence of this development – the ever greater dominance of the United States in world affairs. His guiding principle was of British disengagement from untenable positions and the substitution of international responsibility through defence organizations which would include the Americans.[74] This was not dissimilar from the strategy which Bevin had successfully pursued in a largely European context. Shuckburgh, who worked closely with Eden and whose diary, at least until 1955, gives a far more favourable picture of his master than selected extracts have sometimes implied, insisted that from the time that he came to know him in 1951 Eden was 'constantly preoccupied with the problem of how to get this country into a tenable position, because we were stretched all over the world in places which we had no possibility of continuing to dominate'.[75]

But Eden faced considerable problems, as was obvious in the debate over the future of the Suez Canal Zone, in persuading not only Churchill but a substantial number of his parliamentary colleagues to endorse his analysis. Such figures were more deluded than he was by the mirage of Britain's power-political renaissance in the 1950s. Eden was torn between fighting his corner and biding his time to await the Prime Minister's supposedly imminent departure from the political stage. But his thinking is clearly evident in a cabinet paper on 'Britain's Overseas Obligations' presented in June 1952. Here he argued that if the maintenance of existing commitments would permanently overstrain the economy then the country must reduce its burdens:

> It is clearly beyond the resources of the U.K. to continue to assume the responsibility alone for the security of the Middle East. Our aim should be to make the whole of this area and, in particular, the Canal Zone, an international responsibility. . . . The US have refused to enter into any precise commitments in the Middle East or to allocate forces, and it should be the constant policy of HMG to persuade them to do so.[76]

With her limited resources it was vital for Britain to concentrate on those points where vital strategic needs or the necessities of economic life were engaged. But Eden also recognized that too rapid a withdrawal would undermine the country's international prestige, upon which her voice in world affairs heavily depended. Britain still had a major role to play and she had to be aware that 'the Russians would be only too ready to fill any vacuum'. Like Foreign Secretaries before and after he did not regard the abandonment of a position among the first rank of powers as a serious option.[77]

In many ways Churchill sought to model his 1951 government on the example and the memory of his wartime administration, and the respective positions of himself and Eden as premier and Foreign Secretary, exactly as they had been ten years before, helped sustain this image. But Eden was now entering his third term in the same office and earlier experience made him more than ever intolerant of outside interference in his departmental affairs. Looking back, David Maxwell-Fyfe, now appointed to the Home Office, doubted whether any Foreign Secretary since Rosebery had enjoyed so much freedom of action. It was 'in a sense an abrogation of the role of the Cabinet'.[78] This was an exaggeration but cabinet discussions on foreign policy were comparatively rare and on major issues Eden liked to deal directly with Churchill.[79] None the less the relationship between the two men had undergone an important change since the war years. At that time Eden, though frequently exasperated by Churchill's wilder flights of fancy and general conduct of business, had sometimes had to accept that it was the Prime Minister – the war leader at the height of his powers – who would prevail. By 1951 Eden had grown in stature and, though constrained by an abiding loyalty, he now believed that Churchill was past his best and should already have stepped aside. In any case he understandably regarded diplomacy in peacetime as a more autonomous activity, less subject to the intrusion of others, than in a time of total war. One outgoing Labour minister commented:

> Eden is certainly very well established at the Foreign Office and need not tolerate any interference from the Old Man. He has a much stronger position in the country and an almost equally strong one in the Party. He is, of course, far stronger in the House of Commons.[80]

In practice, Eden never found his position quite that easy. He remained wary at even the possibility of rivals emerging and was always unduly sensitive about his status in the government. Even before his appointment had been officially announced, Shuckburgh overheard him engaged in a heated conversation with Churchill over whether he was entitled to be styled 'Deputy Prime Minister'.[81] Throughout the lifetime of the government Churchill's interventions, in a realm of policy where he at least believed himself to be particularly gifted, caused Eden difficulty and embarrassment. To the latter's understandable irritation it was Churchill who sought to dominate the British delegation when the two men visited America in January 1952.[82] During one particularly awkward cabinet meeting the following December, at which Churchill raised the question of Egypt which was not formally on the agenda, Lord Simonds, the Lord Chancellor, passed Eden a sympathetic note: 'This has been a horrid affair. To have said anything wd. have been to fuel the flames. *But* the whole Cabinet is with you save only the P.M.'[83] 'Nobody can ask me to go on like this,' scribbled Eden in reply. But go on he did.

*

Eden's first months in office gave few grounds for optimism regarding the international situation. Perhaps for this reason he gave serious thought to the possibility of becoming Secretary-General of NATO.[84] The Cold War was now the abiding feature of a diplomatic horizon offering few chances of amelioration. Commenting on the state of affairs created by the Korean War, Eden noted that there was now virtually no contact between East and West on either side of the Iron Curtain. This was 'something new and entirely to be deplored'.[85] Even more depressing was the attitude he found in the United States when he and Churchill made their visit at the beginning of 1952. The Prime Minister was particularly keen to re-establish the wartime Special Relationship which, he believed, had been lost during the years of Labour government. The British ministers soon discovered, however, that President Truman and his Secretary of State, Dean Acheson, were not interested in a purely bilateral Anglo-American partnership, a notion which they attributed to the aged premier's sentimentality and lack of realism.[86] It was impossible, noted Shuckburgh, 'not to be conscious that we are playing second fiddle'.[87] Despite this obvious rebuff Churchill never lost faith in the Special Relationship, even though his final premiership provided much additional evidence that the American government – Democratic or Republican – saw little advantage in pretending that the clock could be turned back to the war years. For Eden, on the other hand, this visit to Washington may have been decisive. His speeches in 1951 suggest that, as in the early war years, he had been ready to approach the American relationship with enthusiasm.

Contact with the US administration, however, made him increasingly sceptical. On his return he admitted that he had been

> forcibly struck – indeed horrified at the way we are treated by the Americans today. They are polite; listen to what we have to say, but make (on most issues) their own decisions. Till we can recover our financial and economic independence, this is bound to continue.[88]

Intrinsic difficulties in the relationship were exacerbated by genuine disagreements over policy, particularly in relation to the Far and Middle East. In the case of the former, Acheson wanted a clear warning to communist China of the consequences of resuming hostilities in Korea, where armistice talks had been proceeding since the previous July. Eden, on the other hand, was cautious about any gestures which risked British involvement in war against China or provided an excuse for further Soviet encroachments in Europe. This matter caused Eden considerable embarrassment on his return to London when he attempted to persuade the Commons that Britain and America were of one mind.[89] Over the Middle East, and particularly the ongoing dispute with Iran, Acheson found Eden no more accommodating than Morrison. Eden was annoyed by America's indulgent attitude towards the Iranian nationalist leader, Muhammad Mussadeq, and Acheson's insistence on seeing him as the only alternative to a communist take-over in Iran.[90]

Relations between Eden and Acheson remained strained throughout the remaining months of Truman's presidency. In later years, particularly once John Foster Dulles had come to occupy a unique position in his personal demonology, Eden seems to have persuaded himself that his relations with Acheson had been warm and friendly. In 1957 he reviewed the course of Anglo-American relations, accepting that the post-war loan had 'marked finally our dependence on U.S.':

> None the less during the Democratic administration this was not so harshly felt. It was only when the Republicans arrived, with their hard headed business methods, that the pattern gradually became clear. I was slow to believe it, because I was loath to believe it.[91]

Eden and Acheson paid one another extravagant compliments in their respective memoirs[92] and went to considerable lengths in their retirements to arrange meetings, especially during Eden's winter retreats to the West Indies.[93] Yet in the wake of the 1952 discussions rumours of an Anglo-American split reached the press. According to *Newsweek*:

> Eden feels the Prime Minister gives in too much to the Americans. . . . His intimates say he is determined to make a show of independence whenever the opportunity arises. Eden's relations with Churchill thus are complicated. . . . Eden and Acheson differ radically in their approach to problems. . . . [Eden] retains some of the English public schoolboy's unconscious rudeness and this is especially hard for Americans to take.[94]

Though Acheson wrote personally to Eden to dissociate himself from this article, it contains more than a grain of truth.[95]

Superficially at least, the two men were very alike, even physically, but their working relationship was never easy. Acheson found Eden condescending and his affected mannerisms tiresome, while Eden was easily wounded by the American's sharpness in debate.[96] A combination of genuine policy differences, especially over the central question of Britain's world position, and the capacity to irritate one another led to a succession of angry confrontations. Lester Pearson of Canada recalled a NATO meeting in Lisbon:

> One evening, I needed every power of persuasion and conciliation I possessed to pacify Dean Acheson and Anthony Eden . . . in a vigorous, post-prandial verbal battle. Acheson had the best of it. He was a master of the brilliant, biting phrase, but Eden's assumption of Oxonian, Foreign Office superiority never faltered.[97]

As Pearson concluded, 'these are two people who shouldn't be left alone in the same room to argue'.[98] In the United States the political tide was moving strongly to the right with a Republican victory in the 1952 presidential election widely anticipated. When Churchill announced that he was 'looking forward to working with Eisenhower [the Republican candidate] after November', it seems likely that Eden shared this sentiment, though not the thought that a Republican victory would provide one more reason for Churchill to postpone retirement.[99]

Anglo-American relations were placed under further strain in June when, without consulting Britain, the Americans bombed power stations on the Yalu River in North Korea. As Acheson was then in London, Eden told the cabinet that he would make it perfectly clear that there should have been prior consultation and that Britain 'would expect to be consulted on any future similar occasion'. Churchill did not dissent, but sounded a significantly different note in stressing the importance of avoiding any public statement which implied a divergence of view between the two governments.[100] It needed nimble footwork from the Foreign Secretary in the Commons to avoid Labour criticism. As Crossman put it, 'twenty minutes proving what we all know: that these were military objectives within Korea; two minutes saying that he had not been consulted; and one minute saying that, despite this, he gave America full support'.[101]

Yet another crisis threatened at the end of the year over the terms of a Korean armistice. Eden found the American stance in the talks in Panmunjom devoid of the flexibility needed for a settlement, but hoped that an open Anglo-American split could still be avoided.[102] Unwell and in pain, Eden arrived in New York on 8 November for a meeting of the UN General Assembly. He found that the American attitude was, if anything, hardening. As he reported to London: 'Acheson himself could not have been more rigid, legalistic and difficult. . . . At times it almost seems as if United States Government were afraid of agreement at this time. This is what some of their own countrymen were beginning to suspect.'[103] The American electorate had rendered Truman's a lame-duck administration, but this made Acheson no more accommodating. Eden could not understand the American attitude – 'It is pretty depressing. And their tone is arrogant.'[104] He welcomed the opportunity to meet the President-elect from whom he clearly expected a more constructive attitude. Acheson, however, remained intransigent. The crucial issue was the Communists' demand for the forcible repatriation of prisoners of war. Eden, taking a more humane line than on a similar matter in 1944, sought support for the compromise proposals of the Indian delegate, Krishna Menon. The latter suggested handing over the prisoners to a Repatriation Committee of neutral powers in order to break the deadlock threatening the armistice negotiations. The Americans deliberately caused Eden further embarrassment by leaking details of a 'serious rift' between London and Washington. Eden found such methods intolerable and was shocked when a semi-inebriated Acheson launched a bitter personal attack on Selwyn Lloyd, Eden's Minister of State.[105] Eden was determined to stand his ground and momentarily considered resignation if he could not get his way.[106] Only when the Russian delegate, Vishinsky, made an all-out attack on the Indian plan did Acheson shift his ground. Wary of appearing to side with the Soviets, the United States now backed the idea of a maximum three-month period of retention of prisoners of war by the international committee after the armistice was agreed. This compromise was embodied in the eventual armistice terms. Just as importantly, an open Anglo-American breach had been narrowly avoided.[107]

It was therefore scarcely against a background of harmonious Anglo-American relations that the Republicans took over the White House in January 1953. In his memoirs Eisenhower relates how, during a meeting in Paris several months earlier, Eden had tried to persuade him not to appoint Dulles as his Secretary of State.[108] John Wheeler-Bennett, Eden's first designated biographer, suggested that it was 'past all belief' that as skilled a diplomatist as Eden would have employed 'so clumsy, tactless and generally "hamfisted" an approach'.[109] By the time, however, that these words were published, Eden had furnished Wheeler-Bennett with an account of the events of May 1952:

> When we were breakfasting together . . . alone and for a confidential talk at his suggestion, he asked me what I knew of various possible candidates for Secretary of State. I praised Dewey and said I felt sure he would be good. I said that I did not know Dulles . . . indicating that I therefore had no view myself, but admitting the difficulties which Morrison had reported to me that he had had with Dulles.[110]

This was one occasion when Eden's real meaning was not concealed by circuitous diplomatic language. He raised the matter again at his meeting with Eisenhower in November. But by this time it was too late and Eden concluded that 'we must do the best we can with him', taking some comfort from the President-elect's hint that Dulles's might be a short-term appointment.[111]

In later years Eden certainly developed a profound hatred of Dulles, not dissimilar to his feelings for Mussolini. It accompanied his perception, for some time also the historical orthodoxy, that Dulles rather than Eisenhower was the real architect of American diplomacy. Yet their working relationship, though seldom easy, was not marked by the degree of animosity Eden later implied, while the contrast he drew in his memoirs with Dulles's predecessors distorted reality:

> My relations had always been cordial and often intimate with the four earlier American Secretaries of State whom I had known as colleagues. . . . My difficulty in working with Mr. Dulles was to

determine what he really meant and in consequence the significance to be attached to his words and actions.[112]

Certainly there was a clash of temperaments. Dulles was more intellectual and more legalistic than Eden. His tedious, didactic monologues designed to take his listener along the logical paths of his own thinking contrasted with Eden's more intuitive style. Even Eisenhower recognized Dulles's curious lack of understanding of how his words and manner might affect another personality.[113] On the other hand, Eden's elegant demeanour and – notwithstanding his inherent professionalism – his cultivation of the ways of the aristocratic English amateur failed to impress his American opposite number. As one of Dulles's staff testified:

> Every time I saw Eden I always felt an overwhelming sense of personal vanity, and Dulles was just the opposite. Dulles may have had intellectual vanity but no personal vanity at all. . . . Just personality wise, they weren't destined to work together. . . . You know, [Eden's] homburg hat and all the rest, and his rather languid air. A calculated lazy manner which is one of the upper-class manifestations of the old English aristocracy. It wasn't Dulles's dish of tea.[114]

But as important as any difference of temperament was Dulles's refusal to see the world as Eden did, with Britain still in the fore-front of world powers:

> The Secretary was convinced that the British throughout the world were a rapidly declining power. He was convinced they no longer had any basic will to meet big international responsibilities, that they were attempting to duck out all over the world, that they were trying to put as good a face on it as possible, but that you simply could not count on the British to carry on in any responsible way, or, indeed, form an effective bulwark with us against anything.[115]

Be that as it may, both Churchill and Eden welcomed the Eisenhower presidency, the premier because of his expectation of rekindling his wartime friendship with the former Supreme Allied Commander, the Foreign Secretary because of the difficulties he had lately experienced with Acheson. Eden was particularly glad that the new President saw him rather than Churchill as the man with whom he wished to work. The Prime Minister's position, judged Eden, was like that of the aged Queen Mary – 'glad she is there, dear old thing'.[116] Yet ironically it may have been advice from Acheson which helped determine Eden's overall attitude to the Republican presidency. Eden heard that the outgoing Secretary of State was deeply concerned at the way the Anglo-American partnership had been drifting seriously apart over the preceding year. Acheson warned that the new administration would not be prepared to take the same positive lead in international affairs or the same diplomatic risks as the Democrats had done. It was now up to Britain 'to take over the torch and show imaginative leadership'.[117]

Churchill determined that he could best revive the sagging Special Relationship through a personal meeting with Eisenhower. He set off for Washington, without Eden, on 30 December 1952. If Churchill seriously believed that he could recreate the sort of partnership he had enjoyed with Roosevelt, this visit showed how badly he had misjudged the American mood. The gap in military and economic terms between Britain and the United States was widening remorselessly. This was inevitably a more fundamental determinant of American policy than Churchill's personal diplomacy. Eisenhower was genuinely fond of the British in general and Churchill in particular, but he did not regard the alliance as the answer to all international ills. He was shocked to see how feeble the elderly premier had become:

> Much as I held Winston in my personal affection, and much as I admire him for his past accomplishments and leadership, I wish he would turn over the leadership of the British Conservation [sic] Party to younger men. . . . he had developed an almost childlike faith that all of the answers are to be found merely in British–American partnership.[118]

Eisenhower and Dulles wanted good relations with Britain, but not to the exclusion of their dealings with other powers. Both Churchill and Eden were slow to grasp this point.

When Eden himself went to Washington in March 1953 he was impressed by his cordial reception. He found Dulles 'very friendly', while Eisenhower told him he could have 'as much time as I wanted [with the President] while I was here, and I think he meant it'.[119] The President regarded Britain as his principal ally, Eden enthusiastically reported, without it seems sensing the precise meaning Eisenhower wished to convey by this remark.[120] Perhaps recognizing that Eden might have misinterpreted him, Eisenhower now sent a letter in which he expressed concern at the way 'we present to the world the picture of British–American association':

> We must, by all means, avoid the appearance of attempting to dom-
> inate the councils of the free world. . . . Our two nations will get
> much further along toward a satisfactory solution to our common
> problems if each of us preserves, consciously, an attitude of absolute
> equality with all other nations.[121]

Eisenhower's warning related to the specific case of policy towards
Egypt, but its application was more general. If Eden understood the
President's message, he seemed not to wish to accept it. A fortnight
later he replied:

> Together we cannot help wielding immense influence, and there is no
> question of domination here. . . . I believe that Egypt is a test case. If
> we can get a settlement there, which the world will see has been
> achieved by our united efforts, the benefits both in the Middle East
> and elsewhere may spread out like ripples on a pond.[122]

There the matter rested, but the difference was genuine and of
considerable significance for the future.

*

One of the most depressing features of the Cold War was the lack of
opportunity to break the deadlock in East–West relations, with its
ever-present threat of escalation to a nuclear war. When in opposition, Churchill had suggested that the way forward lay in the
resumption of meetings at heads of government level, as at Tehran,
Yalta and Potsdam. Such a strategy offered the obvious advantage of
reasserting Britain's claims to a place at the top table of world
diplomacy. But the international climate was hardly conducive to a
summit conference, until, that is, the news came through of Stalin's
sudden death in March 1953. This gave Churchill the chance to
argue that the new Soviet leaders might be more reasonable and
that they should be brought into conference with himself and
Eisenhower at the earliest possible moment. In his memoirs Eden
suggested that he saw no reason to suppose that Stalin's death had
changed anything fundamental in East–West relations. His words
might have been written by as doughty a Cold Warrior as Dulles
himself: 'I did not share the optimism of those who saw in this event
an easement of the world's problems. The permanent challenge of
communism transcends personalities however powerful.'[123] His contemporary minute suggests an unenthusiastic, but less dogmatic,
reaction: 'there is not enough evidence yet. We must go cautiously,
developing contacts when we can and sounding out the Russians.'[124]

He warned Churchill that conciliatory signs from the Soviets might merely indicate a change of tactics, designed to fragment the anti-communist front and bring neutral states over to their side. In the face of 'embarrassing conciliatory manoeuvres', the West should 'respond as freely as we can', without surrendering vital points such as the North Atlantic Treaty.[125] Eden was worried that talk of a summit would only endanger the ratification of the EDC treaty by giving France an excuse for further delay.[126]

Having apparently convinced Churchill on 2 April 1953 that Britain should at least move cautiously, Eden was surprised to receive a telephone call from him the following morning:

> I was taken aback by W . . . saying somewhat challengingly 'So you have given up the idea of meeting Molotov. I don't like that at all.' He had entirely forgotten yesterday's conversation. I went over it again with him carefully and he appeared to agree and to begin to recollect. However his telegram to Ike when it arrived showed that he hadn't really understood our reasons for not asking for early meeting or alarming Ike with report that we were doing so now.[127]

This diary entry reveals Eden's mounting irritation at Churchill's failing powers. His objections to a summit on policy grounds were compounded by the excuse which Churchill's initiative afforded him to delay his retirement into an indefinite future. The half-promise to retain the premiership for no more than a year had not been honoured, and Churchill now presented himself as a man with a mission, to save the world from the incalculable disaster of thermonuclear war. Believing such a prize was within his grasp, he found Eden's hesitant attitude depressing and likely to consign the world to 'years more of hatred and hostility'.[128]

Yet ironically it was not Churchill's but Eden's health which now gave way. He had spent a bad night in considerable pain on 29 March and cancelled his engagements for the following two days. After X-rays revealed gallstones, his doctors advised an immediate operation. In practice Eden would be out of political action for six months. After two unsuccessful operations in England, he left for a third in Boston in early June. His parting words to his junior ministers were that he hoped they would prevent too much appeasement of the Russian bear in his absence.[129] With Eden already *hors de combat*, Churchill made a public call for a summit conference on 11 May. He seemed to envisage precisely the sort of unstructured meeting which Eden had found most dangerous during the war. The conference should 'not be overhung by a ponderous or rigid agenda, or led into mazes and jungles of technical details, zealously contested by hordes of experts and officials drawn up in vast, cumbrous array'. Even if no formal agreements were reached, there 'might be a general feeling among those gathered together that they might do something better than tear the human race . . . into bits'.[130] Churchill's words, offering hope of a 'generation of peace', had wide appeal, but went against everything Eden held important in the conduct of diplomacy. As he later put it:

> I do not believe that anything is to be achieved by meetings at which the burning questions are avoided. As you know, I feel there are serious dangers in having discussions with the bear unless we have some idea what we are to talk about and what we expect to achieve.[131]

Eisenhower was not enthusiastic about Churchill's initiative, but seemed more amenable than his predecessor. Like Eden he had no wish to see a summit meeting getting in the way of Western European military integration. But when Churchill proposed a bilateral meeting with the President to discuss the agenda for a three-power summit, Eisenhower again revealed that this was not the sort of Anglo-American relationship he sought, insisting on 3 June that the French should also be invited. With Eden unable to influence events, Churchill's plans met with less cabinet and Foreign Office opposition than was once thought, and it was American hesitation which kept the situation in check.[132] Even so, Churchill might have made more progress had he not suffered a stroke on the evening of 23 June during a dinner in honour of the Italian premier, de Gasperi. But for his own ill-health, it seems likely that Eden would now have stepped into Churchill's shoes.

After a Mediterranean cruise Eden resumed his duties in early October. Dealing with Churchill proved no easier than before the latter's stroke. Eden found him ready to concede that the Soviet Union appeared not to want talks at any level. But a long after-dinner conversation also revealed that Churchill and he remained at odds about the value of a summit, at least as envisaged by the Prime Minister. It was a 'depressing evening':

> I had to make it clear that I did not regard Four Power talks at the highest level as a panacea. He maintained that in the war it was only the Stalin–Roosevelt–Churchill meetings that had made our Foreign Secretaries' work possible. I said this was not so, nor was it true that to meet without agenda was the best method with Russians. I believe

they liked to have an agenda which they could chew over well in advance. Our most productive meeting with them had been with Hull at Moscow in 1943 before Teheran when we had used just these methods.[133]

Eden clearly preferred the idea of getting Dulles and the French Foreign Minister, Bidault, to London, but he allowed Churchill to approach Eisenhower again, confident that the President would resent being pressed. The argument between the two men was 'firm at times, I telling him that he must not create a situation in which only heads of government could settle anything and they not meet!' Eventually, Churchill sent word from the party conference that he had abandoned his own idea and wished Eden to proceed with plans for a meeting of Foreign Ministers.[134]

Dulles came to London on 16 October and placed further obstacles in the path of a summit conference, citing Eisenhower's difficulty in leaving Washington for any prolonged period. But reports began to appear in the press that Churchill was ready to travel to Moscow without the President. Eden was understandably alarmed: 'I said that if that were so we had better have a talk as we had never agreed anything of that kind.'[135] In the end Eisenhower, having in practice offered such a compromise before Churchill's stroke, had to agree to a three-power Western summit of Britain, the United States and France in Bermuda in early December.

Eden approached this meeting with understandable apprehension. But he could at least see the possibility of progress towards an international conference on the problems of the Far East. His over-riding concern was that Bermuda should produce a demonstration of Western unity. 'It must', he stressed to Churchill, 'be a first principle that the three powers should be seen to be acting together.'[136] Eden hoped to build upon existing agreed policies in relation to Europe in order to extend the area of allied co-operation to Korea and Indo-China.[137] But his cabinet paper of 24 November foreshadowed many of the Anglo-American differences over policy towards the Far East which would surface over the following year, even though there was 'broad agreement with the United States Government that Communism must be contained'.[138]

In many respects Bermuda fulfilled Eden's worst expectations. 'Never again,' he confided to Iverach McDonald of *The Times* on the last morning of the conference.[139] Many problems were a direct result of Churchill's failing powers. Increasingly deaf and following a daily routine appropriate to a man in his eightieth year but hardly conducive to the efficient conduct of diplomatic business, the Prime Minister drove Eden to the point of distraction. 'W hears nothing at the conference which doesn't reduce my problems,' noted Eden of a man who deliberately and ostentatiously turned off his hearing aid when he disapproved of what was being said.[140] Eden's diary entry for 6 December is typical:

> The worst day yet. . . . W late and President annoyed. Bidault a long exposé. Foster replied cautiously and well. W indulged in a tirade about what would happen if E.D.C. failed, a catalogue of calamities. Bidault refrained from easy retort: if it is all so terrible why don't you do a little more. I thought it all in the worst taste. . . . President restored the balance very well. But effect in France may be serious.[141]

Press reports of French annoyance at Churchill's behaviour scarcely fulfilled Eden's hopes for a display of allied unity.[142] But the most interesting aspect of the whole conference was the ease with which Eden and Dulles seemed to co-operate. Early-morning bathing parties led to a wide-ranging and largely satisfactory exchange of views.[143] According to McDonald, Eden said he had got on well with Dulles, especially in their talks on the Far East: 'I came to like him more and more.'[144]

*

By no means all the omens were therefore favourable as Eden embarked upon what many have considered the most distinguished single year of his career. Churchill would not let go of the idea of a heads of government summit and seemed intent on using it as an excuse to hold on to the reins of power. This was despite Eisenhower's protestations at Bermuda that he could discern no real change in Soviet policy since Stalin's death and that Russia remained 'a woman of the streets'. 'Whether her dress was new, or just the old one patched, it was certainly the same whore underneath.'[145] In addition there seemed no area of the globe where easy opportunities existed to reduce East–West tension. And perhaps most worryingly of all – and notwithstanding Eden's more cordial recent dealings with Dulles – significant differences remained just beneath the surface of the Anglo-American alliance. Indeed, in May 1953 Eisenhower had lamented that relations with Britain were at a lower ebb than at any time since the war.[146] Within a fortnight of the Bermuda Conference, Dulles, exasperated by French delays in ratifying the EDC treaty, bluntly stated that America might have to conduct an 'agonizing reappraisal' of its role in Europe, a warning which probably had a greater impact in Britain than in France. While Eden sympathized with the sentiment which underlay Dulles's remarks, he disliked the brutality of their delivery. But he regarded this as merely one aspect of a heavy-handed, unsophisticated American approach to international affairs which had become particularly evident since the outbreak of the Korean War. United States policy in the Far East epitomized the problem. In contrast to America's confrontational attitude towards the Chinese communists, Britain was keen to avoid unnecessary provocation in the hope of steadily moving towards normal trading and diplomatic relations.[147]

With a summit ruled out for the time being, it was agreed at Bermuda that only the three Western Foreign Ministers should meet the Soviets in Berlin in January 1954. Eden emphasized that the main objective of this conference should be the maintenance of Western unity. But this would not be easy granted the conflicting interests of the United States and France. While the Americans were firmly opposed to any formal meeting with China, France was coming to see this as the only way to secure a negotiated exit from her increasingly hopeless colonial war in Indo-China.[148] Whatever the feelings of the Foreign Office as a whole, Eden may not have been as opposed to an eventual five-power conference, including China, as has recently been suggested.[149] He had explained his long-term strategy to Churchill before the meeting in Bermuda:

> If we were to press prematurely for a five-power meeting we should risk increasing our differences with the Americans over the Far East and thus playing into Russia's hands. But if we can agree to work towards such a meeting by stages, beginning with a Korean Conference, there is a real chance of our being able gradually to overcome them.[150]

In fact Berlin gave Eden scope for the sort of diplomatic manoeuvring at which he excelled.[151] An American delegate later paid a revealing tribute:

> Eden was quick, he was skilful, he was eloquent in debate. His rather languid manner concealed a lively, imaginative, perceptive mind . . . he had that almost inbred, instinctive effort, in any conflict, in any collision, great or small, to find a compromise solution. . . . [He saw his role] not as a dictator of events but as a negotiator, as a compromiser, as a mediator.[152]

It soon became apparent at Berlin that no progress would be made in relation to European issues where peace treaties with Germany and Austria had still not been signed. It was, Eden reported, little more than 'a confrontation of opposing views'.[153] Molotov was at his most intransigent – 'the same old gramophone record with scarcely a variant'.[154] Eden was even able to write a letter to his wife while the Soviet Foreign Minister was in full flight, under the subterfuge of taking notes. He was understandably excited by the news that Clarissa Eden was expecting a baby. For some time the conference was little more than an exercise in damage limitation. 'If we could go away with the front of three western powers unbroken', Eden suggested, 'and international tension not increased, I shall not be entirely dissatisfied.' Significantly, he continued, 'if we could add to this some hopes of a conference on the Far East, it would be better still'.[155]

In fact Eden skilfully shifted the emphasis of the conference to Far Eastern issues and won over the United States to the idea of a five-power meeting, involving China. He was able to build on Dulles's concession that he would talk to the Chinese about the situation in Korea, where preliminary discussions had after all been taking place in Panmunjom since the armistice of the previous July. The task was to establish some common ground between Molotov, who wanted to make the agenda for a five-power gathering as wide as possible, and Dulles, who was reluctant to move beyond his concession on Korea. In the end it was Molotov who gave way the most, with agreement being reached that the representatives of the United States, France, Britain, the Soviet Union, China, North and South Korea and 'the other countries the armed forces of which participated in hostilities in Korea' would be invited to Geneva. At the last minute Eden and Dulles also agreed that the restoration of peace in Indo-China should be included on the Geneva agenda. Both men recognized that this was a price they would have to pay to sustain the Laniel government in France, widely seen as the last hope for the ratification of the EDC treaty.

The Berlin Conference thus turned out to be something of an unexpected triumph for the British Foreign Secretary. Yet there was a widespread feeling, even within the British delegation, that this had been partly at the expense of the United States. Just before the conference ended, Shuckburgh noted his concern. He was worried by Eden's attitude and judged him so keen to secure a Far Eastern Conference, 'so as to have some "success" to go home with', that he had forgotten how difficult this topic was for Dulles, particularly in a domestic political context.[156] Eden's private comments show that the favourable impression created at Bermuda had completely disappeared and reveal a near contempt for the Secretary of State's performance in Berlin. Dulles took too much notice of outside advisers and the result was 'amateurish'. His trouble was that, 'while quite reasonable in conversation, he is either clumsy or deliberately provocative in speech at the Conference'.[157] It was not a good augury for Geneva and helps explain the steady deterioration in Anglo-American relations over the next few months.

On 29 March 1954, with news that the key French fortress at Dien Bien Phu was about to fall, Dulles called for 'united action' to stop the subjugation of South-East Asia to communist rule. Eden refused to commit Britain to any measures but agreed that Dulles should visit London to explain what 'united action' might involve. Meetings between the two men on 11 and 12 April produced a genuine misunderstanding which could probably have been avoided if their overall relationship had been warmer. A communiqué issued on 13 April stated that Britain and the United States were 'ready to take part, with the other countries principally concerned, in an examination of the possibility of establishing a collective defence . . . to assure the peace, security, and freedom of Southeast Asia and the Western Pacific'. What Eden envisaged was a NATO-style body for the defence of the Far East which might be organized some time in the future, depending on the outcome of the Geneva Conference. Dulles, however, had in mind an immediate mechanism to internationalize the conflict in Indo-China before the conference even opened.[158] Precisely where the fault of interpretation lay is difficult to ascertain, though it is reasonable to point out that this would not have been the first time that Eden had found it easier to use the vocabulary of diplomacy to slur over an important difference rather than confront it head on. Certainly, one Foreign Office official judged that Britain was 'getting very near having cheated the Americans on this question of starting talks on SE Asia security. . . . When Dulles was in London AE *did* indicate that we should be willing to start talks at once, provided we were not committed to any action in Indochina.'[159] Eden's Commons statement on 14 April hardly suggests that he had deliberately set out to mislead Dulles, even though the latter interpreted it in this sense.

But more important was the impact of this misunderstanding on the course of Anglo-American relations. Back in Washington, and apparently in good faith, Dulles sent out invitations to the ambassadors of nine interested states to a meeting in the US capital to set up a staff-level group for preparatory work on a South-East Asian collective security pact. On hearing this news Eden sent Makins six telegrams in the space of two days forbidding him to attend. His tone suggested mounting animosity towards his American counterpart. He was not aware that Dulles had cause for complaint. Americans might believe that the time had passed when they needed to consider their allies' feelings, but Dulles was creating difficulties for anyone in Britain who wanted to maintain close Anglo-American relations. In the selection of countries invited to Washington, the United States seemed insensitive to Britain's Commonwealth ties.[160] At all events this episode served to poison relations between the two men, perhaps irreversibly. For his part Dulles, confident that Eden had 'gone back on our agreement', was 'absolutely enraged and has been taking it out on the British and Eden ever since'.[161] The two Foreign Ministers met again at a NATO meeting in Paris between 22 and 24 April. With the fall of Dien Bien Phu now imminent, Dulles offered an immediate American air strike against the Vietnamese providing France agreed to fight on. Eden found Dulles in a 'fearfully excited state', while Admiral Radford, chairman of the Joint Chiefs of Staff, spoke of 'bombing China to teach her a lesson once and for all'.[162] 'Put on the spot' and facing the risk of a Third World War, Eden returned to London for instructions. There he secured cabinet backing to refuse to endorse any action in advance of the Geneva Conference. Eden made it clear that Britain's objective was a political rather than a military settlement and that she would work for this at the forthcoming meeting.[163] The Americans were not pleased. As Eisenhower noted:

> Latest reports from Foster Dulles indicate that the British have taken a very definite stand against any collective conversations looking toward the development of an anticommunist coalition in Southeast Asia. Moreover, Eden has apparently gone to the Geneva Conference under strict instructions to press earnestly for a 'ceasefire' in Indochina, possibly with complete indifference to the complex decisions that the French and the Vietnamese will have to make.[164]

Eden's assessment was altogether too optimistic, even complacent: 'It is probably inevitable that the Americans should feel a little sore just now. They will get over it.'[165]

Against this unpromising background the Geneva Conference opened on 26 April. By the beginning of June it was clear that no agreement could be reached on the reunification of Korea, the original *raison d'être* of the conference. But on 21 July three months of negotiations were rewarded by the decision of the French and Viet-Minh to agree to a ceasefire in Vietnam. Though this proved not to be a long-term solution to South-East Asia's problems, Geneva represented a 'singular contribution to international peace' and stands as one of the most considerable diplomatic triumphs of Eden's career.[166] As one of the British delegation later asserted, this turned out to be 'the last example of an independent British policy exercising significant influence in the resolution of a major international crisis'.[167] Yet this achievement was at a heavy price.

By the time that the conference opened, Eden was fearful of the possibility of a complete French military collapse. The cessation of hostilities in Indo-China soon became his 'primary objective'.[168] Seeing Ho Chi Minh more as the leader of a nationalist movement than the instrument of international communist expansion, he aimed to end the war without affording the Americans a pretext to intervene, and could not accept the American premise that the triumph of communism in Vietnam would be but one stage in an unstoppable progression. Indeed, Britain's experience in Malaya made Eden doubtful whether the American government and people would 'ever see the business through' if they did go into Indo-China.[169] Initial progress was slow. Eden's diary reveals that he saw Dulles being as much an obstacle to agreement as either the Russians or the Chinese. 'Foster droned on,' he recorded on 29 April. 'Everybody agreed with me except Foster and Korean.'[170] The Americans seemed apprehensive about reaching any agreement with communists, no matter how innocuous.[171] To Dulles's disappointment Eden refused to adopt the stance of a subservient ally. The Secretary of State was shocked when the communists poured abuse on America's imperialistic actions and the British delegation stood silently by.[172] 'There is no doubt', recorded Shuckburgh, 'that Dulles and A.E. have got thoroughly on each other's nerves and are both behaving rather like prima donnas.'[173] By early May Eden was ready to argue that Britain and the Soviet Union were now occupying positions of moderation with the Americans and Chinese on opposing extremes – 'rather an admission, it sounded, of the divisions in the Western alliance'.[174]

Eden pressed for secret sessions as the best way to conduct serious negotiations. Thereafter the conference slowly gained momentum.

He was at his best in detailed discussions in small groups and his qualities showed to advantage as he sought to build bridges with the communist leaders of China and the Soviet Union. If Eden 'had not set himself out to meet Molotov and Chou En-lai in private as much as possible', judged one observer, 'nothing would have been achieved at all.'[175] Matters improved when Dulles went home on 3 May, leaving the American delegation under his more conciliatory Under-Secretary, Bedell Smith. Eden went to the airport to see him off, but doubted whether the gesture was appreciated. The Americans, he concluded, 'are sore, mainly I suspect because they have made a mess of this conference'.[176] But with Dulles out of the way there was always Churchill to add to his difficulties. Eden had reached the stage where he scarcely expected the Prime Minister to be helpful and the latter's initiative in inviting Bedell Smith to London for talks was fiercely resisted. Eden judged that such a visit would arouse wild speculation that pressure was being applied to the British.[177] 'Position at this conference', he warned Churchill, 'is about as difficult and dangerous to Western unity as anything I have ever seen.'[178]

But as Eden moved slowly towards a diplomatic settlement, emerging thereby as the undisputed star of the conference, he seemed unable and perhaps unwilling to prevent a widening breach in relations with the United States. 'I think they are jealous of our authority and following here,' he concluded. 'They like to give orders, and if they are not at once obeyed they become huffy. That is their conception of an alliance – or Dulles's anyway.'[179] British officials became increasingly worried by Eden's anti-American tone. When the American ambassador in London returned to Washington, he found the atmosphere very bad as far as Anglo-American relations were concerned and sensed an increasing mood for unilateral intervention in Indo-China.[180] Meanwhile Makins, Britain's ambassador in Washington, took it upon himself to warn Eden of the gravity of the situation. The Foreign Secretary should know that, rightly or wrongly, the Americans believed he had gone back on the agreement reached when Dulles was in London. Eden's refusal to enter a Far Eastern collective security pact in advance of Geneva or to support the idea of an air strike at Dien Bien Phu may have been intrinsically sound. But 'the effect . . . was to leave Dulles without a policy at Geneva, since he could not talk to the Chinese, and to make him look rather foolish'. Makins sensed that the Americans were moving towards a historic decision to accept far-reaching commitments on the Asian mainland, carrying the risk of involvement in another war:

> What disturbs me is the effect on Anglo-American relations if, at the time of such a fundamental decision, the British were standing aloof or were opposed to the American policy, and since our interest in the security of that part of the world is possibly greater than the American, and a system of security there with American participation has long been an important objective of our foreign policy, the Americans will not understand why we should be so reluctant in this matter. The decision would be no less historic for us than for them.[181]

After three weeks of negotiations Eden still judged the prospect of an armistice as no better than 'even money'. He had become more than ever convinced of the danger of a Third World War if a settlement were not reached. His problems were not reduced by Churchill's determination to accept Eisenhower's invitation to visit Washington in the second half of June. On the one hand he had no wish to be pulled away from Geneva when discussions there might be reaching a climax. Yet, on the other, he was fearful of allowing a man whom he no longer regarded as mentally or physically fit for high office to make the journey on his own. He warned that he and Churchill might arrive just as the American desire to intervene militarily was at its height. 'The call for us to take part in such an adventure would then be intensified and the strain on Anglo-American relations, when we had to decline, could be all the worse.'[182] By 10 June Eden seemed ready to admit that the conference had failed. His one cause for satisfaction was that Bedell Smith, unlike Dulles, had tried to work for a settlement. 'Though the Americans may not believe it', he told Makins, 'there is all the difference in the world between a Geneva conference which breaks down because the Communists are intransigent and one which breaks down because the Americans won't try.'[183]

Whether either Dulles or Eden would try to make the Anglo-American meeting in Washington a success was another matter. Churchill and Eden flew to the United States on 24 June. The Foreign Secretary's Commons speech on the eve of his departure caused an uproar in Congress. He contrived to pay tribute by name to all the leading actors in the Geneva discussions with the exception of Dulles. More importantly, his call for mutual guarantees was bound to lead to *de facto* recognition of Red China, which remained unacceptable for the Americans.[184] Even on the journey across the

Atlantic Eden poured scorn on the US government. 'Speaking generally they were not so intelligent as Truman's administration; this lack of *nous* was particularly noticeable in the State Department.'[185] Meanwhile Dulles seemed intent on poisoning the atmosphere before Eden's arrival. Makins reported that some of his actions and utterances since returning from Geneva had been 'rather peculiar'. Dulles had told a group of American journalists that differences with Britain were deeper and more serious than was generally realized and by no means limited to South-East Asia. Even in Europe no understanding had been reached on what to do if the EDC collapsed. With the object of alerting Eden that he would be confronted not only with 'intractable political questions, but also with a complicated personal and psychological question as well', Makins summed up Dulles's frame of mind:

> He is a vain fellow, and his amour propre was deeply wounded when on his return from Geneva the American press cheerfully recorded that he had suffered the worst defeat in American diplomatic history, at the same time as they were reporting the bouquets being showered on you for your splendid efforts . . . in bringing the two sides together at Geneva. This is galling to a man whose ambition is to go down in history as one of the great American Secretaries of State.[186]

Once the British party arrived, Eisenhower and Churchill worked hard to smooth over the difficulties. Little had changed from the war years in the respective attitudes of Prime Minister and Foreign Secretary to the American alliance. As Churchill's doctor recorded:

> When there are such great issues at stake the P.M. cannot understand why Anthony should go out of his way to find faults in the Americans. He is particularly annoyed that he should choose such an occasion to have a row with the Secretary of State. Does he not realize that without the United States Britain is alone.[187]

In the event Eden and Dulles avoided a clash and even did something to restore their damaged relationship. Churchill 'relaxed, and subsided heavily into a chair'.[188] If anything, Eden found the premier, glorying in the apparent warmth of Eisenhower's reception, more difficult than his American hosts: 'W. keeps coming in and stopping me by saying the same thing over and over again . . . the old boy loves it. He goes about making the "V" sign and riding in the back of an open car or standing up in it like Hitler!'[189] To Eden's chagrin and the surprise of the British delegation as a whole,

Eisenhower seemed more supportive of Churchill's continuing quest for a summit, or at least an Anglo-Soviet meeting, than might have been anticipated. The importance of this development became apparent after Churchill and Eden left the United States on 30 June to return to Britain on board the *Queen Elizabeth*. It was a sign of Eden's mounting frustration that he seems to have struck an informal bargain whereby Churchill would seek a meeting with the Soviet leader, Malenkov, in August, but agree to retire on or around 20 September. But Eden expressed doubts about Churchill's draft telegram to Molotov, enquiring as to the likely Soviet reaction to a British visit, and was unhappy that such a message should be sent without prior cabinet discussion. Churchill, by contrast, seemed as keen to assert his own prerogatives as Prime Minister as he was to arrange the meeting itself. He sent his draft on to Butler in London – who was heading the government in Churchill's absence – with a request that it be forwarded to Moscow. He could, of course, easily have waited until his return to London and Eden had even offered to deliver the message to Molotov in person once he got back to Geneva.[190]

In London the government was in crisis, with ministers deeply unhappy about Churchill's initiative, especially the way it had been made, and disappointed that Eden had not done more to stop it. The cabinet met on three successive days from 7 July, but failed to resolve the impasse. Churchill seemed determined to go ahead with his mission while several ministers, particularly Salisbury, the Lord President, remained profoundly uneasy. Eden made only limited contributions to the discussions, compromised up to a point by his acquiescence in Churchill's original telegram to Molotov.[191] An open rupture was avoided when the cabinet agreed to delay a decision on an Anglo-Soviet meeting until the outcome of the resumed Geneva Conference was known. Salisbury, however, insisted that he could not agree with Churchill's policy and would have to resign.[192] Yet most of the cabinet now recognized that it would be very difficult to draw back from a Churchill–Malenkov meeting unless some development at Geneva showed up Russian intransigence in an unmistakable light. As one explained:

Otherwise we should be represented as a party of shell-backed Tories headed by a reactionary Marquis who had gone out shooting before September 1st and had shot down the dove of peace as it got up from the turnips. All the Cabinet recognise this half-Nelson.[193]

Over the next fortnight Eden struggled to bring the Geneva Conference to a successful conclusion. In Washington he had hammered out the broad outlines of an agreement with Dulles. But on the journey home he had argued bitterly with Churchill over whether any comment should be made on a speech by Senator Knowland implying that Britain had been pressing America to allow Communist China into the UN. Eden warned that any statement denying the accuracy of Knowland's speech would alienate the Chinese and destroy all chance of success at Geneva.[194] Back in Geneva the outlook again seemed unpromising, but quite suddenly a breakthrough was achieved. On 20 July the Viet-Minh agreed to a partition of their country at the seventeenth parallel, a line further north than expected, and the Americans, admittedly without enthusiasm, agreed to respect the resulting settlement. Eden was probably under few illusions about the terms he had negotiated. He admitted later that there might be a political collapse in Vietnam before elections could be held and accepted that such elections would result in communist gains.[195] He even concluded that the Chinese communists' behaviour could most easily be explained by the fact that they were 'thugs or near thugs . . . who don't really want tension too much relaxed with United States'.[196] Yet in the short term Eden had played the leading part in resolving an apparently intractable diplomatic conundrum. Macmillan concluded that his prestige and authority would be 'much increased'.[197]

Such an enhancement of Eden's position was scarcely evident when the cabinet met again on 23 July to try to resolve the question of Churchill's meeting with Malenkov. In advance of the cabinet Eden adopted a somewhat equivocal position. He told Churchill that he could not back his plans in cabinet, but that he would not oppose the premier's determination to go ahead, providing only that any meeting was not on Russian soil. 'But I can't pretend to the cabinet that I like it.'[198] Eden was coming to the conclusion that letting Churchill have his way had become the essential prerequisite to his ultimate retirement. The cabinet discussion saw much old ground retraced with Churchill and Salisbury exchanging threats of resignation. In a tense and heated atmosphere Eden did his best to effect a reconciliation. Unusually, this put him into the position of the defender of American susceptibilities and of the Anglo-American alliance. He seized upon a Soviet declaration on the Geneva talks issued that morning as a further reason for delay and urged that before any such [further] message was sent [to Molotov], it would be wise to study the announcement which the Soviet Government had issued that morning. . . . If the Russians were about to intensify their propaganda about the aggressive intentions of the United States, it might be difficult to go forward with this project at the present time. It might be expedient that we should take the line that we could not attend a meeting with the Russians while they continued to use their propaganda machine for violent attacks on the policy of our American ally.[199]

Anxious for even a temporary respite, the cabinet concurred.

Over the following weeks events conspired to deprive Churchill of his self-assigned mission to save world peace. On 24 July Moscow called for a conference of European states, along with America and with a Chinese observer, to consider European security questions and German reunification. This initiative was clearly designed to undermine the EDC. It showed 'the mould in which their policy is firmly set' and rendered a bilateral meeting impossible at least for the time being.[200] Churchill, still 'hankering after his solitary pilgrimage', remained a difficult colleague, but Eden now seemed more ready to stand up to him while conceding that the government was 'further every day from being a happy family'. Ministers were growing tired of constant 'slanging matches with the Prime Minister'.[201] Churchill lost some of his enthusiasm for a summit when it was announced in August that a Labour Party delegation had accepted an invitation to visit Moscow. What finally killed off the project, however, was the collapse of the EDC following the French Assembly's vote on 31 August. With the Western alliance in chaos it was impossible to open negotiations with the Soviet leadership and Churchill finally abandoned the idea. Ironically, it was left to Eden himself to see the notion of a summit to fruition. On 25 March 1955, with Churchill's retirement date now settled, Eden gave the cabinet an assessment of the diplomatic situation. The future behaviour of the leading players on the international stage seemed unpredictable. He feared both renewed communist adventurism in the Far East and the growth of American isolationist sentiment. He had now come round to supporting East–West talks as soon as possible. Once installed as Prime Minister, Eden pressed consistently for the summit conference which finally convened in Geneva in July 1955.[202]

*

With Eden able to turn even the collapse of EDC to his diplomatic advantage, skilfully negotiating the creation of the Western European

Union which saved the Atlantic alliance from possible disintegration, 1954 proved to be a year of impressive achievement. Even Dulles paid tribute: 'You gave wonderful leadership to the Lancaster House Conference. . . . I feel that we are, more than ever before, working closely together, both in terms of basic objectives and in terms of their implementation. We must keep this up.'[203] With agreements also reached over the Sudan and the Suez Canal base, and the dispute between Italy and Yugoslavia over Trieste resolved, it would seem churlish to deny that 1954 had indeed been Eden's *annus mirabilis*. His deftness of diplomatic touch had become one of the government's chief assets and Britain's position in the world certainly seemed more secure than it had been in 1951. 'I have watched . . . your superb work all through this year,' wrote Acheson with evident sincerity. 'If there was one thing which you could have done better it has escaped me.'[204] The *Daily Mirror*, which had often attacked Eden in the past, now named him 'politician of the year'. In the circumstances it was entirely natural that the Conservatives should build their successful 1955 election campaign around Eden's image as a man of peace. As even a critic conceded:

> Those middle-of-the-road voters who had earlier felt a certain anxiety at the thought of Sir Winston with his bulldog image leading the nation . . . when world peace seemed to hang in the balance, had no such qualms about Anthony Eden. To them the new Prime Minister epitomised their idea of the perfect peace-maker.[205]

Yet, despite tributes from Dulles and Acheson, underlying tensions in Anglo-American relations remained unresolved. Even the Geneva agreements did not remove fundamental differences in attitude to the Far East in general and Communist China in particular. This became clear when Dulles's quest for a South-East Asian security pact bore fruit in Manila in September 1954. Eden, fearful that the Americans might blunder into a war with the Chinese, announced that he could not go to Manila himself since the collapse of the EDC compelled his presence in London. But he was not anxious to support the United States in a provocative policy in the Far East and repeatedly warned Dulles that British opinion would be unwilling to support America in a war over the so-called offshore islands of Quemoy and Matsu. These islands were under the authority of the Nationalist Chinese authorities on Formosa (Taiwan), but painfully vulnerable to attack from the Communist Chinese mainland.[206] It annoyed Eden that the United States might be willing to gamble with a Third World War – in which British cities would be more vulnerable to nuclear attack than their American counterparts – for the sake of insignificant territories on the other side of the globe:

> I have . . . not forgotten Mr. Dulles's remarks to me last September that Quemoy was indefensible except with the use of atomic weapons. Have the Americans really weighed the dangers of having to use them to defend a useless island simply to support the Nationalist morale by compensating them for the loss of the equally useless Ta Chens?[207]

Eden left London on 19 February 1955 on a tour of the Middle East and Asia. In Bangkok he met Dulles during a three-day meeting of the newly formed South-East Asia Treaty Organization. Their main topic of conversation was Formosa and, while the discussions were friendly, 'their points of view remain[ed] unreconciled'.[208] Eden agreed that it would be 'tragic' if Formosa occasioned a 'serious breach' in the Anglo-American alliance. It was an 'obvious communist manoeuvre and we must not fall into the trap'.[209] Yet he took few steps to heal that breach during his remaining time as Foreign Secretary.

*

Overall, and despite the objectives set out by both Churchill and Eden when still in opposition, the Conservative government had done little to consolidate the Anglo-American alliance. The perception of a common Soviet enemy in the all-pervading Cold War had been insufficient to bring the two countries fully together. 'Sometimes', recalled Shuckburgh, 'there was very strong resentment of the Americans in that government.'[210] Eden's poor personal relations with Dulles were an important factor here, but by no means a total explanation. As has been seen, his relations with Acheson during the first year of Churchill's government were scarcely better. Hopes that with Eisenhower as President it might be possible to return to something like the more intimate partnership of the war years were not fulfilled. But Eden's resentment at what he regarded as the United States' short-sightedness and that country's understandable reluctance to treat Britain on terms of equality reawakened a profound strand of anti-Americanism in his make-up which had also come to the surface during the Second World War. During one moment of exasperation at the 1954 Geneva Conference he gave vent to his feelings: 'All the Americans want to do is replace the French and run Indo-China themselves. They want to replace us in Egypt too. They want to run the world.'[211] Britain's diplomatic 'triumphs' in 1954, for which Eden rightly took credit, seem to have encouraged him to indulge in self-deception about the country's inherent strength and its capacity to pursue a course independent of the United States. In October 1955, by which time he was Prime Minister, Eden insisted that 'the British should not allow themselves to be restricted overmuch by reluctance to act without full American concurrence and support. We should frame our own policy in the light of our interests and get the Americans to support it to the extent we can induce them to do so.'[212] But such feelings were based above all on the failure of Britain and the United States to agree a common policy in an area of the globe which has still to be considered – the Middle East.

*

Even to include the Middle East in a discussion of Eden's role in the Cold War indicates a significant historical statement. Much recent writing on the origins of the Suez Crisis suggests that the fatal divergence between Britain and America at that time had its origins in a fundamental difference of perception as to what was at stake in this area. As one historian has put it:

> Unlike British policy, American objectives rested upon the Cold War perception of the Middle East as a vital link in defence against the Soviet Union. Britain, to obtain U.S. acquiescence and, preferably, support for its traditional position, would use cold war rhetoric in the presentation of its policy, but the Anglo-American 'alliance' in the Middle East could only operate when the cold war interests of Washington and the traditional interests of London could both be fulfilled.[213]

There is some truth in this, as contemporaries recognized, but the distinction between British and American policies was never as clear-cut as is implied. Britain's dependence on Middle Eastern oil obviously added an extra dimension to her attitude to a country such as Iran, which was largely lacking in the United States. Shuckburgh admitted that 'for the United States the Cold War is paramount, whereas for the United Kingdom our economic strength is at the moment fundamental'.[214] But for both countries the Cold War was always a confused amalgam of national self-interest and supposedly disinterested opposition to communist expansion. There was as much self-interest in America's attitude towards Saudi Arabia as in Britain's attitude towards Iraq or Iran. The imagined purity of

354

American motives annoyed Eden just as it had during the war. As he later recalled, 'a trouble with Dulles was that he regarded British and French interests in the Middle East as colonial and American interests in South-East Asia, or anywhere else in the world, as virginal'.[215] Britain and the United States often had different ideas on how to contain the Soviet threat. Understandably these usually coincided with calculations of national advantage, but both were obsessed after 1945 with the problems posed by communist expansion and it was entirely natural, in an area such as the Middle East, that Britain, still a major player on the world stage, should seek to take the lead. This was a part of the globe where she had long-standing interests not yet matched by the United States. In a future war against the Soviet Union the Middle East would remain critically important, both to sustain Western Europe's oil-based economies and as a base for an aerial offensive against southern Russia. Successive British governments saw it as their task to formulate Western policy towards the Middle East, confident that this was to the advantage of the alliance as a whole. As Labour's Herbert Morrison put it in August 1951:

> I should just like to repeat that we do not regard ourselves as being in Egypt simply for the maintenance of our own interests, or those of the Egyptians for that matter, but because we feel that we must bear this responsibility on behalf of all freedom-loving nations. No question of imperialism exists.[216]

The United States, its Cold War priorities fixed in Europe and South-East Asia, was generally amenable to leaving Britain with preponderant influence in the Middle East. This was still the case when Eisenhower became President in 1953.[217] Fairly rapidly, however, Washington came to believe that Britain was not taking sufficient account of the rising tide of Arab nationalism, which would need to be accommodated if the Soviet Union were to be excluded from the area. Such thinking increasingly divided the two Western allies.

If anything, the immediate post-war years saw the importance of the Middle East enhanced in British eyes. The granting of Indian independence in 1947 left Egypt as the main British garrison between Britain and Singapore. A Chiefs of Staff paper drawn up in June 1951, with Labour still in office, judged Europe to be Britain's most important defence commitment in the light of the NATO treaty of 1949, with the Middle East coming second. The loss of the latter would be catastrophic to the country's interests.[218]

It was an axiom of British policy makers at this time that the defence of the Commonwealth and Empire depended on the security of the Middle East and Egypt.[219] Conscious that the imperial edifice was beginning to crumble and aware that the leadership of the democratic world was passing inexorably to the United States, British politicians saw this as the point where the line had to be held. In simple terms Britain had fought in one world war to secure the Middle East and in a second to retain it. Such a prize would not be lightly thrown away. For this reason the nationalization of the Anglo-Iranian Oil Company's installations at Abadan by the Iranian Prime Minister Mussadeq in May 1951 was widely interpreted as a challenge which had to be resisted. From the Conservative right Leo Amery warned Eden in uncompromising terms:

> As for the Middle East the whole position will collapse fatally unless we show that the lion has claws and teeth as well as a tail to be twisted. I believe the time is fast approaching when tough individual action by ourselves will meet with general acquiescence, even from America.[220]

Men like Amery evolved their own version of the domino theory, which Americans tended to apply to the Far East. Later that year he advised Churchill that what had happened at Abadan provided an 'excellent precedent' for the nationalization of the Suez Canal. After that Britain would be expelled from Iraq and lose control over Jordan: 'Obviously there must be some counter stroke. . . . As soon as you are in power it will be essential to act quickly.'[221]

Eden himself agreed with prevailing estimates of the importance of the Middle East. His comments on a Foreign Office minute of March 1945 reveal both this and an interesting attitude towards the United States:

> Certainly we must [secure for the United Kingdom a predominant political position and permanent predominance of defence responsibility]. We must not be too modest about this. Russia must be kept out and I am not over keen to bring any others in. . . . I regard the Suez Canal Committee as too international-minded for this business; not a very proper observation for an aspirant to San Francisco no doubt, but likely to be well understood by the nation that jealously guards the Panama Canal. So should we be for Suez, and for as much of Middle East as our strength and the Palestine millstone will allow.[222]

In the same year he bitterly assailed suggestions that responsibility for the Suez Canal should be transferred to an international authority. The Middle East's defence was a matter of life and death to the British Empire since, as the present war and the war of 1914–18 have proved, it is there that the Empire can be cut in half. . . . We cannot afford to resign our special position in the area – and allow our position to be dependent on arrangements of an international character.[223]

Yet, ironically, by the time that Eden returned to power he was ready to embark upon a more moderate and realistic policy. In many ways this was more acceptable from an American point of view, but it set him at odds with both Churchill and a significant sector of the Tory Party. This struggle was not finally resolved in Eden's favour until 1954 and had a lasting impact on the rest of his career.

As part of his reappraisal of Britain's over-extended commitments Eden concluded that her interests in the Middle East could best be protected by new agreements with the Arab states rather than by holding on desperately to traditional pillars of imperial strength. He looked for multilateral defence pacts to avoid the imperialistic associations of existing treaties. At the heart of this vision lay Egypt and the need to negotiate a new agreement over the Suez Canal. Defence assessments suggested that in the nuclear age a large base was not as necessary as before, especially in view of its irritant effect on Egyptian opinion.[224] Under the existing 1936 Anglo-Egyptian Treaty, for which Eden himself had been largely responsible, Britain would in any case have to evacuate the canal base in 1956. On 9 October 1951 Egypt denounced this treaty, leaving the incoming Conservative government with a difficult problem. But cabinet meetings in February 1952 revealed a significant difference of emphasis between the new Prime Minister and his Foreign Secretary. When Eden suggested that the British forces in the base could be substantially reduced, Churchill warned that they should not be removed until they could be adequately replaced by allied forces under a Middle East command. While Eden wanted to offer concessions to the Egyptians to facilitate agreement, Churchill saw no need for hurried gestures.[225] The latter privately complained that Eden was 'throwing the game away'.[226] At the back of his mind was the opposition within the Tory ranks which would be 'deeply stirred by our moral surrender and physical flight'.[227] But Eden insisted that no Egyptian government would conclude an agreement which provided for land forces from a foreign power being stationed in the country in peacetime.[228]

The Foreign Secretary's position was a delicate one. Always

sensitive to the charge of 'appeasement,' he had no wish to blot his copybook with a substantial section of the party which he expected shortly to lead. But he stood his ground and by April had secured cabinet approval to open negotiations with Egypt.[229] In July a military coup dethroned King Farouk, bringing in a military junta under General Neguib. This offered some amelioration to the situation as the new regime dropped its claim to the Sudan and in August the British ambassador in Cairo raised the idea of a Middle East Defence Organization with the Egyptian government. But trouble rumbled on inside the cabinet throughout the autumn. Each time Eden seemed to have won his point, Churchill – ever ready to take advantage of the Foreign Secretary's absence through illness or diplomatic business – was inclined to step in to undo what had been achieved.[230] Few in the cabinet offered Eden much support.[231] Egypt was one area of foreign policy which Churchill would never allow him to handle alone and their disagreements played an important part in straining their personal relations. Eden was 'rather discredited in the PM's eyes at present,' noted Colville after one episode in the dispute over Egypt.[232]

In the course of 1953 Churchill became more determined than ever not to sanction a policy of 'scuttle'. Shuckburgh recorded the tension at the heart of the cabinet:

[Churchill] described Anthony Eden as having been a failure as Foreign Secretary and being 'tired, sick and bound up in detail'. Jock [Colville] said that the Prime Minister would never give way over Egypt. He positively desired the talks on the Sudan to fail, just as he positively hoped we should not succeed in getting into conversations with the Egyptians on defence which might lead to our abandonment of the Canal Zone.[233]

But Eden was still convinced that only by settling secondary issues such as the Sudan could Britain concentrate 'on those areas where our first interests lie'.[234] Matters came to a head at two cabinet meetings on 11 February. Churchill warned that Eden's proposals represented 'an ignominious surrender of our responsibilities' and 'a serious blow to British prestige throughout the Middle East'. He doubted whether they would command the support of the parliamentary Conservative Party.[235] Such misgivings were not without foundation and the Chief Whip, conscious of a 'sense of general disquiet', advised Eden to win over the 1922 Committee before any formal statement was made to parliament.[236] Eden followed this advice and was then able to win cabinet support at a resumed late-night meeting.[237] Increasingly, however, Eden concluded that Conservative opposition was being orchestrated by Churchill himself.[238]

Eden's problems were not restricted to the domestic political arena. The advent of Eisenhower's administration exacerbated his difficulties. It was perhaps a pity that the American government was not privy to a memorandum which Eden circulated on 16 February 1953. Here he argued that Britain could no longer ignore broad-based nationalist movements: 'In the second half of the twentieth century we cannot hope to maintain our position in the Middle East by the methods of the last century. However little we like it, we must face that fact.'[239] It was a case of 'new times, new methods'.[240] Over one outstanding problem Eden had cause to be pleased with the attitude of the Republican administration. The dispute over the nationalization of the Anglo-Iranian Oil Company had remained a bone of contention through the last months of Truman's presidency. As Eden later insisted, 'just to come to terms at any price with Mussadeq, lest worse befall and he be replaced by communism, would be a policy of despair'.[241] But he could never convert Acheson to his point of view. Indeed the State Department, 'in their anxiety to ward off Communism in Persia, have long desired to assist Mussadeq at the expense of the rights and interests of the AIOC and of Her Majesty's Government'.[242] Eisenhower proved more responsive to the British line and determined to use the CIA in conjunction with British intelligence to restore the Shah to power. The secret Plan Ajax was put into operation when Mussadeq dissolved the Iranian parliament in July 1953 and by late August the Shah was back on his throne. Just over a year later Eden could conclude a new agreement with Iran whereby the country's oil would be produced and marketed by a new international consortium, in which Britain retained nearly 50 per cent of the shares. Surveying his achievements at the end of 1954 Eden judged this, including curing the Americans of their 'Mussadeq mania', 'the toughest of all'.[243]

Elsewhere in the Middle East he had less cause for satisfaction with the Eisenhower administration. Yet, ironically, it is clear that Eden strove throughout 1953 not only to co-ordinate his Middle East policy with the United States but actively to involve that country in a new settlement with Egypt. In April he wrote of 'suck[ing] the Americans into the discussions' and in December confirmed that 'our central purpose . . . should be to get really effective American support in the present negotiations'.[244] He was encouraged by his meeting with Eisenhower in early March – 'at last the two governments are seeing eye to eye regarding the problems of the area' – and reported that the President shared his view of the importance of making satisfactory defence arrangements before Britain began to evacuate the Suez Canal Zone.[245] But Eden probably paid insufficient attention to Eisenhower's concern not to offend Egypt 'by the appearance of joint action'.[246] This was not a throw-away remark by the President. It lay at the heart of a policy reappraisal being undertaken by the new American administration. Dulles decided to tour the Middle East in May to assess the American position at first hand. He concluded that Britain's position was

> rapidly deteriorating, probably to the point of non-repair. . . . Such British troops as are left in the area are more a factor of instability rather than stability. . . . The association of the U.S. in the minds of the people of the area with French and British colonial and imperialistic policies are millstones around our neck.

Dulles saw that the ongoing dispute over the canal base was damaging the prospects of the Arab countries uniting behind the West in a Cold War alignment against the Soviet Union. He suggested a new and independent American policy, abandoning the notion of a Middle East Defence Organization and looking to the creation of a 'Northern Tier' of states along Russia's southern frontier based upon Turkey, Iraq, Iran and Pakistan. But Dulles wanted to proceed cautiously, in a manner which would not make it easy for Britain to grasp the reality of American policy: 'Our efforts with the British must be such as will avoid being placed in a position where we must choose between maintenance of the NATO alliance and action to keep a large portion of the world that is still free from drifting into Soviet hands.'[247] None the less the change of emphasis was of enormous importance. By June a State Department official noted that 'to tie ourselves to the tail of the British kite in the Middle East . . . would be to abandon all hope of a peaceful alignment of that area with the West'.[248]

Eden's prolonged illness in 1953 delayed progress over the Egyptian negotiations, but at the cabinet on 15 October Churchill gave his reluctant backing to the Foreign Secretary's proposals. He was anxious, however, that no final agreement should be signed until ministers had had an opportunity to explain privately to government supporters the situation which had been reached.[249] Churchill's sincerity is open to question as there is considerable evidence that he worked behind the scenes to encourage the government's opponents on the Conservative back benches.[250] 'You see, I am not on our side,' he is once reputed to have said.[251] Not surprisingly, Egypt was an issue Eden got 'really worked up on' and on at least one occasion he drafted a letter of resignation.[252] At one point Churchill suggested sending troops into Khartoum to restore order (which had not in fact been disturbed!) so as to offset the 'disgrace' of withdrawing from the canal. There was no alternative, he warned, 'except a prolonged, humiliating scuttle before all the world without advantage, goodwill or fidelity from those Egyptian usurpers to whom so much is being accorded'.[253] More worryingly still, Eden found the American attitude becoming 'increasingly unhelpful'. He feared that they were 'trying to treat us over Egypt as they treat France over the EDC with less justification! The Americans will have no friends left if they go on in this way.'[254]

None the less, as 1954 began Eden was still arguing for the importance of Britain and the United States co-ordinating their Middle East policies more closely.[255] In his own mind a clear corre-lation existed between the interests of Britain and those of the Western alliance in general:

> The problem confronting us is largely that of meeting our peacetime commitments, while we must retain a right to such facilities as we may need in war. In the present cold war situation, it is through a settlement with Egypt that we could best secure our interests in the area, including our oil interests and the use of the Suez Canal, as assistance in ensuring the stability of the area, thus preventing the spread of communism or neutralism.[256]

But as soon as the Foreign Secretary was out of the country Church-ill was ready to create trouble. The latter telephoned Eden in Berlin in January, warning that he might have to take important decisions about Egypt in his absence.[257] From London Selwyn Lloyd argued that Churchill hated the idea of an agreement and longed for the Egyptians to break off negotiations.[258] Eden could just about cope with the demands of the Berlin Conference. It was the 'sabotage at home that gets me down'. To win over Churchill he needed Amer-ican backing, but Eisenhower had no wish to risk alienating the Egyptians by endorsing Britain's proposals unless and until they themselves were ready to reach an agreement. With opposition mounting inside the Conservative Party Eden began to wonder whether it might not be better to delay a settlement until after Churchill's retirement. 'He is beginning to find the unpopularity of Egypt policy . . . too heavy a burden', noted Shuckburgh, 'and is seeking ways of abandoning it.'[259]

In the spring of 1954 the attitudes of both Churchill and Eden underwent something of an about-turn. The former reluctantly concluded that the development of the H-bomb and the entry of Greece and Turkey into NATO had changed the strategic argument in relation to the Suez base. On 21 June Churchill wrote to Eisenhower, arguing that the residual significance of the canal zone was insufficient to justify the expense and diversion of British troops who, since the war, had been discharging an international rather than a specifically British obligation.[260] Churchill now wanted the Americans to join Britain in negotiating a settlement. He believed that Britain's withdrawal from Egypt could be made more palatable if it could be presented as part of an overall Anglo-American plan to build up a defensive front against communism.[261] Yet on the same day Eden, perhaps exasperated by his problems with the Americans over the Far East, wrote of the 'serious difficulties' which might arise from US involvement. His reasoning was not unlike that used by Dulles himself, with the roles of Britain and America reversed. The Americans were 'not more popular in the Middle East than we are' and it would be no good for Britain's position in the area to be seen to be unable to settle this matter without help from the United States. In addition, it would be difficult for Colonel Nasser, who had replaced Neguib in February, to accept an arrangement of this kind. He would 'be open to the charge of having allowed two great powers into the Canal Zone instead of one'.[262] But at least Churchill and Eden were now both committed to the quest for a settlement. In this situation Antony Head, the Minister of War, went to Cairo in July and secured the outline of an agreement. The Foreign Office then refined the details and a final settlement was signed on 19 October. Britain would evacuate her forces within twenty months, but the base would continue to be maintained by British civilians for a further seven years. Those troops were accorded the right to return if any Arab state or Turkey was attacked by an outside power other than Israel.

Eden presented the outline agreement to the Commons on 29 July 1954. For the time being his right-wing critics in the Suez Group had been thwarted. The Egyptian settlement was achieved against considerable opposition and was typical of his performances throughout the year and of his last Foreign Secretaryship as a whole. Shuckburgh later paid an appropriate and perceptive tribute. As Foreign Secretary Eden was

> an acutely sensitive and skilful player of the diplomatic game. I am not saying that he had original policies (there is probably no such thing) or any particular philosophical approach to foreign affairs; his approach was pragmatic. He was interested in settling problems, in bringing people together and working out agreements. . . . He was, in my opinion, a great Foreign Secretary, and this judgement is not invalidated by the evidence . . . that he had weaknesses and was difficult to work with.[263]

Yet within his triumphs as Foreign Secretary lay most of the seeds of the tragic failure of Eden's premiership. The most important of these seeds lay planted in Egypt and the United States. For the moment, however, Eden could take pride in his achievements. In February 1955 he visited Cairo to meet Nasser for the first and only time. 'It was a glamorous performance,' recalled an Egyptian observer. 'He was the star of Western diplomacy who knew all the answers, talking to an unknown colonel with an uncertain future.'[264] The two men never met again. But their fortunes were now irrevocably intertwined.

12

Suez: the first phase

Americans would not have moved until all was lost. All through
the Canal negotiations Dulles was twisting and wriggling and
lying to do nothing.[1]

A s *Full Circle*, his first volume of memoirs, neared publication,
Eden declared that this would be his last word on the Suez
Crisis, the most controversial aspect of his political career. His position would be that 'I have said what I have got to say in the book
and that I have nothing to add or subtract. Other people can argue as
much as they like.'[2] Other people certainly took up Eden's invitation
and Suez has become the subject of an enormous literature, especially since British government archives were opened in 1987.[3] But
Eden was unable to follow his own self-denying ordinance. Though
Full Circle was his last public statement on the matter, he laboured
for a full twenty years in correspondence and notes, in memoranda
and scribbled jottings, to explain, to defend and to rationalize – to
make sense in fact of what many regarded as his incomprehensible
behaviour in 1956. He made great efforts, as did others on his behalf,
to establish or disprove detailed facts and arguments which might
strengthen his case and weaken that of his critics. Eden knew that,
however history judged the rest of his career, his ultimate reputation
would in most minds depend on assessments of Suez and nothing
else. Did Anthony Nutting stay up half the night with Eden to try to
dissuade him from precipitous action after the dismissal of General
Glubb in March 1956? Was there incontrovertible evidence of Soviet
sponsorship of Nasser's actions? Did Dulles really make a deathbed
repentance of American behaviour towards their British ally? In
relation to these and other questions Eden desperately sought to
amass evidence to place his own conduct in the most favourable
light. As late as the summer of 1976, and seriously ill with cancer, he
angrily confronted Lord Mountbatten whom he held responsible for

briefing an author on the Queen's supposed opposition to the whole Suez enterprise.[4]

Most of what were once called the 'secrets of Suez' have now been uncovered. Documentation was certainly destroyed,[5] some files are still closed and there remains scope for differences of interpretation. But few events of twentieth-century diplomatic history have been subjected to such detailed and exacting scrutiny. Much recent work has rightly stressed the need to view Suez as the product of a regional conflict between Middle Eastern states as well as one between those states and outside powers. An examination of Suez within a biographical study of Anthony Eden must necessarily chart a narrower, more personal path. Yet it remains the case that the whole episode is inexplicable without the dimension provided by Eden himself and that events would have taken a significantly different course had he not been the central player in the drama.

*

At the beginning of 1955 Eden's hopes for a Middle Eastern defence organization seemed at long last ready to bear fruit. Following a Turkish initiative towards Iraq a military agreement was signed in February. Though the Turco-Iraqi Pact came as a surprise to London, the British acceded to it in April, at which time it became known as the Baghdad Pact. This represented the logical culmination of the transfer of Britain's strategic focus in the area from Egypt to Iraq, which made good sense in the light of her recent agreement to evacuate the Suez Canal Zone. The Americans were kept informed and gave their approval to British membership, indicating that they too might join at a later date. As the Baghdad Pact seemed to be a partial fulfilment of Dulles's concept of a Northern Tier – a defensive barrier against Soviet aggression in the Middle East – British and American policies in the area appeared finally to have come together, much as Eden had hoped in the early stages of his Foreign Secretary-ship. But it soon became clear that the new agreement was more successful in entwining the protection of British rather than American interests within an overall Cold War strategy. The Pact could help maintain British influence through the good offices of Iraq, by far the most pro-British state in the Middle East. Since the end of 1953, when Jordan agreed to an increase in British troops on her soil, Eden had been working for an Iraqi-Jordanian axis as the basis of Britain's position. At the same time he had had to accept that the chances of winning over Egypt into the same kind of relationship were slight.[6] As the 1932 Anglo-Iraqi Treaty was due to expire in 1957, the Pact gave obvious opportunities to reinforce Britain's relationship with Baghdad. From an American point of view, however, the Pact threatened to antagonize Nasser and other anti-British Arab leaders. Increasingly, Washington was coming to see Nasser as the one leader of a key Arab state who might have sufficient strength to carry unpopular decisions, including making peace with Israel.[7] Inevitably the American attitude to the Pact began to cool almost as soon as it came into existence. Rather than being united in opposition to Soviet expansion, the Middle East might be bitterly divided by what many would see as a mechanism to perpetuate Britain's quasi-imperial ambitions. Indeed, there was a clear danger that an alliance with Iraq and Turkey would undo most of the gains achieved through Eden's hard-won agreement with Egypt of the previous July.

Eden had taken advantage of his journey to Bangkok in February 1955 to stop over in Cairo to meet Nasser. Evidence conflicts as to whether this meeting did lasting damage to the personal relationship between the two men, though it would have been entirely out of character for Eden to have been gratuitously rude on such an occasion. He reported that the atmosphere was 'most friendly' and that Nasser was not 'entirely negative' on the question of a settlement with Israel. But little progress was made on the question of the Turco-Iraqi Pact. Sir John Harding, the CIGS, set out the strategic arguments in favour of the Pact but to no avail. Nasser professed that his sympathies lay entirely with the West, but also argued that the Pact,

> by its bad timing and its unfortunate content, had seriously set back the development of effective collaboration with the West by the Arab states. We used every argument we could to persuade him at least to restrain his criticism and if the agreement were reasonable in terms to cease his opposition. For instance at one moment I said that he should not treat this pact as a crime to which he replied, laughing, 'No, but it is one'.[8]

There was in fact little likelihood of a meeting of minds. As his 1954 work *The Philosophy of the Revolution* showed, Nasser saw himself as the leader of an Arab bloc independent of any great power and not as one component in a British-led anti-Soviet coalition. His vision of the future was more likely to encompass Britain's expulsion from the Middle East than to include partnership with her.

None the less, Eden was not unduly dismayed by the Egyptian attitude, especially as progress was being made behind the scenes towards a comprehensive Arab–Israeli settlement. Eden and Dulles had agreed to try to bring this about when they met in Paris at the end of 1954. Officials had then set to work on the ultra-secret Plan Alpha, which sought to end the Arab–Israeli impasse by creating a land corridor between Jordan and Egypt, across the Israeli-held Negev. Eden hoped that Alpha, like the Baghdad Pact, would strengthen Britain's influence in the area while checking Soviet penetration. For a while the omens suggested that 1955 might even dwarf 1954 in terms of Eden's achievements. As he told the cabinet in March:

> The United States Government approved of our proposed accession to the Turco-Iraqi Pact and might themselves consider acceding to it at an appropriate time. Jordan and Persia might join later. The Pact was in no sense directed against the Israelis, who might also find it possible to accede to it if their relations with the Arab States could be improved. Complete agreement had recently been reached with the United States authorities upon a plan for settling the present differences between Israel and her neighbours, and this plan would be put into operation as soon as conditions were favourable.[9]

But Eden became increasingly concerned that progress in expanding the Baghdad Pact would be hindered while America remained aloof. His assertion that the overriding need was to work out with the US government 'a common political and strategic concept of the Middle East' got to the heart of the matter, while reflecting his policy failures in this area during his Foreign Secretaryship.[10] He was irritated by America's readiness to court Nasser, which ran the risk of weakening the Turco-Iraqi axis.[11]

On 6 April 1955 Eden finally succeeded Churchill as Prime Minister. If this event gave him cause for satisfaction, there were few developments in the foreign arena over the spring and summer to sustain his optimism. He was increasingly concerned at the possible impact of the disclosure of Plan Alpha. The main danger was of a violent reaction in Iraq against the Baghdad Pact which would sweep the pro-British premier, Nuri es-Said, from power. Eden hoped that the Americans would counteract this by a renewed declaration of their intention to join the Pact and by a commitment to pay for an Iraqi purchase of British Centurion tanks under the US Military Aid programme.[12] He regarded Dulles's insistence on making some public reference to Alpha during a speech on 26 August as yet another example of the Secretary of State's refusal to take the wishes of his leading ally into account. But at least Dulles now said that America would be 'disposed to seek formal adherence' to the Pact once a relaxation of tension between Israel and the Arab world had been achieved.[13] Eden, however, was not impressed. He judged the prospect of America one day joining the Pact as remote and believed that Dulles's conduct was steadily reducing the chances of a Middle East settlement. The scale of American aid to Iraq was also disappointing. The United States agreed in mid-August to pay for ten Centurion tanks if Britain paid for two. In his discussion with Eisenhower Eden had spoken of seventy and ten as the appropriate figures.[14] Significantly, he also began to show irritation at the performance of his own Foreign Secretary, Harold Macmillan. The latter 'had not understood how much our original intentions were being baulked. He showed little fight and is generally too susceptible to a compliment or two . . . from Dulles, who is no doubt glad to be rid of me.'[15]

In the event Dulles's speech did not go beyond general platitudes about the need for a comprehensive peace settlement and caused little stir. Of more importance was the heightening tension between Egypt and Israel. The situation had been worsening since an Israeli raid on the Gaza Strip the previous February. This not only made the achievement of a peace settlement infinitely more difficult, but raised in acute form the question of Egypt's defence and her relations with the great powers. By the end of September, and with negotiations with the West having stalled, Egypt and the Soviet Union had struck a deal whereby a large quantity of arms would be supplied via Czechoslovakia.

This development posed a difficult problem for the British government and one which again tended to separate Eden from Macmillan. Eden was irritated by what he described as Macmillan's 'absence of policy'. The Foreign Secretary 'follows Dulles around like an admiring poodle and that is bad for Foster and worse for British interests in the Middle East'.[16] But the government faced a genuine dilemma. As Shuckburgh explained:

It has long been evident that we were retaining our position and interest in the Middle East only because the Russians were not interfering. Once they start bidding for Arab support as they are now doing we are compelled either to outbid them or to lose the main source of power on which our economy now depends.[17]

Eden's reaction was conditioned in part by his deteriorating political position at home. Little more than four months after leading the Conservatives to victory in the general election he was being criticized for indecision. His failure to reorganize the government or to act against inflation attracted hostile comment. Not surprisingly, Eden initially favoured a forthright response to Nasser's latest initiative. As he told Macmillan, 'we and the Americans should make clear to Nasser the dangers he is running with the Israelis by this Russian deal'.[18] He supported a plan to isolate Egypt by grouping the remaining Arab states firmly around the Baghdad Pact and was interested in the idea of an early approach to Jordan.[19] The reoccupation of the Buraimi Oasis on 16 October seemed to provide evidence that Britain could still take decisive action in the Middle East, even without American support. The most important consideration, he told the cabinet, was to protect British oil interests. The Northern Tier was of greater importance than the attitude of Egypt.[20]

Macmillan, however, was more cautious, influenced perhaps by Dulles's warning in early October that the British should 'try very hard to avoid being sucked into a course of action with the Soviets which we could not or would not share'.[21] The Foreign Secretary did not believe that the time had come to isolate Egypt. Nasser had not yet irrevocably committed himself to the Soviet bloc and might even be anxious to find his way back into Western favour. Jordan's accession to the Baghdad Pact would certainly increase pressure for a guarantee to Israel at a time when, in the light of Nasser's arms deal, it would be difficult to resist such pressure.[22]

In the event British policy managed, somewhat uneasily, to combine the approaches of the Prime Minister and his Foreign Secretary. On the one hand, in response to Macmillan's suggestion that 'we should adopt a policy of moderation in our dealings with Egypt and . . . endeavour to persuade the Americans to do the same', Eden agreed that Britain should indeed, as Shuckburgh had suggested, seek to outbid the Soviets for Arab support, including Egypt's. The means would be an offer to help finance Nasser's dream of a High Dam at Aswan on the Nile, designed to electrify and irrigate the river valley with enormous implications for the Egyptian economy. This would be 'of immense importance in restoring the prestige of the West, and particularly of the older European powers in the Arab world generally. In our dealings with Egypt it could be a trump card.'[23] As Britain could offer no more than a small proportion of the overall cost of this vast project, Eden set to work to persuade

Eisenhower to join him in the endeavour. His language was characteristic of the almost apocalyptic vision of the price of failure which permeated British government circles over the next twelve months. As the British learned of the possibility of an Egyptian–Russian deal over Aswan, Eden warned the President:

> If the Russians were to succeed in this, they would, of course, be ruthless in the Sudan and abuse their control of the Nile waters. The outlook for Africa would then be grim indeed. . . . I hate to trouble you with this, but I am convinced that on our joint success in excluding the Russians from this contract may depend the future of Africa.[24]

But at the same time Britain went ahead with plans to expand the Baghdad Pact. Eden's initial concern that Jordan's accession might have an 'unfortunate' effect on Egypt was soon overcome by the advice of Sir John Glubb, the British commander of the Jordanian army, that the spread of Russian-inspired activities against the Western powers made Jordan's early membership imperative. 'If this country is not soon brought into the Baghdad Pact', reported Britain's ambassador in Amman, Glubb 'fears that she will inevitably gravitate into the Egyptian–Saudi–Syrian and so communist orbit.'[25] At the cabinet on 24 November Macmillan declared that the Pact represented the 'Grand Design' of Britain's Middle East policy within which the solution of the dispute between Egypt and Israel must be sought. Such a settlement would create the conditions in which Russian influence might be halted and progressively reduced. Britain's main effort should now be directed to bringing Jordan into the Pact as quickly as possible. Then, with the Egyptian–Israeli dispute settled, Egypt might gradually and properly resume her leading place among Arab states with the prospect of the Northern Tier alignment spreading 'down towards the Nile'. Eden confirmed that no time should be lost in bringing the Egyptian–Israeli dispute to an end.[26] Indeed, he had already delivered an important speech at the Mansion House, calling upon both Israel and the Arabs to make 'some compromise' between the armistice boundaries won by Israel after the 1948 war and the UN Partition Plan boundaries demanded by the Arabs. Though Eden saw the need for both sides to compromise, his speech was widely interpreted as pro-Arab. He was pleased to receive an encouraging message from Nuri and noted that even Nasser seemed not 'entirely negative'.[27]

Such a wide-ranging strategy needed American support if it were to succeed and Dulles had already indicated serious doubts about at least one aspect of the British scheme. Meeting Macmillan on 9 November, he had warned that the introduction of Israel's neighbours into the Baghdad Pact would create a new problem and make it more difficult for the United States to support it. Unless Lebanon, Syria and Jordan were ready to make peace with Israel, which he doubted, he 'rather wondered whether it was wise to bring them in'.[28] By the end of November Macmillan favoured confronting the Americans with a clear choice: 'Either they back up this attempt to shore up the tottering Middle East area or they will lose it all to communism.'[29] The United States also needed to be persuaded to curb the subversive activities of the Saudi government which, because of the financial arrangements of the Arabian-American Oil Company (ARAMCO), was effectively being subsidized by American money.[30] Not surprisingly, Eden favoured an early visit to Washington. It was, he believed, important to display Anglo-American solidarity before the visit to Britain of the Russian leaders, Khrushchev and Bulganin, scheduled for April 1956.[31] But the prospects for a meeting of minds did not look good. While Macmillan informed Dulles that Britain had decided to 'go all out' in support of the Pact and do everything possible to secure Jordanian adherence – 'for Nuri's position cannot be comfortable while he is alone' – Dulles remained cautious. An immediate move to expand the Pact, he warned, would probably preclude Nasser's co-operation over an Arab–Israeli settlement.[32] Furthermore American adherence to the Pact would now have to be coupled with a security guarantee for Israel which would not be possible before an agreement was reached on Israel's permanent frontiers.[33]

Eden's strategy began to unravel when Sir Gerald Templer, the CIGS, visited Amman at the end of the year to discuss Jordan's possible accession to the Baghdad Pact. The visit turned into a disaster. Pro-Nasser rioting led to a decision by the Jordanian government to stand aloof. Though Pakistan had now joined the Pact, there seemed little immediate prospect of further expansion. Furthermore, Nasser always claimed that the Templer mission broke an assurance he had received the previous April that there would be no attempts to persuade any other Arab states to join.[34] This episode marked the beginning of a period of all-out propaganda by the Egyptians, and in particular the government-controlled Cairo Radio, against the British and their schemes for Middle Eastern defence. Nasser had concluded that Eden could not be trusted.

By the end of 1955, therefore, and seven months before the nationalization of the Suez Canal, the two crucial elements of the crisis which would destroy both Eden's career and his reputation were already in place. In December agreement was reached in principle for a joint Anglo-American offer in association with the World Bank to finance the Aswan Dam scheme. Yet Anglo-Egyptian relations were now set on a downhill path from which they would not recover. Perhaps more importantly, a significant division had begun to emerge in the perceptions of Britain and the United States about what was at stake. The British firmly believed, as Macmillan had told the cabinet, that the Russians had embarked upon a deliberate policy to open up another Cold War front in the Middle East.[35] This perception was reinforced by a series of intelligence reports which began to reach London by November 1955 from a new and supposedly reliable source in Nasser's immediate circle, recruited by an MI6 agent codenamed 'Lucky Break'. These reports, details of which remain classified, appear to have had a significant impact. As Eden's press officer noted, 'It is clear that Nasser has gone further than I had ever supposed towards a tie-up with the communists.'[36] Britain's worries as a major player in the Cold War were heightened by a wholly understandable concern about the vulnerability of her national economy to any interruption in oil supplies. In early October 1955 a Foreign Office report suggested that the United Kingdom's fuel requirements would increase threefold over the next twenty years.[37] To this was added, in the minds of several key figures, the psychological impact of the steady reduction in British power and the feeling that this process should now be halted. The attitude of Eden and others towards the Middle East was not dissimilar to that of Dulles towards the Far East – an attitude which Eden had frequently criticized. In the Middle East it was Eden who was the conventional and somewhat inflexible Cold Warrior, seeing in the Baghdad Pact another NATO. 'Our bitter experience', he noted, 'has convinced us that the ultimate aim of communism is world domination and that all means which may further this end are regarded as good ones.'[38]

Though the US government was also obsessed with the containment of world communism and in one sense welcomed the tougher British line,[39] there were subtle differences in the American outlook. Her indigenous supplies of oil made it harder to view the threat in the Middle East with Britain's degree of urgency. More importantly, though Eden wrote that Britain had decided upon a new approach, basing her policy on 'the concept of partnership between equal and sovereign states',[40] Washington remained suspicious that Britain was motivated as much by residual imperial ambitions as by the common struggle against communism. As Dulles later explained:

> For many years now the United States has been walking a tightrope between the effort to maintain our old and valued relations with our British and French allies . . . and . . . trying to assure ourselves of the friendship and understanding of the newly independent countries who have escaped from colonialism.[41]

As a result, the main American concern was that the inevitable transition of the Middle East from a position of subservience to one of independence should not be accompanied by the advance of communism. America therefore found it difficult to see how Western aims could be achieved without Egyptian co-operation. By contrast Eden, after a period in which he seemed to believe that Nasser's pan-Arabism could be reconciled with a Western-orientated defence system, was now ready to treat Egypt as an enemy.

Clearly much would depend on the outcome of Eden's visit to Washington scheduled for late January 1956. As Prime Minister he stepped easily into the postures vacated by Churchill. The meeting with Eisenhower and Dulles would be one of equals, designed to discuss 'present Soviet tactics and our own in various parts of the world' and also to produce 'in plain and simple terms' a statement of what 'our two countries stand for in world policy; what we believe should be the pattern of international relations and our own beliefs in civil rights and individual freedom'.[42] But the Americans were as opposed to this vision of the Anglo-American alliance as when Churchill had expounded it, and Dulles was particularly wary of any attempt by Eden to establish a Special Relationship with Eisenhower, as this would 'throw an intolerable burden upon the President'.[43] Dulles suspected that the notion of a joint Anglo-American policy and a supposed identity of interest was essentially a device to keep Britain in the top league of world powers. None the less, Shuckburgh led a Foreign Office delegation to prepare for the visit of Eden and his new Foreign Secretary, Selwyn Lloyd, who had replaced Macmillan on 20 December 1955. In the meantime Eden wrote forcefully to Eisenhower urging him to curb the activities of the Saudi government. Saudi money, he warned,

> has been subsidising newspapers in Syria, Jordan and in Lebanon, some of them extremely left and communist or near-communist

papers, which they keep going. . . . Now has come a move, of which you will be aware, to supplant us in Jordan by making payments similar to those which we have been making all these years . . . it becomes increasingly clear that the Saudis, the Russians, the Egyptians and the Syrians are working together. If we don't want to see the whole Middle East fall into communist hands we must first back the friends of the West in Jordan and Iraq. . . . If the Saudis go on spending and behaving as at present there will be nothing left for anybody but the Bear, who is already working in their wake.[44]

The state of Eden's health during the final twelve months of his premiership has been the subject of much comment and speculation. There is evidence that by the time he left for America he was already showing the first signs of the illness which would ultimately compel his retirement. Shuckburgh found him 'nervy and in a curious way frivolous. I don't think he is at all well or at all happy.'[45] During the Atlantic crossing Eden had 'plenty of boring trouble with [his] inside' and contemplated seeking medical attention on arrival in America.[46] The news that he was not going to see Eisenhower as much as he would have like provoked a 'tantrum'. 'I am not going to be treated like this. I will take the next boat home. We shall achieve nothing . . . they cannot treat the British Prime Minister like this. Remember Yalta.' Shuckburgh also noted the first signs of Eden's tendency to equate the present problems in the Middle East with those of the 1930s: 'He compared Nasser with Mussolini and said his object was to be a Caesar from the Gulf to the Atlantic, and to kick us out of it all.'[47] Lloyd, lacking the political stature of his predecessor, was a victim of Eden's moods. 'I would not be Foreign Secretary in such conditions for anything in the world,' recorded Shuckburgh after one display of prime ministerial temper.[48]

In the event the Washington discussions went better than might have been expected. Though it remained clear that the United States would not join the Baghdad Pact until after an Arab–Israeli settlement, the Americans agreed to offer moral support. There were also signs of an improvement in their attitude towards Saudi Arabia. Eden formed a favourable impression of Eisenhower, who was 'really friendly', and it was probably the latter's influence which led to a softening of the British line towards Egypt. As Eden reported:

We agreed that the future of our policy in the Middle East depended to a considerable extent on Nasser. If he showed himself willing to cooperate with us, we should reciprocate. The Americans thought

that present talks about the Aswan dam . . . might indicate his state
of mind. If his attitude on this and other matters showed that he
would not cooperate, we both should have to reconsider our policy
towards him.[49]

Even Eisenhower seemed well pleased, confiding to his diary that he
had never before attended international talks where the spirit of
friendship was more noticeable. 'Even our gravest difficulties could
be discussed in the friendliest debate.'[50]

Nevertheless, while the joint communiqué spoke of the 'unity of
purpose of our two countries', it seems unlikely that underlying
differences had been fully explored. Eden was buoyed up by the
conference's success, influenced perhaps by a secret promise of
'some atomic information which would have saved us many millions
in respect of atomic submarines, and research on atomic propelled
vehicles'. But he had failed to persuade the Americans to put teeth
into the Tripartite Declaration of 1950 by which Britain, France and
the United States had given a theoretical guarantee of the *de facto*
frontiers of the Middle Eastern states pending a final Arab–Israeli
settlement. A year later Eden recalled:

> We failed . . . to get agreement on public and joint warning, with
> French, that we would act together against any aggressor in the Mid-
> East and would prepare plans to do it, making this also clear publicly.
> . . . I pleaded seriousness of situation . . . and that a joint warning
> and the sight of our joint preparations was the one thing which might
> keep the peace. But I made no real progress.[51]

For the time being, however, it did seem that Anglo-American
policies towards the Middle East were more closely aligned than for
several years. In line with the Washington decision that the possibilities of co-operation with Nasser should be further explored, it
was arranged that Lloyd should visit Cairo in early March. In the
resulting discussions the Baghdad Pact remained a stumbling block,
with Nasser emphasizing that any *modus vivendi* with the West
depended on no further Arab states acceding to it. Subject to this,
Nasser seemed willing to revive the idea of an Arab League security
grouping with Iraqi membership and did not rule out its eventual
association with the Baghdad Pact. Following an agreement on these
lines Nasser would call off his propaganda against both the Pact and
British 'colonialism'. On returning to the British embassy, however,
Lloyd learned of the removal of General Glubb from the command of
the Arab Legion on the orders of King Hussein of Jordan, an event

Nasser had not mentioned, but of which Lloyd found it 'difficult to believe' he did not have foreknowledge.[52]

Eden's reaction was less equivocal. According to Nutting, now Lloyd's deputy at the Foreign Office, no arguments could shake the premier's absolute conviction that Glubb's dismissal was Nasser's doing.[53] It seems clear that Eden, perhaps not in the best of health and under considerable domestic political pressure, did react angrily to the news. The need to be seen to be doing something was inevitable and the British ambassador in Amman urged the government to react 'speedily, publicly and effectively against Glubb's removal'.[54] Shuckburgh confirms that Eden was now violently 'anti-Nasser' and that he raised the possibility of a reoccupation of the canal base to counteract what was a serious blow to British prestige.[55] Eden sent word to Hussein that what had happened was 'wholly inexplicable'. Continued British friendship would depend on an immediate public statement that, as proof of the value he attached to that friendship, the King would confirm the remaining British officers in their posts and express confidence in them.[56] But it seems unnecessary to accept Nutting's rather lurid interpretation that the Glubb incident set Eden on an irreversible collision course with Nasser in which he in effect declared a personal struggle to the death against the Egyptian dictator. In fact calmer counsels soon prevailed. Lloyd, now in Bahrein, did not believe that the positive aspects of his Cairo visit had necessarily been lost. Though Glubb's removal was 'a bodyline ball in the middle of the innings', an accommodation with Nasser was 'not impossible to start on a tentative basis with a good deal of mutual suspicion, but possibly broadening out into something more fruitful'.[57] From Jordan Ambassador Duke now argued against taking a more tragic view of the episode than was justified, while Glubb himself, arriving at Chequers on 4 March, calmly pressed the overriding need to preserve good relations with Jordan.[58]

Eden was outraged by what he regarded as an act of treachery on Nasser's part. (Nasser on the other hand professed to believe that Britain had, as a gesture of goodwill, stage-managed Glubb's sacking to coincide with Lloyd's visit to Cairo![59]) Eden also identified a clear challenge to Britain's determination to remain a great power in the Middle East. But his major concerns were probably domestic and political. Shuckburgh's initial reaction to Glubb's removal is instructive: 'For AE it is a serious blow and he will be jeered at in the House, which is his main concern. He wants to strike some blow, somewhere, to counterbalance.'[60] After a very personal triumph in the general election, Eden's popularity had suffered a remarkable slump. In the first weeks of his premiership the press had credited him with outstanding abilities but by the year's end their attitude had undergone an almost total transformation. Criticism was not restricted to left-wing newspapers. Though there had certainly been policy reverses – Macmillan spoke privately of 'Eden's chickens coming home to roost' in the foreign arena[61] – it is not easy satisfactorily to explain the scale or speed of the reversal in his fortunes. Some of Eden's most vehement critics were, it is true, long-term opponents such as Randolph Churchill, anxious to show that their earlier misgivings had been well founded. In part he, like at least one of his successors, fell victim to unfavourable comparisons with the towering figure of his immediate predecessor. At all events, by the closing months of 1955 Eden had managed to acquire an unenviable reputation for indecision and incompetence.

By the autumn the *Spectator* was describing a 'terrifying lack of authority' at the top of the government, which it contrasted with the situation under Churchill. Eden found a leader in the *Daily Telegraph* 'worse than anything the Nazis had ever done'.[62] On 8 January the *Observer* carried the headline 'Eden-Must-Go Move Grows' and predicted that Macmillan might replace him before the year was out, while the *People* reported that Eden was ready to retire to be succeeded by Butler. Such attacks would have been less damaging in relation to a man more immune to criticism. But Eden was a compulsive reader of the political columns and, in the words of his press secretary, showed 'too much vulnerability to press attack', while usually professing indifference to it.[63] As a cabinet colleague noted, 'no one in public life lived more on his nerves than he did, and criticism which Churchill, Attlee or Macmillan would have ignored seemed to wound Eden deeply'.[64] It was a side of public life which Eden had seldom experienced earlier in his career when his standing above the rough and tumble of partisan politics had usually induced a degree of respect, even among those who did not agree with him. As Prime Minister he was less sheltered than as Foreign Secretary and, lacking close friends among his cabinet colleagues, became increasingly lonely and isolated. Those around him did not feel that his stature had grown in step with his political elevation. Shuckburgh, an erstwhile admirer, sadly recorded his impressions:

> He has in my opinion greatly changed in the last two years. He is far away, thinking largely about the effect he is making, not in any way

strengthened in character, as I hoped by the attainment of his ambition. . . . I am wondering whether my extremely critical feelings about him are due to changes in him or in me. Anyhow I can see that there is no prospect of my ever being intimate with him again.[65]

It was as if the premiership had exposed Eden's lack of political skills, which his earlier career as a diplomatist had largely concealed.

The premier's fragile temperament was particularly hurt by a leading article in the *Daily Telegraph* on 3 January, written by the deputy editor, Donald McLachlan. Entitled 'Waiting for the Smack of Firm Government', it read:

> There is a favourite gesture of the Prime Minister which is sometimes recalled to illustrate this sense of disappointment. To emphasise a point he will clench one fist to smack the open palm of the other – but this smack is seldom heard. Most Conservatives . . . are waiting to feel the smack of firm Government.[66]

With press criticism at its height William Clark of the Downing Street Press Office persuaded Eden to hold a series of meetings with newspaper editors, but these achieved little. A public denial that he was contemplating resignation hardly gave the impression that Eden was confident of his own position. Perhaps he recalled a similar gesture by his old mentor, Baldwin, back in 1930.

The problem was that Eden's position was *not* strong. The cabinet reshuffle of December 1955 in which Macmillan left the Foreign Office to replace Butler at the Exchequer may have been justified in political terms but left both men disgruntled. Macmillan's behaviour throughout 1956 is open to the interpretation that he had set his sights on replacing Eden. At the very least it seems likely that Macmillan, always the more adroit political operator, saw a possibility that Eden might not survive a full parliament. A dispute in February during which the new Chancellor threatened to resign over the relatively minor matter of food subsidies left relations between the cabinet's two key members distinctly uneasy. Meanwhile the appointment of Lloyd to the Foreign Office suggested that, after a series of foreign policy mishaps, Eden wanted to resume effective personal control of that department where he felt most at home. Though he had decided upon the change some time earlier, Eden, having once criticized Macmillan's caution, now regarded the fiasco of the Templer mission as indicative of Macmillan's reckless style of diplomacy. [67] Lloyd's were a competent, reliable pair of hands, but he lacked his predecessor's political weight and it struck many as odd that, having himself suffered from the unwanted intrusions of earlier premiers, Eden should now recreate the sort of situation he himself had so often criticized. 'We have no Secretary of State,' judged Shuckburgh in late February. 'We have a rather nervous official who has not the inclination or the courage to take decisions of any kind.'[68]

The situation on the Conservative back benches was little easier. Among the party's right wing Eden was still very much on trial. It was for this reason that the Glubb affair assumed such importance. For many the events of March 1956 represented a vindication of their opposition to Eden over Egypt in 1953 and 1954. Even if the premier had not himself been inclined to draw parallels with the 1930s, his Tory critics would have compelled him to do so. On 6 March Julian Amery, a leader of the original Suez Group, wrote to *The Times* to declare the 'bankruptcy of a policy of appeasement in the Middle East. . . . We are now very close to the final disaster.' Eden had anticipated the charge – one to which, of course, he was always acutely sensitive. Writing to Eisenhower on 5 March he argued that 'a policy of appeasement will bring us nothing in Egypt'. There was, he said, no doubt that the Russians intended to liquidate the Baghdad Pact, in which undertaking Nasser was supporting them. The West's response should show 'that it pays to be our friends'. This meant support for existing Pact members, especially Iraq, while if the Americans could now accede to it the effect would be 'tremendous'.[69]

Such thinking was developed at a cabinet meeting the following day. Britain's Middle Eastern policy must be founded on the need to protect her oil interests in Iraq and the Persian Gulf. The main threat to those interests was the growing influence of Egypt. If America could not be persuaded to join the Pact, then at least she could be encouraged to pursue a policy of greater firmness towards Egypt. Indeed, if Iraq's relative position in the Arab world could be strengthened, it might be that country rather than Egypt which could take the lead in securing an accommodation between Israel and the Arab states.[70] But the Americans remained unconvinced. Eden was dismayed that Dulles was still hoping for a deal with Nasser by which, in return for a moratorium on further Arab membership of the Pact, Egypt would call off its propaganda campaign and accept the so-called Johnston Plan for a Middle East settlement with Israel: 'Whilst we must still try for a Palestine settlement and agreement on Johnston Plan, we shall not achieve them by conciliatory gestures to Egypt, which appear to be abandoning our friends.'[71]

Against this background Eden replied to a debate on the Middle East in the Commons on 7 March. Throughout his career he was a competent rather than a brilliant parliamentary performer. On this occasion, when he particularly needed to re-establish his hold over the Conservative Party, his performance was little short of disastrous, possibly the worst of his political life. William Clark and the editor of *The Times* watched from the press gallery. 'We both froze in horror. It was unreal, a nightmare.'[72] Instead of dealing with the issues Eden indulged in petty attacks on the opposition and was repeatedly barracked. According to Drew Middleton in the *New York Times*, he was subjected to a storm of vituperation unmatched since the last days of Chamberlain's premiership. 'This kind of party polemics by a Prime Minister on a great occasion', judged Richard Crossman, 'has an utterly disastrous effect on the House.'[73] In this instance few disagreed. The range of Eden's Conservative critics extended far beyond the irreconcilables of the Suez Group. Shuckburgh discussed the situation with Clark:

> He is deeply depressed and thinks the 'Eden must go' movement will be redoubled. I really can't wish that it should not. He seems to be completely disintegrated – petulant, irrelevant, provocative at the same time as being weak. Poor England, we are in total disarray.[74]

Eden's outlook might have seemed brighter had there been a realistic prospect, as he had told the Conservative Foreign Affairs Committee on 6 March, of America finally adhering to the Baghdad Pact. From Karachi, however, where he had spoken to Dulles, Lloyd reported that America's reply 'to this particular suggestion must be negative'. US adherence would require Senate approval of which there was no prospect in the present election year. Furthermore, Dulles still believed that this was not the moment to risk a complete break in Anglo-American relations with Egypt.[75] Pressed by Lloyd, Dulles at least conceded that some in the American administration had also reached Eden's conclusion that Nasser was totally unreliable. He himself 'had not yet quite reached it although he was not far off it'.[76] Eisenhower, still hopeful of an Arab–Israeli settlement, struck a similar note. The West might be driven to the conclusion that it was impossible to do business with Nasser. However, 'I do not think that we should close the door yet on the possibility of working with him'.[77] It was perhaps these American hesitations which aroused Eden's interest in the idea of co-operation with the French, very ready as they were to view Nasser as an outright enemy in the light of their mounting difficulties with Arab nationalism in Algeria. At all events, on 12 March Eden spoke to the French premier, Guy Mollet, and for the first time raised the possibility of an Anglo-French alignment against Nasser. 'We are working ourselves up against Nasser', concluded Shuckburgh, 'and deciding that the time has come to overthrow him (if we can) or isolate him.'[78]

At the heart of the problem in Anglo-American relations lay Britain's inability to convince her ally of the seriousness of Nasser's challenge, paralleled by American suspicions that Britain was motivated more by a desire to protect her own quasi-imperial interests than by a supposed altruism in the wider struggle against world communism. Eden was personally convinced that Nasser was not merely an irritating Arab nationalist, but the active, expansionist partner of Russian communism – a threat to specifically British interests, but also to the wider interests of the democratic world. By mid-March he was ready to send Eisenhower details of the latest intelligence findings on Egypt's intentions. These, suggested Eden, 'confirm the wide range of Egyptian ambitions against the Saudis, as well as Iraq and Jordan'.[79] The premier's 'most secret note' remains classified, but Macmillan's contemporary diary record contains a likely guide to its contents. Nasser

> seems to be aiming at a sort of League of Arab Republics (the monarchies are to go) to include Libya, Tunisia, Algeria, Morocco, as well as the Arab States in Asia Minor, etc. Egypt would have a sort of hegemony in this League, which would be a strong and immensely rich affair, especially if the oil revenues were pooled. To start it off, and gain prestige, Egypt will attack Israel in June.[80]

When Lloyd returned from his Middle Eastern tour he too had reached the conclusion that Nasser was an inveterate enemy of the United Kingdom. The Egyptian leader aimed at the leadership of the Arab world and was willing to accept Soviet help to secure it. No basis existed for friendly relations with Egypt and Britain should realign her policy accordingly.[81]

In these circumstances Eden believed it should now be possible to restore a common Anglo-American policy. 'It is essential that we should act together in these matters', he told Eisenhower, 'and I hope that we shall be able speedily to evolve a common line.'[82] To an extent that unity was achieved. Yet an important difference remained. While Eden was in effect committed to Nasser's removal, America still wanted to leave the door ajar for his possible rehabilitation. America's purpose was to let Nasser know that he could not expect to co-operate with the Soviet Union while enjoying 'most favoured nation' treatment from the United States. But 'we would want for the time being to avoid any open break which would throw Nasser irrevocably into a Soviet satellite status and we would want to leave Nasser a bridge back to good relations with the West if he so desires'.[83] It is doubtful whether this narrow but important difference was ever resolved during Eden's premiership. Certainly the United States remained deeply suspicious of Britain's motives, with Dulles warning Congressional leaders early in April that the British had exploited the Baghdad Pact for their own purposes and that America was 'most reluctant publicly to identify ourselves in the area with the U.K.'[84] For the time being, however, as he prepared for the visit of Khruschev and Bulganin, Eden took comfort from the apparent backing of the United States. 'I fully agree with you', wrote Eisenhower, 'that we sd. not be acquiescent in any measure which wd. give the Bear's claws a grip on production or transportation of oil which is so vital to the defence and economy of the Western World.'[85] When, therefore, Eden told his somewhat startled Soviet visitors that oil was vital to Britain's survival and that she would fight to retain it, he did so – at least in his own mind – very much as spokesman for the Western alliance rather than merely as the British leader. He hoped he had succeeded in sending them back to Moscow in a chastened and co-operative mood as regards their Middle East policy.[86]

Mounting Anglo-Egyptian tension made continuing British commitment to the Aswan Dam project something of an anomaly. 'Must we go on with the dam?' asked Lloyd in mid-March with understandable irritation.[87] 'We have no present interest in . . . pressing on with the Dam,' answered Eden.[88] With America moving some way towards the British point of view, it was scarcely surprising that when Lloyd and Dulles met for a NATO meeting in Paris early in May, amid rumours of another Egyptian arms deal with the Soviet bloc, they concluded that the Aswan project should be allowed to 'wither on the vine'.[89] A few days later American resentment towards Egypt increased following Nasser's decision to recognize Communist China. Britain had been inclined to play the matter long but in mid-July Lloyd told the cabinet that the US government was 'likely to share our view that the offer of financial aid . . . should now be withdrawn'.[90] The next day Dulles informed Makins that he would be seeing the Egyptian ambassador on 19 July and would leave him in no doubt that the offer was withdrawn. The British response was that this 'would suit us very well'.[91] When, therefore, Dulles met Ambassador Hussein and the latter asked for an unequivocal answer as to whether Western funding would go ahead, the Secretary of State replied that the offer was no longer on the table. This proved to be a momentous step, triggering the Suez Crisis itself. Understandably, Eden's memoir account sought to place as much responsibility for it as possible upon Dulles's shoulders. Eden 'would have preferred to play this long and not to have forced the issue. . . . We were informed but not consulted and so had no prior opportunity for criticism and comment.'[92] But the record suggests that Eden was less disappointed by Dulles's actions than these words imply. A more accurate summary was given in a contemporary Foreign Office minute: 'Mr. Dulles has taken the decision for us. We were not absolutely in step at the last moment but the difference between us was no more than a nuance.'[93] At all events, on 26 July 1956, during a public speech in Alexandria, Nasser proclaimed the nationalization of the Anglo-French Suez Canal Company. Its revenues would replace the now withdrawn Western funding for the Aswan Dam. The Suez Crisis had begun.

*

For most of a public life spent largely in the resolution of diplomatic conundrums, it was Eden's particular skill to move those around him towards a consensus forming the basis for action. But in the case of Suez Eden began with a near consensus – at least in the domestic context – and had the misfortune to see it fade away in the weeks which followed. When the news came through of Nasser's coup, Eden was hosting a dinner for King Feisal of Iraq and his premier, Nuri es-Said. The Labour leader, Hugh Gaitskell, present as a guest, advised the government to act promptly, 'whatever they did', adding that public opinion would almost certainly support them.[94] At the end of the function Eden hastily called a meeting with the Chiefs of Staff – 'in varying stages of dress, ranging from tweeds to court dress' – the French ambassador, the American chargé d'affaires and those cabinet ministers who were readily available – Lloyd, Salisbury, the Lord President, Kilmuir, the Lord Chancellor, and Home, the Commonwealth Relations Secretary. The emphasis was on a military response and the Chiefs of Staff were asked to draw up appropriate plans as quickly as possible. According to William

Clark, Eden argued that Nasser's action amounted to having 'his thumb . . . in our jugular vein'. His simple question was 'when can we take military action to topple Nasser [and] free the canal?'[95] But this tendency to equate the two objectives ultimately cost Eden dear.

At a full cabinet meeting the following day Eden posed the central issues as he saw them:

> The fundamental question before the Cabinet . . . was whether they were prepared in the last resort to pursue their objective by the threat or even the use of force, and whether they were ready, in default of assistance from the United States and France, to take military action alone.

The cabinet gave an unequivocal answer – 'that our essential interests in the area must, if necessary, be safeguarded by military action and that the necessary preparations to this end must be made'. Though the latter stages of this crisis would be characterized by considerable secrecy and even deception, this early decision gives the lie to any idea that Eden pursued the main goals of British policy in defiance of notions of cabinet government. His problem was that, while he remained committed to this original decision, others sought progressively to distance themselves from it. Significantly, the cabinet also explored what the justification for any British response would be. It would not rest upon international law. As compensation was being offered, Nasser's action amounted to no more than a decision to buy out the shareholders and Eden was clear that the problem should not be treated as a legal issue but as a 'matter of the widest international importance'.[96] Eden confirmed most of these points in conversation with Iverach McDonald later that day. McDonald found him 'quite calm' but determined that Nasser's action could not be accepted. He wanted to go forward in full agreement with France, the United States and the Commonwealth, but emphasized that in the last resort Britain had to prepare for military action, possibly alone. He would be telling Eisenhower that he would not get drawn into the legalities of whether a country could nationalize a company which was technically its own. More than once Eden insisted that weakness in the face of Nasser's aggression would destroy Britain's whole position in the Middle East.[97]

The events of the next few months make little sense unless some effort is made to understand the strength of feeling which the Suez nationalization evoked. Much of the personalized criticism of Eden's conduct focuses on his assumed exaggeration of the menace posed by Nasser's ambitions and upon the parallels drawn with the dictators of the 1930s. 'I thought [Eden] had got that part of it wrong,' concluded Butler, after more than a decade for reflection. 'Nasser was not in politics for the good of our health, but he was no Hitler, no incarnation of evil, no megalomaniac who had to be toppled before free men could rest easy in their beds.'[98] Few would now claim that Nasser had the war potential to do in the Middle East what Hitler had done in Europe. But in 1956 there was less certainty. Too many reputations had been destroyed in the 1930s for the political generation of just two decades later to risk complacency. It was after all the twentieth anniversary of the Rhineland crisis – the moment when, as popular conviction would have it, Hitler should have been stopped before he could embark upon his rake's progress of conquest and pillage.[99] Historical parallels often sound trite in failing to take account of the complexities of individual circumstance. The supposed 'lessons of history' invite simplistic analysis which adds little to genuine understanding. And Eden was not immune to these pitfalls:

> The truth is that if these militant dictators are not dealt with early in their careers, the world has to suffer for it. Apart from Hitler and Mussolini, the classic examples of this are Sukarno and Nasser. If they had been compelled to keep to the rules in their early days, either they would have done so or they would have fallen. In any event, both the Middle East and the Far East would have been saved many millions in expenditure apart from loss of life.[100]

But Eden's views on Nasser were not the preserve of one idiosyncratic extremist, reliving and distorting his own career in the years of appeasement. Even Dag Hammarskjöld, UN Secretary-General, told Selwyn Lloyd in April 1956 that Nasser was comparable to Hitler in 1935.[101] British ministers were convinced, on the basis of intelligence reports which may be one of the most significant but least appreciated elements in the whole crisis, that Nasser should not be viewed in isolation but in the context of global Soviet strategy. 'I was quite clear', recalled Lord Home, 'that Nasser's seizure of the Canal was part of a wider scheme in which the Russians were immensely interested.'[102] From the outset even Dulles recognized that the desire of Britain and France to use force derived less from their intrinsic concern about the canal than their conviction that 'this act should be knocked down or have grave repercussions in North Africa and the British position in other countries'.[103] In the immediate wake of Nasser's coup, Eden judged that the chief danger lay in Arab nationalism sweeping across the whole Middle East.[104] A collective psychology emerged among the British ruling élite that, after a decade of retreat, the country could not afford many more reverses if it were to retain any pretension to great-power status. When Macmillan told Dulles that he would rather pawn the pictures in the National Gallery than accept humiliation at Nasser's hands, there was no doubt an element of hyperbole designed to convince the American of Britain's determination to stand firm.[105] But beneath the rhetoric lay genuine conviction. It was another survivor of the 1930s, Ivone Kirkpatrick of the Foreign Office, stung by the suggestion that Britain was exaggerating Nasser's menace, who gave that feeling its most eloquent expression:

> If we sit back while Nasser consolidates his position and gradually acquires control of the oil-bearing countries, he can and is, according to our information, resolved to wreck us. If Middle Eastern oil is denied to us for a year or two, our gold reserves will disappear. If our gold reserves disappear, the sterling area disintegrates. If the sterling area disintegrates and we have no reserves, we shall not be able to maintain a force in Germany, or indeed, anywhere else. I doubt whether we shall be able to pay for the bare minimum necessary for our defence. And a country that cannot provide for its defence is finished.[106]

In the first days after nationalization the elements of a Suez consensus appeared to fall into place. Eden's brief and factual statement to the Commons before the cabinet meeting on 27 July was followed by a forthright denunciation of Nasser's coup from Hugh Gaitskell and other members of the Labour Party, at least one of whom pointedly warned of the dangers of appeasing dictators. The press reflected the general mood, carrying extravagant condemnations of the Egyptian dictator. In successive editorial headlines *The Times* wrote of a 'Time for Decision', 'A Hinge of History' and 'Resisting the Aggressor', and followed Eden's private thinking in its assertion that 'quibbling' over whether Nasser was 'legally entitled' to take over the canal 'entirely misses the real issues'. Just as striking, however, was the response of the left-of-centre press. The *News Chronicle* suggested that the government would be fully justified in taking retaliatory action, while the *Daily Herald* urged 'No More Hitlers'. Varying this theme the *Daily Mirror* argued that Nasser would be wise to study the fate of Mussolini.[107] Even had Eden not been essentially of the same mind, his scope for manoeuvre would have been extremely limited.

The atmosphere had scarcely changed when the Commons debated the crisis on 2 August. The Conservative mood was not in doubt and Eden's was among the more sober contributions to the debate. Nasser 'will not be argued out of it simply by reason,' insisted Walter Elliot. 'There is a moment when he has to be withstood, and that moment is here.' From the right-wing Suez Group Julian Amery warned against relying too heavily on the United States. That country had 'done us enough harm in the Middle East already. We must make it clear to them that if they will not or cannot join us, then we will go ahead without them. It will not be the first time. Our life is at stake, and we can do no other.'[108] But equally striking were the echoes of such strident sentiments coming from the Labour benches. Gaitskell argued that Nasser's actions were part of a struggle for mastery in the Middle East, which could not be ignored. The Egyptian dictator's tactics were 'all very familiar . . . exactly the same that we encountered from Mussolini and Hitler . . . before the war'.[109] Herbert Morrison, now a backbencher but still carrying some authority as Labour's last Foreign Secretary, went further, insisting that 'if the United States will not stand with us then we may have to stand without them. . . . I ask the Government not to be too nervous.' He favoured acting through the UN, but only if that body was expeditious and effective, otherwise it would be the duty of members such as himself to support the government in the use of force.[110]

Now it is true that on the question of the UN Gaitskell sounded a different note from Morrison, warning that Britain must not get into a position where she could be denounced in the Security Council for aggression. Force could be used 'consistent with our belief in, and our pledges to, the Charter of the United Nations and not in conflict with them'.[111] Having seen Eden on the morning of the debate, Gaitskell believed he had an assurance that force would only be used if Nasser offered further provocation.[112] The passage in his speech on the UN has been cited by those seeking to attribute to Gaitskell a greater degree of consistency than Eden, for one, believed him to have shown during the course of the crisis. Rereading the speech three years later Eden conceded that, despite the 'Hitler simile', it was not as strong as he had believed at the time.[113] But if Eden had paid insufficient attention to the small print of Gaitskell's speech, even the Labour leader's own supporters did not see this as his main message. A fellow member of the shadow cabinet thought he 'rather overdid the Nasser–Hitler line', while the youthful Tony Benn recorded sadly:

> Hugh Gaitskell made a speech of bitter denunciation of Nasser. . . . I felt so sick as I listened that I wanted to shout 'shame'. I very nearly did buttonhole him afterwards and say that his speech had made me want to vomit. Gaitskell's speech . . . ended with a phrase declaring that . . . Britain was bound not to use force without [UN] authority. But the dominating impression of the whole speech was that British prestige and influence required really tough action.[114]

In fact a sufficient measure of agreement probably existed in these first days of the crisis to have sustained an immediate military response to the canal's nationalization. Even a future cabinet waverer such as Walter Monckton later recorded that he was in favour of the tough line adopted by Eden in the wake of Nasser's coup.[115] To Eden's chagrin, however, the capacity to deliver such an immediate response simply did not exist. Soon after his appointment as Minister of Defence in the autumn of 1954 Macmillan had noted that the public would be alarmed to discover that Britain lacked the capacity to fight any war except a nuclear one, but the same revelation in 1956 seems to have taken Eden by surprise.[116] The Chiefs of Staff, meeting on 27 July, noted that no plan existed for taking control of the Suez Canal in the face of Egyptian opposition and that the administrative back-up in the Middle East was inadequate for protracted operations. Several weeks would therefore be necessary to build up British forces so that they could embark on any action with overwhelming force.[117] The problem from Eden's point of view, however, was that the heady atmosphere of late July would be difficult to sustain unless Nasser's control of the canal proved catastrophic for its operation or the Egyptian dictator offered fresh justification for British intervention – something he was determined to avoid. Indeed, once negotiations were begun, military action became ever more difficult to justify without Nasser providing a fresh *casus belli*. As Monckton is said to have put it when first shown the military plans for an attack, 'Very interesting, but how do we actually start this war?'[118] And the longer military action was postponed, the more likely the world was to accept nationalization as a *fait accompli*. This seems to have been the underlying calculation of the Egyptian government. Nasser believed that the period of maximum danger would be in the first few days after nationalization, reducing progressively to something like a 20 per cent risk of an armed attack in the second half of October. 'Thereafter, thanks to the mobilisation of world opinion which Nasser confidently looked forward to, the risk would virtually evaporate.'[119]

The missing element in the otherwise consensual initial reaction to the crisis lay in the United States. Despite the signs of *rapprochement* evident since the beginning of the year, and particularly since March, 'Britain . . . and the United States entered the crisis with divergent interests'.[120] That divergence ultimately proved fatal for British policy and for Eden's premiership, and Suez is best remembered for bringing Anglo-American relations to a post-war nadir. Yet whatever other misjudgements are held against him, Eden does seem to have hoped, in the early stages at least, to co-ordinate British and American policies. But the chances of a single Anglo-American vision emerging were not improved by the nature of this latest Middle East crisis. No one could seriously doubt that the canal was primarily a matter of concern to Britain and France rather than the United States. With its headquarters in Paris the Canal Company was French in character, while the British government was its leading shareholder. For nearly a century this waterway, supposedly vital to the country's prosperity and survival, had symbolized Britain's continuing status as a great imperial power.[121] Its use made a significant difference to the distance between the mother country and its interests in the East – India, Australasia and the Far Eastern Empire. The ever-growing importance of oil underlined its vital strategic role in sustaining the economies of Western Europe. If there were logical arguments to the effect that Britain could cope without the canal, they did not command a sympathetic audience in 1956. Yet Americans found it difficult to summon up the same intensity of feeling. 'In fact', recalled Eisenhower, 'I did not view the situation as seriously as did [Eden], at least there was no reason to panic.'[122] Suez thus threatened to revive suspicions that Britain's interests in the Middle East, notwithstanding all Eden's efforts to convince the United States of the Soviet menace in the area, were at heart selfish and imperial. As Treasury Secretary George Humphrey put it, the British 'were simply trying to reverse the trend away from colonialism, and turn the clock back fifty years', which could not be done.[123] Even to the extent that Americans accepted that Nasser was a Russian puppet whose removal had become a Cold War priority, they were fearful that the principal effect of any Anglo-French military intervention would be to unite the Arab world in passionate hostility and at untold cost to the future of Western influence. As the President later put it, the British 'had gotten themselves into a box in the Middle East. They have been choosing the wrong places in which to get tough.'[124] The fact that Eisenhower was coming up for re-election and, despite the belligerency of the early Republican years, standing as a man of peace, added further compelling reasons for avoiding military conflict.

Eden, of course, understood that Suez was of more immediate concern to Britain than to the United States. But he was sincere in his perception of the overriding Soviet danger (unless, that is, the whole British cabinet was engaged in an exercise of self-deception) and he expected a loyal ally to accept that this was not an issue where Britain could afford to give ground. He recalled occasions when Britain had given way, against her better judgement, to support – or at least not to oppose – America over issues where she felt particularly committed. Intervention in Guatemala in 1954 was an obvious case in point. Surely, he believed, Britain was entitled to reciprocal favours now that her vital interests were at stake. Suez was as important to Britain as Central America was to the United States. As Eden later wrote:

> The course of the Suez Canal crisis was decided by the American attitude to it. If the United States Government had approached this issue in the spirit of an ally, they would have done everything in their power, short of the use of force, to support the nations whose economic security depended upon the freedom of passage through the Suez Canal.[125]

Yet these words display an attitude to the Anglo-American alliance, based on equality, to which no American government had felt it necessary to subscribe since perhaps 1943. That the United States was not prepared to treat its leading ally on the same basis she expected when the roles were reversed may have been irritating – even unfair – but it was a fact of international life and Eden had had plenty of time to assimilate it. 'It would have been good', reflected Eden at the end of the crisis, 'if . . . State Department could have asked itself whenever it found itself in line with Russia against France and U.K., "Am I quite sure I am in the right place and doing what I should be doing?"'[126]

In later years Eden repeatedly blamed Dulles for all Britain's problems over Suez. 'Foster Dulles's diplomatic moves were as tortuous as a wounded snake's', he wrote a decade later, 'and with much less excuse.'[127] His argument was that Dulles had misled him as to how the United States would react to an Anglo-French resort to force. As an Australian observer had noted two years earlier, Dulles 'can be . . . extremely vague in what he says, so that you have little precise idea of what he means to convey'.[128] But at the heart of Eden's misunderstanding lay a fundamental misconception about the distribution of power within Eisenhower's White House. Eden believed that Dulles controlled American policy and that in the last resort he would co-operate in steps to overthrow Nasser. The premier found it difficult to come to terms with the idea that a somewhat inarticulate soldier had become America's head of state.[129] He accepted the view that Eisenhower 'had to be very sedulously briefed in the simplest elements of fact and interpretation, historical and contemporary, if he was to fulfil his constitutional function of assuming responsibility for decisions in foreign policy, and that still in a rather formal and exiguous sense'.[130] Yet historians now agree that Eisenhower and not Dulles shaped American diplomacy, even if the latter endowed it with a moralistic tone which made it all the more unpalatable to a British audience.[131] In anticipating American reactions, therefore, Eden and his government would have been better advised to note Eisenhower's consistent opposition to the use of force than Dulles's somewhat ambiguous utterances. Yet British ministers seem to have been confident – despite warnings from Makins that 'in prevailing conditions we can look for little help'[132] – that the Americans would eventually fall into line. 'We felt that we might argue away like members of a family', recalled Lloyd, 'but at the end of the day would never seriously fall out.'[133] Similarly, Macmillan 'never doubted' the 'tacit assistance of our American friends' in a real crisis.[134] The first diplomatic exchanges between Britain and America after the nationalization gave some grounds for this optimism.

In response to Eden's suggestion of an immediate tripartite conference Eisenhower dispatched to London Robert Murphy, Deputy Under-Secretary at the State Department. Eden impressed upon his American visitor that he was not asking for material support, but 'we do hope you will take care of the Bear'. Murphy's was primarily a fact-finding mission, but he emphasized Eisenhower's desire for a conference of the maritime powers. Ominously he noted that Eden was labouring under the impression that a complete community of interest existed among the allies: 'that was not the American view and I gave no encouragement to the idea'.[135] But Murphy's most important interview was with Macmillan who used the occasion to show his visitor how seriously Britain regarded the situation so as to try to bring US pressure to bear upon Nasser. The Chancellor insisted that Britain would become 'another Netherlands' if she did not respond to the Egyptian challenge.[136] These graphic remarks had the desired effect: 'we have succeeded in thoroughly alarming Murphy. He must have reported in the sense which we wanted and Foster Dulles is now coming over post-haste. This is a very good development.'[137] Dulles brought a long letter from the President whose central theme was clear enough. Not for the last time, however, Eden seemed more capable of assimilating the favourable features of messages from across the Atlantic than those designed to deter him. If Eisenhower and Dulles had at once explicitly excluded force under *any* circumstances, there would have been less scope for later recrimination. This they failed to do. 'We recognise', the letter stated, 'the transcendent worth of the Canal to the free world and the possibility that the eventual use of force might become necessary in order to protect international rights.' But the President had been alarmed to learn from Murphy that Britain might employ force 'without delay or attempting any intermediate and less drastic steps'. The US government remained hopeful that through a conference of the maritime powers such pressure could be brought to bear on the Egyptian government as would ensure the canal's efficient future use. This must be attempted before the action contemplated by the British government could be undertaken. If the situation could only be resolved by drastic methods, then there should be no grounds for the belief that corrective measures had merely been undertaken to protect British interests.[138] Dulles himself explained that Eisenhower's mention of 'intermediate steps' did not imply 'going through the motions of an international conference' but 'a genuine and sincere effort to settle the problem and avoid the use of force'.[139]

In fact, and notwithstanding Macmillan's deliberately alarmist remarks to Murphy, an immediate military strike had been excluded from the British agenda within hours of Nasser's coup – less through a lack of will on Eden's part than as the result of practical restrictions imposed by the government's chief military advisers. In this situation Eden had already concluded that, on balance, it was worthwhile going ahead with a conference. The policy of pursuing diplomatic measures designed to justify subsequent military action was confirmed by the cabinet's newly formed Egypt Committee on 30

July. What was envisaged was a scheme sponsored by Britain, France and the United States, and communicated to Egypt in the form of a virtual ultimatum, to be followed by military operations when Nasser refused to accept it. If by any chance Nasser gave in to the demands of the maritime powers, his domestic prestige would be so damaged as to render his position untenable.[140] British ministers were pleasantly surprised by Dulles's attitude. Certainly the Secretary of State was left in no doubt of the strength of their feelings. Macmillan clearly believed that America could be brought into line and that 'Dulles and Co. [were] moving towards our point of view'. 'We *must* keep the Americans really frightened,' he noted. 'They must not be allowed any illusion. Then they will help us to get what we want, without the necessity for force.'[141]

Though there is some dispute over precisely what Dulles promised, especially during a private meeting with Eden at Hatfield, the latter got the impression that the United States would be prepared to countenance force if all else failed.[142] According to Eden, Dulles assured him that Britain could always count on America's moral support in the event of an Anglo-French military operation against Egypt. When Eden suggested that the situation would probably come to a head in October, Dulles interrupted him to say that he did not want to know anything about Anglo-French military plans.[143] Dulles's use of the phrase that Nasser must be made to 'disgorge' his ill-gotten gains inevitably encouraged Eden to believe that the two governments at least shared a basic objective. 'We did not despair of him,' noted Lloyd.[144] At all events Eden seemed well content. 'At first', he told Iverach McDonald, 'the Americans had been slow in coming along.'

> He had frankly expected Dulles to be rather negative, but Dulles began and continued very well. The talks became better after he had arrived. Undoubtedly the Americans' fear that we and the French might fly off the handle straightaway helped to bring Dulles nearer and faster to our point of view. At any rate, Dulles freely committed himself to the stand that the Canal should not be left in the sole control of Egypt and should be brought under an international authority.[145]

Only later did Eden conclude that the various plans put forward by Dulles were designed less to bring about the results the British desired than to thwart them. The longer military action was postponed the more difficult it became ultimately to resort to it. Yet such thinking seems already to have been in Dulles's mind. In conversation with Hammarskjöld on 10 August he agreed that the longer the delay the less likelihood there was of a resort to force and that this was one reason for advocating a conference of the maritime powers.[146] Similarly, Eisenhower told a meeting of the National Security Council on 9 August that if Nasser showed that Egypt could operate the canal effectively and indicated his intention to abide by the 1888 Convention, 'then it would be nearly impossible for the United States ever to find real justification, legally or morally, for the use of force'.[147]

In the meantime the Egypt Committee had defined British objectives with admirable clarity. 'While our ultimate purpose was to place the Canal under international control', it noted on 30 July, 'our immediate objective was to bring about the downfall of the present Egyptian Government.'[148] Only one of these goals could be openly declared. Britain's case before world opinion was based on the need to secure international jurisdiction over the canal, and not on the destruction of Nasser. But there was no attempt to conceal Britain's underlying objective from American eyes. As Eden wrote to Eisenhower on 5 August:

> I do not think that we disagree about our primary objective . . . to undo what Nasser has done and to set up an International Regime for the Canal. . . . But this is not all. Nasser has embarked on a course which is unpleasantly familiar. . . . I have never thought Nasser a Hitler. . . . But the parallel with Mussolini is close. Neither of us can forget the lives and treasure he cost us before he was finally dealt with. The removal of Nasser, and the installation in Egypt of a regime less hostile to the West, must therefore also rank high among our objectives.[149]

Eden's own papers relating to Suez contain few surprises and revelations. As a busy Prime Minister he had fewer opportunities than hitherto to confide his inner thoughts to his diary and the surviving documentation gives an unmistakable impression of having been sanitized by Eden himself. But there is evidence that he had reason to suppose that the US government would not be shocked by the idea that removing Nasser should be a primary objective of Anglo-American policy. He probably looked to the close ties between the British and American intelligence services to lead to the sort of co-operation which had secured the restoration of the Shah of Iran. Indeed, it is a striking feature of the following months that, while Anglo-American political relations rapidly deteriorated, close collaboration continued in the intelligence field, to the extent that plans for an Anglo-American coup in Syria (Operation Straggle) were still on the table at the time of the Anglo-French military intervention in Egypt in the autumn.[150] Eden records that at one point in the crisis, possibly during his visit to London for the second international conference in September, Dulles suggested that the two countries should work out together how to bring Nasser down:

> It was further agreed that these discussions should take place in Washington. They did so. We were represented at a high official level (in fact Pat Dean [chairman of the Joint Intelligence Committee] and others). Many weeks were spent working out plans. The document at its conclusion was not an impressive one. But it was headed 'Means of bringing about the fall of Colonel Nasser' or something to that effect. Dulles asked that the title should be changed to something more innocuous.[151]

Though the available information is sketchy it seems that MI6, with the knowledge of Eden but possibly behind the back of Selwyn Lloyd, considered plans for Nasser's assassination.[152]

Differences of opinion arose over the composition of the maritime conference, with Dulles insisting that the 1888 Convention must be the allies' legal sheet-anchor and that the conference should include all the signatories of that Convention. The British were inclined to compromise since they wished the conference to convene as rapidly as possible. With every passing day Egypt consolidated her control over the canal. Eventually the British and French governments agreed that invitations should be sent to eight of the original signatories, together with eight of the principal maritime countries and the eight countries most closely interested in the canal by reason of their patterns of trade. Eden's problem, however, was to sustain the sort of atmosphere which had existed in the immediate aftermath of nationalization, which would be necessary to underpin military action if, as expected, the conference failed to enforce an Egyptian climb-down. Military preparations went ahead with all haste with more than 20000 Class A Reservists called up. Yet the appropriate moment might have to be seized whether or not it coincided with the breakdown of the diplomatic process. As the Egypt Committee determined on 9 August:

> Any military action against Egypt should be launched in retaliation against some aggressive or provocative act by the Egyptians. . . . The

government might be compelled to take advantage of any provocative act . . . even though it came at a time when the preparations for military operations were less well-advanced than might have been desired.[153]

The sense of cataclysmic emergency which Eden and others had sought to convey since Nasser's coup was not easy to perpetuate while ships passed without interference through the canal. If Nasser's original action had been outrageous, not everyone found it easy to sustain a sense of injury. Nasser was determined to avoid providing the sort of pretext which Eden hoped for in the form of stopping a British or French ship. He even permitted ammunition to be removed from ordnance stores in Britain's old canal base until a few days before the Anglo-French invasion.[154] Macmillan captured the government's dilemma:

> On what 'principle' can we base a 'casus belli'? How do we get from the Conference leg to the use of force? . . . Yet, if Nasser 'gets away with it', we are done for. The whole Arab world will despise us. . . . Nuri and our friends will fall. It may well be the end of British influence and strength for ever. So, in the last resort, we must use force and defy opinion, here and overseas.[155]

Keeping emotions high was part of Eden's strategy when he broadcast to the nation on 8 August. Arrangements were less than ideal. The tiny studio set aside for him was hot and dusty and situated at the top of three flights of concrete steps lit by naked lightbulbs. 'I felt ashamed', recalled the programme producer, 'that the BBC had not been able to provide a prime minister with more suitable surroundings in which to make so momentous a broadcast.' In view of the importance of the occasion Eden was anxious to keep to the precise wording of a prepared script but found himself unable to do so without using reading glasses which, he feared, would spoil his visual image.[156] But his audience was unaware of these problems. Of more importance were his actual words. Though the language was entirely understandable, the broadcast's effect was increasingly to box Eden into a corner: 'The pattern is familiar to many of us, my friends. We all know this is how fascist governments behave, as we all remember, only too well, what the cost can be in giving in to Fascism.' By equating Nasser with the dictators of the 1930s Eden effectively ruled out a diplomatic solution. This was not the way, as his own experience testified, to deal with such a threat. And by making the dispute very much a personal one – 'our quarrel is not with Egypt, still less with the Arab world; it is with Nasser' – Eden compounded his difficulties, confirming what many already believed. As Clark had noted the previous day, 'If Nasser does get away with it – in fact if Nasser is still dictator of Egypt next year – the Eden government is doomed'.[157]

Despite Eden's efforts the immediate post-nationalization consensus was already beginning to crumble. On the one hand the opinion of the Conservative right had, if anything, hardened. 'These massive preparations and the outburst of nineteenth-century jingoism from the "gun-boat" Toryism is very alarming', judged one observer, 'and the PM with all his knowledge and his former belief in the League and the UN seems to be truckling to them.'[158] But senior Labour figures, including Gaitskell, were busy distancing themselves from too close an association with the government. As early as 9 August, the Cabinet Secretary warned that the idea of using force was growing increasingly unpopular.[159] A Gallup Poll in the second half of August revealed that, if Egypt refused to accept the decision of the maritime conference, only 33 per cent of those questioned would favour military action. Even among Conservative voters only 43 per cent would endorse such measures.[160] Doubts within the Labour Party became apparent when a letter from Douglas Jay and Dennis Healey was published in *The Times* on 7 August, emphasizing that force should only be used in self-defence, in fulfilment of international obligations and the UN Charter, or if Nasser violated the security and freedom of the canal. Gaitskell had clearly modified at least the thrust of his case as set out a week earlier. 'Lest there should still be any doubt in your mind about my personal attitude', he now wrote, 'let me say that I could not regard an armed attack on Egypt by ourselves and the French as justified by anything which Nasser has done so far or as consistent with the Charter of the United Nations.'[161] After a meeting with James Griffiths and Alfred Robens, Eden confirmed that the question of force did not arise at the present time, nor would it while the conference was in being unless Nasser took some further action. 'It is, I feel, only in the light of what happens at the Conference that we can decide our future course.'[162] Technically speaking, this was true enough, but two days later Eden produced a timetable for military action for the Egypt Committee's approval. This assumed that the London Conference would finish its business by 23 August and that Egypt would reject its proposals by 5 September. An assault force would set sail on 7 September and land in Egypt on the 20th. What, of course, Eden could not afford was the postponement of military operations to a winter date which would render them logistically impossible.[163]

For Eden the issues at stake were so patently national in character that he could not comprehend how others could see matters in a different light. Though the claim that he caused plans to be drawn up for the take-over of the BBC is probably exaggerated, Eden found the Corporation's determination to remain 'impartial' quite unacceptable.[164] 'Many people', he told his old colleague Alexander Cadogan, now Chairman of the BBC Governors, 'will judge the strength and determination of Britain by what they hear on the BBC. I do not need to tell you how grave the present crisis is for this country and the whole future of the western world.'[165] As the cross-party consensus fragmented, Eden found it increasingly difficult to confide in Gaitskell, whose donnish intellectualism grated upon him. Certainly he never established the sort of confidential relationship which had existed with his predecessor, Clement Attlee.[166] But divisions of opinion inside Britain made it easier for the Americans to exercise a restraining influence. Eisenhower told a White House meeting on 12 August that he had been greatly encouraged by Gaitskell's stand in opposing force until all possible attempts to reach a peaceful settlement had been exhausted.[167] Eden came to believe that Gaitskell's misgivings, expressed privately to Dulles, gave the latter 'the pretext he wanted to be so unreliable about Suez'.[168]

Notwithstanding these difficulties Eden was still optimistic when the Conference of Maritime Nations met at Lancaster House on 16 August. After lunch with Dulles on the previous day Eden noted that 'F[oster] seemed quite as firm as before and ready to table joint resolution himself. He also seemed not to exclude possibility of joint use of force. I gave him certain details of our plans, in part in order to show him where we stand.'[169] Eden again discussed British military operations with Dulles on 19 August, stressing the difficulty of keeping forces in a state of readiness indefinitely. Dulles accepted the value of Britain's preparations as an indication of her determination to reach a satisfactory settlement, but did not favour provoking Nasser into further action to justify the use of force. He now indicated that if Britain and France became involved in war the United States would be unable to participate in military operations. On the other hand he had warned the Soviet Foreign Minister that, if Britain and France were involved in war, the United States would not stand aside. Eden interpreted this rather ambiguous statement to imply that 'in the event of hostilities the United States would at least provide material help'.[170] Dulles also revealed to Lloyd that America was reluctant to take economic sanctions against Egypt. Yet the outcome of the conference seemed satisfactory from a British point of view. Of the twenty-two nations present (Egypt and Greece had declined the invitation) eighteen subscribed to a resolution put forward by Dulles requiring Egypt to accept that the canal should be run by an international board and not be closed to any nation for political reasons. The one disappointment was that Dulles refused to take these proposals in person to Nasser. That task had to be entrusted to the Australian premier, Robert Menzies.

Ominously, however, Eden was finding it increasingly difficult to maintain the unity of his own cabinet. In later years he insisted that the cabinet was 'remarkably steady and united' throughout the crisis and that

> though the points of view ranged from Harold Macmillan on the right to Walter Monckton on the left, the former being most bellicose and the latter least so, there was never any fundamental difference over any of our major decisions, though there were plenty of them.[171]

Eden was particularly annoyed by Butler's later indiscretions and hints that he had been unhappy with the general thrust of government policy. It was Eden's recollection that Butler 'never once made any criticism of substance at the cabinet nor even to me privately, as he could have done at any time'.[172] But while cabinet unity was preserved to the extent that no senior minister felt obliged to resign, the true picture was one of considerable discord. During August divisions began to emerge whose extent was sometimes concealed by the discreet minutes of the cabinet secretary. On another occasion Eden admitted that differences between senior ministers had been a significant factor in determining policy. As he explained for his biographer:

> We had to adapt our views . . . to allay the qualms of the weaker brethren of whom Rab mattered most. They were not prepared to face direct action . . . by the time that our military preparations were ready. On the other hand they were prepared to go along with more devious arrangements which obscured them. Yet Rab was the one member of the Cabinet who since then had spoken in half truths and criticism of all that happened. In other words, those for whom we had bent our tactics proved the least loyal. This is true because I remember how surprised I was that Rab rallied so readily, even

enthusiastically, to our twisted plans when I explained them to him in detail.[173]

Opinion inside the Egypt Committee began to polarize as early as 2 August. On the one hand Salisbury and others, expressing concern over the legality of Britain's position and proposed actions, were inclined to refer the matter to the UN. At the end of the meeting he felt obliged to write apologetically to Eden about 'our ifs and ans [sic] over the United Nations etc'.[174] By contrast the premier also faced difficulties from another senior colleague who could certainly not be numbered among the 'weaker brethren'. Keen to precipitate military action as quickly as possible, Macmillan told the committee that 'it would be helpful if Egypt were faced with the possibility of a war on two fronts'. In the Chancellor's mind this meant collaboration with Israel, an idea he expounded in more detail the following day. As the Israelis were likely to take action of some kind, Macmillan argued that they should be encouraged to attack Egypt. This scheme, so similar to that ultimately adopted, received little support at this stage, a fact which supports the view that the Suez affair is best seen in two fairly distinct phases, the first of which culminated at the UN in early October 1956. In response to Macmillan's scheme, Selwyn Lloyd argued persuasively that Britain's association with the Israelis would destroy her position in the Arab world and the idea was dropped.[175] Macmillan was also concerned that military planning should reflect the reality of Britain's political objectives, in other words that it should encompass the destruction of Nasser's regime. In a note presented to the Egypt Committee on 7 August he asked what would happen once the allied forces had arrived in the Suez Canal Zone. 'Ought we not to consider a different plan altogether, the purpose of which would be to seek out and destroy Nasser's armies and overthrow his government?' His proposal was for the invasion to come from Libya with a landing west of Alexandria.[176] These ideas enjoyed a mixed reception. Though the original tentative plan for a landing at Port Said to seize the canal was abandoned on 10 August and replaced by one to seize Alexandria and march on Cairo, it seems likely that Macmillan's interventions were ill received by Eden and that thereafter, in the words of his biographer, he 'to some extent retired wounded from the diplomatic and military aspects of the crisis, accepting Eden's reproof and concentrating almost exclusively on Treasury matters'.[177]

The decision of the London Conference to dispatch a five-man mission under Menzies to present its conclusions to Nasser brought the use of force distinctly closer, while heightening tensions within the British government. At the Egypt Committee on 24 August Eden reported that he had told Dulles that, if Britain could not get what she wanted by diplomatic pressure, she had no alternative but to resort to force. Dulles, he stressed, had not seemed shocked.[178] Others were less certain. Increasingly a division was emerging between those ministers who favoured postponing force until genuine negotiations had been exhausted and those who simply looked to the diplomatic process to establish a pretext for invasion.[179] The increasingly extravagant language of the latter – which often reads curiously after an interval of forty years – bore witness to the magnitude of the issues which they believed to be at stake.

On 23 August Clark had lunched with Lord Home, the Commonwealth Secretary. Home was personally still firm about the need for force but regretted the speed with which Eden had pushed the original decision through the cabinet without time for a proper discussion. Clark gathered that the whole cabinet was 'a bit weak and searching almost desperately for a moral basis for action'. Butler, in particular, was 'discouraged by the whole outlook' and had come back from holiday 'a very damp influence'.[180] He feared that 'we have got ourselves into a position where we shall press the button before we have a moral basis for action which will carry conviction in the country, the free world and the Conservative party'.[181] But it was Monckton's intervention at the Egypt Committee on 24 August which really took the lid off the situation. The Defence Minister asked whether the government would get the backing of its own public opinion and that of the world in general for the use of force if no further 'incident' took place. Eden, Salisbury, Home and Macmillan countered that Nasser's defeat must be secured by one means or another. 'If not, we would rot away.'[182] But the flurry of correspondence unleashed by his remarks gives a clearer impression of the impact of Monckton's intervention than do the bland minutes of the meeting. Both Home and Salisbury wrote of Monckton's 'outburst', while Lennox-Boyd, the Colonial Secretary, admitted to being 'horrified' by his remarks. While personally convinced that Britain was 'finished if the Middle East goes and Russia and India and China rule from Africa to the Pacific', Home warned Eden of the extent of ministerial unrest:

Even before Walter's outburst . . . I had thought that I had better warn you that I see a definite wavering in the attitude of some of our colleagues towards the use of force. . . . Derry Amory [Minister of Agriculture], for instance, . . . feels the deepest anxieties but I think would be ready to face up to it if all the processes of UNO had been exhausted. . . . The anxiety of some, Rab for instance, might be removed if we didn't have to go on thinking in terms of button pushing and dates and had plenty of time for diplomatic manoeuvre.[183]

Lennox-Boyd confirmed that if 'Nasser wins or even appears to win we might as well as a government (and indeed as a country) go out of business'. He was annoyed that Monckton had waited a month before voicing his objections: 'All these difficulties stood out miles when we first embarked on our policy.' If such uncertainty existed within the cabinet, he concluded, 'we can't be surprised if it exists in the country'.[184]

But it was Eden's friend and long-time colleague Lord Salisbury who best captured the genuine dilemma confronting the whole government. The Lord President warned Eden that Monckton's doubts might find a measure of support inside the full cabinet. Butler was clearly 'not happy' and had been making enquiries which revealed that a number of others, especially the younger members, had not yet made up their minds. 'I thought you might like to know this', wrote Salisbury, 'as the case for force will clearly need to be closely and cogently argued by those of us who agree with it.' His personal difficulty, however, was that he could see no means by which Britain could take action within the terms of the UN Charter 'to which we have solemnly adhered'. His somewhat bizarre proposal indicated the impasse threatening to paralyse the government's scope for manoeuvre:

I suppose that you haven't considered the possibility of us and the French putting the Suez problem to the United Nations, and, if we get no response adequate to the needs of the situation, saying that we have no option but to leave UNO, on the ground that it is clear that it is quite ineffective for the purpose for which it was created, the maintenance of justice and the sanctity of treaties. We must therefore resume our liberty of action.[185]

Irrespective, therefore, of the constraining hand of the United States, the hesitations of the opposition and the fact that Britain would certainly have to await the outcome of the Menzies mission, the advocates of a forceful solution now faced considerable obstacles even inside the cabinet and the Egypt Committee. The cabinet secretary summed up the position:

> some in varying degrees think that, before we resort to force, we must be able to show that we have made an honest effort to reach a settlement by peaceful means and have exhausted all the 'other methods'. We have had the Conference, and we now have a plan for making at least a bow to the United Nations . . . but there may still be a feeling in some quarters that, even so, there should be some further provocative acts by Nasser before we take the final step.[186]

Such a taut situation was unlikely to react well on Eden's always brittle temperament. Monckton found him in a 'highly emotional state and making life very difficult all around him', while Eden's private secretary recounted 'terrible tantrums' over a weekend at Chequers.[187] Clark's diary describes a man living on his nerves. Inaccurate or unhelpful press comments led to 'a violent explosion' and 'screaming at the F[oreign] S[ecretary] over the phone'.[188] An inability to delegate responsibility was more of a handicap to Eden as premier than earlier in his career. 'He wants to be Foreign Secretary, Minister of Defence and Chancellor,' concluded Norman Brook.[189] But Eden retained the capacity to let the storms blow themselves out and to present an image of charm and equanimity to the watching world. At the full cabinet on 28 August he tried to regain the initiative. Over the previous week discussions had turned to the possibility of taking Britain's case to the Security Council, though Dulles had indicated his opposition. Eden too was unenthusiastic and, somewhat unrealistically, canvassed the alternative idea of a reference to NATO or WEU, where Britain would be 'among friends', to ensure a more favourable reception.[190] The cabinet discussion, however, left him with little alternative. Monckton rehearsed the arguments he had deployed at the Egypt Committee and warned that military action would be condemned by opinion abroad and would leave even British opinion divided. In like vein Butler stressed that Britain would need to show that all practicable steps had been taken to secure a satisfactory settlement by peaceful means before a resort to force. In this situation Eden secured general agreement that Britain should go to the Security Council before taking military action.[191]

It was now clear that, whatever the outcome of the Menzies mission, there would be no immediate military strike. Few expected the Australian premier to meet with success. Menzies found Nasser courteous but unyielding. While the mission was still in Cairo, Eisenhower, at a press conference, declared, somewhat unhelpfully, that America was 'committed to a peaceful settlement of this dispute, nothing else'. Menzies reported that these words were 'received with glee by the Egyptians and were undoubtedly treated by Nasser as indicating that he could safely reject our proposals'.[192] In all probability Eisenhower's remarks had little effect. When it came, Nasser's response represented an unequivocal rejection of international control and would probably have been the same without Eisenhower's intervention. But the episode can only have increased Eden's doubts as to whether he and the Americans were really working for the same end. As at other times in his career this sort of situation awakened a latent yearning to be free of American tutelage. A few days earlier he had written to Monckton about the dependence of British military forces on American equipment:

> Have we in fact got ourselves into the position that we cannot send British units to Kenya or to Singapore without first seeking American consent? Is it tolerable that our right – indeed, our duty – to maintain law and order in our colonial possessions should be open to question – and indeed to veto – by the United States? This is a serious matter; and, when our immediate preoccupations . . . are less, I should be obliged if you would consider it and put a paper to the Defence Committee.[193]

Eisenhower was more convinced than ever of the importance of a peaceful resolution of the dispute. His special envoy to the Middle East, Robert Anderson, reported that Egypt might accept a negotiated settlement safeguarding the rights and interests of the maritime nations. Anticipating the failure of the Menzies mission, on 4 September Dulles proposed the idea of a Suez Canal Users' Association (SCUA) by which the users of the canal would employ their own pilots and pay dues to the new association rather than to Egypt direct, with Egypt receiving only an appropriate share of the revenue. The British cabinet accepted this proposal on 11 September. Dulles argued that the 1888 Convention gave the users all the rights they needed. If Egypt interfered with these rights, her position would become untenable. Almost certainly, however, Dulles saw SCUA primarily as a negotiating device which could be useful in delaying and probably avoiding military operations.[194] Eden, by contrast, viewed it as a 'cockeyed idea but if it means the Americans are with us then I think we can accept it'.[195] He was keen to re-establish a close working partnership with the United States but not for quite the disinterested motives suggested in his memoirs.[196] Though it would certainly occasion further delay, SCUA carried the hope of provoking an incident which would justify military action in such a way as would leave America implicated at the side of her allies. One British official had already argued that the concerted denial of canal dues should have a high priority since 'we shall need any luck we can get in the way of provocation by Colonel Nasser'.[197] At the cabinet on 11 September Macmillan used the same argument, insisting that SCUA would not in itself provide a solution. He regarded its establishment as a step towards the ultimate use of force – it would help bring the issue to a head.[198]

While Eden hoped that SCUA would tie the United States into an eventual decision to recover the canal by force, Eisenhower was busy distancing himself ever more clearly from the British position. An exchange of letters in early September revealed the gulf now separating the two men. In a wider context the correspondence exposed the failure of Britain and America in the post-war era to define the nature of their relationship and the authority which the greater could exercise over the lesser power. At the end of the Suez Crisis an official report concluded that its effect had been to define the conditions within which British foreign policy must henceforth operate. 'We could never again resort to military action, outside British territories, without at least American acquiescence.' This was undoubtedly true, but it formed no part of Eden's basic assumptions during 1956.[199]

Eisenhower's first letter on 3 September began by detailing the areas where Britain and America were in agreement. It then continued:

> I am afraid, Anthony, that from this point onward our views on this situation diverge. As to the use of force or the threat of force at this juncture, I continue to feel as I expressed myself in the letter Foster carried to you some weeks ago. . . . I must tell you frankly that American public opinion flatly rejects the thought of using force, particularly when it does not seem that every possible peaceful means of protecting our vital interests has been exhausted without result.[200]

As several commentators have pointed out, this message appeared to place a total ban on a British resort to force of which Eden could have been in no doubt.[201] Its effect, noted Clark, was to bring 'the P.M. racing back, almost in despair'.[202] Three days later the President secretly vetoed CIA plans to depose Nasser because of the risk that this would be counter-productive as regards Western influence in the Middle East.[203] As an American diplomat later explained, the US government opposed a military solution not because it wanted to save Nasser but because it believed such action would unify the Arab world behind him.[204] Eden's only hope was to try once again to convince Eisenhower of the magnitude of the threat posed by Nasser, something he had been attempting throughout the year without ever fully succeeding. His reply, drafted by Ivone Kirkpatrick, Permanent Under-Secretary at the Foreign Office, put forward an apocalyptic vision of Nasser's future designs, under-pinned by analogies with 1936, when the West (including Eden) had acquiesced in Hitler's remilitarization of the Rhineland, and 1948, when Soviet aggression in Berlin had been successfully resisted. The seizure of the canal, warned Eden, was 'the opening gambit in a planned campaign designed by Nasser to expel all Western influence and interests from Arab countries. . . . When that moment comes Nasser can deny oil to Western Europe and we here shall all be at his mercy.' If this assessment was correct, Eden insisted, and if the only alternative was to allow Nasser's plans to evolve until all of Western Europe was held to ransom by Egypt 'acting at Russia's behest', then Britain's duty was plain. 'We have many times led Europe in the fight for freedom. It would be an ignoble end to our long history if we tamely accepted to perish by degrees.'[205]

On receipt of this letter Eisenhower consulted Dulles. The latter felt that Eden's fears were exaggerated, while the President disliked his tendency to polarize the debate between two dire alternatives – 'the immediate use of force . . . or an inevitable demise'.[206] 'Permit me to suggest', Eisenhower now wrote to Eden, 'that when you use phrases in connection with the Suez affair, like "ignoble end to our long history" . . . you are making Nasser a much more important figure than he is.' Eisenhower outlined his preferred course of action. The results desired by both nations could best be assured by slower and less dramatic processes than military force. It ought to be possible gradually to isolate Nasser and gain a victory which would be not only bloodless but more far-reaching than anything brought about by force. This situation would only change if Nasser himself resorted to violence in clear disregard of the 1888 Convention. 'That would create a new situation and one in which he and not we would be violating the United Nations Charter.'[207] All this was reasonable enough – except that it took no account of the position, logistical and political, in which Eden found himself. Playing the matter long would involve the abandonment of Britain's existing military plans. By 15 August the Anglo-French commanders had settled on a plan of operations, eventually condenamed 'Musketeer'. Nine days later their staffs produced plans for an aerial attack to begin on 15 September with seaborne landings two days afterwards.[208] Eisenhower's proposals would have thrown such timetables into confusion. As Eden later explained:

> The weather was likely soon to be a limiting factor. Therefore, the alternative to action was in effect capitulation. We would have to bring our forces home and release our ships. As a result the pressure we had used effectively on Nasser by the mere existence of our armed forces would have ended and seen to be so. . . . It would only have been a matter of time before all were compelled by him to pay dues and the effect on all our friends, if they still survived, from Libya to Baghdad would have been disastrous.[209]

Whether Eden's premiership could have survived such a visible reverse is open to question.

Eisenhower's message needed no clarification. But during the first two weeks of September the British ambassador in Washington, Roger Makins, also stressed that London could not expect either moral or material support from America for military action. A go-it-alone policy of military intervention would deal the United States 'a body blow'.[210] Quite simply the Americans 'did not believe that the method and the tempo which we were advocating were the right ones'.[211] If all this was not sufficiently disheartening for Eden, Dulles wasted little time in undermining the effectiveness of SCUA, his own brainchild. When consulted on 7 September on canal dues, he said that the first result of the user nations announcing that they intended to deny payment to Egypt would be the denial of the right of passage to their ships. The second would be an oil shortage in Europe. Dulles now presented SCUA as a means of co-operating with Egypt rather than as a means of pressure. He also introduced the notion that a substantial part of the dues should be paid over to Egypt. On 10 September he suggested that the user nations should be prepared for some interference by Egypt with the passage through the canal and that in this situation American tankers would turn round and go via the Cape of Good Hope.[212]

Eden, however, in the light of Eisenhower's admonitions, had no real alternative to pursuing Dulles's proposals which still offered the best prospect of implicating the Americans more deeply and of denying dues to Egypt except on the users' terms. Somewhat astonishingly, the cabinet meeting on 11 September revealed that the option of force had still not been abandoned. Monckton again warned that any premature recourse to force, especially if taken without American support and approval, would precipitate disorder throughout the Middle East, while alienating a substantial body of public opinion in Britain and elsewhere. But the crucial intervention was that of Macmillan, who argued that matters should be brought to a head:

> This was of great importance from the point of view of the national economy. If we could achieve a quick and satisfactory settlement . . . confidence in sterling would be restored; but if a settlement was long delayed the cost and the uncertainty would undermine our financial position. He therefore hoped that Parliament would be persuaded to give the Government a mandate to take all necessary steps, including the use of force, to secure a satisfactory settlement of this problem.[213]

Those who sense that Macmillan played a devious role through the Suez Crisis, aimed at the ultimate removal of Eden from the premiership, have understandably focused on these remarks. Only four days earlier the Treasury Permanent Secretary had advised Macmillan in a very different sense. Sir Edward Bridges had referred to the 'vital necessity from the point of view of our currency and our economy of ensuring that we do not go it alone and that we have the maximum United States support'.[214] Macmillan understood the significance of these words, appending the comment, 'Yes: this is just the trouble. The US are being very difficult.' But his contribution to the cabinet discussion was scarcely consistent with the caution of his senior civil servant. It enabled Eden to sum up the mood of the meeting:

> It would be a difficult exercise in judgement to decide when the point had been reached when recourse must be had to forceful measures. In determining this, we should weigh not only the state of public opinion in the United States but also the views of the French who were eager to take firm action to restore the situation and were increasingly impatient of delay.[215]

Lloyd later suggested that no one who had been at that cabinet could have been in any doubt that if Britain tried SCUA without success, and then went to the Security Council without success, 'the use of force would be the policy'.[216]

The collapse of the domestic consensus was more than ever apparent when the Commons reassembled on 12 September. In a speech frequently interrupted by opposition MPs Eden announced the setting up of SCUA but confirmed that if Egypt interfered with its operation Britain would be free to take such further steps either through the UN or by other means as would safeguard her rights.[217] But each time Eden tried to show that the ultimate use of force had not been excluded Dulles stepped forward to undermine him. In a further press conference the Secretary of State now insisted that American ships would not shoot their way through the canal – 'it may be we have the right to do it but we don't intend to do it'. Almost certainly these words severed any remaining strands of trust between the two men. As Eden later commented, it 'would be hard to imagine a statement more likely to cause the maximum allied disunity and disarray'.[218] When Eden wound up the parliamentary debate the following day, Gaitskell understandably asked whether the British would 'shoot their way through the canal'. Cornered, and under pressure from Tory backbenchers, Eden replied that if Egypt rejected SCUA Britain would refer the whole question to the Security Council.[219] It struck many as odd that this concession had had to be wrung from the man whose early career had been inextricably linked with the League of Nations and who had been one of the architects of the United Nations Organization. As one disillusioned observer recorded: 'to put it in an irritated aside confirms the impression that he has lost his hold and that although he plays the right words, he plays them in the wrong order and in the wrong manner'.[220] The use of force had been delayed still further. It was now difficult to see how SCUA could provide the pretext for military action as Eden had hoped when first accepting Dulles's scheme. 'I may be too optimistic', noted Gaitskell, 'but my feeling is that we are probably over the hump now.'[221] Even when the Western canal pilots walked out, their Egyptian replacements proved – against expectation – fully able to cope with the weight of traffic artificially increased by the British and French in a desperate attempt to show the new staff to be incompetent. A record fifty-seven ships passed through the canal on 7 October, by which time Harold Watkinson, the Minister of Transport, had conceded the failure of Operation 'Pile Up'.[222] Like so many 'experts' the Suez Canal Company's pilots had overvalued their own importance.

Eden could no longer realistically expect united domestic support for military action. The Chief Whip advised that even within the

Conservative Party some opposed force even as a last resort.[223] Others were still ready to put their faith in the premier's sureness of touch in the diplomatic arena. 'They believe that if military action is taken it will be because Sir Anthony really believes that such action is absolutely necessary.'[224] But Eden himself had by no means abandoned the option of a solution by force. He had, he told Churchill, been struggling to keep a firm and united front, but 'firm is even more important than united'. Though the Americans had a mental caveat about action before their presidential elections on 6 November, this was not something he could accept.[225] The SCUA Conference duly opened in London on 19 September with an innocuous agreement reached two days later. Eden was now anxious to refer the matter to the UN as quickly as possible, 'otherwise we shall appear to drift – and in fact be drifting'.[226] He clearly believed that until this step had been taken the justification for military action would not exist to the satisfaction of all. He did not want anyone to think that going to the Security Council represented a change of policy or that it was likely to produce a satisfactory solution – 'everyone knows what the difficulties are with the power of the veto'.[227] But even this decision caused further friction within the Anglo-American alliance. When Eden and Dulles dined together on 20 September there was no indication that reference to the UN was imminent. Yet two days later Lloyd sent Dulles, now back in Washington, an urgent message that Britain intended to have the issue inscribed on the Security Council's agenda. Dulles was considerably upset by the way the matter was handled, but Eden's explanation suggested that he no longer placed American susceptibilities at the top of his priorities: 'I should add that no kind of undertaking was given on Thursday evening. To best of my recollection just before Dulles left Foreign Secretary was arguing for earlier meeting and Dulles for a later one. I do not recall any decision being asked for or taken.'[228]

Dulles continued to make somewhat ambiguous statements on the use of force. On American television he seemed to put Eden's case very fairly: 'You can't always count on nations not using force unless there is some alternative which conforms to international peace and justice. Unless we can work it out, some system which is just, then I don't think you can expect people to go on for ever asking people not to resort to force.'[229] Clark sensed that the government was near to despair. He judged that Eden and others would still like to destroy Nasser but that they dared not do so against

American wishes and without a pretext.[230] Eden was keen that
Macmillan, in Washington for a meeting of the International Monetary Fund, should impress upon Dulles that Britain's position
remained what it had been from the beginning, with force the last
resort. But some way or other Nasser must be compelled to 'disgorge'.[231] In the event Macmillan gave astonishingly optimistic
reports after meeting both Eisenhower and Dulles. The President
was understandably anxious about the forthcoming elections, but
otherwise supportive. According to Macmillan, Eisenhower was
'really determined, somehow or another, to bring Nasser down'.
Macmillan explained Britain's economic difficulties in 'playing the
hand long' and Eisenhower seemed to understand. 'I also made clear
that we must win, or the whole structure of our economy would
collapse.'[232] Macmillan gained a similar impression from Dulles,
subsequently telling Eden that Dulles 'quite realised that we might
have to act by force'. Indeed, he thought the threat of force was
vital, whether or not Britain used it, to keep Nasser worried. Dulles's
concern too was with the election:

> He reminded me of how he and the President had helped us in May
> 1955 by agreeing to the four-power meeting at the top level, which
> had undoubtedly been of great benefit to us in our electoral troubles.
> Could we not do something in return and try and hold things off until
> after November 6th.[233]

But these reports gave a misleading impression of the attitude of
the American leaders. Makins, who was well placed to judge, saw no
basis for Macmillan's optimism. The striking thing about the interview with Eisenhower was the lack of discussion of Suez itself. The
two men, who had first got to know one another in North Africa
more than a decade earlier, seemed more interested in exchanging
wartime reminiscences. It seems that Macmillan did not even raise
the circumstances in which Britain might use force against Nasser
nor what the US reaction would be.[234] The Chancellor's motives for
his clear misrepresentation must remain a matter of speculation. The
most favourable interpretation is that he was trying to use his
wartime friendship with Eisenhower to strengthen Eden's domestic
position against cabinet waverers. Alternatively, it is possible that he
was deliberately shepherding Eden towards a political disaster from
which he hoped to be the beneficiary.

If Eden was not confused enough as to what American leaders
really believed, Dulles, perhaps in retaliation for Britain's failure to

consult him over the reference to the UN, was ready with one
further contortion. Not content with again undermining SCUA –
'there is talk about teeth being pulled out of the plan, but I know of
no teeth: there were no teeth in it, so far as I am aware' – he now
launched a gratuitous attack upon the underlying motivation of
Britain and France:

> The United States cannot be expected to identify itself 100 per cent
> either with the colonial powers or the powers uniquely concerned
> with the problem of getting independence as rapidly and as fully as
> possible. . . . I hope that we shall always stand together in treaty
> relations covering the North Atlantic, [but] any areas encroaching in
> some form or manner on the problem of so-called colonialism find the
> United States playing a somewhat independent role.[235]

Though Dulles quickly tried to make amends for his ill-judged
remarks (which did, in fact, represent his views), trying to persuade
Makins that he had not intended to link Suez and the colonial
question and had been drawn into an 'undesirable' line of discussion, Eden was understandably outraged. The charge of 'colonialism'
was one he never forgot.[236] Nutting recalled that Eden read a report
of Dulles's words and then threw it across the table, angrily demanding, 'Now what can you say for your American friends?'[237] Eden
could be forgiven for concluding that this was but the latest example
in a pattern of American perfidy which made a nonsense of any
notion of friendship or alliance:

> We leant over backwards to keep in line with US – Users Club was
> classic example. We did not like it. It was US project. We would have
> preferred to go direct to UN. We endorsed it. That was beginning of
> our Parliamentary troubles. It became project for collecting dues for
> Nasser. Dulles press conference during our debate. His outburst about
> colonialism. His purpose to smash us.[238]

Precisely what Eden now intended remains a matter of controversy. According to Nutting, the reference of the dispute to the UN
was never meant as a serious move, but rather as a device to set the
stage for war by inviting a Russian veto in the Security Council and
so demonstrating that redress would have to be sought by other
means.[239] Eden himself lent some credence to this interpretation.
Writing in 1970 he argued: 'Critics of Suez seldom tell us what other
course should have been pursued. There was only one – to go to
Geneva to negotiate and to come to an agreement which I knew
Nasser would not observe for more than a few weeks.'[240] But it must

have been obvious to everyone that the combination of faltering domestic support, the impotence of SCUA, clear signs of unrest among the military reservists, the inexorable advance towards autumn and the unhelpful American attitude had rendered intervention a very difficult course.[241] On balance it seems likely that Lloyd was authorized to negotiate in good faith in New York, though the military option had not been abandoned. It was, however, necessary in mid-October for the British Chiefs of Staff to recommend the replacement of the existing military plan, Musketeer Revise, by the so-called Winter Plan. This retained many of the features of Musketeer Revise, but extended the timetable to allow for shorter daylight hours and a prolonged phase of aerial bombardment.

Meeting Dulles on 5 October Lloyd insisted that the British and French were indeed making a genuine effort for a peaceful solution, but that the Middle Eastern situation was deteriorating and that, if no result emerged from the Security Council, force would probably be the lesser evil. To this Dulles gave the characteristically ambiguous response that America did not rule out force as an ultimate sanction, although in this particular case they thought it would be a fatal mistake.[242] According to Lloyd it was the French Foreign Minister, Christian Pineau, rather than himself who was unenthusiastic about a peaceful resolution.[243] This confirmed the impression that Eden had formed when meeting the French leaders on 26 September: 'Mollet [the French premier], as I believe, would like to get a settlement on reasonable terms if he could. I doubt whether Pineau wants a settlement at all.'[244]

Correspondence exchanged between Eden and Lloyd while the latter was at the UN confirms that both men were moving towards a negotiated settlement. As an official report later concluded, 'thus we came near to abandoning our dual objective' of recovering the canal and removing Nasser from power.[245] Between 9 and 12 October, Lloyd, Pineau and the Egyptian Foreign Minister, Mahmoud Fawzi, engaged in direct negotiations. Against expectations, an outline agreement was soon reached. It was based on 'six principles' which largely derived from Dulles's speech to the first Suez conference in the early days of the crisis. These were:

1 There should be free and open transit through the canal without discrimination, overt or covert.
2 The sovereignty of Egypt should be respected.

3 The operation of the canal should be insulated from the politics of any country.
4 The manner of fixing tolls and charges should be decided by agreement between Egypt and the users.
5 A fair proportion of the dues should be allotted to development.
6 In case of disputes, unresolved affairs between the Suez Canal Company and the Egyptian government should be settled by arbitration.

That part of the Anglo-French resolution containing the six principles was passed unanimously by the Security Council. The remainder, which endorsed the proposals of the eighteen powers as corresponding to these principles and invited Egypt to formulate means to implement them, secured a positive nine to two vote but was vetoed by the Russians. Although vaguely expressed, the six principles were an improvement on anything which Britain might have expected to obtain had the canal concession expired in the normal way in 1968.

Though Lloyd later suggested that the UN discussions did not come close to a settlement, his initial response was enthusiastic:

> I think it can fairly be said that the results are better than we were justified in expecting before we came. . . . The suspicion that we were treating the United Nations simply as a formality has been dissipated. . . . I think we have convinced both delegates and the UN Secretariat that we have been making a genuine attempt to get a peaceful settlement and that we have been gravely wronged by Egypt.

He was not, however, ready to exclude force altogether. Britain's very presence at the UN

> to some extent imposes on us the settlement of the dispute by processes of negotiation. This presents difficulties for a resort to force unless it is shown to be the last resort. On the other hand I think that with the changed atmosphere here we can count on a more understanding reaction if we have to take extreme measures.[246]

The tone of Eden's telegram suggested that he too was prepared to see the Security Council proceedings as much as a basis for further discussion as they were a springboard for military action:

> Should not we and the French now approach the Egyptians and ask them whether they are prepared to meet and discuss in confidence with us on the basis of the second half of the resolution which the Russians vetoed? If they say yes, then it is for consideration whether

we and the French meet them somewhere, e.g. Geneva. If they say no, then they will be in defiance of the view of nine members of the Security Council and a new situation will arise.[247]

With an understandable, if over-optimistic, sense of relief, Eisenhower told a press conference that 'it looks like here is a very great crisis that is behind us'. Events proved otherwise.

13

Suez: the crisis

I knew how warlike we must appear, how much acting against
the spirit of UN and of all my life's work. Had to be accepted.[1]

B Y 14 October 1956 the odds probably favoured a peaceful resolution of the Suez affair. Eden's handling of the crisis had not
been without fault or misjudgement, but his room for manoeuvre,
particularly in a domestic political context, had never been great.
Whatever line he had taken would have alienated a substantial
section of opinion. Acquiescence in Nasser's coup would have raised
the spectre of appeasement, especially in the light of Eden's moderate policy towards the canal base since 1951. As the French Prime
Minister, Guy Mollet, later asserted, the instinctive reaction to the
coup was an anti-Munich reflex.[2] On the other hand, after the first
days of universal outrage, a 'straight bash' not endorsed by the UN
had been outside the bounds of political possibility, granted Britain's far from compelling legal case and Nasser's scrupulous avoid-
ance of further provocation. Trapped within what was dangerously
close to a no-win situation, Eden had until mid-October been guilty
of no single act which would permanently damage his reputation.
The crisis then entered its second and in some ways distinct phase.
Yet there was a greater measure of continuity than some recent
historians have implied. For if it is misleading to picture Eden as a
man who, from the time of Nasser's coup or even the dismissal of
General Glubb, began to act 'like an enraged elephant charging
senselessly at invisible and imaginary enemies in the international
jungle',[3] declaring a personal war to the death against the Egyptian
dictator, it is also unrealistic to suggest that he had become committed to a peaceful settlement by 14 October and yet was recon-
verted to military action later that same day. His remark to Lloyd
about a 'new situation' arising if Egypt rejected further negotiations,
and his realization that whatever was decided in New York would

leave unresolved the question of Nasser himself as a Soviet-backed threat to the peace, suggests that force remained a possible option in Eden's mind.

Indeed, the evidence suggests that a military response was always his preferred solution. But the possibility of securing it had become ever more remote, not least as a result of American obstruction. In such circumstances Eden *was* prepared to envisage a negotiated settlement, but only on terms which represented a clear defeat for Nasser. A letter to Robert Boothby on 2 October captures the balance of his thinking:

> Of course what we have been and are trying to do is to settle this matter by negotiation. But our American as well as our French allies agree that the negotiation must be on our terms. It would be disastrous if the result of negotiation was a settlement which strengthened Nasser's position and increased his prestige. . . . It seems to me, therefore, that we stand a chance of achieving our objective by negotiation only if we show the greatest firmness and resolution. But it is no more than a chance. . . . The essential thing is that Nasser should not be judged by the world to have got away with it.[4]

The second phase of the Suez Crisis centres upon the collusive agreement between Britain, France and Israel to attack Egypt. The details of this plot, which seem more suited to the pages of a Buchan novel than the chronicle of British diplomacy, are now well known, and an outline of events will suffice here. On 14 October, with Lloyd still in New York, Eden received General Maurice Challe, Deputy Chief of Staff of the French air force, and Albert Gazier, Mollet's Minister of Labour, at Chequers. The visitors suggested that Israel might be encouraged to attack Egypt in such a manner as to enable Britain and France to send in forces to separate the combatants and protect the canal. Eden was, at the least, intrigued. On 16 October, with Selwyn Lloyd hastily summoned back from the UN, Eden discussed the idea with a small group of senior ministers. Later that day he and Lloyd flew to Paris for a meeting with Mollet and Pineau from which even the British ambassador was excluded.[5] Discussions seem to have been deliberately based upon the contingent possibility of an Israeli attack rather than a formal conspiracy to incite that attack, though all must have understood the reality which underlay their cautious words. At a full cabinet meeting on 18 October Eden indicated that an Israeli strike against Jordan or Egypt might be imminent. The cabinet agreed that Britain would have to defend Jordan under existing treaty commitments, but that no support should be offered to Egypt notwithstanding Britain's obligations under the 1950 Tripartite Declaration. Eden made two important points. The first was that it would be far better that Israel should attack Egypt rather than Jordan; the second that, while Britain was still working for a peaceful settlement, 'it was possible that the issue might be brought more rapidly to a head as a result of military action by Israel against Egypt'.[6] Under Israeli pressure for more concrete British assurances, Lloyd travelled incognito to Sèvres on the outskirts of Paris on 22 October. There, still talking in terms of possible eventualities, he sought to assure Israeli leaders, including the premier David Ben-Gurion, that if Israel did attack Egypt in such a way as to threaten the canal then Britain and France would issue an ultimatum demanding that both sides withdraw their forces from the area of the canal. It could be anticipated that the Egyptians would reject this ultimatum, affording Britain and France the pretext for military intervention so conspicuously lacking hitherto.

The Israelis were less than enthusiastic about the role they were asked to play – accepting the opprobrium of aggression followed by the indignity of capitulating to an ultimatum – and on 23 October the cabinet was told that 'from secret conversations which had been held in Paris with representatives of the Israeli Government, it now appeared that the Israelis would not alone launch a full-scale attack against Egypt'.[7] Later that day, however, Pineau arrived in London with the message that the Israelis were indeed keen to co-operate providing they were assured of an early attack on the Egyptian air force by the RAF. Patrick Dean, Deputy Under-Secretary at the Foreign Office, and Donald Logan, Lloyd's assistant private secretary, were now sent to Sèvres to finalize arrangements. These officials seem to have blurred the always questionable distinction between contingency planning and a collusive plot even though Dean began by handing over a letter from Lloyd which emphasized that Britain was not asking the Israelis to take any action. It was merely a question of stating what reactions there would be if certain things happened. Israel now agreed to invade on 29 October and to threaten the canal with a major offensive. On the following day the British and French governments would demand that Egypt and Israel cease fire and withdraw their forces ten miles either side of the canal, while Anglo-French forces began a temporary occupation of key positions on the canal. An Egyptian rejection would result in an Anglo-French attack early on 31 October. Dean signed a record of the discussions, the notorious Protocol of Sèvres, on Britain's behalf. This, however, was not what Eden wanted. He had not expected a written record and seemed to think the officials should have realized this. He now sent Dean and Logan back to Paris on a futile quest to recover the French and Israeli copies of the document and have them destroyed.[8] At the cabinet on 25 October no mention was made of the Sèvres Protocol, though ministers were informed that the Israelis were, 'after all, advancing their military preparations with a view to making an attack on Egypt', probably on 29 October.

Exactly what ministers understood remains uncertain. Too much significance can be attached to the precise wording of the cabinet minutes which do not, of course, purport to give a verbatim record of the discussions. A few senior ministers including Macmillan were aware of the full story, but others may simply have believed that Britain had received advance intelligence of Israeli military plans. As a subsequent official report noted, 'the Cabinet agreed in principle, *and on a hypothetical basis*, that in the event of an Israeli attack on Egypt, HMG should join with the French Government in calling upon the belligerents to stop'.[9] Yet Eden, constrained by the secrecy of the Sèvres agreement, did go so far as to mention the word 'collusion' and the judgement that he was 'specifically untruthful' to the cabinet would seem harsh:[10]

> We must face the risk that we should be accused of collusion with Israel. But this charge was liable to be brought against us in any event; for it could now be assumed that, if an Anglo-French operation were undertaken against Egypt, we should be unable to prevent the Israelis from launching a parallel attack themselves; and it was preferable that we should be seen to be holding the balance between Israel and Egypt rather than appear to be accepting Israeli cooperation in an attack on Egypt alone.[11]

At all events the character of the Suez crisis had been completely transformed.

Perhaps the most astonishing feature of the whole collusion episode was Eden's belief that he could erase the record of its existence. Apart from his abortive attempts to have all copies of the Sèvres Protocol destroyed, he even seems to have scratched out General Challe's name from the visitors' book at Chequers.[12] The details of the Anglo-Franco-Israeli agreement were hidden from all but a handful of civil servants and senior ministers; formal meetings of the Egypt Committee were apparently suspended between 17

October and 1 November; and written records were either kept to a minimum or subsequently destroyed.[13] Even so, enough people in the three countries were aware of what had gone on to make it unlikely that the truth would not somehow enter the public domain. Additionally, the cover story and the military arrangements needed to sustain it were inherently implausible. This was odd for a man like Eden who prided himself on his political antennae. As early as mid-November Harold Nicolson recorded that he and his son, then MP for Bournemouth, had 'always believed that there was some collusion between the French and the Israelis to which we were a consenting party'. If the story got out, he added, 'I do not see how the Government can survive. It is an utterly disgraceful tale.'[14] Get out it did. The enterprising journalists Merry and Serge Bromberger published an account of the collusion in their book *Les Secrets de l'Expédition d'Egypte* as early as March 1957. Over the next few years both Pineau and General Dayan, the Israeli Chief of Staff, confirmed the essentials of what had happened.[15]

Meanwhile Eden and the small group of colleagues who had been kept informed continued to indulge in an act of collective amnesia, maintaining the fiction that no collusion had taken place and even preferring to ignore the charge rather than deny it.[16] A confidential official report drawn up in 1957 at the request of the cabinet secretary made no reference to this feature of the crisis. 'I used my own judgement,' recalled Guy Millard, the report's author. 'I left that out because I didn't think it would be discreet at that time to write about that aspect of it.'[17] After his famous denial to the Commons on 20 December – 'there were no plans to get together to attack Egypt . . . there was no foreknowledge that Israel would attack Egypt' – Eden made no subsequent public reference to the matter. Only occasionally, in his most private papers, was the question discussed and then only obliquely. 'Collusion is a new term of abuse for concerting foreign policy between free allies,' he wrote in 1957. 'Only Moscow and Nasser may plan with secrecy and impunity.'[18] If Eden engaged in an act of deception in 1956, his tendency in later years was to indulge in self-deception. As work proceeded on the preparation of *Full Circle*, the promptings of his research assistant, Robert Blake, that the matter at least merited discussion, fell on stony ground: 'I feel that you ought somewhere to deal with the allegation quite specifically. It might conveniently come in chapter six when you describe the meeting in Paris on October 16th, for that is the occasion when, according to the left, you concerted some allegedly mysterious and sinister plans with the French ministers!'[19] Collusion was a secret to be taken to the grave. In his account of Suez, published posthumously, Lloyd at least touched upon the question. But he continued to evade the real issues, engaging in a semantic debate on the word 'collusion' and still insisting that, although discussions with the Israelis had certainly taken place, they had been on the basis of what Britain would do in particular circumstances and involved no binding agreements between the British and Israeli governments.[20]

Some, even among his political opponents, could not believe that Eden had become involved in this sort of deception. 'I have never believed in the rather crude version of collusion between Britain, France and Israel,' insisted the Labour MP and security expert George Wigg in 1958. 'I believe that the truth about Sir Anthony Eden is that he was, and remains, an honourable man.'[21] Yet at the end of the day the charge remains and nothing has done more to damage Eden's standing at the bar of history. Under his leadership Britain pretended to move into Egypt to stop a war when in fact she had connived at starting that war to create a pretext to move in. He was therefore guilty of misleading – or at least trying to mislead – British and world opinion as to what British troops were being asked to do. 'Collusion implies something dishonourable', pleaded Lloyd, 'and we were all honourable men.'[22] Perhaps so, but Eden, whose career had been founded on integrity and honour, stands hopelessly besmirched in the eyes of majority opinion. How did this situation arise? Why did he accept the bait which his French visitors dangled before him on 14 October?

Four main factors need to be considered. The first is that the domestic political situation seemed to be closing in on Eden. The Conservative Party Conference at Llandudno debated Suez on 11 October. For the Suez Group, Captain Waterhouse successfully proposed an amendment to the anodyne platform motion. As Waterhouse explained, 'at all costs and by all means Nasser's aggression must be resisted and defeated'.[23] The conference rose to Julian Amery when he announced that

If the discussions at the Security Council do not bring Nasser to his senses, I believe the process of negotiation will then be exhausted. . . . Our hands will be free to use any and every measure . . . necessary to achieve our ends, including if necessary . . . force. We must go forward with the Americans' approval if we can get it, without it if they withhold it, and against their wishes if needs be.[24]

Eden appeared before the conference two days later. It was not an environment in which he usually felt at ease. The typical conference delegate was considerably to the right of his own political instincts. Eden was awaiting the outcome of the Security Council vote, but must have noted that the passage in his speech which emphasized that the use of force could not be excluded was greeted with enormous enthusiasm by the party rank and file. At the very least he would have realized that a compromise scheme, such as that being sketched out in New York, would not have been well received by his own party.[25]

The state of Eden's health must also be considered. The evidence is conflicting. According to some who knew him well, including his former PPS Robert Carr, Eden was never the same man again after his three operations in 1953. Lord Carr has testified that the fevers, which were symptomatic of Eden's failure to make a complete recovery after the eight-hour Boston operation of June 1953, had begun to recur some months before the start of the Suez Crisis.[26] By contrast, Eden's official biographer, who had the advantage of consulting the medical records, insists that 'he made an astonishing recovery [after 1953], and it is wholly incorrect to ascribe subsequent events to his health which, until October 1956, was excellent'.[27] What is incontrovertible is that on 5 October, while visiting his wife in University College Hospital, London, Eden suffered a sudden attack of fever with his temperature reaching 105°F. He stayed in hospital over the weekend, but was back in Downing Street by 8 October, admitting that he was still 'pretty weak'.[28] After this time there seems little question but that Eden was not a fit man. He admitted to the strain imposed by the diplomatic crisis upon his physical reserves: 'the burden was much increased . . . by the differences in time across the Atlantic, with the result that it was difficult to get much sleep, having to discharge one's responsibilities during the day and often be available to UNO during the night'.[29] Precisely what drugs he was prescribed and what their effect upon his behaviour may have been remains uncertain, but many who knew him have testified to a concoction of painkillers and amphetamines which would be consistent with erratic behaviour and changes of mood.[30] Several contemporaries noted that in the later stages of the drama Eden displayed a calm serenity, in marked contrast to his usually excitable and nervous demeanour. An assistant described him as 'almost boyish, reminiscent of a young officer of the First World War, very calm, very polite, the captain of the first eleven in a critical match'.[31] Jebb recalled Eden confessing that he had been 'practically living on benzedrine' and in the last year of his life Eden read, apparently without comment, an account of his premiership suggesting that he had been taking that drug since July.[32] At all events his American specialist judged in the spring of 1957 that his condition then was such that his judgement might well have been impaired over the preceding six months.[33]

Yet there is some evidence that Eden's incapacitation in early October was only the culmination of a period of ill-health. He made few entries in his personal diary for 1956 compared with earlier years, but pain and tiredness are recurrent themes in those few. Entries in August and September see Eden complaining of a lack of sleep.[34] That of 21 August, a full six weeks before his collapse at University College Hospital, is particularly revealing:

> Felt rather wretched after a poor night. Awoke 3.30 onwards with pain. Had to take Pethidine in the end. Appropriately the doctors came. Kling was more optimistic than Horace [Evans]. We are to try a slightly different regime. Agreed no final decision until a holiday had given me chance to decide in good health.[35]

As Rhodes James explains the 'final decision' related to the possibility of yet another operation, an indication surely that Eden had been concerned about his state of health for some while. At the time of his final resignation he admitted that he had been taking a combination of drugs and stimulants since his 1953 operations and that, 'since Nasser seized the Canal in July', he had been obliged to increase the drugs considerably and also the stimulants necessary to counteract them.[36] Fevers recurred at infrequent intervals during the last two decades of Eden's life and, if he had indeed been having such problems in 1956, later experience suggests that each attack would have left him debilitated for several days. Be that as it may, Eden's engagement diary indicates that, apart from his weekend in hospital, he consulted Dr Evans or Dr Kling on at least ten occasions between the canal nationalization and the end of October.[37]

Arguments based upon Eden's medical condition must not be taken too far. The miscalculations of Suez — and miscalculations there certainly were — cannot be fully explained in terms of one man's ill-health. After Eden's death Butler asserted that he had become a 'one-man band' during the crisis, 'much nearer to being a dictator than Churchill at the height of the war'.[38] This was the culmination of a process of Butler trying to distance himself from the government's policy which began during the crisis itself.[39] It was something which Butler's conduct in 1956 only partially justified and for which Eden felt considerable distaste, regarding his former colleague as a source of 'many fairy tales'.[40] But it has been central to the present account to show that Eden did not abandon the conventional norms of cabinet government. Even though the full truth about collusion was kept to a more restricted audience, with the majority of senior civil servants in particular being excluded, there is no evidence that Lloyd's or Macmillan's conduct was affected by illness. As one recent writer puts it, 'if their collective judgement was clouded, it was not by collective painkillers'.[41] All the same, Eden's personal commitment to the collusion plan was a seminal moment in the evolution of the Suez affair and it seems plausible to suggest that the weeks of tension, exacerbated by a lack of sleep and the strain of ill-health, may have left him more suscep-tible to the simple attractions of the French plan and less conscious of its equally obvious drawbacks than he would otherwise have been. Lord Carr strikes the right balance:

> I find it difficult to accept the judgement that Anthony's health did not have a decisive influence at least on the conduct of his policy. I agree that he might well have pursued the same basic policy had he been well, but I find it very hard to believe that he would have made such obvious miscalculations in its execution both in the political and the military spheres.[42]

Some account must be taken of pressure from Britain's French allies. Until Suez the two countries had rightly tended to see one another as rivals in the Middle East. As recently as October 1955 Eden had minuted angrily that the 'French continue to attack our policy in Mid East publicly and sabotage it privately'.[43] Suez brought the two governments together in opposition to Nasser, but the French could always take a straightforward and uncompli-cated attitude towards the canal's recovery which was denied to Britain. They did not have to take account of Commonwealth opi-nion, placed no overriding importance on their relationship with America and did not have a network of alignments in the Arab world to consider. This absence of complications made France increasingly irritated by the postponement of military action and by Britain's insistence on the need for a further pretext to justify a strike against Egypt. As Pineau recalled:

> I was struck by the fact that the English sought above all else a method of justifying their action in the eyes of the Arabs and before world opinion. I thought it would be much simpler to say that the Egyptians had taken over the canal illegally, that this was an aggression under international law, and that we were acting purely and simply to recover it. That was not sufficient for the English.[44]

France's freedom of manoeuvre enabled her to enter into an increasingly close relationship with Israel, something which, as has been seen, had been specifically excluded by Eden and Lloyd because of its effect upon Arab opinion. The military alliance between the two countries began soon after Israel's raid on Gaza in February 1955 and by October of that year the French Defence Ministry had agreed to supply Israel with seventy-two Mystère IV fighters, AMX tanks and other equipment. Eden appeared to regard the arming of Israel with equanimity,[45] but by the late summer of 1956 a critical situation was arising. In early September the French approached Israel to discuss joint military planning against Egypt. They needed considerable persuasion to go along with SCUA and their approach to the UN was unenthusiastic and unhelpful. Eden sensed that the French, particularly Pineau, were anxious to avoid further delay and were in the mood to blame everyone, including Britain, if military action was not taken before the end of October.[46] With Lloyd apparently making progress in his discussions with the Egyptian Foreign Minister, Fawzi, Reuters on 11 October announced Pineau's imminent departure from New York and quoted him as saying that no advance had been made and that no basis existed for further negotiation.[47] The French were concerned that Egypt would be able to drag out talks on the UN resolution indefinitely.[48] By the end of September, in fact, the leaders of France and Israel had begun putting together the details of the plan placed before Eden at Chequers a fortnight later.

It thus seems plausible to argue that Eden was coming under considerable pressure from across the Channel. The French were discussing with the Israelis an attack on Egypt, with or without British support. For the first time in many years in the troubled history of Anglo-French relations the French were making the running. The Fourth Republic was not renowned for political stability and Eden expressed concern that the lack of progress over Suez might bring about the fall of Mollet's government.[49] In addition, 'the French had Algiers on their hands and could not make their [military] contribution available indefinitely'.[50] By early October Eden had become genuinely worried that the French and Israelis might go ahead with an invasion on their own. One witness to the meeting in Paris on 16 October recorded that Bourgès-Maunoury, the French Defence Minister, asked Eden three times what he would do in the case of a Franco-Israeli attack and each time Eden pretended not to have heard. This prospect may have made him more sympathetic to the collusion plan which at least allowed the burden of aggression to fall squarely on Israel's shoulders.[51] At all events, at the cabinet of 23 October, when Lloyd was still saying that it might be possible to negotiate an agreement for the effective international supervision of the canal, he also added that the French would not give their full co-operation to such a policy.[52]

But French pressure must be viewed in conjunction with a fourth and perhaps decisive factor which, by mid-October, was seriously restricting Eden's options. Recent interpretations of Suez have tended to present the Jordanian factor as the missing element without which the premier's conversion from the possibility of a peaceful settlement to the willing participant in a collusive plot to resolve matters by force is inexplicable. There is some truth in this, but the importance of Jordan in Britain's calculations has long been apparent. According to Pineau, from the point of view of the British cabinet the whole Middle Eastern problem turned on the Jordanian question.[53] Eden admitted in his memoirs that he was acutely aware of the consequences of action by Israel against Jordan and that at his Paris meeting on 16 October he had asked the French ministers to do all they could to impress upon Israel that an attack on Jordan would have to be resisted by Britain.[54]

By the late summer of 1956 a tense military and diplomatic situation existed between Israel and Jordan. Fedayeen raids, instigated and organized by Egypt, were being mounted from Jordanian territory into the heart of Israel. On 14 September King Hussein asked for Iraqi troops to be sent immediately to bolster his position, but Israel threatened retaliation if a single Iraqi soldier moved into Jordan. On 10 October very heavy Israeli artillery and mortar fire was directed against the Jordanians at Qalqilya. Britain estimated that Israel would be able to overrun Jordan in quick time and Eden feared that this would precipitate a general Middle Eastern conflagration which would wipe out British influence. The existence of the

Anglo-Jordanian Treaty, together with Britain's determination to maintain the Iraqi-Jordanian axis as the linchpin of her power in the area, thus created a very difficult situation. Britain certainly had an operational plan – Cordage – to come to Jordan's aid against Israel, but a new Arab–Israeli war might even see her on the same side as Egypt and opposed by France. As Eden later explained:

> So this nightmare haunted me. Jordan folly, Israeli retaliation, British commitment and our airforce Meteors called upon to fight French Mystères. . . . The danger of all this became worse when Nasser completed joint command with the foolish Hussein and Syria and stepped up his verbal attacks against Israel. . . . It was at this stage that we repeated our warnings public and private about Jordan and got the French, in the course of our visit to Paris, to undertake to press them privately on Israel. We were determined not to fight for Jordan with support of Nasser and Soviet equipment against Israel supported by French. The only way to make sure of this was to warn Israel off Jordan. *We had no reason to weep if she then concentrated her thoughts all the more firmly elsewhere.*[55]

When, therefore, Eden received Challe and Gazier on 14 October, the most attractive part of their proposals was the proposition that the British premier could have the decisive voice in deciding whether Israel, feeling herself threatened by the Arab world in general, attacked Jordan or Egypt. The possibility of the former provided the most compelling short-term justification for collusion. As Eden explained to the cabinet, 'the political situation in Jordan was unstable, and there were signs that Israel might be preparing to make some military move. If the Israelis attack Jordan, we should be in a position of very great difficulty. . . but it would be contrary to our interests to act on this alone, in support of Jordan against Israel.'[56]

Eden's fatal decision begins to look comprehensible. In poor health, under pressure from the more vocal sections of his own party as well as from impatient French allies, disillusioned by American behaviour and aware that his military plans could not be kept in place much longer, he now faced the prospect of an unwanted Middle East war in which the alignments would be in defiance of Britain's interests and objectives. The collusion plan arrived at the right psychological moment. Eden perhaps took comfort from those who seemed ready to place an unquestioning confidence in his judgement. One junior minister wrote:

Especially do I sympathise with your sense of frustration and dismay at the want of resolution shown by many of our people. It makes it all the more important for us in the Conservative Party under your leadership to build up again the sense of loyalty, dignity and national purpose without which nothing can go right for us in the world.[57]

At the crucial cabinet meeting on 25 October several objections were raised. In particular it was noted that Britain's proposed course of action would offend the US government and might do lasting damage to Anglo-American relations. But, in so far as ministers fully understood what was going on, none felt strongly enough to resign.[58]

By the last week of October, therefore, the die was cast. While Nasser, in accordance with his original calculations, probably assumed that the danger of an Anglo-French military strike had all but passed, the secret plans were now in place. But Nasser was not the only one to be deceived. Those in the know maintained a complete blackout of information. Few public servants in the Foreign Office, the Treasury or even the senior diplomatic corps knew anything of the Sèvres agreement. Several key ministers were kept in the dark. 'I was told nothing of the contingency plans', recalled Lord Hailsham, the First Lord of the Admiralty, 'and when ships were actually beginning to move I was misled as to their true purpose.'[59] Not until British planes appeared over Cairo did Britain's ambassador to Egypt, Sir Humphrey Trevelyan, know what was afoot.[60] In the interests of secrecy even General Sir Charles Keightley, Commander-in-Chief of the allied forces about to go into Egypt, was left unaware of Israeli military plans.[61] 'I knew absolutely nothing,' recalled Makins, now back in London as Permanent Secretary to the Treasury, 'was told absolutely nothing, but I knew enough to realise that something very big was in the wind.'[62] Meanwhile the Washington embassy was kept curiously vacant pending the leisurely arrival of Makins's successor, Harold Caccia. Eden even seems to have concealed the truth from Salisbury, his oldest political friend, who had been absent from the key cabinet meetings because of heart trouble.[63]

Eden now sailed serenely on towards the denouement of the whole Suez affair. There was something of the young crusader about him again. Meeting him on 26 October, as the first stages of the Hungarian revolt seemed to indicate a split in the communist monolith, John Colville found Eden 'cheerful and apparently exhilarated'.[64] The Air Force Chief of Staff, Sir William Dickson, noted that in these final days he was like a 'prophet inspired, and he swept the Cabinet and Chiefs of Staff along with him, brushing aside any counter-arguments and carrying all by his exaltation'.[65] On the evening of 29 October the Anglo-Franco-Israeli plan began to unfold. Israel launched an attack on the Sinai peninsula, her main objective, while dropping parachutists at the eastern end of the Mitla Pass in conformity with the need to appear to be threatening the canal. The US government, aware of the Israeli military build-up but continuing to anticipate an attack on Jordan, immediately suspected Anglo-French involvement. As the transcript of a White House meeting that evening records:

> The President thought that the British are calculating that we must go along with them (he thought they were not banking too heavily on our being tied up in the election, but are thinking in longer range terms). He thought we should let them know at once of our position, telling them that we recognise that much is on their side in the dispute . . . but that nothing justifies double-crossing us.[66]

The British cabinet met at 10 a.m. the following day to hear Eden announce that, in accordance with the earlier 'contingency' plan, Britain and France would, subject to French agreement, send notes to Israel and Egypt requiring them to withdraw ten miles either side of the canal and to consent to an Anglo-French force temporarily occupying the canal zone. Even at this early stage there was some appreciation of the difficulties which Britain's actions might create for the Anglo-American alliance:

> Our reserves of gold and dollars were still falling at a dangerously rapid rate; and in view of the extent to which we might have to rely on American economic assistance we could not afford to alienate the United States Government more than was absolutely necessary.[67]

After a visit from Mollet and Pineau had given the appearance of consultation, ultimatums were presented to the Israeli and Egyptian ambassadors in London. Eden then made an announcement to the Commons. But Britain's sincerity was patently unconvincing. The supposedly even-handed ultimatums clearly favoured Israel by allowing her to remain ninety miles inside Egyptian territory. The mere twelve hours allowed for a reply suggested that Britain was keener to take action herself than to bring the Israeli-Egyptian conflict to a swift conclusion. William Clark, who had been sent on an unwanted holiday during the crucial period of secret planning, found himself 'completely flabbergasted by this idea which was wholly new to me, and seemed at once to prove that all this was a deep-laid plot of some age'.[68] Harold Nicolson, who twenty years earlier had seen himself as an 'Eden man', recorded his changed feelings:

> I do not see that it was at all necessary for Anthony to indulge in such hypocrisy. He went out of his way to do so. He is a rotten creature, vain and purposeless, and I hope he is soundly defeated at the next election. A Prime Minister should give some idea of principle and consistency; Anthony is just all over the place all the time.[69]

Of more importance, however, was the reaction of the US government. A first telegram from Eden to Eisenhower on 30 October made no mention of the ultimatum, stressing the imminent meeting with the French. This message crossed with one from the President asking for help in clearing up what exactly was happening between the United States and her European allies. America knew that France had provided Israel with considerable quantities of equipment, including aircraft, and her UN ambassador, Cabot Lodge, had been struck by the unhelpful attitude of his British opposite number, Pierson Dixon. The latter had argued that the Tripartite Declaration – which might have been expected to provide the basis for any Western response to the present crisis – was 'ancient history' and 'without current validity'. Eisenhower concluded:

> The possible involvement of you and the French in a general Arab war leaves your government and mine in a sad state of confusion. . . . if the United Nations find Israel an aggressor Egypt could very well ask the Soviets for help. Then the Middle East fat would really be in the fire. It is the latter possibility that has led us to insist that the West must ask for a United Nations examination and possible intervention, for we may shortly find ourselves confronted with a *de facto* situation that would make all our present troubles look puny indeed.[70]

Eden's second telegram, announcing that the ultimatums had been issued, went on to express regret that Britain had not been able to include America in her plans: 'Of course my first instinct would have been to ask you to associate yourself and your country with the declaration. But I know the constitutional and other difficulties in which you are placed.'[71] This second message failed, however, to reach the President before he had had the opportunity to learn of the ultimatums from press reports. Eisenhower's response to this news was couched in a noticeably cooler and more formal style than anything he had previously written to Eden, even to the extent of addressing him 'Dear Mr. Prime Minister':

> I have just learned from the press of the 12 hour ultimatum. . . . I feel I must urgently express to you my deep concern at the prospect of this drastic action even at the very time when the matter is under consideration . . . by the Security Council. It is my sincere belief that peaceful processes can and should prevail to secure a solution which will restore the armistice conditions as between Israel and Egypt, and also justly settle the controversy with Egypt about the Suez Canal.[72]

At the Security Council that evening Britain and France, to the dismay of world opinion, vetoed an American resolution demanding that Israel withdraw from Egyptian soil and calling on all UN members to refrain from the use or threat of force. Dixon argued that no action which the Security Council could take would contribute to the objectives of stopping the fighting and safeguarding the free passage of the canal. A second resolution from the Soviet Union, reproducing the draft of the American resolution but omitting the preamble which condemned Israel, was also vetoed by Britain and France. With Soviet assistance the United States then secured the transfer of the matter to the General Assembly under the terms of the rarely invoked 'Uniting for Peace' resolution. Not surprisingly, the Egyptians now rejected the Anglo-French ultimatum. After a delay of twelve hours designed to allow American civilians to be evacuated, Eden authorized the start of military action. RAF bombing of Egyptian airfields began on the night of 31 October/1 November.

One of the most telling charges laid against Eden's conduct has been his total misjudgement of the American reaction to Anglo-French intervention. 'I have always thought', Lloyd later admitted, 'that the most damning criticism . . . was that we misjudged the United States reaction.'[73] Even Eden, to whom self-criticism did not come easily, made a similar concession as he prepared for a BBC interview many years later: 'I certainly made mistakes. One of these was my judgement of what the United States Government's reaction would be to the Anglo-French intervention.'[74] Yet this charge is only partially correct. The very existence of the collusive plot was a recognition of the likely American response to the use of force. Had he taken the United States into his confidence Eden knew that the most likely result would have been for Dulles to come hurrying across the Atlantic with yet more delaying tactics and obfuscation.[75] Eisenhower was in no doubt that Eden had been made fully aware of his attitude. 'It was undoubtedly because of his knowledge of our bitter opposition to using force in the matter', the President confided, 'that when he finally decided to undertake the plan he just went completely silent.'[76] Eden had no reason to believe that the Americans would welcome military action. Kirkpatrick, one of the few civil servants taken into his confidence, later testified that Suez had been 'a hideous choice of difficulties' and that he, like Eden and Lloyd, fully realized that whatever course was adopted, great risks had to be run.[77] Where Eden miscalculated was on the scale and intensity of American opposition. 'I believed that the Americans would issue a protest,' conceded Macmillan, 'even a violent protest in public; but that they would in their hearts be glad to see the matter brought to a conclusion. They would therefore content themselves with overt disapproval, while feeling covert sympathy.'[78] This might even have been the case but for the timing of the intervention and the implausible reasons devised to justify it. Eden and Macmillan probably believed that it would be to Britain's advantage, notwithstanding Dulles's clear warning about the importance of 6 November, to launch operations before the American elections.[79] They calculated on Eisenhower's preoccupation with domestic politics and the difficulty for any presidential candidate in doing anything to upset the powerful Zionist vote. Yet Eisenhower was furious at being double-crossed only days before Americans went to the polls and could not but recall his co-operation a year earlier in fixing the date of the Geneva summit to assist Eden's electoral prospects.

In determining the American reaction, moreover, British efforts to secure a viable pretext for intervention only made matters worse. Eisenhower's memoirs indicate that by as early as 2 November the US government had heard that Pineau had told their ambassador in Paris 'the whole history of French collusion', including British involvement.[80] The feeling of betrayal, of having been deceived and cheated by their closest ally, lay at the heart of the American reaction and explains why much of the subsequent retribution was directed personally towards Eden. Dulles told Britain's minister in Washington on 31 October that what rankled most was the deliberate concealment on their ally's part.[81] It did not take Eisenhower long to reach a more understanding view of Britain's underlying objectives. 'It is difficult for us to put ourselves in their shoes', he explained to Senator Knowland, 'but don't condemn the British too bitterly.'[82] He realized that Eden and his colleagues had become convinced that Nasser's seizure of the canal was the last straw for Britain which 'simply *had* to act in the manner of the Victorian period'.[83] The President might even have acquiesced in an Anglo-French military strike had it occurred immediately – something which the need to adhere to the cover story rendered virtually impossible. In a 1964 interview he confirmed that 'We assumed that, if the three nations did attack, they would all move at one time, and it would be over in almost twenty-four hours. . . . Had they done it quickly, we would have accepted it.'[84] Certainly there is plenty of evidence that leading figures in the US administration subsequently had second thoughts about their outright opposition, at least as regards the Anglo-French aim of dealing with Nasser once and for all.[85] Granted that the Americans had concluded earlier in the year that Nasser was indeed a threat to the Western position in the Middle East, this is not surprising. 'I do *not* differ from you in your estimate of his intentions and purposes,' Eisenhower had insisted to Eden in September.[86] Most famously, when Lloyd visited Dulles at the Walter Reed Hospital on 17 November, the latter is reported to have asked his astonished visitor, 'Selwyn, why did you stop? Why didn't you go through with it and get Nasser down.'[87] Lloyd recalled that this, above all others, was the moment when he could have been knocked down by the proverbial feather. Yet Dulles's remark was less inconsistent than it might appear, though it in no way obviated American resentment at the stealth and deception with which Britain had gone about her task. While Americans never looked upon the Suez Crisis as the right issue upon which to deal with Nasser, they might have regarded his fall from power as some compensation for Britain's otherwise ill-judged initiative.

American opposition was only part of a worldwide reaction cutting across traditional loyalties and alignments. Among the dominions only Menzies for Australia and Holland for New Zealand were genuinely supportive. The Canadian premier's reaction was typical. 'I had never before seen him in such a state of controlled anger,' recorded a governmental colleague. 'I had never seen him in a state of any kind of anger.'[88] The response of the Arab world was entirely predictable, while Britain's NATO allies followed the American lead. In the febrile atmosphere of the General Assembly Britain faced a hostile majority of anti-colonialist states, and with no power of veto.

On 2 November Britain was outvoted by sixty-five to five, in a call for a ceasefire and the withdrawal of British, French and Israeli forces from Egypt. Only Australia and New Zealand joined Britain, France and Israel in opposition. Later that day the cabinet agreed, on Eden's recommendation, that Britain and France would be prepared to transfer responsibility for the separation of the combatants and the protection of the canal to a UN peacekeeping force as soon as one could be effectively established.[89] The international outcry against Britain was far greater than against France, greater indeed than that against the Soviet Union for invading Hungary. The world seemed to expect a higher code of conduct from Britain – and from Eden in particular – than that involved in the attack on an African state and the subterfuge employed to justify it. Aneurin Bevan spoke for many when he declared in the Commons on 1 November:

> I am bound to say that I have not seen from the Prime Minister in the course of the last four or five months . . . any evidence of that sagacity and skill that he should have acquired in so many years in the Foreign Office. . . . I have been astonished at the amateurishness of his performance. There is something the matter with him.[90]

Granted that no domestic consensus in favour of force had existed since the first days of the crisis, it was hardly surprising that British military action in defiance of the UN and against a country with whom she was not officially at war occasioned violent opposition. In the Commons the scenes were the worst since before the First World War. On 31 October Gaitskell denounced what the government had done as 'an act of disastrous folly whose tragic consequences we shall regret for years'.[91] Passions became so heated that the Speaker had to suspend the sitting for thirty minutes. Yet in response to an opposition motion of 'No confidence', Eden produced a brilliant reply which, according to a cabinet colleague, was the best speech of his life, silencing his critics at least momentarily.[92] It would be entirely misleading, however, to suggest that Eden had no support. In parliament and in the country there were intensely held feelings on both sides of the debate. But what was entirely lacking was the sense of national unity characteristic of a country embarking upon military action. A substantial body of opinion believed that the government had behaved so discreditably as to forfeit all right to support, even with British forces about to go into action. Richard Crossman captured the prevailing atmosphere:

The fact is that the Government's case is so terribly unconvincing, so ingeniously disingenuous, such a palpable tissue of prevarications that I began to realise why the House had turned into a bear-garden – because it's almost impossible to listen to this nonsense and stay still.[93]

There was considerable unrest inside the Conservative Party with the whips' office under Edward Heath working furiously to maintain a semblance of unity. Two junior ministers, Anthony Nutting and Edward Boyle, resigned. William Clark of the Downing Street press office also left his post. Had the conflict gone on much longer, internal dissension within the party would have become a significant factor. On the morning of 6 November, the day Eden announced the ceasefire, a letter signed by sixteen Tory MPs was delivered to 10 Downing Street, calling for a ceasefire and hinting at the withdrawal of their support if satisfactory assurances were not offered.[94] Ministerial broadcasts by Eden on 3 November and by Gaitskell the following evening probably had the effect of consolidating the division of opinion along largely party lines. Eden sought to confirm his credentials as an internationalist and man of peace. Even a critic who deemed it an 'odious performance' conceded that the broadcast was none the less effective.[95] Eden believed that, at a time of military conflict, he had the right to put the national point of view without the right of reply by the opposition. Once British forces were committed they were fighting the Queen's enemies and criticism was tantamount to subversion.[96] But Gaitskell miscalculated. Having secured the right of reply from a somewhat reluctant BBC, he used his broadcast to launch a bitter personal attack on the premier. Eden was 'utterly, utterly discredited in the world'. Only one thing, Gaitskell insisted, could save country's reputation and honour. 'Parliament must repudiate the Government's policy. The Prime Minister must resign.' But, as many Labour supporters recognized, their leader's appeal made it more rather than less likely that the bulk of Tory MPs would rally to Eden's defence.[97]

Eden could not even rely on the unquestioning obedience of those whose role it was to carry out the bidding of their political masters. Motivated by a combination of principled opposition and hurt feelings at being deliberately deceived, many public servants considered their positions. At the Foreign Office, Ivone Kirkpatrick, as influenced as Eden himself by the supposed lessons of the 1930s, never wavered in his support. Indeed, it seems likely that

Kirkpatrick proved the worst type of adviser for Eden at this time, feeding his 'dubious historical analogies'.[98] As has been seen, Eden leaned heavily throughout his career on the advice of others before making up his mind on important matters. His decision for secrecy, however, had severely restricted the number of those to whom he could turn during this most fateful episode. Most of Kirkpatrick's senior colleagues had been systematically excluded from the government's plans. Harold Nicolson found the Foreign Office 'enraged'. Middle East specialists in particular felt badly let down, their patient work of recent years destroyed at a stroke.[99] The Foreign Office African Department had been enthusiastic supporters of the 1954 agreement on the canal zone and doubted whether military action would lead to anything more than another costly occupation of Egypt.[100] After consulting various senior colleagues, Paul Gore-Booth, a future Permanent Secretary, sent Kirkpatrick a confidential note on 2 November:

> In the course of the week's business I have seen a lot of members of the Office of all ranks, and have been deeply impressed with the dismay caused throughout our ranks by HMG's action. People are doing their duty but with a heavy heart and a feeling that, whatever our motives, we have terribly damaged our reputation. I have not sought this opinion, but it is only honest to add that I myself, with my USA, UN and Asian background, have been appalled by what has been done – even granted the gravity and the imminence of the Nasser menace.[101]

Even though Anglo-French co-operation lay at the heart of the Suez venture, Britain's ambassador in Paris, Gladwyn Jebb, had only an inkling of what was going on. His increasingly vigorous protests at being kept in the dark were ignored except, at the last moment, 'for a vague hint of what was impending'.[102] At a Kremlin reception Khrushchev accused the British and French governments of behaving like bandits. Though this particular example of Soviet hypocrisy was hard to swallow, William Hayter, Britain's ambassador, privately considered that the assessment was not far from the truth and drafted several letters of resignation during the crisis.[103] 'I should have found things much more difficult', he later admitted, 'if I had been British representative at the United Nations.'[104] There, Pierson Dixon did his best to carry out his responsibilities, but with a heavy heart, his task made no easier by subtle shifts in the British position during the first days of November. The effort of concealing his true feelings, Dixon confessed, 'and putting a plausible and confident face on the case was the severest moral and physical strain I have ever experienced'.[105] On the whole the discipline and commitment to duty of the military held firm, though Eden did face a challenge from the always unpredictable First Sea Lord, Earl Mountbatten. With the invasion force already on the high seas, Mountbatten wrote to urge Eden to accept the General Assembly resolution and 'to beg you to turn back the assault convoy before it is too late'. He warned that the actual landing of troops would spread the war with 'untold misery and world-wide repercussions'.[106]

In many ways, then, this was very unlike war, even of an undeclared kind. Eden faced as many domestic as external enemies. It was an unusual spectacle, and one which did Eden's international credibility no favours, to see a British premier vehemently condemned by respected newspapers, just as British forces went into action. 'We had not realised', an *Observer* editorial confessed, 'that our Government was capable of such folly and such crookedness':

> It is essential that the world should know that the Eden administration no longer has the nation's confidence. Unless we can find means of making that absolutely clear, we shall be individually guilty of an irresponsibility and a folly as great as that of our Government.[107]

Eden's one hope was to secure a quick military success – to silence domestic critics and present international opponents with a *fait accompli*. 'I'm not the kind of moralist who can believe that an action such as Eden and Mollet planned would not be justified by entire success,' conceded Crossman. 'I just knew it wouldn't succeed.'[108] The requirements of the collusion plan made a quick victory impossible. The prisoner of Eden's contorted agreement with France and Israel, British forces could not give the impression of merely awaiting the opportunity to invade Egypt. In fact the invasion force left Malta on 29 October and could not reach Port Said until 6 November with airborne landings on Port Said and Port Fuad on 5 November. This delay proved fatal, allowing time for world opinion to mobilize and for internal doubts within the government to multiply. Military constraints made political failure inevitable. The longer the Anglo-French force took to sail across the Mediterranean, the clearer it became that it would lose the race against political pressure. On the day British troops landed Dixon sent an urgent telegram from New York:

These developments make it absolutely certain . . . that at the inevitable Assembly meeting later today and tonight, delegations will concentrate their attention on our failure to comply with the cease-fire and in particular on the reported bombings on populated areas in apparent contradiction to our declared policy. They will be in a very ugly mood and out for our blood and I would not be surprised if the Arab-Asians and the Soviet bloc did not try to rush through some resolution urging collective measures . . . against us.[109]

With hindsight it seems strange that, after agreeing to the evacuation of the Suez Canal Zone in 1954, Britain had taken no steps to provide herself with deep-water harbour facilities in Cyprus, which would have considerably reduced the sailing time. 'I had to admit I had not thought of it', Eden later recorded, 'and Admiralty never suggested it.'[110] To his credit, moreover, Eden was determined that British military action should keep civilian casualties to a minimum, even if this lessened the chances of a quick victory. On the twentieth anniversary of Suez and in the last months of his life, he reacted angrily to the 'ruthless, cruel and extraordinary' suggestion that Britain should have 'smashed Egypt up on the first day and inflicted maximum damage and casualties if we were going to do anything at all'.[111] Yet Britain was involved in a war situation and war, as Eden well knew, meant casualties. A more ruthless man would have shown fewer scruples. At all events, someone of Eden's experience might have been expected to show greater awareness of the military aspects of the whole venture.

Even if military action had been successful, it is by no means clear precisely what the government hoped to achieve. Surprisingly little thought had been given to the post-invasion situation. It was certainly open to question whether Nasser would simply have disappeared within forty-eight hours of the start of military operations and 'everything would [then] be O.K.'[112] Lloyd spelt out British hopes in conversation with the Australian Foreign Minister on 31 October. The government, he insisted, did not propose to reoccupy Egypt. Their aim was to seize the canal and there was 'at least a good chance' that Nasser would be overthrown 'from within'. The United Kingdom would then, before leaving Egypt, make a satisfactory deal with whatever regime succeeded him. The British had more far-reaching ideas about a wider Middle East settlement, including new arrangements for the area's oil resources, but these were longer-term proposals. Not surprisingly, Lloyd conceded that the whole enterprise was a considerable gamble, but judged that it was better than a 'slow death', a metaphor on the lips of several ministers at this time.[113]

By Sunday, 4 November Britain's problems were fully apparent, especially when news arrived that both Israel and Egypt were ready to accept a ceasefire. The day saw two meetings of the Egypt Committee followed by a full cabinet in the evening. At the second meeting of the committee it was announced that there had been some discussion at the UN about oil sanctions. Macmillan, hitherto the government's leading 'hawk', is said to have thrown up his arms, exclaiming 'Oil sanctions! That finishes it.'[114] Eden informed the cabinet that he had received a letter from Hammarskjöld. This placed a heavy onus of responsibility on Britain and France implying that, unless a ceasefire were accepted, they alone would be responsible for the continuation of hostilities which it had been their professed intention to stop. Eden had become the prisoner of his own cover-story: 'If both the Israelis and the Egyptians were . . . willing to accept a ceasefire, it would be difficult to deny that the purpose of our intervention had been achieved.' If both sides were willing to lay down their weapons, it could scarcely be argued that further military action was necessary to separate the combatants. Furthermore, if the initial phase of the Anglo-French landing force encountered opposition and had to be reinforced from the air or by naval bombardment which resulted in heavy civilian casualties, 'we might well be unable to sustain our position in the face of world opinion. We should run a grave risk that the United Nations would . . . adopt collective measures, including oil sanctions, against the United Kingdom and France.'

In this situation Eden proposed three possible courses of action and asked each minister to express his opinion. The first was to proceed with the landing at Port Said, while repeating the offer, now reinforced by a Canadian proposal at the General Assembly, to hand over to the UN as soon as an effective force could reach the canal. The second was to postpone parachute landings for twenty-four hours and the third to defer all further military action on the grounds that Britain and France had succeeded in ending the Israeli-Egyptian conflict. Eden favoured the first option, for which he received majority but by no means unanimous support. With varying degrees of conviction, four ministers would have preferred the second course and two the third. Only Monckton, however, who favoured suspending all further military action indefinitely, felt strongly enough to insist on reserving his position.[115] Many years later Butler suggested that Eden, confronted by a cabinet split, temporarily left the meeting and in effect threatened to resign. This greatly annoyed Eden, occasioning a characteristic flurry of correspondence with surviving participants to confirm his recollection of events.[116] Yet according to Clarissa Eden's diary, based presumably on information provided by Eden, the latter did take Butler, Macmillan and Salisbury to one side to tell them that, if they were not prepared to continue military operations, he would indeed have to resign.[117] It is certainly clear that Eden had to fight hard to carry his colleagues with him. As he later recorded, 'I wrestled for hours with Cabinet not to halt us from getting ashore when there was first report of ceasefire between parties'.[118]

At all events, towards the end of the meeting news came through that, contrary to earlier reports, Israel was not prepared to end hostilities, at least under the conditions specified in the UN resolution. 'As a result of this', recalled Eden, 'I think that we were all convinced that a ceasefire having not yet been achieved in the area, together with Israel's refusal to withdraw from the Egyptian territory she had occupied, was a sufficient ground for going ahead with our plans.'[119] Lord Home seemed euphoric at the outcome: 'We are not out of the wood, but we have won a decisive round. If our country rediscovers its soul and inspiration, your calm courage will have achieved the miracle.'[120] Eden felt sufficiently confident to launch a final appeal to Eisenhower for sympathy, if not approval, for Britain's actions. Once again he sought to raise the stakes. Nasser might become a 'kind of Moslem Mussolini'. This was the moment to curb his ambitions and secure a final settlement of the problems of the Middle East. 'If we let it pass, all of us will bitterly regret it.'[121]

On 5 November British and French paratroops landed at the northern end of the canal. Seaborne forces landed at first light the following day after a naval barrage. Resistance was considerable though British casualties were light. But at 5 p.m. with only about a third of the canal under Anglo-French control the Commander-in-Chief was ordered to cease fire. 'Why we stopped', mused Eden as he prepared his memoirs, 'is going to be more difficult to explain than why we started.'[122] Yet on this occasion *Full Circle*, not renowned for its candour, probably provides the fundamental answer. 'Fighting between Israel and Egypt ceased on November 6th and our plans were therefore never completed. . . . We had intervened to divide and, above all, to contain the conflict. The occasion for our intervention was over, the fire was out.'[123] Eden was the victim of the myth devised to justify British intervention. The end of Israeli-Egyptian conflict and the decision of the UN Secretary-General to raise an international force meant that he could not continue military action without forfeiting what credibility remained to him. To have gone on would have confirmed that the ostensible reason for intervention had not been the real one.

Of course, there were other factors. On the morning of 6 November Dixon said he thought he could hold the position at the UN a little longer, though it would not be easy beyond the end of the week unless the Americans, 'who were leading the pack, could be induced to slow down the hunt'.[124] Even so, Dixon had already warned that there was no chance of Britain pursuing her objectives, 'without alienating the whole world'.[125] The menacing presence of the American Sixth Fleet symbolized the complete collapse of the Special Relationship. The plain fact was that the overwhelming majority of world opinion was lined up against Britain and 'we were forced to bow to world opinion, as reflected in the United Nations'.[126] In writing his memoirs this was something Eden was loath to accept, despite the promptings of the Cabinet Office, but it was true none the less.[127]

Two further considerations helped determine the outcome of events. On 5 November the Soviet premier Bulganin sent threatening notes to Britain, France and Israel hinting at military action, while inviting Eisenhower to join him to put an end to Anglo-French aggression. In their respective memoirs Eden, Lloyd and Macmillan all denied that this threat significantly influenced the cabinet's decision, but, if the prospect of missile attacks on British cities was quickly dismissed, ministers did consider 'the possibility of a Soviet invasion of Syria or some other area in the Middle East, and possibly a direct Soviet attack on the Anglo-French forces in the Canal area'.[128] Moreover, the Chiefs of Staff subsequently admitted that the threat of Russian intervention had necessitated a British realignment alongside America, while recently declassified papers suggest that the Soviet menace was taken seriously, at least by the British intelligence community.[129]

Some controversy also surrounds the final factor which helped determine the British decision to halt operations – pressure on the pound. Sterling's position as a reserve currency made it particularly vulnerable to speculation. The unusually reticent minutes of the cabinet meeting of the morning of 6 November, at which the decision was taken, make no mention of Britain's financial position as a factor in moving ministers towards a ceasefire, nor does the subsequent review presented by the Chiefs of Staff. Eden's memoirs, however, like those of several cabinet colleagues, emphasize that 'a run on the pound, at a speed which threatened disaster to our whole economic position, had developed in the world's financial markets'.[130] The 'increasing drain on the gold and dollar reserves' was also highlighted by the secret Cabinet Office report drawn up in 1957.[131] The cabinet secretary agreed that the financial crisis 'certainly played a part', though he believed that Eden's account 'exaggerates it slightly'.[132] Macmillan seems to have become convinced by news arriving from America during the night of 5/6 November that all operations would have to be stopped and to have informed Lloyd of this before the cabinet meeting opened.[133] During that meeting the Chancellor received confirmation that George Humphrey, the US Treasury Secretary, was obstructing Britain's efforts to make sufficient withdrawals from the IMF to protect sterling from speculation. Without an end to operations Macmillan could 'not any more be responsible for Her Majesty's Exchequer'. If sanctions were imposed on Britain, he argued, the country was finished.[134]

Macmillan may even have exaggerated the threat to sterling as it is striking that there is no documentary evidence of warnings on 5 or 6 November from his Treasury officials in the extreme terms of his advice to the cabinet.[135] Even more significantly, in the weeks before the invasion, Macmillan took no action, such as drawing a tranche of funds from the IMF, to ensure financial stability, despite the advice of his department. Indeed, apart from his 'wobble' at the Egypt Committee on 4 November he remained a decisive advocate of intervention until his startling volte-face two days later. On 21 September Leslie Rowan, head of the Treasury's Overseas Finance Branch, had reported that there was 'a very dangerous outlook for sterling in the coming months' and that 'unless something . . . is done, both soon and successfully, sterling will be in the greatest danger, and our other resources – IMF, dollar securities, etc. will not do much to put off the day'.[136] Eden may not even have been shown the Treasury's warnings.[137] If Macmillan's duplicity during the Suez Crisis remains unproven, it is hard not to question his competence. Be that as it may, his transformation from leading 'hawk' to chief advocate of a ceasefire – 'first in, first out' in Harold Wilson's biting phrase – may well have been crucial.

The cabinet decided before lunchtime on 6 November that, subject to agreement with France, military operations should halt that day. A telephone call to Mollet brought reluctant French acquiescence[138] and at 6 p.m. Eden announced in the Commons to enthusiastic opposition applause that a ceasefire would come into force at midnight, London time. As members left the Chamber, recorded Crossman, 'I think the general Labour view was that this was the greatest climbdown in history and that Eden couldn't survive'.[139] Yet this was not how Eden saw matters himself. The *Manchester Guardian* described him as being 'as self confident as he had been throughout the whole affair'.[140] He still believed that Port Said could be held as a bargaining counter to ensure the prompt clearance of the canal, which Nasser had deliberately blocked, and to secure a general settlement. As regards Anglo-American relations Eden assumed that any damage done by Britain's conduct would soon be forgiven and forgotten. Indeed it would now be up to the United States to seize the unique opportunity created by Anglo-French military intervention to propose its own settlement for the canal and for Arab–Israeli relations.[141] Whatever their recent difficulties, Britain and America were after all on the same side in the wider issues of global politics, as Eisenhower's rebuttal of Bulganin's proposal for joint action had shown. The Chairman of the Joint Chiefs of Staff tried to explain the new situation to General Gruenther, the Supreme Allied Commander at SHAPE:

> Our action had ungummed UNO and HMG felt that it was up to UNO to seize the first real opportunity offered to her to reach a lasting solution to the world's greatest danger spot. Nevertheless we greatly fear that UNO will not grasp this fleeting opportunity. If UNO does not act promptly and handle both Egypt and Israel firmly the Soviets will quickly turn the situation to their advantage. . . . As HMG sees it everything depends on America being willing and able to act quickly. Unless she uses all her influence at once the UN plan will not be effective. And unless she puts her full sanction behind the plan . . . to ensure real effect to the will of UNO, this fleeting opportunity will be lost forever and the Soviets will win.[142]

Eden was to be sorely disappointed. Some months later he discussed matters with Pierson Dixon: 'Eden said that he did not mind about the ceasefire. By that time we were ashore in Port Said and thus had a "gage". What he did object to was the way in which the United States had pushed us into an immediate withdrawal.'[143] His optimism in the wake of the ceasefire seems curiously misplaced, symptomatic perhaps of mounting ill-health. Yet Eden's confidence was no doubt increased by his first communications with Eisenhower after the crucial cabinet meeting on 6 November. As he was about to enter the Commons a telephone call came through from the President. The latter

> was, of course, delighted. He said that we had got our objectives, the fighting was over and it had not spread. After some more agreeable expressions, he remarked 'This is a very good line. I suppose that it is the new cable.' I said I thought it was. He added we can talk freely on this and we will use it. Don't hesitate to ring at any time. I thanked him and said I would, but that I must go into the House now.[144]

Though Eisenhower emphasized that he did not want troops of any of the great powers, including Britain and France, included in a UN peacekeeping force – the first sign that the United States was not ready uncritically to adopt Eden's vision of the way ahead – Eden was clearly delighted by the President's cordial tone. That evening he telegraphed Mollet to assure him that 'the friendship between us all is restored and even strengthened'. As a result of the Anglo-French initiative 'we have laid bare the reality of Soviet plans in the Middle East and are physically holding a position which can be decisive for the future'.[145]

The following day Eden took up Eisenhower's invitation and telephoned Washington. His aim was to sound out the possibility of a visit to America by Mollet and himself. Eisenhower, now confirmed in the White House by the American electorate, seemed sympathetic, suggesting that an early visit would suit him best. It was agreed that Eden and Mollet should fly out the following day. This was the authentic Eisenhower, keen to restore the traditional working friendship between Britain and America and not prepared to allow hurt feelings to stand in the way of a meeting.[146] In two subsequent calls on the same day, however, Eisenhower progressively backtracked from this position. In the first he sought an assurance that Eden and Mollet would not use their visit to argue about the terms of UN resolutions since this might only underline Anglo-American differences. In the second, only half an hour before a formal announcement was due to be made, he explained that he would be too busy with Congressional business to see his European partners for the time being.[147] Eden blamed Eisenhower's change of heart – as he did most other things which went wrong during the crisis – upon Dulles, then in hospital recuperating from an emergency operation for cancer.[148] In fact, although Eisenhower visited

Dulles that day and the latter did support postponing Eden's visit at least until all Anglo-French forces had been withdrawn, the crucial influence seems to have been that of the acting head of the State Department, Herbert Hoover, a longstanding Anglophobe with a particular dislike of Eden.[149] The fact was that a visit to Washington would give out the wrong signals as far as the Middle Eastern priorities of American diplomacy were concerned, linking British and French interests to those of the United States.[150]

But Eisenhower's change of heart was a severe blow to Eden. It came, of course, just when a policy with which he had been personally and inextricably associated was being judged by the vast majority of British opinion – if not by Eden himself – to have been a failure. But the problem went deeper from Eden's point of view. The events which followed made it clear that the US administration, dominated for the time being by Hoover and Treasury Secretary Humphrey, was determined to keep Eden personally at arm's length. Between the second week of November and his final resignation the following January the British premier was effectively ostracized by the American government. Yet it was striking that, when the new British ambassador, Harold Caccia, presented his credentials on 9 November, Eisenhower seemed warm and forgiving, keen to emphasize the enduring importance of the Anglo-American alliance irrespective of recent differences.[151] It is inconceivable that Eden's senior colleagues did not begin to wonder whether his continued presence at the head of the government might not act as a long-term impediment to the restoration of good relations. By early December even the Israeli premier had heard that the Americans 'will not make up with the English until Eden goes'.[152]

Over the next fortnight the reality of Eden's position – that in fact he held no bargaining counters – became fully apparent. It amounted to a personal humiliation from which his standing would have been unlikely to recover, even had his health not once more intervened. He later reflected bitterly:

> Now if we could have held on to what we had taken until Canal was cleared and Canal issue was agreed, that would have been a useful lever. As it was US insisted in rubbing our noses in it and inserting the word 'forthwith' [in the UN resolution of 23 November].[153]

On 7 November the General Assembly passed a resolution calling for the immediate withdrawal of British, French and Israeli troops. Eden's inclination was to resist and to insist that Britain and France, who had the expertise and the equipment, should take the lead in clearing the canal. There was still some optimism in the air. The cabinet agreed that Britain should devote her efforts to inducing the Americans to acknowledge the existence of a dangerous situation in the Middle East which they had consistently refused to recognize over the previous decade.[154] Momentarily Eden may have believed that the way ahead, now that America had shown her unreliability, lay in closer association with Europe.[155] The government comfortably survived a vote of confidence on 8 November, though eight Conservatives – all opponents of the Suez expedition – abstained. One of these even wrote to express his admiration for Eden's courage 'in these desperate days'.[156] Those MPs who were disappointed that Eden had called a premature halt to military operations trooped obediently into the government lobby.

Lloyd set off for New York on 12 November, expressing confidence that the situation was beginning to thaw. 'I shall continue to drive into their heads that the Anglo-French forces in Port Said are the only effective bargaining counter for either the United Nations or the Americans with the Egyptians and the Russians.'[157] At the UN Lloyd tackled Cabot Lodge, impressing upon him that Soviet penetration of the Middle East had gone much further than the United States believed and that, but for American opposition, Britain would have achieved a 'brilliant success' and the elimination of Nasser. Lloyd found Lodge friendly but reticent, not disputing his arguments but insisting that what Britain had done was wrong.[158] On 15 November Eden told the Egypt Committee that Britain would be prepared to remove her troops, unit by unit, in a planned operation as UN forces arrived. She could not, however, accept that her forces would have to be totally withdrawn before any agreement for clearing the canal became operative.[159] Lloyd found the situation in Washington less hopeful and it left him feeling 'rather depressed'.[160] Eisenhower was unwilling even to meet him. Dulles was 'friendly', but still recovering from his operation and anxious to evade responsibility. Hoover, by contrast, was 'quite negative':

It is clear that the most antagonistic elements are the second rank in the State Department. . . . In the absence of an effective Secretary of State they are more powerful than they should be. . . . There is no desire here to see Nasser built up, but there is no grip in the administration as to what is immediately involved, or what should speedily be done. It will be 1917 and 1941 all over again.[161]

The plain fact was that the US administration saw no need to bargain with Britain about the terms of her withdrawal. Their position was strengthened by Hammarskjöld who reported that Nasser had refused to negotiate about the canal until British and French troops left his country. Lloyd's suggestion that, if Nasser were allowed to determine the shape and functions of the UN peace-keeping force, Britain would 'seriously consider resuming our operations' was patently unrealistic.[162] The economic crisis, perhaps exaggerated by Macmillan on 6 November, was now very real. In the first half of the month Britain's reserves fell by 200 million dollars. With the canal blocked Britain would face an enormous bill to pay for oil from the western hemisphere. But at a meeting in Paris on 15 November the Chancellor's request for help with oil supplies was bluntly refused. American co-operation was dependent on Britain's unconditional withdrawal.[163] As an official at the Washington embassy noted, the Americans were 'hurt and piqued at our action which they look on as a blunder and they are determined to treat us as naughty boys who have got to be taught that they cannot go off and act on their own without asking Nanny's permission first'.[164] It was a harsh judgement, but not without validity.

*

By mid-November there was renewed concern about Eden's health. His physician, unsure whether the problem was a recurrence of Eden's old bile duct trouble or simply the result of physical exhaustion, advised a period of complete rest. A public announcement was made on 19 November and Eden flew to Jamaica to stay at Goldeneye, the holiday home of the thriller writer, Ian Fleming, leaving Butler as acting Prime Minister. It was an unfortunate decision. If Eden were to keep in touch with London, messages would need to be transported a considerable distance over rough roads from the Governor's residence to Goldeneye. Notwithstanding his poor physical condition, it struck many as insensitive that, as Britain prepared for petrol rationing, Eden retreated to an island popularly regarded as a holiday resort for the idle rich. Even Ann Fleming judged that 'Torquay and a sun-ray lamp would have been more peaceful and patriotic'.[165] At all events Eden's departure made his political survival increasingly problematical.

It is now well established that, some days before Eden's departure, Macmillan, Butler and Salisbury had opened unofficial contacts with

Washington through the intermediary of the US ambassador in London, Winthrop Aldrich.[166] These senior ministers seem to have concluded that Eden's 'no surrender' attitude was untenable in the light of all that had happened and that his illness offered an opportunity to change the direction of policy, even if this also meant a new Prime Minister. The twin bases of their discussions were that Britain would accept American terms for withdrawing from Egypt and that Eden was unlikely to survive as premier. Whether Eden's removal was something they were prepared actively to work for, or whether they expected it to happen of its own accord, remains unclear. The American aim seems to have been to keep a Conservative government in power, but one probably not headed by Eden. Not surprisingly, it was the Chancellor who took the lead. Despite the impression given in his memoirs, Macmillan now believed that Eden 'could never return and remain Prime Minister for long'.[167] While Macmillan kept the nature of his American contacts secret, he did tell the cabinet on 20 November:

> although any formal approach to the United States would be premature at the present time, we should endeavour to establish informal contact with them through the Treasury Delegation in Washington, in order gradually to enlist their support for the loan we should have to raise.[168]

Secluded in Jamaica, Eden remained unaware of what was happening, but the correspondence which passed across the Atlantic shows both that he was now considerably out of step with his senior colleagues and that he was no longer being treated by them as head of the government. On 1 December he cabled London to insist that, despite Britain's difficulties with her 'American friends', the only option was to stand firm 'on the ground that we have chosen and I believe that they will come round'. While admitting that the financial situation might become 'difficult', he added cryptically that 'we have resources which I would rather not put in a telegram but which I would rather use than yield'.[169] But the cabinet was moving in a different direction. In the last days of November the decision to capitulate was taken. All indications from America suggested there was no alternative. In conversation with Caccia, 'Humphrey more than once used the simile that the United Kingdom was an armed burglar who had climbed in through the window while Nasser was the house-holder in his nightshirt appealing to the world for protection'.[170] Lloyd returned from the United States to admit that he had made no progress. The Americans 'expected us to purge our contempt of the President in some way, and were temporarily beyond the bounds of reason. They had no intention of lifting a finger to preserve us from financial disaster until they were certain that we were removing ourselves from Port Said quickly.'[171]

At the cabinet on 28 November Lloyd offered his resignation, which was declined.[172] Instead it was agreed that he should make a formal announcement to the Commons on 3 December of Britain's intention to withdraw.[173] Told of this development, Eden sought an assurance that any decision would be consistent with his own earlier stipulation about the need for British participation in the clearance of the canal. Butler's reply that 'the policy on which we have decided is consistent with the course which you set for us' was patently untrue.[174] Eden even contemplated an early return to London to take part in the parliamentary debate but was dissuaded from this course. Equally alarming from his point of view was the news of growing dissension inside the Tory Party. His PPS warned that the Suez Group was talking in terms of his resignation.[175] In the event, though fifteen right-wing MPs abstained in the division, Eden was reassured that his own position in the party and the country remained strong.[176]

Feeling considerably better, Eden was understandably keen to return as quickly as possible. But the debate which took place about the statement he would make on arrival at London airport showed how dramatically his authority had been diminished within the government he still nominally led. In effect Eden was overruled by his subordinate ministers. He had already sought – without success – to persuade his colleagues that they should take the offensive in their parliamentary statements, criticizing US policy during the course of the Suez Crisis and, in particular, their behaviour over SCUA.[177] Now, for all practical purposes, Eden's statement was composed for him by Butler, Salisbury and Lloyd, omitting the combative remarks he wanted to make at the expense of the United States, the Soviet Union and the UN.[178] The tone of Butler's critique of Eden's own draft vividly illustrates the reversal in the latter's fortunes:

The draft which Millard sent you was, we thought, the sort of line you might take in the light of both national and international situation now obtaining. . . . People throughout the country are looking forward rather than back. In particular there is a growing wish to end the breach with the United States. It is important that your first pronouncement should be in tune with this.

For this reason Butler objected to a passage in which Eden seemed to equate the Soviet Union and Communist China with the United States and the UN. In addition Butler was getting ready to abandon the pretence surrounding British policy since the ultimatums to Israel and Egypt at the end of October: 'we suggest the deletion of reference to our Middle East action in terms of fire-prevention. This has been laboured a lot.'[179] Such sentiments reflected a growing mood. 'What decent Tories mind', noted Harold Nicolson, 'is that [Eden] has placed them in a false position by obliging them to say things . . . which they now know to have been untrue.'[180]

During the last fortnight of 1956 the weakness of Eden's position became ever more apparent. Brendan Bracken heard that, on arrival in England, he was almost immediately confronted by a deputation headed by Salisbury and Butler informing him that while the cabinet was willing to carry on until the following Easter, if it was then clear that his health was not fully restored, a change of leadership would be necessary.[181] Eden himself remained optimistic. Assuring the Queen that his stay in Jamaica had had a most beneficial effect on his health, he again took up the theme that British intervention in Egypt had created a unique opportunity if only the Americans would step into the vacuum before the Russians did so.[182] He even began making tentative plans for his visit to the Soviet Union scheduled for May 1957.[183] But Eden's incursions into the parliamentary arena were little short of disastrous. His first reappearance in the Commons was greeted with stony silence. Embarrassingly, just one loyal Tory rose to wave his order paper before resuming his seat.[184] On 18 December Eden appeared before the 1922 Committee of backbench Conservatives. Though the chairman wrote to offer his congratulations, many of those present judged his performance to be unconvincing.[185] Eden confessed himself unable to answer a point about the Tripartite Declaration which he would have to look up. Crossman heard that the occasion had sealed his fate, though uncertainty about who should succeed him – 'the Suez rebels will not take Butler and the Tory progressives will not take Macmillan' – probably meant that he would not be removed until defeated at a general election.[186] After the 1922 meeting one MP wrote urging Eden to make an unequivocal denial of the charge of collusion.[187] Two days later, under severe pressure in the House, Eden did just that:

I want to say this on the question of foreknowledge, and to say it quite bluntly to the House, that there was not foreknowledge that Israel would attack Egypt – there was not. But there was something else. There was – we knew it perfectly well – a risk of it, and, in the event of the risk of it, certain discussions and conversations took place, as, I think, was absolutely right, and as, I think, anybody would do.[188]

As it happened, this was the last speech which Eden made in the Commons. Yet his words would haunt him for the rest of his life.

The Christmas vacation gave Eden time to ponder his future at Chequers. To Kilmuir, the Lord Chancellor, he expressed doubts about being able to carry on.[189] Though continuing health worries were his principal concern, there were clearly others. Domestic political pressure and the attitude of the United States also counted as a letter from Patrick Buchan-Hepburn, the former Chief Whip, makes clear:

> I hope you will not dwell too much on what the critics say. I do not think that many of them would relish it for long – and I include colleagues – if you were to take them at their word here and now. . . .
> I feel pretty certain that it would be a grievous blow to us, in every particular, if it appeared that the U.S. could influence the tenure of office of the British Prime Minister.[190]

Buchan-Hepburn's advice was equivocal. While urging Eden to stay at his post, he also discussed the possibility of a later 'more dignified and appropriate' resignation. One old friend, Lord Coleraine, the former Richard Law, suggested three important prerequisites to Eden's continuance in office. The first was a 'massive reconstruction' of the government; the second that he should pay more attention to the Commons; and the third that he reconsidered his own functions as premier – 'it's true that he shouldn't be a Baldwin, but it's not necessary that he should be a Churchill either'.[191]

As 1957 opened, Eden's inclination remained to soldier on. With the United States preparing to issue the Eisenhower Doctrine, indicative of a tougher approach to Soviet incursions into the Middle East, he took comfort that the Americans were 'now coming our way'.[192] Above all he would still not admit that Suez had been a mistake or even devoid of success. He was stung by the first sentence of a paper which Macmillan submitted to the cabinet on the state of the economy, a sentence which in effect admitted the expedition's failure. 'I do not think that the events of Suez can be reckoned as a tactical defeat,' Eden insisted. It was too early to pronounce on an operation of that kind. There were tangible gains. The Egyptian air force had been destroyed, Jordan and Syria kept from active alignment with Nasser, the extent of Soviet penetration in the Middle East exposed and the UN given a chance to take effective action.[193] In private he mused: 'Diplomacy is a continuing process. The consequences travel on. You cannot just close a chapter if you want to and leave it at that saying this has failed, let's forget it. It may not have failed: no one can tell that at once.'[194] In fact, of course, it was truer to say that Britain had 'nothing to compensate for a political defeat of the first magnitude'.[195] Eden's inability to come to terms with reality was the most disturbing feature of his continued premiership.

Eden's health possibly accounted for his poor judgement. He was now seeing Evans on a daily basis. The latter's advice and that of two other specialists was that the period of rest in Jamaica had brought only temporary improvement, that the bouts of fever were likely to recur and that Eden's constitution was no longer equal to the demands of his office. Though reluctant to admit that his political life was over, not least because his departure might be interpreted as an admission that Suez had been a mistake, he sought an audience with the Queen on 8 January to announce his decision to resign. The cabinet expressed shock and dismay, but several ministers must have anticipated this outcome. Macmillan's capacity to dissemble came to his aid and increased with the passage of the years. Asked in 1971 whether the question of Eden's resignation had even crossed his mind, he replied:

> Never thought of it. I see him now – it was in that little room, the only room in No 10 that gets the western sun – in January, it was – and he still looked just the same: elegant, gallant, always with a rather Elizabethan figure, he looked just like a young officer in the first war. And he told me this and I couldn't believe it, and when he told the cabinet afterwards, they just couldn't believe it.[196]

In fact Macmillan, recognizing that he would be a strong candidate for the succession, had been subtly distancing himself from Eden for some time. After a meeting in Paris on 12 December Dulles recorded that the Chancellor was 'very unhappy' at the way Suez had been handled, but that Eden had 'taken this entirely to himself and he, Macmillan, had no choice except to back Eden'. Macmillan hoped that a change of government would leave himself or Butler as Prime

Minister, if not immediately on Eden's return, then within six months.[197] One MP recalled a striking scene in the Commons at the end of the year with Macmillan walking to a seat on a bench below the gangway to chat to Tory backbenchers to avoid taking his front-bench place next to Eden.[198] One well-informed observer judged that Macmillan's real intention was 'to push his boss out of No. 10'.[199]

Yet there is little doubt that health was the genuine reason for Eden's resignation. The evidence that ministers including Butler helped exaggerate the seriousness of his condition is unconvincing.[200] Though Eden lived on for two more decades, his constitution was never robust, he tired easily and without question he was physically unequal to the demands of modern political life. That of course leaves open the rather different question of whether Eden could have survived in office for any length of time but for his illness. This seems unlikely, though the procedure for removing a sitting Prime Minister against his will was by no means as clear in 1957 as it has since become. By his failure to accept the facts of the situation as they existed after 6 November Eden had lost much credibility. The feeling, even among cabinet members, was for a new start and the most obvious way to mark it was by a change at the top. In letting Suez be seen as a personal contest between himself and Nasser Eden had merely contributed to this conclusion. Whether senior colleagues had been gently suggesting that resignation would be in the best interests of the party and the country remains imponderable. Certainly, Eden now recognized that his departure might help to break the impasse in Anglo-American relations.[201]

Unusually, the Queen made no formal request for Eden's advice as to his successor, but 'enabled [him] to signify' his debt to Butler during his premiership, especially during his time in Jamaica.[202] That he should have revealed, however obliquely, a preference for Butler over Macmillan must seem somewhat strange. He was not close to either man and may have wished that there had been a viable third candidate available.[203] Macmillan had been more at one with Eden over Suez, at least until his volte-face on 6 November, and Eden can have had no confidence that a cabinet headed by Butler would strive to capitalize on the Anglo-French intervention in the way he still believed was possible. But Eden resented Macmillan for having 'about turned too quickly' and it is open to speculation that

453

news of the Chancellor's recent disloyalty had already reached him.[204]

*

Eden's political career was over. He resigned not only the premiership but also his parliamentary seat of Warwick and Leamington which he had held for thirty-three years. On 18 January 1957 he and his wife set sail for New Zealand in search of rest and recuperation.

As Anthony Nutting famously concluded, Suez had been 'no end of a lesson'. Yet it is open to question how many lessons Eden himself had learned. In one of his last papers as Prime Minister he assessed the lessons as he saw them. He made interesting points about Britain's need to secure her financial and economic independence by reducing defence commitments and curbing the spiralling costs of the welfare state. He mentioned playing a world role 'on a more modest scale than we have done heretofore' and even hinted at a closer association with Europe. But at no point did he consider the future need to secure world, or at least American, support for major foreign policy initiatives.[205] In fact Eden left office, and in due course went to his grave, convinced that, as far as Suez was concerned, he had been more sinned against than sinning. If the requirements of true Greek tragedy include catharsis and self-knowledge then Eden does not qualify. In transit to New Zealand he wrote to Lloyd of the need to stand by what had been done and maintain that it was right. 'I feel far more strongly about this than I did in 1938. It was just an open question then, I thought, whether appeasement might succeed. Now it cannot possibly.'[206] In later years he developed ever more extravagant claims for the Suez venture and exaggerated assessments of the price paid for American opposition – 'three wars, two between Israel and her Arab neighbours, and the third and bloodiest between Arabs and the Yemen'.[207] Eden was convinced that he had been badly let down by his American allies, and for the most unworthy reasons. He became understandably bitter:

> The US government took its decision. Militant nationalism must be supported. The United Nations as a reactionary organisation must be supported because it was also anti-colonial. The greatest opportunity for a Middle-East settlement must be spurned. The USSR must be given unlimited opportunities to fish in trouble waters. . . . Britain and France must be ordered out immediately and the Egyptian dictator picked up, dusted down and put back in his place again.[208]

Yet the overwhelming weight of historical opinion remains against him. Suez stands as a towering personal and political tragedy in Eden's career. Constrained by the secrets of Suez, and particularly the secrets of Sèvres (which in fact were unlikely to remain secret), he did not even use his memoirs to present the best case on his own behalf. Instead Eden resorted to a series of platitudinous analogies with the 1930s which, though present in his thinking in 1956, were by no means his only calculations.[209] The cabinet secretary sympathized with his predicament:

> I think . . . you pass rather lightly over the considerations which ultimately led the government to intervene. I realise only too clearly the difficulties which you have had in dealing with this. But . . . it may strike some readers that, as compared with the many pages devoted to your account of the negotiations and conferences . . . the narrative runs pretty rapidly over the period immediately before our actual intervention.[210]

After forty years the task of rehabilitating Eden's conduct still appears daunting. The case for the prosecution remains very much as Harold Nicolson stated it in late October 1956: 'To risk a war with more than half the country against you, with America and UNO opposed, and even the Dominions voting against us, is an act of insane recklessness and an example of lack of all principle.'[211] But if historical revisionism remains inappropriate there is always scope for historical balance. What does seem unjust is the proportion of the indictment which Eden personally has had to bear. The episode is, of course, irrevocably associated with his name. Yet Suez seemed altogether out of character. Having made his reputation on the basis of high-minded integrity, his conduct in 1956 appeared doubly culpable. Launched on his career by the League of Nations, it was ironic that the UN should have helped to bring him down.[212] By contrast, the French premier, Mollet, not only remained in power but saw his parliamentary position strengthened.[213] In Britain Eden's leading colleagues managed to ride out the storm. Lloyd continued as Foreign Secretary until 1960, later served as Chancellor and Leader of the House, and ended his career as a respected Speaker of the Commons. Macmillan was the ultimate beneficiary of Suez, enabled by Eden's retirement to embark upon a seven-year premiership which is still held to be among the more impressive of post-war history. Yet Macmillan was party to all the key decisions of the crisis, more intent perhaps than Eden himself to bring Nasser down, and merits particular mention for his misinterpretation, wilful or otherwise, of US attitudes that September. The prediction that Eisenhower would 'lie doggo' was the most grievous mistake of the whole affair.[214] If, therefore, Suez still stands condemned, there is a good case for a more equitable distribution of the burden of responsibility.

The Suez Crisis contained much deception, some of it self-deception, but Eden and the majority of informed British opinion sincerely regarded Nasser as a major threat not only to specifically British interests but to the wider interests of the Western world in the global conflict with Soviet communism. To this extent the excessive personalization of the crisis into a struggle between Eden and Nasser over the precise issue of the nationalization of the canal – a tendency increased by Eden's notorious fits of temper and unguarded remarks – conceals the reality of the situation. Eden's reaction to Nasser was just part of the initial cross-party consensus which greeted the nationalization coup. Such an assessment of the Egyptian leader's menace may have been exaggerated, but it was widely shared. And if Nasser was an incipient Mussolini then only the threat, or perhaps the use, of force would stop him. Granted the paralysis of the UN because of the Soviet veto, the attitude of the Labour opposition seems in retrospect untenable.[215] Confronted by the nationalization crisis and at a time when his own domestic political position was not strong, a passive reaction on Eden's part was simply not possible. The problem was that there was no easy solution. 'I don't believe', reflected Lord Home, 'given the military limitations in the early days, that there was a perfect way of dealing with the Suez situation.'[216]

One distinguished historian has written of Eden's career that 'no one who tried to guess the area in which disaster would occur, if it was to occur at all, would have guessed at foreign affairs'.[217] Yet the reverse may be the case, for it was a singular failure of Eden's last foreign secretaryship that he did not bring British and American strategies in the Middle East fully into line nor, in a broader context, succeed in defining clearly Britain's position as the subordinate partner in the alliance. The difficulty was exacerbated by a long-term unease on Eden's part about America's role in international affairs and her replacement of Britain as the leading world power. At the heart of Suez lay a series of fundamental Anglo-American differences and misunderstandings. The United States, which so readily equated the forces of nationalism with those of the Kremlin in

South-East Asia, was ready to draw a nice distinction between those same forces in the Middle East. But even the US government reached the conclusion during 1956 that Western interests would best be served by Nasser's removal. The Americans, however, never believed that the canal nationalization provided the occasion to secure this end. While Eden viewed military action as the sort of pre-emptive operation which should have been used against Hitler and Mussolini before 1939, the Americans believed that the use of force would be an inflammatory act likely to unite the Arab world in opposition and drive it into Moscow's arms.

Recent accounts have stressed that there was no excuse for Eden's misjudgement of the American position. If he and others believed that Eisenhower would acquiesce in British military action, 'this was their wishful thinking – not deception by the United States'.[218] The President had made his position perfectly clear – though Macmillan had also succeeded in misrepresenting him. But about Dulles's remarks and behaviour there was a tortured ambiguity. His many-faceted character did not allow for easy assessment. Eden never understood the balance of power inside the US government. 'Ike . . . don't [sic] count much more than Pétain did,' he remarked to Churchill in 1957. 'Laval did the work.'[219] This, of course, was incorrect. But it is difficult for the historian to be censorious, since it was a misjudgement shared by a generation of his own profession.

By mid-October the complexities of the crisis had multiplied. On the one hand, in the absence of a fresh pretext for intervention, Eden was tending, albeit reluctantly, towards a peaceful settlement. On the other, he faced new anxieties in terms of renewed domestic political pressure and a worsening situation in the Middle East, including the possibility of an unwanted Israeli-Jordanian war. He calculated that Nasser – and therefore, in his mind, Russia – would move against Israel by March or April of the following year. The advance of autumn meant that a decision on using force could not be long delayed. In this situation and with his judgement almost certainly affected by chronic tiredness and poor health, Eden seized at the collusive plan dangled before him by his French visitors on 14 October. The characteristic Eden would have agonized long and hard over such a difficult decision, seeking reassurance and support from those around him. But the secret nature of the Sèvres Protocol ruled this out. Even so, he probably managed to convince at least himself that the agreement with France and Israel – in 1976 he brought himself to admit that there had been 'understandings'[220] – was on a contingency basis, a diplomatic subtlety which the leaders of the other countries involved understandably regarded as unnecessarily obtuse. The Sèvres plan did not offer an instant panacea. It involved enormous risks and Eden knew it. He was certainly gravely mistaken in thinking that the cover-story would hold. Thereafter he was the prisoner of his own pretence, even at the cost of misleading the Commons in an age before the public had become used to their leading politicians being 'economical with the truth'. Yet if Suez had been judged a success, it seems unlikely that so much attention would have focused on such issues. At the end of the day the problem which Eden could never overcome was that Britain was cast as an international pariah and her expedition ended in the most humiliating of circumstances before it had achieved its ill-defined aims. Eden gambled – and he lost.

14

Eden: an assessment

Anthony Eden is not an easy man to know. Added to a natural
reserve, he sets high store upon good manners and they sometimes concealed the depth of his convictions. His elegant
appearance and ease of manner, his disregard of men's clubs,
his appreciation of painting, his liking of female society and his
hobby of gardening perhaps concealed from the superficial
observer the inflexible nature of his principles on high matters
of State.[1]

A T the meeting called to elect Eden's successor as Conservative
leader, the senior backbencher Walter Elliot paid tribute to the
outgoing Prime Minister. 'A man's life', he insisted, 'is the whole of
his life, and not this or that particular patch of shadow or of
sunshine.'[2] This was Elliot's attempt at the difficult task of remind-
ing his audience that there was far more to Eden's long career than
the recent sorry tale of Suez. But behind even the totality of Eden's
political life there lies the story of a private human being. The
private man was not well known. Many observers commented
upon his lack of friends, particularly among political colleagues.
He had 'absolutely no friends', asserted Iain Macleod with patent
exaggeration.[3] More realistically Evelyn Shuckburgh 'soon discovered that Eden did not have many close friends, or seem to want
them'.[4] Those closest to him were drawn from that narrow circle of
political and professional associates with whom he had forged his
career in the later 1930s. As these men progressively left the stage,
Eden tended to eschew his remaining political contemporaries and
seek the companionship of a new generation of younger acolytes
including Robert Carr, Allan Noble and, for a time, Anthony Nutting,
a group contemptuously described by other Tory backbenchers as
the 'kissing ring' and the 'inner kissing ring'.[5] The British premiership, probably because it has no equal in the domestic government,

is an inherently lonely office. For Eden it was particularly so. Even his relationship with Salisbury was never quite as close as in pre-war days, while J. P. L. Thomas – 'the only one I can remember coming to see him voluntarily, as it were, and without a business reason'[6] – made clear his wish to retire from politics soon after Eden became Prime Minister.

Eden found many aspects of political life uncongenial. Its partisan nature often appalled him and he never believed that one side in the political divide possessed a monopoly of rectitude and wisdom. Political tolerance figured strongly in his make-up – 'to be civilised is to tolerate,' he once insisted[7] – and the concept of national unity always exercised a pronounced appeal. As a result he was never a good party man. He viewed the Conservative right with distaste and for some while toyed with the possibility of a major realignment of British politics. This became a problem for him in his quest for the premiership and as Prime Minister, for it meant that he never possessed a strong power-base in the parliamentary party. In general Eden was 'unclubbable'. As his official biographer has recorded: 'Eden loathed men's clubs, refused Churchill's invitation to join The Other Club, was never seen in the [Commons] Smoking Room . . . and, even when he became Prime Minister, refused the offer of honorary membership of the Athenaeum'.[8] In 1941 Eden found himself happily involved in a meeting of a Trade Union Club with Ben Tillett in the chair: 'It is no doubt unfortunate for a Tory M.P. but I am infinitely happier among these folk than in the Carlton Club, and they like me better than does the C.C.! I am in the wrong pen it seems.'[9] A visit to his constituency a year later produced different emotions:

> Cannot pretend that we enjoyed it much. Crowded journey both ways. Luncheon of people very smugly middle class – timid but patronising! Forty minute wait for return train. . . . Gossiped to some soldiers standing in the corridor. That was best part of the day.[10]

This preference for military company is striking and characteristic. Eden's outlook had been profoundly affected by service in the Great War. As he wrote towards the end of his life, 'war promoted working together into something good and true and rare, the like of which was never to be met with in civil life'.[11] Shortly after the 1918 armistice he confided to his sister that he was contemplating standing for parliament. It was 'a choice of that or staying in the Army'. He concluded that 'a semi-military semi-civil job would suit me best'.[12] Such an occupation was denied him, but he probably achieved greatest professional satisfaction during his short spell as War Secretary in 1940. His diary, particularly during the Second World War, contains numerous references of unaffected sincerity to the simple pleasure of being with ordinary soldiers.

Small wonder that most of those with whom he mixed in political life knew little of the real Eden. For this reason he was not a particularly popular minister. His three periods at the Foreign Office never succeeded in creating that mixture of admiration and affection inside the Civil Service which was associated with Ernest Bevin. Though 'Anthony' was the name by which most of the staff referred to him amongst themselves, few would have thought of so addressing him to his face.[13] 'He either gives you his full confidence', noted Lockhart, 'or does not tell you anything.'[14] Essentially a shy man, Eden had the unfortunate knack of advertising his trivial qualities, while hiding more substantial attributes. Most political contemporaries tended to judge him by the same surface characteristics as did the general public, reaching a less favourable conclusion than did the world at large. Especially during his time as a junior minister, the public may be forgiven for having formed their assessments as much on the basis of external appearance as upon ill-informed ideas about Eden's contribution to the workings of the British government. As an early biographer put it, the public, 'finding him politically sincere, physically attractive and socially beyond reproach, . . . conferred on him a testimonial from its own immediate impressions'.[15] He was, after all, the most handsome and best-dressed politician of his era. Slim, elegant and with good looks which lasted into old age, Eden's was an instant appeal. He had 'the talent for looking wonderful no matter what he wore, so that women and haberdashers swooned as he passed'.[16] Eden 'moved always with the fluent ease of a river between its banks, almost with a suggestion of bravura'.[17]

But the very attributes which contributed to his wide popular appeal condemned him in the eyes of many who saw him at somewhat closer quarters as light-weight and insubstantial. The fact that he was good-looking and well dressed encouraged others to describe him as superficial and something of a poseur.[18] 'Robert Taylor' and 'Miss England' were among the irreverent and derogatory nicknames he attracted.[19] Vanity was a significant factor in his character. Sunbathing played an important part in his routine. An American diplomat remembered the Geneva Summit of 1955:

Eden had just [in fact in August 1952] married his second wife, who was much younger. . . . Mr. Eden, I think, was quite interested in keeping this impression of glamour, as much to her as to the public. He was obviously taking sunbaths or a sunlamp every day, because throughout the entire conference he had a blooming, vigorous tan. He was quite chagrined, I think, because he didn't come out of the conference as the outstanding diplomat.[20]

Eden's, judged a Tory backbencher at the height of the Second World War, was 'purely a press-made reputation . . . mere platitudes and amiable generalities'.[21] To many MPs he was simply the 'spoilt darling of fortune'.[22] 'Foreigners find him frivolous and light-metal,' remarked 'Chips' Channon.[23] John Colville concurred: 'many as are his qualities, they are not very solid'.[24] Labour colleagues in the wartime coalition tended to agree. 'I am afraid Eden is rather a light-weight,' concluded Hugh Dalton after a War Cabinet meeting in 1941.[25] Dalton's party leader was of the same mind: 'He's a funny little bird. He's got no status of his own. He's only a Private Secretary to the P.M.'[26]

Eden was not helped by what many regarded as an unfortunate personal manner. His violent displays of temper, in such marked contrast to the suave exterior of the public figure, have been noted several times already. It was subordinates who suffered most. Indeed, he could be unduly deferential to those above him. Perhaps there is something in the remark of an unidentified colleague that 'Eden lives in a no-man's-land inhabited by superiors and inferiors. He acknowledges no equals.'[27] Certainly, Eden's outbursts took many by surprise. His press secretary recalled:

The defect was one of the best-kept secrets about Eden. I heard not a whisper of its existence until I went to Downing Street; he enjoyed a public reputation, of course, as the most considerate, charming and calm of men. But in private, under pressure and hostility, he used to become perfectly terrible.[28]

This characteristic was perhaps inherited from his notoriously irascible father, a man capable of destroying his greenhouse plants simply because they were the wrong colour. As Eden's brother noted, their father's 'rudeness was offensive to the most patient. His constant and exaggerated irritation with the minor details of life . . . became quickly wearisome.'[29] So too with Eden himself:

An official car that broke down in Piccadilly, the importunity of a bevy of flash-light photographers, a secret telephone that refused to 'scramble', sometimes threatened to assume the proportions of a hideous departmental *faux pas* or a major diplomatic reverse.[30]

Indeed the fact that Eden's highly strung outbursts were as likely to be sparked off by an apparently trivial incident as by a matter of substance meant that his anger was often dismissed as mere petulance. 'He is like a child,' judged Shuckburgh. 'You can have a scene with a child of great violence with angry words spoken on both sides and ten minutes later the whole thing is forgotten. This is not possible with grown-ups, but it is the regular thing with A.E.'[31] Eden's rage would spend itself quickly. If he ended a conversation by slamming down the telephone, he was just as likely to ring back the following day to apologize with all the charm for which he was equally renowned.[32] As far as Eden was concerned, no damage had been done. But his cavalier rudeness could give lasting offence. Intensely sensitive himself, he could be strangely insensitive to the feelings of others. As William Clark remembered: 'The private secretaries sometimes refused to take things up with him because they knew it would worry him and cause an explosion. Always at the back of everything was the fear that he would lose his temper and we should be sworn at.'[33] Clark felt the strain 'of being polite on paper to someone whose sole manifestations to me are snappy notes via private secretaries, or saccharine remarks to my face'.[34] Almost certainly, this trait in Eden's character was exacerbated in his later career by mounting ill-health. As Prime Minister he was dangerously thin-skinned.

A striking number of observers drew attention to a feminine streak in Eden's make-up. By this few meant to imply that he was effeminate, though his habit of addressing male colleagues as 'my dear' was disconcerting, especially for Americans.[35] Such verbal affectations – 'bless you' was another[36] – did not endear him to everyone. Shuckburgh related his experience of a couple of days in the country with his master:

> Eden took him for a walk in the garden one afternoon and to Shuckburgh's astonishment Eden talked to the flowers like a little boy. He would say looking at a flower bed: 'Peony dear, you need a little water, don't you?' Then he would turn to Shuckburgh and say: 'Evelyn dear, do get a little water for Peony.'[37]

More usually 'feminine' was used to describe Eden's sensitivity to criticism, his jealousy – particularly of potential rivals – and his mental processes which seemed heavily dependent on intuition and quickness of perception. He did not 'brood over things', concluded Macmillan.[38] The last, which might well be considered a virtue, merely underlined the curious perception of frivolity which Eden never quite lost.

But Eden did not allow his instincts to act as a substitute for hard work. He was not blessed with intellectual brilliance. Highly intelligent and with a good brain, he was not really a man of ideas, at least original ones. 'His approach to problems often strikes one as instinctive rather than based on long, intellectual process,' recorded Oliver Harvey, who knew him better than most. 'But he always safeguards himself by a very firm grasp of the facts.'[39] This characteristic lay at the heart of his diplomatic triumphs. His forte was in patient, painstaking and sometimes plodding negotiations, seeking out the ground for compromise and conciliation where none seemed to exist. It was not the most spectacular of political skills, but it frequently proved invaluable. A European observer noted the way he could 'change the atmosphere of a conference from deep pessimism and general distrust to mild optimism and willingness to cooperate'.[40] Eden had a tremendous appetite for work — a virtue, but one which played upon his nerves and reacted upon his health. This was the man who allowed himself just one day in Sussex for his honeymoon during the 1923 general election campaign. Probably no figure in the wartime government, Churchill included, carried a comparable load. In 1941 his doctor diagnosed a duodenal ulcer: 'He urged bed and treatment, but agreed in the end on latter alone and see how matters were a fortnight hence. I have a shrewd idea that this is really my danger signal for over-work.'[41] By the end of the war he was on the verge of exhaustion. The struggle against Hitler compelled many to exceed their powers of endurance. Yet throughout his life, even as a comparatively young man, Eden tended to drive himself too hard for his own good. In a perceptive article published in 1954, with Eden's reputation at its height, an American journalist noted:

It is his constant desire to see everything, to have a hand in any verdict, which leads to crises of nerves — the kind of crises that marred his first experience in Cabinet office. . . . Slowly he would become so immersed in what he was doing and so confident that there was nothing more important in the whole field of British activity that any check or criticism, however slight, would produce a passionate outburst.[42]

Eden thrived on detail. His speed with paperwork was impressive.[43] This made him a good complement to someone like Churchill whose strength lay in the grand, the sweeping and the visionary.[44] But it also encouraged the perception – not entirely justified, but too frequently noted to be ignored – that his was a mind which could sometimes not see the wood for the trees. 'Everywhere this fantastic attention to detail,' recorded Clark early in 1956. 'Preparing his speech on Cyprus he wants detailed proof of Makarios's guilt – no lifting up of his eyes to the long term.'[45] Eden was not prone to great leaps of imagination. His capacity to inspire was at its greatest in the 1930s when it was based upon a popular image of what he stood for which never quite coincided with reality. This shortcoming showed particularly in his public speeches. On occasions he could be quite effective. He became, particularly in the years after 1945, adept at winding up a Commons debate and responding to points made by his opponents. But his more formal speeches were too obviously manufactured to be truly impressive, and were frequently marred by elaborately worded platitudes and clichés. Of one Eden speech Churchill reputedly remarked that 'it contains every cliché except "Please adjust your dress"'.[46] 'This is not the result of lack of application,' insisted Drew Middleton. 'He alters and polishes his speeches repeatedly, but the gift is not there.'[47] It was the same with his writings. Whatever their merits, Eden's memoirs represent no literary masterpiece. Yet ironically, in the last years of his life, he managed to find an altogether lighter touch when writing of his youth and experience on the Western Front. *Another World* was published in 1976 to almost universal acclaim. Perhaps Eden took the advice of his friend, Richard Law: 'I hope that you will not dictate it. I have noticed over the years that when you write in your own hand the English is alive and flexible, but when you dictate it becomes stiff and formal.'[48]

Notwithstanding his intuitive streak, decision making did not come easily to Eden. Over matters such as the Indian Viceroyalty in 1943 or the question of whether to give up the Foreign Office, either to lighten his workload or to broaden his experience of domestic politics, he agonized long and hard before reaching his conclusion. An image of indecision was easily created. 'He had been discussing this with all his friends for months', wrote Lockhart of Eden's crushing wartime ministerial burden, 'and cannot make up his mind.'[49] Yet this impression was somewhat misleading. On reflection the same observer concluded that Eden's greatest weakness was 'to make his own decision . . . and then to consult others without telling them what he had decided.'[50] Indeed he did not like others to press their ideas upon him unless he had asked them to. Shuckburgh experienced this problem: 'It was only a matter of time before he started saying this sort of thing about me, as he always used to about others who gave him unpalatable advice. It is like the ancient kings who slew the bearers of bad news.'[51] The point was that Eden, perhaps showing the impact of an over-domineering father, lacked a certain self-confidence and needed reassurance. 'How do you think I did?' he would ask a colleague, or 'I hope I didn't overdo it; but I did feel very strongly about it.'[52] Paradoxically, therefore, the man who had few really close friends liked to have his staff around him. According to Brendan Bracken, Eden 'could not live without an array of secretaries'.[53] The writing of a telegram or a speech required an audience:

> Wavell and Dill in their dressing-gowns sat side by side on a sofa in the drawing room. Longmore arrived in uniform, and watched the two weary soldiers, looking like a couple of teddy-bears, trying to give the Foreign Secretary's eloquence the attention it demanded.[54]

Such habits grated on those of calmer temperament. Macmillan was one who never reconciled himself to the rather frenetic atmosphere in which Eden seemed to exist: 'He'd have a mass of people round when he was trying to write a speech. How he ever wrote it I can't imagine. Everybody talking; a tremendous flap went on. . . . He was always very excitable, very feminine-type, very easily upset, easily annoyed.'[55]

Macmillan, who found time during the Suez Crisis to read the whole of George Eliot together with sundry other literary works, attributed many of Eden's failings to his obsessional working habits. He was 'basically not an interesting man. Unlike Winston, he never had a chance to read.'[56] This was unfair, an indication perhaps that Macmillan never really knew well a man with whom much of his own career was intertwined. Eden was among the most civilized and cultured figures to hold high office in Britain this century. He read widely with tastes ranging from Shakespeare to modern novels. For sport he preferred tennis to the country pursuits usually associated with his class. Though he had little feeling for music, he enjoyed the theatre and, at one time, the cinema. But his particular passion, from his student days, was for painting and his own collection included works by Corot, Monet, Derain, Degas, Picasso, Braque and Gwen

John.[57] He greatly enjoyed gardening, telling Lockhart that when he was tired or bored in cabinet he would think, or try to think, of his garden and that this soothed him.[58] John G. Winant, America's wartime ambassador in London, described a weekend at Binderton, Eden's home in Sussex:

> I have never known anyone who cared more about flowers or vegetables or fruit trees, or wind blowing across wheatfields or the green pastures which marked out the Sussex Downs. We used to get our fun weeding the garden. We would put our despatch boxes at either end, and when we had completed a row we would do penance by reading messages and writing the necessary replies. Then we would start again our menial task, each in some subconscious fashion trying to find a sense of lasting values in the good earth.[59]

Yet the presence of despatch boxes among the flower beds is revealing. If there is an element of truth in Macmillan's strictures, it is that Eden found it harder than some of the more successful political figures to switch off completely from his public duties. 'Except for a few hours of sleep', recalled Nutting, 'there was never one moment in the day or night when he ever seemed to want to relax.'[60] Separated from his work, he complained that he felt like Robinson Crusoe until his papers arrived by messenger in the Foreign Secretary's box.[61] As Valentine Lawford put it:

> For him even a walk on the Downs was no more than a pretext for a tour of infinitely vaster and less green horizons; and when he picked and offered one an apple from his favourite tree or a fig from the southern, stableyard wall, one had the strong impression that the hospitable gesture, though it was as natural as that of any Happy Countryman, only momentarily interrupted a political train of thought.[62]

Almost certainly the qualities which accounted for his successes as a diplomat made Eden less effective as a politician. As a patient seeker after compromise he tended to avoid any premature direct reference to issues which might offend those he was trying to conciliate. Like a good poker player Eden knew not to reveal his hand too soon. But the blurred edge of diplomatic parlance, couched in subtle ambiguities and earnest homilies, was not always appropriate to the rough and tumble of domestic politics. The French diplomat René Massigli believed that Eden's urbanity created its own problems. 'Misunderstandings could arise because he expressed himself so politely, or because he stated his views so moderately that his hearers might think that he only held them moderately.'[63] At his worst he tended to tell people what they wanted to hear, rather than what he really thought. It was possible to end a conversation with Eden with a curiously empty feeling, 'thinking how reasonable, how agreeable and how helpful he has been, and then [one] discovers that in fact he has promised nothing at all!'[64] Political life demanded a tougher, sometimes rougher, approach. Several contemporaries noted that Eden was not a particularly effective operator in cabinet. 'He is not very good at standing up for himself,' judged Lockhart. 'He is nothing like as good as Brendan [Bracken] at standing up for his staff.'[65] Letting it be known how strongly he felt about a matter was a problem which bedeviled Eden's relations with John Simon and also latterly with Neville Chamberlain. Most obviously it weakened his bargaining position with Churchill after 1945, when others were keener than Eden himself to confront the issue of the old man's retirement even though it was Eden who had most to gain. 'The whole problem', noted Lockhart in December 1945, 'is who is to bell the Winston cat.'[66] It was still the whole problem nearly a decade on. As late as the summer of 1954 Colville found Eden 'feeling bashful about choosing the right moment' to raise the question with Churchill.[67] 'He does not know how to put it to Winston,' confirmed Lord Moran.[68] In fact Eden preferred to leave this delicate task to others. 'He wants me to tell [Churchill] this,' recorded Macmillan that October, 'which I agreed to do.'[69]

Eden might have been a more effective politician, or at least less obsessional about his work, had his private life been happier and more secure. Again the public image contrasted with private reality. Eden did not wear his private grief or anxiety on his sleeve. 'Anthony seems to bear all his troubles easily,' noted one in whom he came to confide. 'Certainly no one would guess from his outside manner that he was worried.'[70] He married Beatrice Beckett, daughter of the part-owner and chairman of the *Yorkshire Post* in 1923. She was attractive and vivacious, an ideal partner it seemed for an ambitious young politician. Two sons, Simon and Nicholas, were born in 1924 and 1930 respectively. But from a fairly early date the marriage started to go wrong, afflicted by an incompatibility of interests and temperaments. Beatrice preferred jazz to politics, flying to gardening. Both parties agreed to go their separate ways and both had their affairs. Such unconventional relationships were more common among Eden's political generation than the general public realized. 'No word of Ursula [Filmer-Sankey] so far!' recorded an anxious Eden in 1930. 'Out of sight, out of mind?'[71] A commitment to their children and a continuing mutual affection kept the couple together through the 1930s, but by the end of the war the marriage had ceased to have much meaning. The death of Simon in a plane crash in the Burmese jungle in the last weeks of the conflict was a shattering blow for both parents. Eden carried on his public duties with a stoicism which few could have matched. With his son still listed only as missing, Eden confided his anxiety to Lockhart: 'He asked me to say nothing about this, as he did not wish the news to come out before the election was over in case people might say that he was trying to get sympathy out of his son's misfortune.'[72]

Throughout his first marriage Eden was beset by financial worries. Poverty is a relative concept, but Eden certainly found it difficult to sustain the lifestyle to match his career and social standing. 'If we never fare worse', he noted at the end of 1929, 'we shall have no cause to complain, even though we never be either millionaires or prime ministers.'[73] Though the premiership awaited him, wealth did not. In his early career he had had to 'write and write – and write' to make ends meet.[74] The extravagance and irresponsibility of his unpredictable mother, who lived on until 1945, added to his difficulties. Not until the Conservatives lost office that year did he have the opportunity to take on company directorships and earn money in the City. It was one more factor which tempted him on occasion to give up the whole political game.

By the end of the war Beatrice had entered into a long-term relationship with the American, C. D. Jackson, who had been senior vice-president of Time Incorporated before becoming Eisenhower's political warfare representative in England. This liaison seems to have been investigated by the security services and knowledge of it, or Eden's own flirtations, might have had a damaging impact on his career had it entered the public domain.[75] According to J. P. L. Thomas, husband and wife reached a bargain at the end of the war that if Eden remained in politics and had a chance of becoming premier she would not seek a divorce or create a scandal. If, on the other hand, as he seriously considered at this time, Eden became UN Secretary-General, that would be a different matter.[76] This possibility passed and at Christmas 1946 Beatrice decided to leave for the New World, thus giving Eden the option of a divorce on the grounds of desertion after three years. She was 'obviously bored with him'.[77] But Eden clearly hoped that a reconciliation might still be possible. The late 1940s were probably the most unhappy years of his career:

He made it quite obvious that he would like her back, kept referring all the time to her belongings. She had left everything here: her jewellery, her clothes, etc. Asked pathetically what he was to do with them. . . . [He] thinks that the C.D. Jackson affair is over and says that Jackson, who is married himself, has behaved very badly. But the hard fact remains that Beatrice herself is adamant, refuses to answer all entreaties from her own relations, and has told Anthony that she simply cannot face political life.[78]

By 1949 Beatrice had transferred her affections from Jackson to a New York surgeon, whom she wanted to marry. Though Eden now had to give serious thought to divorce, he did so without enthusiasm. 'I feel that he is much unhappier about the whole business than anyone would realise from his external manner,' remarked Lockhart.[79]

The divorce became effective in 1950. Two years later Eden married Clarissa Churchill, daughter of Winston Churchill's younger brother John. In his second marriage to a woman more than twenty years his junior Eden found true happiness and contentment for perhaps the first time in his life. Though some observers believed that Clarissa was excessively protective and tended to exacerbate Eden's natural volatility,[80] she remained his devoted companion for the rest of his life. When, in the early years of their marriage, Churchill repeatedly put off his promised retirement, Eden was genuinely tempted to give up politics altogether – 'some way out of this awful impasse' – to enjoy the happiness for which he had waited so long. In the darkest days of the Suez Crisis, Clarissa was at his side, supportive throughout. The very last entries in his diary in 1976, by which time Eden was mortally ill, reveal a man who seemed finally to have put his political worries behind him and to be at peace with himself: 'Very beautiful day. Soft sunshine, gentle light and breeze. Enjoyed it. . . . Exquisite vase of crimson glory and mignorette from beloved C[larissa].'[81]

*

There is a curious shape to Eden's career. Reaching senior office early on, he occupied the place of heir apparent for an uncomfortably long time before his disappointingly brief premiership. Overall, the present assessment has sought to portray a career of distinction. Oddly, the least impressive period of his political life was probably the 1930s, at least when judged against his own criteria. Though he uttered the right phrases in public, Eden acquired a reputation for rectitude which his actual performance never quite justified. A ministerial colleague asserted that 'at the centre of every crisis you could always find Anthony, personifying the struggle of Good against Evil'.[82] Churchill too gave his valuable imprimatur to Eden's record: 'It might almost be said that there was not much difference of view between him and me, except of course that he was in harness.'[83] Yet Eden was always closer to government policy – whether it was conducted by Simon, Hoare or Chamberlain – than these simple conclusions implied. Even the supposed internationalist and League of Nations man was more usually a conventional exponent of traditional Foreign Office thinking on Britain's national self-interest. If the central failure of British diplomacy in this decade was not to understand that Hitler's assumption of power transformed not only the objectives of German foreign policy but also the very parameters of international relations and that the German dictator could not be reconciled to the existing order through a policy of concessions, with or without corresponding gestures by Germany, then Eden was part of that failure. In lesser details of policy his attitude was more distinctive – his greater stress upon the need for rearmament, his determination to seek out possible allies, particularly the United States, and his doubts whether Mussolini could ever be appeased. It was an honourable enough record, but not quite the one Eden himself came to portray. Indeed he seems sincerely to have convinced himself of the accuracy of the image which grew up around him. Perhaps he took to heart an after-dinner conversation with Churchill in October 1941. According to the latter:

> In time all the Munich men would be driven out. Neville was lucky to have died when he did. Edward [Halifax] could not have stayed at F.O. The public had forgotten its own errors, would only remember its former leaders' and would take vengeance.[84]

At all events the perception of Eden's pre-war career proved as important as the reality. Lacking a secure power-base in party-political terms, he could use his national standing which derived from this reputation as the foundation for further advance. It remained important for years to come. A local newspaper observed his election tour in 1951 and recorded some questionable but revealing judgements among the voters: '"You were the best Foreign Secretary we have ever had!" shouted a man at a Birmingham meeting last night. "There would not have been a war if you had had your road", shouted another, in homely local idiom, at Acocks

Green.'[85] From Eden's point of view his experience of the 1930s, and in particular of what that decade did to the reputations of so many fellow ministers, made him peculiarly sensitive throughout the rest of his career to the charge of 'appeasement'. To say that he became resolute in the need to stand up to all dictators, whencesoever they came, would be going too far. Eden's wartime career showed a clear grasp of power-political reality, especially in relation to what could or could not be denied to the Soviet Union. But he became capable of making facile historical analogies when they suited his purpose. For one whose strength lay in a mastery of intricate detail, Eden could be curiously prone to simplify complex historical and political situations.

Eden's performance as a leading figure in Britain's war directorate appears altogether more impressive. Carrying a burden of responsibility which was probably beyond the capacity of any one man, he none the less played a vital role in shaping British and Allied diplomacy. His interaction with Churchill was the key factor. Becoming a sort of Alan Brooke of the diplomatic arena, Eden could steer the premier into what were usually wise and realistic lines of policy. However difficult he had sometimes found it with Chamberlain and others to clarify his own policy positions, and notwithstanding his perennial difficulty in confronting personal questions with Churchill himself, Eden exercised an influence over the Prime Minister unmatched by any other member of the War Cabinet. Whatever his undoubted qualities as a war leader, Churchill was not the best business manager ever to head the British government. Working with him could be an exasperating experience. Eden recorded the events of an evening in August 1941:

> Winston suddenly decided he would rather see 'Emma Hamilton' than have a meeting and the film people were summoned. They had already left for London and could not be caught on road despite repeated demands. So eventually we had our meeting, but did not sit down to it until near midnight nor get up until near 2 a.m! Max [Beaverbrook] and I then motored off to our respective Surrey homes. I was in bed by 4 a.m., and felt it was all rather like a Russian evening.[86]

On another occasion Churchill himself conceded that he 'would end by killing us all by these late hours, which may well be true'.[87] His personal intervention into key areas of wartime diplomacy caused Eden immense irritation, which he did not seek to conceal from those around him. Yet, whatever the strain, Eden remained Church-ill's 'loyal (if complaining) lieutenant'.[88] Where the situation demanded it, however, he would wage a war of attrition to ensure that his own views prevailed. In so doing he had more of an eye for the longer term than critics implied or his own working habits tended to suggest. The question of de Gaulle and post-war France is probably the outstanding example of his clarity of thought.

By the end of the conflict Eden's standing was considerably enhanced and the war years may indeed represent the pinnacle of his achievement. But, partly because he was already designated as Churchill's successor, critics were ever ready to point to his weak-nesses, particularly of character and political skill. Many, such as Vansittart, felt strangely ambivalent about him:

> He was not quite sure whether in these new difficult circumstances Anthony was big enough for the job. Previously luck had always been on his side. He had always had a good start given to him. . . . Now he had to start from nothing. He was like a sprinter who had achieved wonderful success from starting holes and now was deprived of these holes. Could he make the running? Van agreed that Anthony had grown greatly in stature during the war. Never-theless, he remained uncertain.[89]

Freed from the constraints of office, Eden gave more attention than hitherto to domestic politics. In a vague way he had already culti-vated a progressive image and during the war had sought to associ-ate the Foreign Office idea of international reconstruction with the concept of domestic reform.[90] But the intellectual climate after 1945 was not one in which he was likely to shine, he remained uncertain of himself on economic matters and he was always happier when reverting to his chosen field of diplomacy. By this stage Eden firmly believed that the Conservative leadership should already be his and at best he could only display a 'controlled impatience' at Churchill's reluctance to stand aside.[91] Indeed, granted that Eden's character weaknesses became more rather than less evident as the years went by, it seems beyond question that the great war leader did his chosen successor a disservice in making him wait so long. That delay periodically caused Eden to wonder whether it was worth his while to stay on in political life. During the war he genuinely questioned his own suitability for the highest office of state.[92] Gradually, how-ever, such doubts were overcome and, behind repeated outbursts about 'how beastly' politics were,[93] there developed an intensely ambitious streak and a feeling that the premiership was his logical destiny.

When he was reinstated as Foreign Secretary in 1951, Eden's renewed ministerial partnership with Churchill proved altogether less fruitful than it had been a decade earlier. The premier's failing powers made him keener than ever to focus his energies on his principal interests – foreign and defence policy. Both men suffered from the delusion of their political generation that Britain's natural role was that of a great world power. Churchill was probably less confident than Eden of the country's independent power, but mistaken in thinking he could recreate anything approaching the relationship he had enjoyed with Roosevelt in the early stages of the war. Eden seemed more sure of continuing British independence and autonomy, but also saw the importance of the Anglo-American partnership, providing this was founded on terms acceptable to Britain.

Eden's perception of the United States was probably the most important single element in his diplomatic outlook. Intellectually, he understood the importance of good Anglo-American relations. Early in his career he described the co-operation of the two countries as 'the most important safeguard for world peace in the years that are to come'.[94] Only weeks before his death he was still insisting that the alliance remained 'the key to the preservation of our and Western security'.[95] But the practical experience of working with the United States left him deeply troubled. He viewed the Churchill–Roosevelt relationship with mixed feelings. Though it represented a useful means of exerting influence at the top level of the US government, it also, Eden believed, made Churchill unnecessarily susceptible to American pressure.[96] He never therefore wanted to see the American partnership become the only prop in Britain's international standing. To such misgivings was added the largely unspoken resentment of one whose adult life witnessed the progressive replacement of his own country by the United States as the dominant force in international affairs. This made him particularly concerned that Anglo-American relations should be conducted on a basis of equality, even when he was only too aware that the two countries were not in fact equals. Shortly before the nationalization of the Suez Canal Company Eden wrote to Eisenhower about the future of NATO:

I am quite sure . . . that we shall go forward together to shape the future as we have done the past. We have had many more difficult problems than this and as long as we are in step I have no doubt that we can handle this one without causing disarray.[97]

Such phraseology helped to massage the damaged ego of a declining imperial power, but it implied an interpretation of the Special Relationship which successive US governments were loath to accept.[98]

In his final period as Foreign Secretary – and with little positive assistance from Chuchill – Eden achieved a series of considerable diplomatic successes, particularly during 1954. As before, his great strengths were in negotiation and conciliation. Yet Eden's triumphs may well have encouraged him to think that Britain could still take the lead on the world stage, with or without American backing. 'Eden was determined to go ahead,' noted William Clark of the decision to warn the Russian leadership against supplying arms to Egypt. '"If we'd waited on the Americans", [Eden] said, "I'd never have gone round Europe after EDC smashed; we'd never have had a Geneva conference. They're always unprepared to act." '[99] The reality, of course, was that ears 'were now bent more toward the United States than Britain, and a Dulles monosyllable (or even his silence) could mean more than Eden's most eloquent peroration'.[100]

But if Eden's preference as a diplomat was for a tactical advance on a limited front, his longer-term goals were more clearly defined and adhered to than critics have allowed. Notwithstanding his determination that Britain should remain a great power, he understood that her resources were severely overstretched and that the burdens of responsibility needed to be shed or at least shared. In the Middle East which he, like most of his contemporaries, saw as a vital symbol of continuing British power, he wanted to develop a system of informal influence based on new agreements with nationalistic governments and to draw the United States into the resulting arrangements. As a long-term strategy it was not unlike that for which in Europe his predecessor, Bevin, received considerable praise. But Eden's failure was to convince the Americans that Britain's national interests were synonymous with those of the Western alliance. America tended to see the respective priorities of Britain and herself as incompatible.

Eden's reluctance to embrace Europe has been singled out as indicating his inability to see beyond the immediate problems of day-to-day diplomacy. He is thus portrayed as an old-fashioned figure – a Commonwealth man or even, and despite his misgivings about the United States, an Atlanticist – when it was vital to be a

European. While Eden concentrated his attention on the Commonwealth – doomed to go its disparate ways – and the Special Relationship – which was never going to be as special for America as it was for Britain – wiser men turned to the third and most neglected of Churchill's interlocking circles as the country's only long-term salvation in the second half of the twentieth century. But if Eden was guilty of a grave misjudgement it is at least easy to understand why he made it and clear that it was one shared with the vast majority of his political generation. The effective collapse of the Commonwealth as a force in world affairs and the rapid decline of Britain, at least in relative terms *vis-à-vis* rivals on the continent, were developments few foresaw at the time. If a major reappraisal of Britain's world role was called for, then this should have been easier in the late 1940s, when most economic indicators seemed to compel such a reassessment, than in the early 1950s, when the country's economic performance seemed to take a decisive turn for the better and offer some justification for the quest to sustain great-power status. Yet an unfavourable comparison is sometimes drawn between Eden the tactician and Bevin the strategist.[101] In fact successive British leaders, at least until the end of the 1960s, including those such as Macmillan who laid claim to 'European' credentials, all placed undue stress upon the notion of Britain as a great power and the need to play a world role out of proportion to the country's inherent capability. The vision Eden is supposed to have lacked was held by very few during the period when he held responsibility for the direction of British foreign policy.

With hindsight his attitude towards the processes of European integration appears to have been consistent, honourable and in its own way far-sighted. He believed, like most of the ruling élite of the 1950s, that Europe was moving in a direction unacceptable to Britain. Yet, while some pretended that Britain could divert the course of European integration into more acceptable channels, Eden considered that the country should be made aware from the outset of the probable destination of the European movement. As he put it in his retirement: 'It is neither a secret nor an exaggeration to say that most of those on the continent of Europe who have been eager enthusiasts for the Common Market are also convinced believers in a European federation.'[102] From the perspective of the mid-1990s Eden's approach seems not without merit.

Eden's tenure of the Foreign Office after 1951 appears all the more impressive when set in the context of the continued postponement of Churchill's retirement. This delay did little for Eden's self-confidence or his equanimity. Moran recorded a conversation with Harold Macmillan a few weeks before Churchill's retirement:

> For fifteen years, according to Harold, Winston has harried Anthony unmercifully, lectured him and butted in on his work, until poor Anthony is afraid to make a decision on his own. Anthony apparently has taken this a good deal to heart, and has been very nervy lately.[103]

Thus, while Eden's reputation as a diplomat continued to prosper, doubts mounted as to his political toughness. If the 'instinct for the jugular' be a characteristic of political success, then Eden's make-up was indeed defective. He could surely have engineered Churchill's resignation at an earlier date had he been prepared to act decisively, as decisively perhaps as Churchill himself would have done in a similar situation. Moran drew an interesting contrast between the two men in 1953:

> '[Churchill] was so much against my leaving the F.O. that I [Eden] gave in.' There it is in a sentence: the essential difference between the two men is exposed. If the roles were reversed it is impossible to picture Winston meekly accepting such a handicap in the climb to power. . . . Day in, day out, [Eden] lacks the hard core which in Winston is hidden by his emotional nature and by his magnanimity.[104]

However exasperated Eden became at Churchill's behaviour, he would never conspire against him nor use the old man's patent physical and mental decline to his own advantage.[105]

With Churchill clinging tenaciously to office in the summer of 1953, Shuckburgh and Salisbury discussed Eden's prospects. Perhaps the latter's real role, pondered Salisbury, was to be a great Foreign Secretary. 'But that is not how he sees it,' countered Shuckburgh. 'That is the trouble with them all,' Salisbury concluded. 'They are so ambitious.'[106] Eden, the Foreign Secretary, succeeded as a diplomat; his failure as Prime Minister was as a politician. He had great merits, but in the last resort was not well equipped for the highest office of state. This is not merely a retrospective judgement, a convenient later rationalization of Eden's ill-fated premiership. His final arrival at 10 Downing Street stirred conflicting emotions. The judgement of one of his ministers that no Prime Minister since Pitt had entered office with higher expectations of success may have been true of the public at large.[107] But while some applauded Eden's belated promotion as the just reward for a ministerial career of considerable distinction, many, particularly in the upper levels of the Tory Party, still doubted his suitability. 'He is a diplomat but Prime Ministers have to give orders,' warned Walter Elliot only a day after Eden assumed his new role, while a perceptive article in the American magazine *Life* questioned whether he had the character or capacity to deal with the difficult issues facing him on the domestic front.[108]

As Prime Minister Eden never really managed to stamp his own personality and identity on the British government. His failure to carry out a full-blooded reconstruction of Churchill's cabinet was widely condemned. Eden was aware that comparisons would inevitably be made with his predecessor and generally not to his advantage. Yet in a curious way he sought to ape Churchill, even to the extent of conducting some governmental business from his bed.[109] Perhaps in order finally to remove himself from the great man's shadow he pointedly excluded Churchill from the Conservatives' 1955 general election campaign. That election was the highspot of Eden's premiership and very much a personal triumph. One particular success was his appearance on television. Eden took naturally to the camera – of which Churchill had been extremely wary – and became the first Prime Minister to use the ministerial broadcast as a weapon of political debate.[110] After the election, however, his standing underwent a rapid decline and the feeling, so prevalent in the campaign, that he was in control quickly evaporated. A senior colleague described the slump in Eden's fortunes as one of the most inexplicable phenomena he had witnessed.[111] The accusation of indecisiveness was only partially justified, but its impact on Eden's brittle temperament seems beyond question.

As a subordinate minister Eden had been a difficult colleague. Many who now served under him as Prime Minister found him fussy and overbearing. The problem was not really apparent at cabinet meetings where Eden proved to be an efficient chairman, setting a marked contrast with Churchill's declining years. But in the day-to-day conduct of government business his reluctance to delegate became a particular handicap. It was a curious failing in one who had suffered in similar manner at the hands of Chamberlain and Churchill. 'He kept on sending me little notes,' recalled Macmillan, 'sometimes twenty a day, ringing up all the time. He really should have been both PM and Foreign Secretary.'[112] Macmillan's Foreign

Office successor, Selwyn Lloyd, fared no better – thirty telephone calls from Chequers over the Christmas weekend of 1955. 'He *cannot* leave people alone to do their job,' judged Shuckburgh.[113] Sensing himself to have been over-promoted, Lloyd found it difficult to cope with Eden's interference, admitting that on one occasion the latter had kept him on the telephone between 10 p.m. and 2 a.m.[114] Indeed, the telephone became almost a drug for Eden. To an extent this was the natural reaction of a former Foreign Secretary to separation from his area of particular expertise. In part Eden was seeking compensation for the isolation of the highest office of state. But as Prime Minister he seemed unable to free himself from the trivia of departmental administration. 'I hear we now consult NO 10 about the appointment of Embassy guards in Amman,' recorded one exasperated official.[115] Nor was the Foreign Office the only ministry to incur this excessive attention:

> He insisted, for example, on having submitted to him proposed appointments to the boards of nationalized industries. And as he knew nothing of the men or the requirements of the jobs, and anyhow hadn't time to deal with them, the only effect was that they were then held up in his office for embarrassingly long periods.[116]

Though no doubt moved by his personal tragedy, there were probably few of Eden's colleagues who were genuinely sorry at his resignation in January 1957. It was one of Macmillan's singular achievements to replace the frenzied tone of Eden's premiership by one of studied calm.[117] Duncan Sandys, David Kilmuir and Quintin Hailsham were among ministers who commented on the transformation of the atmosphere at the heart of government.[118] Hailsham's description of Macmillan as 'unflappable' probably contained an implied contrast with his predecessor. But it was Iain Macleod who best captured the difference between the two regimes:

> He said it was summed up beautifully for him when the government faced a major Ford strike at Dagenham. The phone rang and his private office said the Prime Minister wanted to speak to him. He thought, 'Oh God, he is going to be just like Anthony. Whenever anything like this happens, he will be phoning me every five minutes to ask what is going on and what I am doing about it, making suggestions.' He wearily picked up the phone: 'Good morning, Prime Minister.' Harold said, 'Dorothy tells me there is a terribly good film on . . . and we wondered if you and Eve would like to join us.'[119]

Even without the intrusion of Colonel Nasser and Eden's own ill-health, it remains difficult to envisage that an Eden premiership could have lasted its full term.

There was of course more to Eden's premiership than Suez. He had, for example, begun to pay more attention to the problems of the domestic economy than has generally been allowed. But Suez dominated the last six months of Eden's career and its shadow falls heavily not only across his time as Prime Minister but upon his overall historical reputation. The crisis invites a series of questions to which no conclusive answers can be given. Would Eden the Foreign Secretary, operating in his favoured environment, surrounded by officials and trying out policy options upon them, have pursued the same policy as did Eden the Prime Minister? Would an Eden not burdened by the domestic political pressures of the premiership and the need to be seen to apply the smack of firm government have followed a more conciliatory course? Would a fully fit Eden have taken the desperate gamble of collusion with France and Israel? Yet two overriding points stand out. In the first place the seeds of the Suez disaster clearly go back to his time as Foreign Secretary. Despite his expertise in foreign affairs Eden did have his diplomatic blindspots. His failure to come to terms with the nature of the Anglo-American alliance was definitely one. His general approach to the problems of the Middle East may well have been another. Though, from his student days, he regarded himself as a specialist in the region's affairs, his understanding was oddly defective. Perhaps he always viewed the subject in a narrow academic light. Certainly, like most of his contemporaries, he found it difficult to come to terms with the transience of the era of British imperialism. Be that as it may, his remark in 1940 that all Egyptians were governed by fear to a greater extent than most other peoples conveys no particular depth of understanding.[120] In the second place Eden's policies in 1956 were determined by an assessment of the gravity of the crisis, the enormity of the issues at stake and the impossibility of accepting defeat which was widely held, and in some cases sustained even after the November débâcle. This assessment related both to specifically British interests and to those of the whole Western world. Nearly two months before Nasser's coup Eden was speaking graphically of the consequences of any interruption in the supplies of Middle Eastern oil in terms of unemployment and hunger in Britain.[121] In the wake of the November ceasefire the CIGS could still write defiantly of the positive achievements of Anglo-French intervention:

> Some people in England today say that what we've done in the Middle East will have terrible effects in the future. . . . The reality is that we've checked a drift. With a bit of luck we've not only stopped quite a big war in the Middle East, but we've halted the march of Russia through the Middle East and on into the African continent.[122]

If Suez involved major misjudgements, these were not Eden's alone. Yet whatever assessment is finally reached of the Suez affair, it remains but one episode in a far longer career. Forty years on, nothing that Eden did during the crisis appears so reprehensible as to colour an assessment of his whole public life. It is the lot of statesmen to face appallingly difficult decisions. Sometimes they get things wrong. Those who subsequently chronicle their careers are usually spared such ordeals and find it relatively easy to pass judgement. Yet it is time for the events of Suez to loom less large. Over more than two decades Eden exercised a decisive impact on British politics and, in particular, over Britain's external relations. The management of the foreign policy of a nation which was, despite its traditions and obligations as a world power, in manifest decline was no easy task. Overall Eden did as well as, if not better than, any of his contemporaries. His many achievements make it difficult to deny him a place among the top rank of Foreign Secretaries. But he was also – and notwithstanding the mistakes of Suez – a man of integrity. Though he could become obsessive in his attitude towards those he perceived as implacable enemies – Mussolini, Nasser and latterly Dulles – he was none the less remarkably free of malice. There was a patent honesty about the man which is attractive. In a meeting with Dulles in November 1956 Eisenhower expressed disappointment at having 'continually to downgrade' his estimate of Eden.[123] The final verdict of history may yet turn, full circle, towards an opposite conclusion.

Notes

ABBREVIATIONS

1. Government manuscript sources

CAB   Cabinet Office Papers
FO   Foreign Office Papers
PREM   Prime Minister's Papers
PRO   Public Record Office (Kew)

2. Private papers

Avon Papers (AP)	Papers of Lord Avon, Birmingham University Library
Baldwin MSS	Papers of Stanley Baldwin, Cambridge University Library
Beaverbrook MSS	Papers of Lord Beaverbrook, House of Lords Record Office
Brabourne MSS	Papers of Lord Brabourne, India Office Library
Butler MSS	Papers of R. A. Butler, Trinity College, Cambridge
Cecil MSS	Papers of Lord Cecil of Chelwood, British Library
Chamberlain MSS	Papers of Austen (AC) and Neville (NC) Chamberlain, Birmingham University Library
Cilcennin MSS	Papers of J. P. L. Thomas, Carmarthenshire Record Office
Davies MSS	Papers of Clement Davies, National Library of Wales, Aberystwyth
Derby MSS	Papers of Seventeenth Earl of Derby, Liverpool City Library
Evans MSS	Papers of Paul Emrys-Evans, British Library
Lloyd George MSS	Papers of David Lloyd George, House of Lords Record Office
Halifax MSS	Papers of First Earl of Halifax, Borthwick Institute of Historical Research, York
Hankey MSS	Papers of Maurice Hankey, Churchill College, Cambridge
Harvey MSS	Papers of Oliver Harvey, British Library
Linlithgow MSS	Papers of Lord Linlithgow, India Office Library
Selborne MSS	Papers of Third Earl of Selborne, Bodleian Library, Oxford
Simon MSS	Papers of Sir John Simon, Bodleian Library, Oxford
Templewood MSS	Papers of Sir Samuel Hoare, Cambridge University Library
Woolton MSS	Papers of Lord Woolton, Bodleian Library, Oxford

3. Published primary sources

DBFP	E. L. Woodward et al (eds), *Documents on British Foreign Policy, 1919–39*, 2nd and 3rd series (London, HMSO)
DBPO	R. Bullen and M. Pelly (eds), *Documents on British Policy Overseas* (London, HMSO)
FRUS	*Foreign Relations of the United States* (Washington, Department of State)
H of C Debs	*House of Commons Debates* (Hansard), 5th series

CHAPTER 1

1. Lord Butler, *The Art of Memory: Friends in Perspective* (London, 1982), p. 87.
2. T. Barman, *Diplomatic Correspondent* (London, 1968), pp. 175–6.
3. H. Brandon, *Special Relationships* (London, 1988), pp. 133–4.
4. 'Cato', *Guilty Men* (London, 1940). The influence of *Guilty Men* is by no means exhausted. See, for example, S. Aster, '"Guilty Men": The Case of Neville Chamberlain', in R. Boyce and E. M. Robertson, eds, *Paths to War* (London, 1989), pp. 233–68.
5. D. Bardens, *Portrait of a Statesman* (London, 1955), p. 112.
6. Quoted in Bardens, *Portrait of a Statesman*, p. 132.
7. In fact, of course, far more than 'a few' survived. But the myth of the 'lost generation' was itself an important historical reality.
8. W. Clark, *From Three Worlds* (London, 1986), pp. 147–8.
9. Quoted G. McDermott, *The Eden Legacy and the Decline of British Diplomacy* (London, 1969), p. 23.
10. A. J. Sylvester, *Life with Lloyd George* (London, 1975), p. 196.
11. D. Jay, *Change and Fortune: A Political Record* (London, 1980), p. 74.
12. W. S. Churchill to George VI 16 June 1942, Avon Papers (hereinafter AP) 19/1/10.
13. W. S. Churchill, *The Gathering Storm* (London, 1948), p. 190.
14. Churchill, *The Gathering Storm*, p. 188.
15. Churchill, *The Gathering Storm*, p. 201.
16. House of Commons Debates, 5th series, vol. 408, cols 1293–4: 'His constant self-preparation for the tasks which had fallen to him, his unequalled experience as a Minister at the Foreign Office, his knowledge of foreign affairs and their past history, his experience of conferences of all kinds, his breadth of view, his powers of exposition, his moral courage, have gained for him a position second to none among the Foreign Secretaries of the Grand Alliance.'
17. E. Shuckburgh, *Descent to Suez: Diaries 1951–56* (London, 1986), p. 255.
18. J. Margach, *The Abuse of Power* (London, 1978), p. 100.
19. Margach, *The Abuse of Power*, pp. 100–1.
20. N. Nicolson, ed., *Harold Nicolson: Diaries and Letters 1945–1962* (London, 1968), p. 329.
21. H. Thomas, *The Suez Affair* (London, 1986 edn), p. 13.
22. *The Times*, 15 Jan. 1977.
23. S. Aster, *Anthony Eden* (London, 1976), p. 165.
24. A. Nutting, 'Sir Anthony Eden', in H. Van Thal, ed., *The Prime Ministers*, vol. 2 (London, 1975), p. 330. See also R. Jenkins, 'Anthony Eden in the Thirties', in *Essays and Speeches* (London, 1967), p. 39. 'Even amongst Lord Avon's most fervent admirers there must now be few who wish that this disposition of his time [ten

years as Foreign Secretary; twenty-one months as Prime Minister] had been reversed.'

25. Among those for whom jealousy was an important motivation was Randolph Churchill. His longstanding antipathy towards Eden, dating from the 1930s, was no doubt intensified by the feeling that Eden had usurped the place at his father's side which was rightfully his. See R. S. Churchill, *The Rise and Fall of Sir Anthony Eden* (London, 1959).

26. J. Chauvel, *Commentaire*, vol. 3 (Paris, 1973), p. 216.

27. Sir O. Mosley, *My Life* (London, 1968), p. 273.

28. Sir P. Harris, *Fifty Years in and out of Parliament* (London, n.d.), p. 128.

29. J. Vincent, ed., *The Crawford Papers* (Manchester, 1984), p. 590.

30. S. Ball, ed., *Parliament and Politics in the Age of Baldwin and MacDonald* (London, 1992), p. 288.

31. D. Dutton, 'Simon and Eden at the Foreign Office, 1931–1935', *Review of International Studies*, 20 (1994), pp. 35–52.

32. R. F. V. Heuston, *Lives of the Lord Chancellors 1885–1940* (Oxford, 1964), p. 488.

33. Bardens, *Portrait*, p. 194.

34. R. Cockett, ed., *My Dear Max: The Letters of Brendan Bracken to Lord Beaverbrook 1925–1958* (London, 1990), p. 91.

35. Eden to Churchill 8 Oct. 1947, AP 19/4/5.

36. M. Gilbert, *Winston S. Churchill*, vol. 5 (London, 1976), p. 696.

37. Gilbert, *Churchill*, p. 904; J. Ramsden, *The Age of Balfour and Baldwin* (London, 1978), p. 366.

38. D. Middleton, *The Supreme Choice* (London, 1963), p. 46.

39. R. R. James, ed., *Chips: The Diaries of Sir Henry Channon* (London, 1967), p. 26.

40. A. Calder, *The People's War: Britain 1939–45* (London, 1969), p. 105.

41. A. Forbes to Eden 10 Oct. 1946, AP 20/43/109.

42. M. Amory, ed., *The Letters of Ann Fleming* (London, 1985), p. 217.

43. J. F. Naylor, *A Man and an Institution: Sir Maurice Hankey, the Cabinet Secretariat and the Custody of Cabinet Secrecy* (Cambridge, 1984), pp. 294–5.

44. Sir A. Eden, *Full Circle* (London, 1960), foreword.

45. D. Dutton, 'Living with Collusion: Anthony Eden and the Later History of the Suez Affair', *Contemporary Record*, 5, 2 (1991), pp. 201–16.

46. W. R. Louis and R. Owen, eds, *Suez 1956: The Crisis and its Consequences* (Oxford, 1989), p. 128.

47. Chandos to Eden 29 Sept. 1959, AP 23/17/37.

48. Eden to Chandos 1 Oct. 1959, AP 23/17/37A.

49. K. Young, ed., *The Diaries of Sir Robert Bruce Lockhart 1939–1965* (London, 1980), p. 762.

50. S. Lloyd to N. Brook 8 Aug. 1959, quoted R. Lamb, *The Failure of the Eden Government* (London, 1987), p. 307.

51. It is true that the word 'Katyn' does not appear in the index of Eden's volume of memoirs dealing with the Second World War and the fate of Poland generally is given short shrift. But this issue was less central to the author's ultimate historical reputation.

52. Eden to R. Hankey 6 Dec. 1961, AP 23/35/19A.

53. Eden to F. Manor 3 April 1963, AP 33/6.

54. Eden to E. Heath 9 June 1966, AP 23/38/16B.

55. H. Macmillan, *Tides of Fortune 1945–55* (London, 1969); Lord Boothby, *My Yesterday, Your Tomorrow* (London, 1962); Lord Kilmuir, *Political Adventure* (London, 1964); I.

Macleod, *Neville Chamberlain* (London, 1961); A. Nutting, *No End of a Lesson* (London, 1967); R. S. Churchill, *The Rise and Fall of Sir Anthony Eden* (London, 1959); E. Spier, *Focus: a Footnote to the History of the Thirties* (London, 1963); Lord Moran, *Winston Churchill: The Struggle for Survival 1940–1965* (London, 1966); N. Nicolson, ed., *Harold Nicolson: Diaries and Letters 1930–39* (London, 1966); H. Thomas, *The Suez Affair* (London, 1967); A. J. P. Taylor, *English History 1914–1945* (Oxford, 1965); H. Grisewood, *One Thing at a Time* (London, 1968); N. Bethell, *The Last Secret* (London, 1974); F. S. Northedge, *The Troubled Giant* (London, 1966); M. Cowling, *The Impact of Hitler* (Cambridge, 1975).

56. Lord Lambton to Eden 21 July 1966, AP 23/43/76.

57. Eden to S. Lloyd 24 April 1967, AP 23/44/82.

58. Eden to O. Harvey 26 Oct. 1965, AP 23/36/50.

59. Eden to E. Heath 7 Nov. 1968, AP 23/38/26; F. Bishop to Eden 20 Feb. 1969, AP 20/51/7; R. Allan to Eden 16 Sept. 1969, AP 20/51/8. Allan provided an interesting insight into how Eden's notorious temper might have given rise to the story in the first place: 'It is not inconceivable that you may have said something like this to a member of your personal and confidential staff, even perhaps adding " . . . you can pass this on to your friend Grisewood if you like". You would naturally not expect that person to have done this on an official basis.' Allan to Eden 15 Jan. 1969, AP 23/6/57.

60. Halifax diary 6 June 1940, Halifax Papers A 7.8.4.

61. A. J. P. Taylor, *English History 1914–1945* (Oxford, 1965), p. 754.

62. Lord Moran, *Winston Churchill: the Struggle for Survival 1940–1965* (London, 1966), p. 592.

63. Note by D. Dilks 2 Aug. 1962, AP 20/50/3.

64. Pimlott's article appeared in *Economic Age*, Nov. 1969.

65. Allan to Sir C. Curran 4 March 1975, AP 23/6/142C. The programme's original title, 'We cannot afford to be sentimental about this', was taken from one of Eden's wartime minutes. In the end it was changed to 'Orders from Above'.

66. N. Nicolson to Eden 15 June 1965, AP 23/55/2.

67. Eden to A. Mann 6 Sept. 1961, AP 24/47/37.

68. Eden to Sir N. Brook 18 Aug. 1961, AP 23/56/84A.

69. Eden to A. Mann 6 Sept. 1961, AP 24/47/37.

70. Eden to J. A. McCracken 11 Aug. 1974, AP 24/18/4.

71. Eden to A. F. Phillpotts 26 Nov. 1966 and 7 Jan. 1967, AP 23/57/9,12.

72. Eden to Lord Lambton 18 April 1964, AP 23/43/53; Eden to I. MacDonald 20 April 1964, AP 23/46/70.

73. Sir J. Wheeler-Bennett and A. Nicholls, *The Semblance of Peace* (London, 1972) p. 34.

74. Sir J. Wheeler-Bennett, *Munich: Prologue to Tragedy* (London, 1948), p. 170.

75. See Appendix C of Wheeler-Bennett and Nicholls, *Semblance of Peace*, 'Britain and Europe in 1951'.

76. Sir J. Wheeler-Bennett, *Knaves, Fools and Heroes: Europe Between the Wars* (London, 1974), pp. 13–14.

77. Cf. W. F. Deedes, 'Honourable to the Point of Folly', *Daily Telegraph*, 18 Oct. 1986.

78. See correspondence in *Sunday Telegraph*, 9 Nov. 1986.

79. R. R. James, *Anthony Eden* (London, 1986), p. 603.

80. Eden diary 4 Sept. 1966, AP 20/2/11. Yet Sir Robert Rhodes James may have striven too hard – perhaps inadvertently – to create an appropriate picture of Eden's retirement. The photograph in the official biography of Eden sharing a drink

with three elderly men and captioned 'with Wiltshire friends in retirement' dates in fact from a campaign tour of the Midlands in October 1951. Eden's main interest in his colleagues on this occasion was, presumably, their votes. See *Birmingham Gazette*, 23 Oct. 1951.

81. Eden diary 8 Sept. 1975, AP 20/2/20.
82. Brandon, *Special Relationships*, p. 134.
83. See, especially, James, *Eden*, pp. 3–110.
84. Extracts from the unpublished memoirs of Lady Sybil Eden (Eden's mother), AP 22/14/12.
85. Diary 31 Dec. 1930, AP 20/1/10.
86. Diary 27 Aug. 1931, AP 20/1/11.

CHAPTER 2

1. Lord Avon, *The Reckoning* (London, 1965), p. 277.
2. K. Young, ed., *The Diaries of Sir Robert Bruce Lockhart 1939–1965* (London, 1980), p. 42.
3. A. Shlaim, P. Jones and K. Sainsbury, *British Foreign Secretaries since 1945* (Newton Abbot, 1977), p. 82.
4. R. Bassett, *Democracy and Foreign Policy* (London, 1968), p. 5.
5. J. Strachey in *Left News*, cited Bassett, *Democracy and Foreign Policy*, p. 625.
6. Viscount Cecil, *All the Way* (London, 1949), p. 199.
7. Lord Avon, *Facing the Dictators* (London, 1962), p. 40.
8. Diary 19–20 Nov. 1931, AP 20/1/11.
9. House of Commons Debates, 5th series, vol. 193, cols 1105–9.
10. H of C Debs, 5th series, vol. 210, col. 2165. The Locarno treaties had limited Britain's formal defence commitments in Europe to the protection of the Franco-German and German-Belgian frontiers, while confirming the demilitarization of the Rhineland. The Geneva Protocol, designed to strengthen the League Covenant, had implied an open-ended commitment to all member states.
11. H of C Debs, 5th series, vol. 210, cols 2162–3.
12. H of C Debs, 5th series, vol. 235, cols 697–8.
13. C. Thorne, *The Limits of Foreign Policy: The West, the League and the Far Eastern Crisis of 1931–1933* (London, 1972), p. 191.
14. *Documents on British Foreign Policy* (hereinafter *DBF*), 2nd series, vol. 10, no. 545.
15. Thorne, *Limits*, p. 241.
16. See, however, his speech in the House of Commons on 29 February 1932, H of C Debs, 5th series, vol. 262, cols 917–20.
17. D. Dutton, *Simon: A Political Biography of Sir John Simon* (London, 1992), pp. 138–9.
18. Bassett, *Democracy and Foreign Policy*, p. 371, quoting Harold Laski.
19. Avon, *Facing the Dictators*, p. 23.
20. Harvey to Eden 24 May 1961, AP 7/24/64.
21. R. Jenkins, 'Anthony Eden in the Thirties', *Daily Telegraph* 19 Nov. 1962.
22. Jenkins, 'Anthony Eden in the Thirties'.
23. See correspondence in AP 33/1.
24. Simon to L. S. Amery 19 March 1932, FO 800/286.
25. Simon to Baldwin 11 Oct. 1933, Baldwin MSS 121 fo. 86.
26. Simon to MacDonald 30 Dec. 1932, FO 800/287.
27. The best account is probably still J. Wheeler-Bennett, *The Pipe Dream of Peace: The Story of the Collapse of Disarmament* (London, 1935). See also D. Richardson and C.

Notes

Kitching, 'Britain and the World Disarmament Conference', in P. Catterall and C. J. Morris, eds, *Britain and the Threat to Stability in Europe 1918–45* (Leicester, 1993), pp. 35–56.

28. Lord Vansittart, *The Mist Procession* (London, 1958), p. 486.
29. A. C. Temperley, *The Whispering Gallery of Europe* (London, 1939), pp. 280–1.
30. Temperley, *Whispering Gallery*, p. 248.
31. S. De Madariaga, *Morning without Noon* (Farnborough, 1974), p. 273.
32. H. Macmillan, *Winds of Change* (London, 1966), p. 393.
33. Temperley, *Whispering Gallery*, p. 228.
34. In 1958 Eden told Selwyn Lloyd that 'he thought he himself had one outstanding quality and that was his capacity to gauge public opinion'. R. Lamb, *The Failure of the Eden Government* (London, 1987), p. 9.
35. Diary 30 Jan. 1932, AP 20/1/12.
36. Diary 6 July 1932, AP 20/1/12.
37. Eden to Baldwin 9 May 1932, AP 14/1/81.
38. Eden to Simon 23 Nov. 1932, AP 14/1/120.
39. Eden to D. Dilks 30 April 1971, AP 33/2, writing of the Cadogan diaries.
40. Simon to W. Runciman 24 Dec. 1932, Runciman MSS 254.
41. Dutton, *Simon*, p. 337.
42. Diary 26 July 1932, AP 20/1/12.
43. Diary 28 Oct. 1932, AP 20/1/12. Eden's impressions were confirmed by Sir Alexander Cadogan who wrote, 'To navigate difficult seas you must have both a chart and a captain. I had hoped to get the latter, but you know the difficulties that have arisen there. I thought at least we were going to get a chart, but even that seems doubtful.' Cadogan to Eden 14 Jan. 1933, AP 14/1/153.
44. Diary 31 Oct. 1932, AP 20/1/12.
45. D. Dilks, ed., *The Diaries of Sir Alexander Cadogan 1938–45* (London, 1971), p. 8.
46. *Manchester Guardian*, 1 March 1933.
47. S. Ball, ed., *Parliament and Politics in the Age of Baldwin and MacDonald: The Headlam Diaries 1923–1935* (London, 1992), p. 264.
48. A. Chamberlain to Ivy Chamberlain 4 March 1933, AP 33/8iii.
49. See Eden's acrimonious correspondence with Lord Londonderry, the Secretary of State for Air, in July 1933, AP 14/1/194.
50. Eden to S. Baldwin 22 Feb. 1933, AP 14/1/142. In his memoirs Eden suggested that he now began the practice of bypassing Simon and confiding his ideas first to Baldwin (*Facing the Dictators*, p. 31). This assertion has been accepted by several historians, but Eden's own papers suggest that he presented his plan for a draft convention to Simon and Baldwin at the same time.
51. Cadogan to Eden 6 March 1933, AP 14/1/154.
52. Eden to Simon 6 March 1933, AP 14/1/224.
53. Avon, *Facing the Dictators*, p. 34.
54. H of C Debs, 5th series, vol. 276, cols 511–626
55. N. Chamberlain to I. Chamberlain 25 March 1933, Chamberlain MSS, NC 18/1/821.
56. Cabinet 17 May 1933, CAB 23/76, CC(33)35.
57. Eden to Baldwin 1 May 1933, AP 14/1/144.
58. Eden to Simon 1 May 1933, Simon MSS 76 fos 120–3.
59. Simon to Baldwin 12 May 1933, Baldwin MSS 121 fos 33–4.
60. Eden to Baldwin 22 June 1933, Baldwin MSS 121 fo. 50.
61. Avon, *Facing the Dictators*, p. 38.
62. Diary 23 June 1933, AP 20/1/13.

63. W. Rees-Mogg, *Sir Anthony Eden* (London, 1956), p. 48.
64. W. Ormsby-Gore to Baldwin 8 Oct. 1933, Baldwin MSS 121 fos 83–4.
65. Eden to Ormsby-Gore 25 Aug. 1933, AP 14/1/206C.
66. Diary 14 Oct. 1933, AP 20/1/13.
67. Avon, *Facing the Dictators*, p. 47.
68. R. R. James, *Anthony Eden* (London, 1986), p. 119.
69. N. H. Gibbs, *Grand Strategy: Rearmament Policy* (London, 1976), p. 85.
70. Note by Simon after Germany's withdrawal 14 Oct. 1933, Simon MSS 77 fo. 62.
71. Memorandum 20 Oct. 1933, CAB 24/243, CP(33)240; Cabinet 23 Oct. 1933, CAB 23/77.
72. Eden to Irwin 23 Oct. 1933, AP 14/1/188.
73. *DBFP*, 2nd series, vol. 5, no. 509, and vol. 6, no. 32.
74. Eden to Simon 5 Dec. 1933, AP 14/1/231.
75. Simon to Vansittart 23 Dec. 1933, Baldwin MSS 121 fos 129–35.
76. Diary 17 Nov. 1933, AP 20/1/13.
77. Diary 8 Jan. 1934, AP 20/1/14.
78. Diary 14 Jan. 1934, AP 20/1/14.
79. Ball, ed., *Headlam Diaries*, p. 289.
80. *The Times*, 3 Jan. 1934.
81. H of C Debs, 5th series, vol. 286, cols 2061–74.
82. K. Middlemas and J. Barnes, *Baldwin* (London, 1969), p. 751.
83. Eden to Simon 18 Feb. 1934, AP 14/1/858.
84. Eden's later comment in his memoirs that 'Hitler had not made the grim impression upon me which I recorded a year later' (*Facing the Dictators*, p. 70) scarcely reflects his contemporary reaction.
85. Eden to Baldwin 21 Feb. 1934, Baldwin MSS 122 fos 31–3.
86. James, *Eden*, p. 135.
87. Eden to Lord Tyrrell 22 Feb. 1934, AP 14/1/372A.
88. Eden to Baldwin 21 Feb. 1934, Baldwin MSS 122 fos 31–3.
89. Eden to R. MacDonald 22 Feb. 1934, AP 14/1/338.
90. M. Gilbert, *Horace Rumbold: Portrait of a Diplomat 1869–1941* (London, 1973), p. 390.
91. *DBFP*, 2nd series, vol. 6, no. 308.
92. Diary 24 Feb. 1934, AP 20/1/14.
93. R. Lamb, *The Drift to War 1922–1939* (London, 1989), p. 87; A. Peters, *Anthony Eden at the Foreign Office 1931–1938* (Aldershot, 1986), p. 49.
94. *DBFP*, 2nd series, vol. 6, no. 303.
95. Eden to O. Harvey 26 May 1961, AP 7/24/64A.
96. Eden to Simon 27 Feb. 1934, AP 14/1/859; Eden to Baldwin 27 Feb. 1934, AP 14/1/255 (not sent).
97. *The Times*, 6 March 1934.
98. Avon, *Facing the Dictators*, p. 87.
99. CAB 23/78.
100. CAB 23/78, cabinet 19 March 1934.
101. Eden to W. Elliot 8 May 1934, AP 14/1/301A.
102. Simon to Eden 8 June 1934, AP 14/1/861.
103. Murray to Eden 6 July 1934, Murray MSS 316 fo. 12.
104. Eden to Cranborne 10 Oct. 1934, AP 14/1/290A.
105. H of C Debs, 5th series, vol. 296, cols 213–14.
106. Simon diary 11 Dec. 1934, Simon MSS 7 fo. 9.

107. Temperley, *Whispering Gallery*, p. 298.
108. Avon, *Facing the Dictators*, pp. 100–7.
109. Simon diary 9 Dec. 1934, Simon MSS 7 fo. 8; Simon to Eden 20 Nov. 1934, FO 800/289.
110. CAB 23/80, CC(34)43.
111. N. Chamberlain to H. and I. Chamberlain 9 Dec. 1934, Chamberlain MSS NC 18/1/898.
112. Avon, *Facing the Dictators*, p. 119.
113. Eden to R. Cecil 11 Dec. 1934, Cecil MSS Add. MS 51083.
114. Eden to A. Cadogan 15 March 1935, AP 14/1/405B.
115. Diary 12 Nov. 1934, AP 20/1/14.
116. Neville Chamberlain diary 17 Dec. 1934, NC 2/23A.
117. Chamberlain to H. and I. Chamberlain 9 Dec. 1934, NC 18/1/898.
118. Chamberlain diary 11 Dec. 1934, NC 2/23A.
119. Eden to Baldwin 11 Jan. 1935, Baldwin MSS 123 fos 170–4.
120. Diary 28 Feb. 1935, AP 20/1/15.
121. Chamberlain to H. Chamberlain 9 March 1935, NC 18/1/908.
122. Diary 4 March 1935, AP 20/1/15.
123. Beatrice Webb diary 16 March 1935.
124. Avon, *Facing the Dictators*, p. 129; N. Rose, *Vansittart: Study of a Diplomat* (London, 1978), p. 120.
125. Diary 24 March 1935, AP 20/1/15.
126. P. Schmidt, *Hitler's Interpreter* (London, 1951), pp. 18, 26.
127. Diary 26 March 1935, AP 20/1/15.
128. Avon, *Facing the Dictators*, p. 142.
129. Avon, *Facing the Dictators*, p. 178.
130. Simon to George V 27 March 1935, FO 800/290.
131. Simon diary 27 March 1935, Simon MSS 7 fos 24–5; Simon to Dr S. Berry 5 April 1935, Simon MSS 82 fos 38–9.
132. Chamberlain to I. Chamberlain 30 March 1935, NC 18/1/911.
133. CAB 24/253, CP(35)41.
134. *DBFP*, 2nd series, vol. 12, nos 655 and 659.
135. *DBFP*, 2nd series, vol. 12, nos 669, 670, 673. 'Impression left upon us was of a man of strong oriental traits of character with unshakable assurance and control whose courtesy in no way hid from us an implacable ruthlessness.'
136. Eden to A. Chamberlain 10 April 1935, AC 41/1/10.
137. Diary 24 April 1935, AP 20/1/15.
138. It fell to Eden to assure the House of Commons on 11 July that the Anglo-German Naval Agreement was not incompatible with the Stresa Front. H of C Debs, 5th series, vol. 304, col. 616.
139. Diary 24 April 1935, AP 20/1/15.
140. Vansittart to Eden 20 May 1935, AP 14/1/536.
141. Lumley to Eden 4 April 1935, AP 14/1/460.
142. Ormsby-Gore to Eden *c.* April 1935, AP 14/1/491.
143. Sassoon to Eden *c.* April 1935, AP 14/1/509.
144. R. James, ed., *Chips: The Diaries of Sir Henry Channon* (London, 1967), p. 34.
145. R. James, *Victor Cazalet: A Portrait* (London, 1976), p. 167.
146. R. James, *Memoirs of a Conservative: J. C. C. Davidson's Memoirs and Papers 1910–37* (London, 1969), pp. 406–7; H. M. Hyde, *Baldwin: The Unexpected Prime Minister* (London, 1973), p. 384.

147. Note by A. J. Sylvester 21 May 1935, Lloyd George MSS G/22/1/10.

148. Middlemas and Barnes, *Baldwin*, p. 822.

149. Lord Templewood, *Nine Troubled Years* (London, 1954), p. 136.

150. Hoare to A. Chamberlain 22 July 1935, AC 41/1/62.

151. A. J. P. Taylor, ed., *Off the Record: Political Interviews 1933–43* (London, 1973), p. 48, (interview with Leslie Hore-Belisha).

152. *Manchester Guardian*, 19 June 1935.

153. Chamberlain to Hoare 26 July 1935, Templewood MSS VIII:3.

154. H of C Debs, 5th series, vol. 304, cols 539–40.

155. N. Chamberlain to H. Chamberlain 1 Sept. 1935, NC 18/1/930.

156. Eden to V. Cazalet 28 Aug. 1935, AP 14/1/409.

157. *DBFP*, 2nd series, vol. 14, no. 138.

158. Avon, *Facing the Dictators*, p. 199.

159. Avon, *Facing the Dictators*, p. 179.

160. *DBFP*, 2nd series, vol. 14, no. 253 (CP(35)98).

161. Diary 15 May 1935, AP 20/1/15.

162. Avon, *Facing the Dictators*, p. 204.

163. Eden to Drummond 15 May 1935, AP 14/1/435.

164. Eden to Simon 21 May 1935, AP 14/1/515B.

165. Avon, *Facing the Dictators*, p. 214.

166. Simon diary 27 May 1935, Simon MSS 7 fo. 30.

167. Eden to Drummond 3 June 1935, AP 14/1/436A.

168. *DBFP*, 2nd series, vol. 14, no. 301; Peters, *Anthony Eden*, p. 154; Lamb, *Drift to War*, p. 139; R. Lamb, *The Ghosts of Peace 1935–1945* (Salisbury, 1987), p. 12.

169. Avon, *Facing the Dictators*, p. 221.

170. V. H. Rothwell, *Anthony Eden: A Political Biography 1931–57* (Manchester, 1992), p. 21; see also M. Toscano, 'Eden's Mission to Rome on the Eve of the Italo-Ethiopian Conflict', in A. O. Sarkissian, ed., *Studies in Diplomatic History and Historiography in honour of G. P. Gooch* (London, 1961), pp. 126–52.

171. Peters, *Anthony Eden*, p. 124.

172. Avon, *Facing the Dictators*, pp. 243, 245.

173. Avon, *Facing the Dictators*, p. 242.

174. *DBFP*, 2nd series, vol. 14, no. 366.

175. CAB 23/82, CC(35)40.

176. Eden to Lord E. Percy 30 Aug. 1935, AP 14/1/500A.

177. Cabinet 31 July 1935, CAB 23/82, CC(35)41.

178. Eden to Ormsby-Gore 12 Aug. 1935, AP 14/1/493A.

179. Avon, *Facing the Dictators*, p. 254.

180. Emphasis added.

181. Meeting of ministers 21 Aug. 1935, CAB 23/82.

182. Hoare to Chamberlain 18 Aug. 1935, NC 7/11/28/24.

183. *DBFP*, 2nd series, vol. 14, no. 493.

184. Avon, *Facing the Dictators*, p. 261.

185. CAB 23/82, CC(35)43.

186. Cabinet 2 Oct. 1935, CAB 23/82, CC(35)44.

187. Cabinet 16 Oct. 1935, CAB 23/82, CC(35)47.

188. Hoare to Eden 16 Oct. 1935 and Wigram to Hoare Oct. 1935, FO 800/295.

189. Avon, *Facing the Dictators*, pp. 267, 286.

190. Record of meeting 1 Nov. 1935, AP 20/4/5A.

191. Rose, *Vansittart*, p. 171.

192. Lamb, *Ghosts of Peace*, pp. 26–7; Lamb, *Drift to War*, pp. 152–6.
193. Neville Chamberlain diary 29 Nov. 1935, NC 2/23A.
194. Young, ed., *Bruce Lockhart Diaries 1939–1965*, p. 525.
195. CAB 23/82, CC(35)50. Compare James, *Eden*, p. 154: 'Eden had no instruction to negotiate with Laval.'
196. Diary note 1946, AP 20/1/26.
197. Hoare to Eden 8 Dec. 1935, AP 14/1/450 J.
198. Hoare to Baldwin 8 Dec. 1935, AP 20/4/12B.
199. James, *Eden*, p. 154.
200. CAB 23/82, CC(35)52.
201. Eden to R. Hankey 15 Nov. 1960, AP 23/35/6.
202. R. A. C. Parker, 'Great Britain, France and the Ethiopian Crisis 1935–1936', *English Historical Review*, LXXXIX (1974), p. 313.
203. 'December 1935 Crisis', Templewood MSS XIX:6.
204. CAB 23/82, CC(35)53.
205. CAB 23/82, CC(35)54.
206. *DBFP*, 2nd series, vol. 15, no. 363.
207. Macmillan, *Winds of Change*, p. 445.
208. R. Hankey to 'Vincent' 17 Dec. 1935, AP 20/4/18.
209. Chamberlain diary 15 Dec. 1935, NC 2/23A.
210. Simon to Eden 17 Dec. 1935, AP 14/1/517.
211. L. Pearson, *Memoirs 1897–1948: Through Diplomacy to Politics* (London, 1973), p. 100.
212. 'Invitation to join Mr Baldwin's government, December 1935', AC 41/1/68.
213. B. Bond, ed., *Chief of Staff: The Diaries of Lieutenant-General Sir Henry Pownall 1933–1940* (London, 1972), p. 93; *Daily Express* 20 Dec. 1935.
214. *Daily Herald*, 19 Dec. 1935.
215. A. Campbell Johnson, *Anthony Eden: A Biography* (London, 1938), p. 297.
216. *The Spectator*, 18 Oct. 1935.
217. Strang to Lord Halifax 13 Jan. 1957, Halifax MSS A2.278.107.

CHAPTER 3

1. Note in 1946 diary, AP 20/1/26.
2. A. Roberts, *Eminent Churchillians* (London, 1994), pp. 137–210.
3. As late as the general election of 1983 the then leader of the Labour Party, Michael Foot, castigated the Lord Chancellor, Lord Hailsham, as 'a man of Munich'.
4. U. Bialer, 'Telling the Truth to the People: Britain's decision to Publish the Diplomatic Papers of the Inter-War Period', *Historical Journal*, 26, 2 (1983), p. 355.
5. Sir J. Wheeler-Bennett, *Munich: Prologue to Tragedy* (London, 1948), p. 268.
6. Bialer, 'Telling the Truth', p. 355.
7. J. Harvey, ed., *The War Diaries of Oliver Harvey 1941–1945* (London, 1978), p. 61 (8 Nov. 1941).
8. D. Dilks, ed., *The Diaries of Sir Alexander Cadogan, 1938–1945* (London, 1971), p. 415 (1 Dec. 1941).
9. Bialer, 'Telling the Truth', p. 363.
10. Eden's statement in the foreword to *Full Circle* (London, 1960) that the themes of his memoirs were 'the lessons of the 'thirties and their application to the 'fifties' is well known. But the extreme statement of his obsession with this idea came in a

syndicated newspaper article, 'The Lessons of Munich', published in September 1963. See copy in AP 7/9/1.

11. For example I. Macleod, *Neville Chamberlain* (London, 1961).

12. Undated note for memoirs, AP 20/1/26.

13. R. Maudling, *Memoirs* (London, 1978), p. 63; Lord Avon, *Facing the Dictators* (London, 1962), p. 319.

14. T. Jones, *A Diary with Letters* (Oxford, 1954), p. 309.

15. Note by Hoare May 1937, Templewood MSS X/5, H07.

16. Dilks, ed., *Cadogan Diaries*, p. 378 (13 May 1941).

17. J. Harvey, ed., *The Diplomatic Diaries of Oliver Harvey 1937–1940* (London, 1970), p. 56 (3 Nov. 1937).

18. V. Lawford, 'Three Ministers', *The Cornhill Magazine*, 1010 (1956–7), p. 83.

19. J. Colville, *Fringes of Power* (London, 1985), p. 350.

20. *New Statesman*, 28 Dec. 1935.

21. R. R. James, ed., *Chips: The Diaries of Sir Henry Channon* (London, 1967), p. 49.

22. *Spectator*, 27 Dec. 1935.

23. M. Gilbert, *Winston S. Churchill*, vol. 5, companion part 2 (London, 1981), p. 1363, see also *Saturday Review*, 28 Dec. 1935.

24. Avon, *Facing the Dictators*, p. 324.

25. N. Nicolson, ed., *Harold Nicolson: Diaries and Letters 1930–39* (London, 1966), p. 243.

26. *Documents on British Foreign Policy* (hereinafter *DBFP*), 2nd series, vol. 15, no. 460; Avon, *Facing the Dictators*, p. 323.

27. Cabinet 29 Jan. 1936, CC(36)3, CAB 23/83.

28. Memorandum by Eden 11 Feb. 1936, CAB 24/260.

29. Avon, *Facing the Dictators*, p. 323.

30. B. Bond, ed., *Chief of Staff: The Diaries of Lieutenant-General Sir Henry Pownall 1933–1940* (London, 1972), p. 103.

31. *DBFP*, 2nd series, vol. 15, no. 529.

32. Avon, *Facing the Dictators*, p. 335.

33. *DBFP*, 2nd series, vol. 15, no. 517.

34. Cabinet Committee on Germany 17 Feb. 1936, cited M. Cowling, *The Impact of Hitler* (Cambridge, 1975), p. 105.

35. *DBFP*, 2nd series, vol. 15, no. 541.

36. House of Commons Debates, 5th series, vol. 308, col. 918.

37. D. Bardens, *Portrait of a Statesman: The Personal Life Story of Sir Anthony Eden* (London, 1955), p. 147.

38. H of C Debs, 5th series, vol. 309, cols 76–87, esp. col. 83.

39. Cabinet 5 March 1936, CC(36)15, CAB 23/83; R. A. C. Parker, 'Great Britain, France and the Ethiopian Crisis, 1935–1936', *English Historical Review*, 351 (1974), p. 328; R. Lamb, *The Drift to War 1922–1939* (London, 1989), p. 171.

40. A. R. Peters, *Anthony Eden at the Foreign Office 1931–1938* (Aldershot, 1986), p. 177.

41. Avon, *Facing the Dictators*, p. 340.

42. *DBFP*, 2nd series, vol. 16, no. 48.

43. Avon, *Facing the Dictators*, p. 341.

44. Cabinet 9 March 1936, CC(36)16, CAB 23/83.

45. Compare Eden's statement in the Commons later that day: H of C Debs, 5th series, vol. 309, cols 1808–13.

46. R. A. C. Parker, *Chamberlain and Appeasement* (London, 1993), p. 70.

47. *DBFP*, 2nd series, vol. 16, no. 37.

48. *Documents Diplomatiques Français* 1932–1939, 2e série, tome I, no. 316.

49. Vansittart to Eden 7 March 1936, AP 14/1/628.

50. A. Adamthwaite, *France and the Coming of the Second World War* (London, 1977), pp. 39–40.

51. Avon, *Facing the Dictators*, pp. 350, 354.

52. Cabinet 11 March 1936, CC(36)18, CAB 23/83.

53. Nicolson, ed., *Nicolson Diaries and Letters*, p. 252; J. T. Emmerson, *The Rhineland Crisis, 7 March 1936* (London, 1977), pp. 177–200.

54. Cabinet 16 March 1936, CC(36)20, CAB 23/83.

55. Vansittart to Eden 22 March 1936, AP 14/1/629.

56. *DBFP*, 2nd series, vol. 16, no. 122; see also Eden's summary of reports from the Air Staff and General Staff on the strategic importance of the demilitarized zone. Its disappearance was 'likely to lead to far-reaching political repercussions of a kind which will further weaken France's influence in Eastern and Central Europe, leaving a gap which may be eventually filled either by Germany or by Russia'. C. Barnett, *The Collapse of British Power* (London, 1972), p. 408.

57. K. Feiling, *The Life of Neville Chamberlain* (London, 1946), p. 280.

58. H of C Debs, 5th series, vol. 310, col. 1446.

59. Avon, *Facing the Dictators*, p. 346.

60. H. Macmillan, *Winds of Change* (London, 1966), p. 463.

61. E. Evans to Eden 30 July 1963, AP 23/31/12.

62. W. S. Churchill, *The Gathering Storm* (London, 1948), pp. 170–1.

63. P. M. H. Bell, *The Origins of the Second World War in Europe* (London, 1986), p. 211.

64. N. Chamberlain diary 12 March 1936, NC 2/23A.

65. Pierson Dixon diary 25 May 1957, cited in P. Dixon, 'Eden after Suez', *Contemporary Record*, 6 (1992), p. 179.

66. Avon, *Facing the Dictators*, p. 366. Self-criticism did not come easily to Eden. On another occasion he insisted to Churchill that 'if the French were not in fact prepared to march into the Rhineland, and I am convinced that they were not, then I do believe that we salved all that we possibly could out of this very difficult situation.' Note on Churchill's draft memoirs, sent to Churchill early 1948, AP 19/4/10A.

67. Emmerson, *Rhineland Crisis*, p. 241.

68. G. Peden, *British Rearmament and the Treasury 1932–1939* (Edinburgh, 1979), p. 83.

69. Diary 30 April 1936, AP 20/1/16.

70. Lamb, *Drift to War*, p. 181.

71. Eden to Lord Tweedsmuir 9 April 1936, AP 14/1/626A.

72. *DBFP*, 2nd series, vol. 16, no. 121.

73. Parker, 'Britain, France and Ethiopia', p. 325.

74. Cabinet 5 March 1936, CC(36)15, CAB 23/83.

75. M. Toscano, 'Eden's Mission to Rome on the Eve of the Italo-Ethiopian Conflict', in A. O. Sarkissian, ed., *Studies in Diplomatic History and Historiography in honour of G. P. Gooch* (London, 1961), pp. 126–52.

76. Harvey, ed., *Diplomatic Diaries*, p. 65 (19–23 Dec. 1937).

77. *The Times*, 18 Jan. 1936.

78. Cabinet 26 Feb. 1936, CC(36)11, CAB 23/83.

79. *DBFP*, 2nd series, vol. 15, no. 442.

80. *DBFP*, 2nd series, vol. 16, no. 16.

81. *DBFP*, 2nd series, vol. 16, no. 246.

82. Cabinet 22 April 1936, CC(36)30, CAB 23/84.

83. *DBFP*, 2nd series, vol. 16, no. 343.

84. Bond, ed., *Pownall Diaries 1933–1940*, p. 110.
85. Cabinet 22 April 1936, CC(36)30, CAB 23/84.
86. Lumley to Eden 23 April 1936, AP 14/1/598.
87. Diary 27 April 1936, AP 20/1/16; Chamberlain diary 27 April 1936, NC 2/23A.
88. Cabinet 29 April 1936, CC(36)31, CAB 23/84.
89. FP(36)1, CAB 27/622; diary 30 April 1936, AP 20/1/16.
90. T. Jones, *A Diary with Letters* (Oxford, 1954), p. 193.
91. Diary 20 May 1936, AP 20/1/16.
92. *DBFP*, 2nd series, vol. 16, no. 272.
93. *DBFP*, 2nd series, vol. 16, no. 339.
94. Diary 5 June 1936, AP 20/1/16.
95. H of C Debs, 5th series, vol. 311, cols 1734–7.
96. Diary 27 May 1936, AP 20/1/16.
97. *DBFP*, 2nd series, vol. 16, no. 347.
98. Cabinet 29 May 1936, CC(36)40, CAB 23/84.
99. Cabinet 10 June 1936, CC(36)41, CAB 23/84.
100. Chamberlain to H. Chamberlain 14 June 1936, NC 18/1/965.
101. Parker, 'Britain, France and Ethiopia', p. 331.
102. Neville Chamberlain diary 17 June 1936, NC 2/23A; Vansittart to Eden 15 June 1936, AP 14/1/630.
103. Cranborne to Eden 16 June 1936, AP 14/1/574.
104. Cabinet 17 June 1936, CC(36)42, CAB 23/84; *DBFP*, 2nd series, vol. 16, no. 360.
105. Cabinet 6 July 1936, CC(36)50, CAB 23/85. Emphasis added.
106. D. McLachlan, *In the Chair: Barrington Ward of The Times 1927–1948* (London, 1971), p. 125.
107. L. R. Pratt, *East of Malta, West of Suez* (Cambridge, 1975), p. 41; *DBFP*, 2nd series, vol. 17, no. 157; cf. R. R. James, *Anthony Eden* (London, 1986), p. 168.
108. J. Connell, *The 'Office': A Study of British Foreign Policy and its Makers 1919–1951* (London, 1958), p. 245.
109. Adamthwaithe, *France and Second World War*, p. 43; D. Carlton, 'Eden, Blum and the Origins of Non-Intervention', *Journal of Contemporary History*, 6 (1971), pp. 40–55. Yet Blum's biographers continue to imply that, if Britain did not actually force France to adopt non-intervention, she created a climate in which the French had little choice: J. Lacouture, *Léon Blum* (Paris, 1977), pp. 355–6.
110. D. Carlton, *Anthony Eden* (London, 1981), p. 92; J. Edwards, *The British Government and the Spanish Civil War, 1936–1939* (London, 1979), p. 29.
111. Edwards, *Spanish Civil War*, p. 61.
112. Cabinet 29 July 1936, CC(36)55, CAB 23/85.
113. Cranborne to Eden 20 Oct. 1936, AP 14/1/578.
114. Peters, *Eden at Foreign Office*, p. 228.
115. CP(36)233, 31 Aug. 1936, CAB 24/264.
116. *DBFP*, 2nd series, vol. 17, nos 356 and 291.
117. Cabinet 4 Nov. 1936, CC(36)63, CAB 23/86.
118. W. J. Mommsen and L. Kettenacker, eds, *The Fascist Challenge and the Policy of Appeasement* (London, 1983), p. 273; *DBFP*, 2nd series, vol. 17, no. 352.
119. J. Barnes and D. Nicholson, eds, *The Empire at Bay: The Leo Amery Diaries 1929–1945* (London, 1988), p. 429.
120. Nicolson, ed., *Nicolson Diaries and Letters*, p. 279.
121. Minute by Eden 16 Nov. 1936, *DBFP*, 2nd series, vol. 17, no. 350; Eden to Baldwin 27 Dec. 1936, Baldwin MSS vol. 124.

122. Peters, *Eden at Foreign Office*, p. 225.
123. Cadogan unpublished diary 24 Sept. 1936, AP 23/15/19; Peters, *Eden at Foreign Office*, p. 223; D. Morton to W. Churchill 17 Oct. 1936, cited M. Gilbert, *Winston S. Churchill*, vol. 5 (London, 1976), p. 790.
124. See, for example, cabinet 4 Nov. 1936, CC(36)62, CAB 23/86.
125. Cabinet 16 Nov. 1936, CC(36)66, CAB 23/86; Mommsen and Kettenacker, eds, *Fascist Challenge*, p. 368; P. Haggie, *Britannia at Bay: The Defence of the British Empire against Japan 1931–1941* (Oxford, 1981), p. 105.
126. Diary 16 Nov. 1936, AP 20/1/16.
127. *DBFP*, 2nd series, vol. 17, no. 440.
128. *DBFP*, 2nd series, vol. 17, no. 527; Mommsen and Kettenacker, eds, *Fascist Challenge*, p. 275.
129. Diary 5 Jan 1937, AP 20/1/17.
130. CP(37)6, 8 Jan. 1937, CAB 24/267; meeting of ministers 8 Jan. 1937, CAB 23/87; *DBFP*, 2nd series, vol. 18, nos 32–3; Peters, *Eden at Foreign Office*, p. 246.
131. Diary 9 Jan. 1937, AP 20/1/17.
132. Cabinet 13 Jan. 1937, CC(37)1, CAB 23/87.
133. Diary 13 Jan. 1937, AP 20/1/17.
134. Cadogan unpublished diary 10 Jan. 1937, AP 23/15/19.
135. H of C Debs, 5th series, vol. 319, cols 93–108.
136. Eden to Leith-Ross 19 Jan. 1937, cited A. Crozier, *Appeasement and Germany's Last Bid for Colonies* (London, 1988), p. 196.
137. Cabinet Committee on Foreign Policy 18 March 1937, CAB 27/622; see also Eden to Lord Cecil of Chelwood 5 Feb. 1937, AP 33/4.
138. Memorandum by Eden 29 April 1937, FO 954/1.
139. *DBFP*, 2nd series, vol. 18, nos 307 and 327; Peters, *Eden at Foreign Office*, p. 242.
140. Cabinet 3 Feb. 1937, CC(37)5, CAB 23/87.
141. *The Times*, 21 Nov. 1936.
142. *The Times*, 15 Dec. 1936.
143. Diary 6 March 1937, AP 20/1/17.
144. D. Reynolds, *Britannia Overruled: British Policy and World Power in the Twentieth Century* (Harlow, 1991), p. 119.
145. 'We had no cause to crave negotiation with Facist Italy, our navy and air force were far more powerful than hers': Avon, *Facing the Dictators*, p. 425.

CHAPTER 4

1. D. Dilks, ed., *The Diaries of Sir Alexander Cadogan 1938–1945* (London, 1971), p. 55 (21 Feb. 1938).
2. Diary 20 May 1936, AP 20/1/16.
3. Diary 24 Aug. 1936.
4. Lord Avon, *Facing the Dictators* (London, 1962), p. 445; but see also note of interview with Lord Harvey 2 Nov. 1954, AP 7/11/25B: 'Baldwin gave AE no support whatever.'
5. J. Harvey, ed., *The Diplomatic Diaries of Oliver Harvey 1937–1940* (London, 1970), p. 34 (26 March 1937).
6. Avon, *Facing the Dictators*, p. 445.
7. Note of interview with Lord Harvey 2 Nov. 1954.
8. Hoare to Chamberlain 17 March 1937, NC 7/11/30/74.
9. Chamberlain to Ida Chamberlain 4 July 1937, NC 18/1/1010.

10. Diary 5 Jan. 1937, AP 20/1/17.
11. M. L. Roi, '"A Completely Immoral and Cowardly Attitude": The British Foreign Office, American Neutrality and the Hoare–Laval Plan', *Canadian Journal of History*, 29 (1994), pp. 333–51.
12. Eden to R. Lindsay 18 Jan. 1937, cited S. Roskill, *Naval Policy Between the Wars: The Period of Reluctant Rearmament 1930–1939* (London, 1976), p. 360.
13. A. Campbell-Johnson, *Sir Anthony Eden: A Biography* (London, 1955), p. 137.
14. Lindsay to Eden 22 March 1937, cited M. Murfett, *Fool-Proof Relations* (Singapore, 1984), p. 22.
15. Murfett, *Fool-Proof Relations*, p. 27.
16. C. MacDonald, *The United States, Britain and Appeasement 1936–1939* (London, 1981), p. 24.
17. MacDonald, *United States, Britain and Appeasement*, p. 30.
18. Avon, *Facing the Dictators*, p. 531; Eden to R. Lindsay 21 July 1937, FO 954/29.
19. Minute 30 July 1937, cited P. Lowe, *Great Britain and the Origins of the Pacific War* (Oxford, 1977), p. 19.
20. Cabinet 29 Sept. 1937, CC(37)35, CAB 23/89; A. Peters, *Anthony Eden at the Foreign Office 1931–1938* (Aldershot, 1986), p. 323.
21. N. Chamberlain to H. Chamberlain 28 Aug. 1937, NC 18/1/1018.
22. Murfett, *Fool-Proof Relations*, p. 63.
23. Cabinet 6 Oct. 1937, CC(37)36, CAB 23/89.
24. J. Charmley, *Chamberlain and the Lost Peace* (London, 1989), p. 37.
25. N. Chamberlain to H. Chamberlain 9 Oct. 1937, NC 18/1/1023.
26. R. Dallek, *Franklin D. Roosevelt and American Foreign Policy 1932–1945* (New York, 1979), p. 150.
27. Harvey, ed., *Diplomatic Diaries*, p. 55 (2 Nov. 1937); MacDonald, *United States, Britain and Appeasement*, p. 44; Murfett, *Fool-Proof Relations*, pp. 73, 78; C. Hull, *Memoirs*, vol. 1 (London, 1948), p. 553.
28. Eden to Lindsay 10 Nov. 1937, cited Murfett, *Fool-Proof Relations*, p. 82.
29. Minute of 26 Aug. 1937, cited L. R. Pratt, *East of Malta, West of Suez* (Cambridge, 1975), p. 70.
30. *Documents on British Foreign Policy*, 2nd series, vol. 18, no. 365.
31. Defence Plans (Policy) Sub-Committee meeting 19 April 1937, CAB 16/181.
32. Minute of 24 April 1937, cited Peters, *Eden at Foreign Office*, p. 240.
33. Minute of 15 April 1937, cited D. Dilks, '"We must hope for the best and prepare for the worst": the Prime Minister, the Cabinet and Hitler's Germany, 1937–1939', *Proceedings of the British Academy*, LXXIII (1987), p. 315.
34. *DBFP*, 2nd series, vol. 18, no. 566; see also cabinet meeting 2 June 1937, CAB 23/88.
35. W. J. Mommsen and L. Kettenacker, *The Fascist Challenge and the Policy of Appeasement* (London, 1983), p. 164.
36. C. Barnett, *The Collapse of British Power* (London, 1972), pp. 465–6.
37. Hertzog to Smuts 24 May 1937, cited J. van der Poel, *Selections From the Smuts Papers*, vol. 6 (Cambridge, 1973), p. 83.
38. Cabinet 14 July 1937, CC(37)30, CAB 23/89.
39. Eden to Chamberlain 16 July 1937, PREM 1/276; Avon, *Facing the Dictators*, p. 452.
40. Chamberlain diary 19–27 Feb. 1938, NC 2/24A.
41. Chamberlain to H. Chamberlain 1 Aug. 1937, NC 18/1/1014.
42. Chamberlain to H. Chamberlain, 24 Oct. 1937, NC 18/1/1025.
43. Eden to Halifax 1 Aug. 1937, Halifax MSS A4.410.21.1.
44. Eden to Vansittart 4 Aug. 1937, AP 13/1/58D.

45. Avon, *Facing the Dictators*, p. 455.
46. Eden to Halifax 11 Aug. 1937, AP 20/53/88.
47. Halifax to Vansittart 15 Aug. 1937, FO 954/13.
48. Halifax to Chamberlain 15 Aug. 1937, NC 7/11/30/64.
49. Halifax to Chamberlain, 19 Aug. 1937, PREM 1/276.
50. Chamberlain to Lord Weir 15 Aug. 1937, NC 7/11/30/141.
51. Cabinet 8 Sept. 1937, CC(37)34, CAB 23/89.
52. Eden to Chamberlain 9 Sept. 1937 and note by Chamberlain, PREM 1/210.
53. Chamberlain to H. Chamberlain 12 Sept. 1937, NC 18/1/1020.
54. Harvey, ed., *Diplomatic Diaries*, p. 48 (22 Sept. 1937).
55. Eden to Chamberlain 14 Sept. 1937, FO 954/2; Eden to Churchill 14 Sept. 1937, cited M. Gilbert, *Winston S. Churchill*, vol. 5 (London, 1976), p. 869.
56. Note by Eden 14 Sept. 1937, AP 13/1/58M.
57. W. C. Mills, 'The Nyon Conference: Neville Chamberlain, Anthony Eden and the Appeasement of Italy in 1937', *International History Review*, 15 (1993), pp. 2–3.
58. Cabinet 29 Sept. 1937, CC(37)35, CAB 23/89; Pratt, *East of Malta*, p. 89.
59. Chamberlain diary 18 Feb. 1938, NC 2/24A.
60. Cabinet 29 Sept. 1937, CC(37)35, CAB 23/89.
61. See revealing entry in Harvey, ed., *Diplomatic Diaries*, p. 57 (7 Nov. 1937).
62. Harvey, ed., *Diplomatic Diaries*, p. 50 (15 Oct. 1937).
63. Chamberlain to H. Chamberlain 6 Nov. 1937, NC 18/1/1027.
64. Harvey, ed., *Diplomatic Diaries*, p. 56 (3 Nov. 1937).
65. Eden to Chamberlain 3 Nov. 1937, PREM 1/210; see also discussion at the cabinet on 7 Nov. 1936, CC(36)64, CAB 23/86, when Hoare drew the conclusion that the state of British military preparedness meant that her foreign policy would 'have to proceed very quietly'.
66. F. R. Gannon, *The British Press and Germany 1936–1939* (Oxford, 1971), p. 132.
67. Diary 8 Nov. 1937, AP 20/1/17.
68. D. Dutton, *Simon: A Political Biography of Sir John Simon* (London, 1992), pp. 263–4.
69. Avon, *Facing the Dictators*, p. 509.
70. Note by Halifax 6 May 1946, Halifax MSS A4.410.3.3; see also Lord Halifax, *Fulness of Days* (London, 1957), p. 184; Lord Templewood, *Nine Troubled Years* (London, 1954), pp. 281–2; N. Chamberlain to H. Chamberlain 24 Oct. 1937, NC 18/1/1025.
71. Avon, *Facing the Dictators*, p. 512; R. Cockett, *Twilight of Truth: Chamberlain, Appeasement and the Manipulation of the Press* (London, 1989), pp. 33–42.
72. Harvey, ed., *Diplomatic Diaries*, p. 57 (7 Nov. 1937).
73. Halifax MSS A4.410.3.3; Cabinet 24 Nov. 1937, CC(37)43, CAB 23/90A.
74. Cabinet 24 Nov. 1937, CC(37)43, CAB 23/90A.
75. N. H. Hooker, ed., *The Moffat Papers* (Cambridge, Mass., 1956), p. 182.
76. *Documents Diplomatiques Français*, série 2, tome 7, p. 576.
77. *DBFP*, 2nd series, vol. 19, nos 311 and 348.
78. CID meeting 2 Dec. 1937, cited Murfett, *Fool-Proof Relations*, pp. 99–100.
79. Chamberlain to Ida Chamberlain 26 Nov. 1937, NC 18/1/1030.
80. B. H. Liddell Hart, *The Memoirs of Captain Liddell Hart*, vol. 2 (London, 1965), p. 137.
81. N. H. Gibbs, *Grand Strategy: Rearmament Policy* (London, 1976), p. 305.
82. 'Defence Expenditure in Future Years', CP(37)316.
83. Cabinet 22 Dec. 1937, CC(37)49, CAB 23/90A. On reflection, but not it would appear at the meeting itself, the Foreign Secretary argued that Britain should not be content with less than parity in the air and that bombers should play an important

part as 'it is not good military doctrine to design a force for purely defensive action alone'. Eden to Swinton 31 Dec. 1937, AP 24/31/14E.

84. Eden to Chamberlain 31 Dec. 1937, PREM 1/314.

85. Harvey, ed., *Diplomatic Diaries*, p. 65 (19–23 Dec. 1937).

86. L. Pratt, 'The Anglo-American Naval Conversations on the Far East of January 1938', *International Affairs*, 47 (1971), p. 753.

87. P. Haggie, *Britannia at Bay: The Defence of the British Empire against Japan 1931–1941* (Oxford, 1981), pp. 118–19.

88. Lowe, *Pacific War*, p. 34.

89. Murfett, *Fool-Proof Relations*, p. 125.

90. See diary 5 Jan. 1938, AP 20/1/18, after lunch with Lloyd George and Churchill in the South of France: 'I told them something of our efforts to ensure co-operation with the U.S. Both seemed impressed with the progress that had been made.'

91. Chamberlain diary 19–27 Feb. 1938, NC 2/24A.

92. Eden to Chamberlain 1 Jan. 1938, PREM 1/276; Harvey, ed., *Diplomatic Diaries*, p. 67 (1–13 Jan. 1938).

93. Chamberlain to Eden 7 Jan. 1938, AP 20/6/4.

94. Chamberlain to Eden 13 Jan. 1938, AP 20/6/7.

95. I. Colvin, *The Chamberlain Cabinet* (London, 1971), p. 82; compare Eden's later assertion that rumours of closer relations between Germany and Italy from the end of 1937 tended to 'influence my own attitude towards conversations with either dictator': *Facing the Dictators*, p. 476.

96. Eden to Chamberlain 9 Jan. 1938, PREM 1/276; cf. Harvey, ed., *Diplomatic Diaries*, p. 65 (23 Dec. 1937): 'He said he regarded Muss as anti-Christ.'

97. Eden to Chamberlain 31 Jan. 1938, PREM 1/276.

98. Eden to Chamberlain 9 Jan. 1938, PREM 1/276.

99. Eden to Chamberlain 9 Jan. 1938, NC 7/11/31/100.

100. Phipps to Hankey 9 Jan. 1938, cited N. Rose, *Vansittart: Study of a Diplomat* (London, 1978), pp. 210–11.

101. Vansittart to Eden 14 Sept. 1936, AP 14/1/631.

102. W. Fisher to Chamberlain 15 Sept. 1936, NC 7/11/29/19; diary 21 Dec. 1936, AP 20/1/16.

103. Eden to Baldwin 5 Jan. 1937, AP 14/1/641B; Cadogan unpublished diary 10 Jan. 1937, AP 23/15/19.

104. Eden to Chamberlain 12 Dec. 1937, NC 7/11/30/46; Cadogan unpublished diary 13 Dec. 1937, AP 23/15/19.

105. Chamberlain to Ida Chamberlain 12 Dec. 1937, NC 18/1/1031.

106. Peters, *Eden at Foreign Office*, p. 307.

107. A. J. P. Taylor, ed., *Off the Record: W. P. Crozier Political Interviews 1933–1943*, (London, 1973), p. 120; see also B. Pimlott, ed., *The Political Diary of Hugh Dalton* (London, 1986), pp. 230–1; J. Bright-Holmes, ed., *Like it Was: The Diaries of Malcolm Muggeridge* (London, 1981), p. 256.

108. Eden to D. Dilks 24 June 1971, AP 33/2.

109. Note by Simon 20 Jan. 1938, AP 8/1/6.

110. Chamberlain to Roosevelt 13 Jan. 1938, cited D. Reynolds, *The Creation of the Anglo-American Alliance 1937–41* (London, 1981); DBFP, 2nd series, vol. 19, no. 430.

111. Chamberlain diary 19–27 Feb. 1938, NC 2/24A.

112. It was at Eden's prompting that Robert Blake included this phrase in his review of I. Macleod, *Neville Chamberlain* (London, 1961), AP 33/3.

113. Dilks to Eden 20 Aug. 1966, AP 33/2.

114. Salisbury to Eden 20 Aug. 1961, AP 23/60/70.

115. Harvey, ed., *Diplomatic Diaries*, p. 68 (14 Jan. 1938).

116. Eden to Chamberlain 17 Jan. 1938, AP 20/6/8.

117. Dilks, ed., *Cadogan Diaries*, p. 37 (15 Jan. 1938).

118. Harvey, ed., *Diplomatic Diaries*, p. 72 (17 Jan. 1938).

119. Harvey to Eden 18 Jan. 1938, AP 8/2/104.

120. Harvey, ed., *Diplomatic Diaries*, p. 69 (note for Eden).

121. Harvey, ed., *Diplomatic Diaries*, p. 70 (note for Eden).

122. Dilks, ed., *Cadogan Diaries*, pp. 39–40 (21 Jan. 1938).

123. Harvey, ed., *Diplomatic Diaries*, pp. 76, 78 (20 and 24 Jan. 1938).

124. Halifax to Eden 21 May 1943, Halifax MSS A 4.410.4.15; note by Halifax, June 1948, A 4.410.4.31; G. Bennett, 'The Roosevelt Peace Plan of January 1938', *Foreign and Commonwealth Office Occasional Papers*, 1 (1987), pp. 33–4.

125. Templewood, *Nine Troubled Years*, pp. 262–3.

126. J. P. L. Thomas to Eden 22 Jan. 1938, AP 14/1/822; R. R. James, *Victor Cazalet: A Portrait* (London, 1976), p. 197.

127. Eden to Chamberlain 31 Jan. 1938, PREM 1/276.

128. Notes of meeting of Foreign Policy Committee 19 Jan. 1938, by Mrs. Scott, Eden's literary secretary, cited S. Roskill, *Hankey: Man of Secrets*, vol. 3 (London, 1974), pp. 299–300.

129. Harvey, ed., *Diplomatic Diaries*, p. 81 (31 Jan. 1938); see also undated memorandum by Eden, AP 13/1/64C.

130. *DBFP*, 2nd series, vol. 19, no. 469.

131. Harvey, ed., *Diplomatic Diaries*, p. 85 (6 Feb. 1938).

132. Cabinet 9 Feb. 1938, CC(38)4, CAB 23/92.

133. D. Dilks, '"We must hope for the best"', p. 322; Barnett, *Collapse of British Power*, p. 471.

134. Harvey, ed., *Diplomatic Diaries*, p. 91 (16 Feb. 1938).

135. Cabinet 16 Feb. 1938, CC(38)5, CAB 23/92.

136. *DBFP*, 2nd series, vol. 19, no. 534. Compare the note written for Eden's research assistant as his memoirs were being prepared: '[Lord Avon] thinks that there ought to be something in resignation chapter to show that as a result of the threats to Austria he wanted to stiffen in his attitude towards Hitler and that he feared that his colleagues would not be willing to do this.' Note for David Dilks 30 Oct. 1961, AP 33/2.

137. Colvin, *Chamberlain Cabinet*, p. 98.

138. Dilks, ed., *Cadogan Diaries*, p. 44 (4 Feb. 1938).

139. Diary note 1946, AP 20/1/26.

140. Ivy Chamberlain to N. Chamberlain 16 Dec. 1937, NC 1/17/5.

141. I. Macleod to Eden 27 Sept. 1961, AP 23/47/2A.

142. Memorandum by Lord Cilcennin, AP 7/24/81.

143. Eden to Chamberlain 8 Feb. 1938 and Chamberlain to Eden 8 Feb. 1938, AP 13/1/64K, L.

144. Harvey, ed., *Diplomatic Diaries*, p. 83 (6 Feb. 1938); Dilks, ed., *Cadogan Diaries*, p. 45 (8 Feb. 1938).

145. Chamberlain to H. Chamberlain 13 Feb. 1938, NC 18/1/1039; cf. Dilks, ed., *Cadogan Diaries*, p. 47 (14 Feb. 1938).

146. A. Duff Cooper, *Old Men Forget* (London, 1953), p. 211.

147. Simon diary 12 Feb. 1938, Simon MSS 7 fo. 64.

148. Chamberlain to Eden 8 Feb. 1938, AP 13/1/64L.
149. Simon diary 12 Feb. 1938, Simon MSS 7 fo. 64.
150. Eden to Chamberlain 8 Feb. 1938, AP 13/1/64K; Avon, *Facing the Dictators*, p. 574.
151. Cabinet 16 Feb. 1938, CC(38)5, CAB 23/92.
152. Harvey, ed., *Diplomatic Diaries*, p. 92 (17 Feb. 1938); R. Boyce and E. M. Robertson, eds, *Paths to War: New Essays on the Origins of the Second World War* (London, 1989), pp. 180–1.
153. Eden to Chamberlain 17 Feb. 1938, AP 20/6/9.
154. Chamberlain diary 19–27 Feb. 1938, NC 2/24A.
155. 'Chamberlain and self with Grandi, February 1938', note in 1946 diary, AP 20/1/26.
156. M. Muggeridge, *Ciano's Diplomatic Papers* (London, 1948), p. 183.
157. Diary 18 Feb. 1938, AP 20/1/18.
158. Chamberlain to H. Chamberlain 27 Feb. 1938, NC 18/1/1040.
159. Roskill, *Hankey*, vol. 3, p. 303.
160. 'Record of events connected with Anthony Eden's resignation 19–20 Feb. 1938', Halifax MSS A4.410.4.11.
161. Simon diary 19 Feb. 1938, Simon MSS 7 fo. 67.
162. For the views of Leslie Hore-Belisha and Walter Elliot see R. J. Minney, *The Private Papers of Hore-Belisha* (London, 1960), p. 101, and N. Rose, *Baffy: The Diaries of Blanche Dugdale 1936–1947* (London, 1973), p. 84.
163. Harvey, ed., *Diplomatic Diaries*, p. 94 (19 Feb. 1938).
164. Halifax, 'Record of events'.
165. Halifax, 'Record of events'.
166. Halifax, 'Record of events'.
167. Simon diary 20 Feb. 1938, Simon MSS 7 fos. 68–9.
168. Roskill, *Hankey*, vol. 3, p. 303.
169. Halifax, 'Record of events'.
170. Harvey, ed., *Diplomatic Diaries*, pp. 96–7 (20–21 Feb. 1938); memorandum by Lord Cilcennin AP 7/24/81.
171. Halifax, 'Record of events'.
172. Halifax diary 29 April 1941, Halifax MSS 7.8.8; D. Carlton, *Anthony Eden* (London, 1981), pp. 129–30; Harvey, ed., *Diplomatic Diaries*, p. 96 (20 Feb. 1938).
173. Note exchanged at cabinet 20 Feb. 1938, AP 13/1/64.
174. Chamberlain diary 19–27 Feb. 1938, NC 2/24A.
175. Caccia to Eden 20 Sept. 1961, AP 23/14/7.
176. House of Commons Debates, 5th series, vol. 322, col. 242.
177. R. A. C. Parker, *Chamberlain and Appeasement* (London, 1993), p. 95.
178. Harvey, ed., *Diplomatic Diaries*, p. 57 (7 Nov. 1937); Peters, *Eden at Foreign Office*, p. 299.
179. N. Thompson, *The Anti-Appeasers* (Oxford, 1971), p. 58.
180. Avon, *Facing the Dictators*, p. 508.
181. James, *Cazalet*, p. 197; cf. Pratt, *East of Malta*, p. 86: 'Reduced to its essentials, Eden's dispute with Italy was not: "Shall fascism or democracy prevail?" Rather it was "Who shall rule the Mediterranean?"'
182. But note Halifax's view that Eden greatly exaggerated the extent to which Chamberlain intervened in the conduct of diplomacy. Halifax to Strang 7 Jan. 1957, Halifax MSS A 2.27.8.107.
183. Charmley, *Chamberlain*, p. 29.

Notes

184. Chamberlain to Annie Chamberlain 3 March 1938, cited Dilks, ed., *Cadogan Diaries*, p. 52.
185. Roskill, *Hankey*, vol. 3, p. 303.
186. Cabinet 16 Feb. 1938, CC(38)5, CAB 23/92.
187. W. S. Churchill, *Step by Step 1936–1939* (London, 1939), pp. 322–5.
188. T. Jones, *A Diary with Letters* (Oxford, 1954), p. 369.
189. These potential friends and allies did not at this stage include the Soviet Union: Hooker, ed., *Moffat Papers*, p. 174.
190. Eden to Chamberlain 31 Jan. 1938, PREM 1/276; see also Harvey, ed., *Diplomatic Diary*, p. 89 (14 Feb. 1938).
191. Eden to Lindsay 25 Jan. 1938, cited Murfett, *Fool-Proof Relations*, pp. 148–9.
192. Chatfield to Inskip 25 Jan. 1938, cited Pratt, 'Anglo-American Naval Conversations', p. 758.
193. Bennett, 'Roosevelt Peace Plan', p. 31; F. S. Northedge concluded that if Chamberlain was credulous in relation to the dictators, 'so was Eden in relation to Roosevelt's power to render effective assistance' (*The Troubled Giant: Britain Among the Great Powers 1916–1939* (London, 1966), p. 485). This remark pained Eden and led him to contemplate legal action. Eden to Strang 12 Oct. 1966, AP 24/66/6. But it does contain a grain of truth.
194. Chamberlain to F. Morton Price 16 Jan. 1938, cited Dilks, '"We must hope for the best"', p. 322.
195. Dilks to Eden 23 Jan. 1967, AP 33/2.

CHAPTER 5

1. C. Coote, *Editorial* (London, 1965), p. 184.
2. R. Blake, in J. Mackintosh, ed., *British Prime Ministers in the Twentieth Century: Churchill to Callaghan* (London, 1978), p. 77.
3. S. Roskill, *Hankey: Man of Secrets* (3 vols, London, 1970–4), vol. 3, p. 305.
4. N. Nicolson, ed., *Harold Nicolson: Diaries and Letters 1939–1945* (London, 1967), p. 75.
5. N. Fisher, *Harold Macmillan* (London, 1982), p. 172.
6. N. Chamberlain to Ida Chamberlain 23 April 1939, Chamberlain MSS, NC 18/1/1095.
7. H. Ickes, *The Secret Diary of Harold L. Ickes: The Inside Struggle 1936–1939* (London, 1955), pp. 370–1.
8. R. R. James, *Anthony Eden* (London, 1986), p. 222.
9. W. Churchill, *The Gathering Storm* (London, 1948), p. 191.
10. Note for biographer 9 May 1968, AP 33/5.
11. 'Account of a meeting of the Foreign Affairs Committee of the Conservative Party on February 17th 1938', AP 33/9.
12. J. Connell, *The 'Office': A Study of British Foreign Policy and its Makers 1919–1951* (London, 1958), p. 269.
13. A. Adamthwaite, *France and the Coming of the Second World War* (London, 1977), p. 83.
14. J. Harvey, ed., *The Diplomatic Diaries of Oliver Harvey 1937–1940* (London, 1970), p. 99 (22 Feb. 1938). See also N. Rose, 'The Resignation of Anthony Eden', *Historical Journal*, 25,4 (1982), p. 930.
15. Harvey, ed., *Diplomatic Diaries*, p. 104 (27 Feb. 1938).
16. *Spectator*, 25 Feb. 1938.

17. N. Rose, ed., *Baffy: The Diaries of Blanche Dugdale 1936–1947* (London, 1973), p. 84.
18. Eden to Harvey 18 June 1965, AP 23/36/49; C. Coote, *Companion of Honour* (London, 1965), p. 158.
19. R. R. James, *Victor Cazalet: A Portrait* (London, 1976), pp. 208–9.
20. J. Harvey, ed., *The War Diaries of Oliver Harvey* (London, 1978), p. 143 (24 July 1942).
21. Rose, ed., *Baffy*, pp. 85–6.
22. S. Olson, ed., *Harold Nicolson: Diaries and Letters 1930–1964* (London, 1980), p. 120. This edition contains several passages not published in Nigel Nicolson's three-volume edition. Several of the earlier omissions followed objections by Eden.
23. N. Nicolson, ed., *Harold Nicolson: Diaries and Letters 1930–39* (London, 1966), p. 324.
24. Earl Winterton, *Orders of the Day* (London, 1953), p. 230.
25. M. Gilbert, *Winston S. Churchill*, vol. 5 (London, 1976), p. 904.
26. House of Commons Debates, 5th series, vol. 332, cols 45–50.
27. Harvey, ed., *Diplomatic Diaries*, p. 97 (21 Feb. 1938).
28. J. Barnes and D. Nicholson, eds, *The Empire at Bay: The Leo Amery Diaries 1929–1945* (London, 1988), p. 456.
29. Nicolson, ed., *Nicolson Diaries 1930–39*, pp. 324–5; H. Macmillan, *Winds of Change* (London, 1966), pp. 536–7.
30. Connell, *Office*, p. 273.
31. A. Duff Cooper, *Old Men Forget* (London, 1953), p. 214. 'The debate', noted one minister, 'did a good deal to put things straight.' Winterton, *Orders*, p. 231.
32. R. R. James, ed., *Chips: The Diaries of Sir Henry Channon* (London, 1967), pp. 146–7.
33. Nicolson, ed., *Nicolson Diaries 1930–39*, p. 327.
34. Hankey to Robin Hankey 1 March 1938, Hankey MSS HNKY 3/43; K. Feiling, *The Life of Neville Chamberlain* (London, 1946), p. 339.
35. Eden to Halifax 5 March 1938, Halifax MSS A 4.410.11.2.
36. J. Vincent, ed., *The Crawford Papers* (Manchester, 1984), p. 587.
37. Harvey, ed., *Diplomatic Diaries*, p. 102 (26 Feb. 1938).
38. D. Birn, *The League of Nations Union 1918–1945* (Oxford, 1981), pp. 188–91.
39. Eden to P. Emrys-Evans 13 Aug. 1965, AP 23/31/15.
40. Ball to N. Chamberlain 21 Feb. 1938, NC 7/11/31/10. The *Manchester Guardian* (24 Feb. 1938) noted the 'curiously distorted' view offered by the Conservative press.
41. Emrys-Evans to Eden 27 June 1961, AP 23/31/46.
42. Diary 27 Feb. 1938, Simon MSS 7 fo. 70.
43. Earl of Avon, *The Reckoning* (London, 1965), pp. 3, 11.
44. Avon, *Reckoning*, p. 7.
45. D. Carlton, *Anthony Eden* (London, 1981), p. 133.
46. J. P. L. Thomas to Emrys-Evans March 1938, AP 23/31/4A.
47. H of C Debs, 5th series, vol. 332, cols 235–47.
48. N. Chamberlain to H. Chamberlain 13 March 1938, NC 18/1/1041.
49. M. Patrick to Eden 15 March 1938, AP 14/1/795.
50. Evans to J. P. L. Thomas 19 March 1938, AP 14/1/739.
51. A. Mann to Eden 17 March 1938, AP 14/1/770B.
52. James, ed., *Chips*, pp. 151–2.
53. Patrick to Eden 25 March 1938, AP 14/1/797.
54. R. Tree, *When the Moon was High* (London, 1975), p. 74.
55. P. Harris, *Forty Years in and out of Parliament* (London, n.d.), p. 137.
56. James, *Eden*, p. 203.
57. D. Margesson to Chamberlain 17 March 1938, NC 7/11/31/188.

58. Rose, ed., *Baffy*, pp. 90–1.
59. J. Margach, *The Abuse of Power: The War between Downing Street and the Media from Lloyd George to James Callaghan* (London, 1978), p. 101.
60. Butler to Lord Brabourne 23 Feb. 1938, Brabourne MSS Eur. F97/22B.
61. A. Mann to Eden 4 March 1938, AP 14/1/770A.
62. Nicolson, ed., *Nicolson Diaries 1930–39*, p. 328.
63. R. Boothby, *I Fight to Live* (London, 1947), p. 148.
64. James, ed., *Chips*, p. 147.
65. Harvey, ed., *Diplomatic Diaries*, p. 128 (22 April 1938); Avon, *Reckoning*, pp. 16–17.
66. Halifax in conversation with Thomas Jones about Eden's future, April 1938, E. Ellis, *T. J.: A Life of Dr Thomas Jones* (Cardiff, 1992), pp. 425–6.
67. Nicolson, ed., *Nicolson Diaries 1930–39*, p. 334.
68. Thomas to Eden 9 April 1938, AP 14/1/823.
69. Harvey, ed., *Diplomatic Diaries*, p. 131 (22 April 1938).
70. Eden to Duchess of Atholl 23 April 1938, AP 14/1/678B.
71. Harvey, ed., *Diplomatic Diaries*, p. 130 (22 April 1938).
72. James, *Eden*, p. 207.
73. Harvey, ed., *Diplomatic Diaries*, p. 129 (22 April 1938).
74. Eden to R. Beckett 17 April 1938, AP 20/6/1.
75. Tree to Eden 12 July 1938, AP 14/1/834.
76. Ellis, *T. J.*, pp. 425–6.
77. Harvey, ed., *Diplomatic Diaries*, p. 147 (1 June 1938).
78. Clarke to J. Ball 27 Jan. 1939, cited R. A. C. Parker, *Chamberlain and Appeasement: British Policy and the Coming of the Second World War* (London, 1993), p. 324.
79. Sandys to Eden 28 April 1938, AP 14/1/803.
80. Barnes and Nicholson, eds, *Empire at Bay*, p. 506.
81. Nicolson diary 15 Nov. 1938 (unpublished); James, *Cazalet*, p. 208.
82. Eden to A. Mann 27 May 1938, AP 14/1/772A.
83. Cranborne to Eden 13 April 1938, AP 14/1/710.
84. Eden to M. Patrick 14 April 1938, AP 14/1/800; to P. Emrys-Evans 28 April 1938, AP 23/31/4A; to Churchill 24 April 1938, AP 20/6/13A.
85. Avon, *The Reckoning*, p. 8.
86. A. Roberts, *The Holy Fox: A Biography of Lord Halifax* (London, 1991), p. 105.
87. Harvey, ed., *Diplomatic Diaries*, p. 129 (22 April 1938).
88. Harvey, ed., *Diplomatic Diaries*, p. 141 (20 May 1938).
89. Rose, ed., *Baffy*, pp. 90–1.
90. Harvey, ed., *Diplomatic Diaries*, p. 153 (14 June 1938).
91. Harvey, ed., *Diplomatic Diaries*, p. 158 (2 July 1938).
92. Eden to Cranborne 8 June 1938, AP 14/1/714.
93. Eden to Countess of Warwick 2 July 1938, AP 14/1/840.
94. Rose, ed., *Baffy*, p. 93; cf. Simon diary 3 Aug. 1938, Simon MSS 7 fo. 75.
95. Olson, ed., *Nicolson*, p. 132.
96. Baldwin to Eden 6 Sept. 1938, AP 14/1/685; Harvey, ed., *Diplomatic Diaries*, p. 172 (8 Sept. 1938).
97. Harvey, ed., *Diplomatic Diaries*, p. 172 (9 Sept. 1938). Yet when Henderson showed himself reluctant to deliver just such a warning and the government withdrew its instruction to do so, Halifax noted that Eden, who visited him again on 11 September, 'thought we could not have done otherwise'. Note by Halifax, FO 800/314.
98. Cranborne to Eden 9 Sept. 1938, AP 14/1/718.

99. Lord Davies to Eden 9 Sept. 1938, AP 14/1/725.

100. B. Liddell Hart, *Memoirs* (2 vols, London 1965), vol. 2, pp. 162–3.

101. *The Times*, 12 Sept. 1938.

102. Eden to Emrys-Evans 16 Sept. 1938, AP 23/31/4A.

103. Harvey, ed., *Diplomatic Diaries*, p. 186 (18 Sept. 1938); Nicolson, ed., *Nicolson Diaries 1930–39*, pp. 360–1.

104. C. Coote to Eden 20 Sept. 1938, AP 13/1/66C.

105. Rose, ed., *Baffy*, p. 99; Liddell Hart, *Memoirs*, vol. 2, p. 165.

106. Harvey, ed., *Diplomatic Diaries*, p. 196 (25 Sept. 1938).

107. Eden to Chamberlain 28 Sept. 1938, NC 13/11/655. Eden later claimed that he had not written to Chamberlain at this time and that the message of support was probably from the journalist, Guy Eden: note for biographer 20 June 1975, AP 33/4. The letter in the Chamberlain papers is, however, in Eden's own hand.

108. Harvey, ed., *Diplomatic Diaries*, p. 202 (30 Sept. 1938).

109. Nicolson, ed., *Nicolson Diaries 1930–39*, p. 372; E. Spier, *Focus: A Footnote to the History of the Thirties* (London, 1963), p. 11; Rose, ed., *Baffy*, p. 108.

110. Eden to Spier 6 Dec. 1961, AP 19/2/44B.

111. James, *Eden*, p. 211.

112. *Daily Telegraph*, 12 March 1965.

113. M. Gilbert, *Winston S. Churchill*, vol. 5, companion part 3 (London, 1982), p. 1189.

114. Rose, ed., *Baffy*, pp. 109–10.

115. Anthony Crossley diary 30 Sept. 1938, cited Gilbert, *Churchill*, vol. 5, 3, p. 1170.

116. H of C Debs, 5th series, vol. 339, cols 77–88; Winterton, *Orders*, p. 244.

117. Eden to S. Flower 4 Oct. 1938, AP 13/1/66J.

118. Barnes and Nicholson, eds, *Empire at Bay*, p. 525; Nicolson, ed., *Nicolson Diaries 1930–39*, p. 375; Harvey, ed., *Diplomatic Diaries*, p. 212 (11 Oct. 1938).

119. Amery to Eden 10 Oct. 1938, AP 14/1/671. Though it was Amery who told Chamberlain about this, Eden readily confirmed his version of events.

120. Barnes and Nicholson, eds, *Empire at Bay*, pp. 527–8.

121. Eden to Baldwin 30 Sept. 1938, Baldwin MSS vol. 124, fos 136–7. Interestingly Eden objected to Maurice Cowling quoting from this letter in the 1970s. See correspondence in AP 24/18.

122. Hoare to Chamberlain 5 Oct. 1938, NC 7/11/31/133.

123. Halifax to Chamberlain 11 Oct. 1938, NC 7/11/31/1244.

124. Chamberlain to H. Chamberlain 15 Oct. 1938, NC 18/1/1073.

125. Avon, *The Reckoning*, p. 31; Eden to Major J. J. Astor 29 July 1938, AP 14/1/675A.

126. Boothby, *I Fight*, p. 164.

127. Lord Wolmer is usually included in lists of the Eden Group. In his draft memoirs, however, Wolmer noted: 'As far as I was concerned there never had been an Eden Group. I had been supporting the line Winston Churchill had taken.' Selborne MSS, Eng. hist. c 1017 fo.128.

128. Tree, *Moon was High*, p. 75.

129. Emrys-Evans to Amery 1 July 1954, Evans MSS, Add. MS 58247.

130. Margach, *Abuse*, pp. 102–3.

131. R. Cockett, *Twilight of Truth: Chamberlain, Appeasement and the Manipulation of the Press* (London, 1989), pp. 101–2.

132. L. S. Amery, *My Political Life* (3 vols, London, 1953–5), vol. 3, p. 298.

133. Harvey, ed., *Diplomatic Diaries*, p. 117 (15 March 1938).

134. B. Pimlott, *Labour and the Left in the 1930s* (Cambridge, 1977), pp. 166–7; H.

Notes

Dalton, *The Fateful Years: Memoirs 1931–1945* (London, 1957); Hugh Dalton diary 6 Oct. 1938.

135. Hugh Dalton diary 12 Oct. 1938.
136. Barnes and Nicholson, eds, *Empire at Bay*, p. 512.
137. Rose, ed., *Baffy*, p. 115.
138. Cecil King, quoted in P. Ziegler, *Diana Cooper* (London, 1981), p. 189.
139. Barnes and Nicholson, eds, *Empire at Bay*, p. 532.
140. Harvey, ed., *Diplomatic Diaries*, pp. 210–11 (8 Oct. 1938).
141. Lindsay to Eden 25 Oct. 1938, AP 14/1/765.
142. Bartlett to Eden 31 Oct. 1938, AP 14/1/689.
143. Bartlett to Eden n.d., [Nov. 1938], AP 14/1/690.
144. Harvey, ed., *Diplomatic Diaries*, p. 221 (18 Nov. 1938).
145. Undated memorandum by Sir T. Eden, AP 14/1/737A.
146. B. Pimlott, *Hugh Dalton* (London, 1985), p. 259.
147. Nicolson diary 30 Nov. 1938 (unpublished).
148. Liddell Hart, *Memoirs*, vol. 2, pp. 210–11.
149. V. Bonham Carter to Eden 21 Oct. 1938, AP 14/1/697.
150. *News Chronicle*, 28 Nov. 1938.
151. Cranborne to Thomas 30 Oct. 1938, AP 14/1/722.
152. Harvey, ed., *Diplomatic Diaries*, pp. 221–2 (18 Nov. 1938).
153. Gilbert, *Churchill*, vol. 5, p. 1019; Nicolson diary 15 Nov. 1938 (unpublished).
154. Duchess of Atholl to Eden 14 Nov. 1938, AP 14/1/682.
155. T. Eden to Eden 25 Oct. 1938, AP 14/1/737.
156. Harvey, ed., *Diplomatic Diaries*, pp. 217–18 (13 Nov. 1938); Barnes and Nicholson, eds, *Empire at Bay*, p. 535; Olson, ed., *Nicolson*, p. 141.
157. Harvey, ed., *Diplomatic Diaries*, p. 219 (15 Nov. 1938).
158. Tree, *Moon was High*, p. 80.
159. T. Jones, *A Diary with Letters* (Oxford, 1954), p. 422.
160. Harvey, ed., *Diplomatic Diaries*, p. 229 (25 Dec. 1938); Gilbert, *Churchill*, vol. 5, 3, p. 1323; D. Hacking to Lord Derby 20 Dec. 1938, Derby MSS 920 DER (17) 31/6.
161. Harvey, ed., *Diplomatic Diaries*, p. 230 (25 Dec. 1938).
162. Harvey, ed., *Diplomatic Diaries*, p. 234 (2 Jan. 1939).
163. Harvey, ed., *Diplomatic Diaries*, p. 236 (6 Jan. 1939).
164. Harvey, ed., *Diplomatic Diaries*, p. 236 (4 Jan. 1939).
165. Harvey, ed., *Diplomatic Diaries*, p. 236 (6 Jan. 1939); Gilbert, *Churchill*, vol. 5, 3, p. 1340.
166. Eden to Baldwin 15 Jan. 1939, AP 20/7/57.
167. Harvey, ed., *Diplomatic Diaries*, pp. 253–4 (Harvey to Halifax 14 Feb. 1939).
168. Cranborne to Halifax 25 Feb. 1939, Halifax MSS A2.278.64.4.
169. N. Chamberlain to Ida Chamberlain 12 March 1939, NC 18/1/1089.
170. N. Chamberlain to Ida Chamberlain 12 March 1939, NC 18/1/1089.
171. H of C Debs, 5th series, vol. 345, cols 461–2; Nicolson diary 15 March 1939 (unpublished).
172. James, ed., *Chips*, p. 186.
173. Harvey, ed., *Diplomatic Diaries*, p. 263 (18 March 1939).
174. Harvey, ed., *Diplomatic Diaries*, p. 264 (19 March 1939); Cranborne to Eden 22 March 1939, AP 14/2/22D.
175. Nicolson diary 24 and 27 March 1939 (unpublished).
176. Draft of speech, AP 20/7/12; Eden to Chamberlain 1 April 1939, cited D. Carlton, *Anthony Eden* (London, 1981), p. 148.

Notes

177. H of C Debs, 5th series, vol. 345, col. 2515; Liddell Hart, *Memoirs*, vol. 2, p. 218.
178. Thomas to Eden n.d., AP 14/2/97D. Channon noted that Thomas was annoyed that 'Winston is stealing all Anthony's thunder': James, ed., *Chips*, p. 204.
179. Harvey, ed., *Diplomatic Diaries*, p. 289 (8 May 1939).
180. James, ed., *Chips*, p. 200.
181. Harvey, ed., *Diplomatic Diaries*, pp. 280–1 (16 April 1939).
182. N. Chamberlain to H. Chamberlain 15 April 1939, NC 18/1/1094.
183. Harvey, ed., *Diplomatic Diaries*, pp. 280–1 (16 April 1939).
184. Harvey, ed., *Diplomatic Diaries*, p. 289 (8 May 1939); Inskip diary 4 May 1939.
185. Eden to Cranborne 27 June 1939, AP 14/2/27A.
186. *Yorkshire Post*, 12 June 1939, reporting Eden's speech at Grove Park, Warwick.
187. N. Chamberlain to Ida Chamberlain 10 June 1939, NC 18/1/1102.
188. Gilbert, *Churchill*, vol. 5, p. 1081.
189. Harvey, ed., *Diplomatic Diaries*, pp. 300, 302 (30 June 1939; 4 July 1939).
190. James, *Cazalet*, p. 212; see also James, ed., *Chips*, p. 204.
191. Eden to Cranborne 12 July 1939, AP 14/2/28.
192. Cranborne to Eden 13 July 1939, AP 14/2/28A.
193. Cranborne to Eden 16 July 1939, AP 14/2/28A.
194. Nicolson, ed., *Nicolson Diaries 1930–39*, p. 406. Eden later complained of Nicolson's diaries: 'These continue good from my point of view, or at least satisfactory, until about the end of November 1938. Then there is not much of any kind till June 1939, when it becomes very critical. . . . Three weeks later is a little better; July, more critical.' Eden to D. Dilks 28 May 1965, AP 33/2. In fact the diaries serve as an interesting barometer of Eden's standing among those who looked to him for a lead.
195. Eden to Thomas 12 Aug. 1939, AP 14/2/103.
196. Eden to R. Law 12 Aug. 1939, AP 20/7/60.
197. Nicolson diary 1 Aug. 1939 (unpublished).
198. Cranborne to Eden 17 Aug. 1939, AP 14/2/32A.
199. Cooper, *Old Men Forget*, p. 259.
200. Churchill to Chamberlain 2 Sept. 1939, cited K. Feiling, *The Life of Neville Chamberlain* (London, 1946), p. 420.
201. Nicolson, ed., *Nicolson Diaries 1930–39*, p. 420; Duff Cooper diary 2 Sept. 1939.
202. Avon, *The Reckoning*, p. 62.
203. A. Eden, *Full Circle* (London, 1960), p. 266.
204. Harvey, ed., *Diplomatic Diaries*, p. 115 (12 March 1938).
205. R. Law to Eden 13 July 1939, AP 14/2/69.
206. Dilks to Eden 15 May 1965, AP 33/2.
207. Rose, ed., *Baffy*, p. 141.
208. Cranborne to Eden 21 Feb. 1939, AP 14/2/22.
209. Diary 25 Nov, 1941, AP 20/1/21.

CHAPTER 6

1. P. H. Spaak, *The Continuing Battle: Memoirs of a European 1936–1966* (London, 1971), p. 83, quoting Eden in July 1944. This chapter was written before the publication of John Charmley's controversial book, *Churchill's Grand Alliance* (London, 1995). Somewhat to my surprise I find myself in broad agreement with Dr Charmley's analysis, at least as far as it concerns Eden's attitude towards the Special Relationship.

2. Diary 12 Feb. 1957, AP 20/2/5. Eden's post-Suez diary is an important and neglected source. But it was used as a notebook rather than a daily journal and cited dates must only be regarded as indicating the place of the entry within the manuscript.

3. Diary 2 Jan. 1957, AP 20/2/5.

4. Eden to Lord Hailes 27 Sept. 1958, AP 23/34/3A.

5. Eden to Salisbury 28 Dec. 1957, AP 23/60/14A.

6. Enclosure with Normanbrook to Eden 24 Aug. 1959, AP 23/56/46A.

7. Lord Boothby, *Recollections of a Rebel* (London, 1978), p. 211.

8. Diary 2 Jan. 1957, AP 20/2/5.

9. P. M. H. Bell, *A Certain Eventuality: Britain and the Fall of France* (Farnborough, 1974), pp. 50–1.

10. WP(40)276, 18 July 1940, CAB 66/10.

11. W. Kimball, *Churchill and Roosevelt: The Complete Correspondence*, vol. 1 (London, 1984), pp. 6–7.

12. Churchill to Eden 5 Nov. 1942, AP 20/9/257B.

13. Churchill to Eden and Attlee 24 Sept. 1943, FO 954/22.

14. Churchill to Cranborne 3 April 1945, FO 954/20.

15. Earl of Avon, *The Reckoning* (London, 1965), foreword.

16. Avon, *The Reckoning*, p. 364.

17. Churchill to Eden 7 Sept. 1943, FO 954/1.

18. P. Gore-Booth, *With Great Truth and Respect* (London, 1974), p. 123. Hull later paid tribute to Eden as possessing 'an agreeable personality and a high order of intelligence': C. Hull, *The Memoirs of Cordell Hull*, vol. 2 (London, 1948), p. 1474.

19. V. Lawford, 'Three Ministers', *Cornhill Magazine*, no. 1010 (1956–7), p. 88.

20. D. Dilks, ed., *Retreat from Power: Studies in British Foreign Policy of the Twentieth Century*, vol. 2 (London, 1981), p. 79.

21. Lord Blake, in J. P. Mackintosh, ed., *British Prime Ministers in the Twentieth Century*, vol. 2 (London, 1978), p. 78.

22. B. Bond, ed., *Chief of Staff: The Diaries of Lieutenant-General Sir Henry Pownall*, vol. 2 (London, 1974), p. 189.

23. E. Barker, *Churchill and Eden at War* (London, 1978), p. 308.

24. Memorandum 8 Nov. 1942, WP(42)516, CAB 66/30.

25. Robert Eden (1741–84) became Governor of Maryland in 1768.

26. On reading a draft of *The Reckoning* Robert Blake wrote: 'I was very interested in the picture of Roosevelt which emerges from your book – a very different one from Winston's.' Blake to Eden 7 Sept. 1963, AP 33/3.

27. Amery to Lord Linlithgow 23 Dec. 1942, Linlithgow MSS vol. 11.

28. B. Lapping, *End of Empire* (London, 1985), p. 119.

29. D. Dilks, ed., *The Diaries of Sir Alexander Cadogan 1938–45* (London, 1971), p. 632.

30. J. Lash, *Roosevelt and Churchill 1939–1945: The Partnership that Saved the West* (London, 1977), p. 331; W. Armstrong, ed., *With Malice toward None: A War Diary by Cecil H. King* (London, 1970), p. 196.

31. Note by Isaiah Berlin, AP 20/10/696.

32. Sir R. Bruce Lockhart, *Comes the Reckoning* (London, 1947), p. 362.

33. Diary 21 July 1941, AP 20/1/21.

34. R. Sherwood, *Roosevelt and Hopkins: An Intimate History* (New York, 1950), p. 237.

35. Eden to Churchill 12 June 1940, AP 20/8/15.

36. Avon, *The Reckoning*, p. 372.

37. Lash, *Roosevelt and Churchill*, p. 331; A. Eden, *Freedom and Order* (London, 1947), pp. 103–11.

38. D. Reynolds, *The Creation of the Anglo-American Alliance 1937–41* (London, 1981), p. 170.
39. Barker, *Churchill and Eden*, p. 147.
40. Minute of 23 May 1941, cited Reynolds, *Anglo-American Alliance*, p. 200.
41. J. Harvey, ed., *The War Diaries of Oliver Harvey 1941–1945* (London, 1978), p. 31 (12 Aug. 1941).
42. Harvey, ed., *War Diaries*, p. 31 (11 Aug. 1941).
43. Diary 21 July 1941, AP 20/1/21.
44. N. Chamberlain to H. Chamberlain 17 Dec. 1937, Chamberlain MSS NC18/1/1032.
45. Memorandum 5 Feb. 1941, WP(41)24, CAB 66/14.
46. WP(41)230, CAB 66/19.
47. Sherwood, *Roosevelt and Hopkins*, pp. 258–9.
48. D. Day, *Menzies and Churchill at War* (London, 1986), p. 165.
49. WP(41)172, CAB 66/17.
50. Reynolds, *Anglo-American Alliance*, p. 237.
51. Eden to Halifax 22 July 1941, cited Lash, *Roosevelt and Churchill*, p. 381.
52. Eden to Churchill 2 Aug. 1941, AP 20/8/516.
53. Eden to Churchill 12 Sept. 1941, AP 20/8/547.
54. WP(41)230, CAB 66/19.
55. Eden to Halifax 18 Oct. 1941, cited Lash, *Roosevelt and Churchill*, p. 457; Eden to Hankey 10 Oct. 1941, AP 20/53/81.
56. Harvey, ed., *War Diaries*, p. 68 (3 Dec. 1941).
57. Note by Eden on Hankey to Eden 3 Oct. 1941, AP 20/53/80.
58. J. Colville, *The Fringes of Power: Downing Street Diaries 1939–1955* (London, 1985), p. 213.
59. Colville, *Fringes of Power*, p. 404.
60. Dilks, ed., *Cadogan Diaries*, p. 356.
61. F. Kersaudy, *Churchill and de Gaulle* (London, 1981), p. 131.
62. Lord Moran, *Winston Churchill: The Struggle for Survival* (London, 1966), p. 501.
63. Dilks, ed., *Retreat from Power*, vol. 2, p. 74.
64. Eden to Churchill 31 Aug. 1941, AP 20/8/537.
65. Churchill minute on Eden to Churchill 28 Nov. 1941, AP 20/8/604.
66. R. Sherwood, *The White House Papers of Harry L. Hopkins* (London, 1948), vol. 1, p. 454.
67. G. Kolko, *The Politics of War* (New York, paperback edn, 1990), p. 73.
68. G. Warner, 'Franklin D. Roosevelt and the Postwar World', in D. Dutton, ed., *Statecraft and Diplomacy in the Twentieth Century* (Liverpool, 1995), p. 162.
69. V. H. Rothwell, *Anthony Eden: A Political Biography 1931–57* (Manchester, 1992), p. 82.
70. Diary 4 March 1944, AP 20/1/24.
71. Kolko, *Politics of War*, p. 83.
72. R. Dallek, *Franklin D. Roosevelt and American Foreign Policy 1932–1945* (Oxford, 1979), p. 408.
73. C. Giuliani, 'Eden, de Gaulle and the Free French: un bienfait inscrit dans la mémoire?' in Dutton, ed., *Statecraft and Diplomacy*, pp. 115–16.
74. Avon, *The Reckoning*, pp. 316–17.
75. C. Thorne, *Allies of a Kind: The United States, Britain and the War against Japan, 1941–1945* (Oxford paperback edn, 1979), p. 245.
76. J. Barnes and D. Nicholson, eds, *The Empire at Bay: The Leo Amery Diaries 1929–1945* (London, 1988), pp. 767–8; Harvey, ed., *War Diaries*, p. 90 (3 Feb. 1942).

77. K. Young, ed., *The Diaries of Sir Robert Bruce Lockhart 1939–1965* (London, 1980), p. 140.
78. Diary 6 Feb. 1942, AP 20/1/22.
79. Reynolds, *Anglo-American Alliance*, p. 261.
80. Avon, *The Reckoning*, p. 323.
81. Diary 25 Feb. 1942, AP 20/1/22.
82. Eden to Churchill 17 June 1942, AP 20/9/139.
83. Eden to Churchill 5 Feb. 1942, AP 20/9/13.
84. R. T. Thomas, *Britain and Vichy: The Dilemma of Anglo-French Relations 1940–42* (London, 1979), p. 134.
85. Memorandum for cabinet 8 July 1942, WP(42)285, AP 20/9/515A.
86. Churchill to Eden 30 May 1942, AP 20/9/115.
87. Avon, *The Reckoning*, p. 347.
88. Avon, *The Reckoning*, pp. 397–8.
89. Avon, *The Reckoning*, p. 341.
90. Harvey, ed., *War Diaries*, p. 168 (15 Oct. 1942).
91. Churchill to Eden 18 Oct. 1942, FO 954/7.
92. Churchill to Eden 18 Oct. 1942, FO 954/7.
93. Note by Eden 20 Oct. 1975, AP 7/23/590.
94. R. M. Hathaway, *Ambiguous Partnership: Britain and America 1944–47* (New York, 1981), p. 46.
95. Reynolds, *Anglo-American Alliance*, p. 261.
96. Thorne, *Allies of a Kind*, p. 220.
97. WP(42)375, CAB 66/28. See also Eden to Halifax 12 Aug. 1944, PREM 4/30/11: 'China would be so dependent on the other Great Powers that she would not be likely to pursue any very independent policy . . . in matters affecting international peace and security.'
98. Thorne, *Allies of a Kind*, pp. 178–9.
99. Thorne, *Allies of a Kind*, p. 226.
100. Young, ed., *Lockhart Diary*, p. 198; Harvey, ed., *War Diaries*, p. 170 (19 Oct. 1942).
101. Eden to Churchill 19 Oct. 1942, AP 20/9/229.
102. Eden to Churchill 22 Sept. 1942, AP 20/9/185.
103. Diary 15 Nov. 1942, AP 20/1/22.
104. Thomas, *Britain and Vichy*, p. 141.
105. Cranborne to Eden 16 Nov. 1942, AP 20/39/30.
106. C. de Gaulle, *The Complete War Memoirs* (New York paperback edn, 1984), p. 362.
107. Diary 20 Nov. 1942, AP 20/1/22.
108. Eden to Churchill 26 Nov. 1942, AP 20/9/286.
109. Cranborne to Churchill 17 Dec. 1942, AP 20/39/40.
110. Eden to Cranborne 21 Dec. 1942, AP 20/39/40A.
111. Diary 24 Dec. 1942, AP 20/1/22.
112. Dilks, ed., *Cadogan Diaries*, p. 504 (17 Jan. 1943).
113. Churchill to Eden 18 Jan. 1943, CAB 120/76.
114. Moran, *Winston Churchill*, p. 81.
115. Churchill to Eden 27 Feb. 1943, AP 20/10/458.
116. Diary 13 June 1942, AP 20/1/22.
117. Eden to Churchill 2 March 1943, AP 20/10/41.
118. Halifax to Cranborne 19 April 1943, Halifax MSS A4.410.4.19.
119. Sherwood, *White House Papers*, vol. 2, p. 719.
120. Harvey, ed., *War Diaries*, p. 225 (3 March 1943).

121. Avon, *The Reckoning*, p. 360.
122. Avon, *The Reckoning*, p. 372.
123. Hull, *Memoirs*, vol. 2, p. 1213; Dallek, *Roosevelt and Foreign Policy*, p. 408.
124. Thorne, *Allies of a Kind*, p. 281.
125. Harvey, ed., *War Diaries*, p. 229 (13 March 1943).
126. Avon, *The Reckoning*, p. 374.
127. W. D. Leahy, *I Was There* (London, 1950), p. 187.
128. Dallek, *Roosevelt and Foreign Policy*, pp. 389–90; J. Wheeler-Bennett and A. Nicholls, *The Semblance of Peace* (London, 1972), p. 98.
129. Sherwood, *Roosevelt and Hopkins*, p. 716.
130. Sherwood, *White House Papers*, vol. 2, pp. 706–7.
131. WM(43)53, CAB 65/38.
132. Diary 23 May 1943, AP 20/1/23.
133. Eden and Attlee to Churchill 23 May 1943, AP 20/10/706.
134. Churchill to Eden and Attlee 24 May 1943, AP 20/10/709.
135. Kersaudy, *Churchill and de Gaulle*, p. 285.
136. Diary 11 June 1943, AP 20/1/23.
137. Diary 18 June 1943, AP 20/1/23.
138. Kersaudy, *Churchill and de Gaulle*, p. 290; Eden to Churchill 7 July 1943, AP 20/10/208.
139. Diary 8 July 1943, AP 20/1/23.
140. Harvey, ed., *War Diaries*, p. 274 (13 July 1943).
141. Diary 13 July 1943, AP 20/1/23; Avon, *The Reckoning*, pp. 397–8.
142. PREM 3/81/8, cited D. Carlton, *Anthony Eden: A Biography* (London, 1981), p. 220.
143. See, for example, B. Pimlott, ed., *The Second World War Diary of Hugh Dalton 1940–45* (London, 1986), p. 642: 'Lockhart says that Eden's relations with the P.M. are good, but "because he pays the price" of always giving in.'
144. Diary 20 July 1943, AP 20/1/23.
145. Churchill to Eden 30 July 1943, AP 20/10/579.
146. Barker, *Churchill and Eden*, p. 162; diary 31 July 1943, AP 20/1/23.
147. Note by Eden 5 Aug. 1943 on Eden to Churchill 3 Aug. 1943, AP 20/10/265.
148. Diary 12 Aug. 1943, AP 20/1/23.
149. Eden to Churchill 3 Oct. 1943, AP 20/10/305.
150. Diary 20 Aug. 1943, AP 20/3/4; Hull, *Memoirs*, vol. 2, p. 1237.
151. Diary 21 Aug. 1943, AP 20/3/4.
152. Wheeler-Bennett and Nicholls, *Semblance of Peace*, p. 102.
153. Hull, *Memoirs*, vol. 2, p. 1241.
154. Eden to Churchill 12 July 1943, AP 20/10/201.
155. Diary 10 Sept. 1943, AP 20/1/23.
156. Diary 26 Sept. and 2 Oct. 1943, AP 20/1/23.
157. Eden rated Moscow 'the most successful of our three power meetings, with Teheran less so and Yalta least so'. Eden to Wheeler-Bennett 13 Aug. 1966, AP 33/4.
158. Diary 20–21 Oct. 1943, AP 20/3/5.
159. K. Sainsbury, *The Turning Point* (Oxford, 1985), pp. 18–19.
160. Avon, *The Reckoning*, p. 416; Eden to Wheeler-Bennett 13 Aug. 1966, AP 33/4.
161. F. W. Deakin *et al.*, eds, *British Political and Military Strategy in Central, Eastern and Southern Europe* (London, 1988), p. 54.
162. Diary 25 Nov. 1943, AP 20/3/5.
163. Diary 26 Nov. 1943, AP 20/3/5.

Notes

164. Sainsbury, *Turning Point*, p. 215.
165. Diary 29 Nov. 1943, AP 20/3/5.
166. War Cabinet 13 Dec. 1943, WM(43)169, CAB 65/40.
167. Diary 28 Nov. 1943, AP 20/3/5.
168. Eden to Churchill 27 Dec. 1943, AP 20/10/755.
169. Eden to Halifax 28 Jan. 1944, cited Thorne, *Allies of a Kind*, pp. 392–3.
170. Harvey, ed., *War Diaries*, p. 348 (15 July 1944).
171. Eden to Churchill 21 Feb. 1944, AP 20/11/91.
172. A. Verrier, *Through the Looking Glass* (London, 1983), p. 104; Eden to Beaverbrook 12 June 1944, cited Thorne, *Allies of a Kind*, p. 391.
173. Note by Eden on Halifax to Eden 30 May 1944, AP 33/9.
174. Eden to Churchill 5 Feb. 1944, AP 20/11/43.
175. Avon, *The Reckoning*, p. 464.
176. Diary 4 March 1944, AP 20/1/24.
177. Eden to Churchill 6 April 1944, AP 20/11/210.
178. Churchill to Eden 10 May 1944, AP 20/12/252.
179. Pierson Dixon diary 13 June 1944, AP 23/25/24A.
180. Dilks, ed., *Cadogan Diaries*, pp. 634–5 (5 June 1944).
181. Diary 5 June 1944, AP 20/1/24.
182. Note by Eden 5 June 1944, AP 20/12/883; Eden to Churchill 6 June 1944, AP 20/11/412.
183. Diary 6 June 1944, AP 20/1/24.
184. Diary 7 June 1944, AP 20/1/24.
185. P. Dixon, *Double Diploma: The Life of Sir Pierson Dixon, Don and Diplomat* (London, 1968), p. 92.
186. De Gaulle, *Complete War Memoirs*, p. 557.
187. See, for example, Eden to Churchill 26 June 1944, PREM 3/182/2, and 8 July 1944, AP 20/11/506; Churchill to Eden 9 June 1944, AP 20/12/326.
188. Avon, *The Reckoning*, p. 457.
189. Dilks, ed., *Cadogan Diaries*, p. 675 (23 Oct. 1944).
190. Eden to Churchill 10 Sept. 1944, AP 20/12/894.
191. Moran, *Winston Churchill*, p. 179.
192. Hathaway, *Ambiguous Partnership*, p. 113.
193. Diary 15 Sept. 1944, AP 20/3/8.
194. Eden to J. Colville 8 Nov. 1967, AP 24/16/2.
195. Hathaway, *Ambiguous Partnership*, p. 113.
196. Note for Cadogan 5 Jan. 1945, FO 954/22.
197. Diary 4 Jan. 1945, AP 20/1/25.
198. Eden to Churchill 4 Jan. 1945, AP 33/9.
199. Diary 1 Feb. 1945, AP 20/3/11.
200. Sherwood, *White House Papers*, vol. 2, p. 849.
201. Eden to Churchill 16 Jan. 1945, AP 20/13/19.
202. US Dept of State, *Foreign Relations of the United States: The Conferences at Malta and Yalta, 1945* (Washington, 1955), p. 498.
203. Diary note, 1946, 'Far East and Yalta', AP 20/1/26.
204. Diary 2 Feb. 1945, AP 20/3/11.
205. Avon, *The Reckoning*, p. 518.
206. W. F. Kimball, *The Juggler: Franklin Roosevelt as Wartime Statesman* (Princeton, 1991), p. 122.
207. W. Millis, ed., *The Forrestal Diaries* (London, 1952), pp. 63–4.

208. Diary 3 July 1945, AP 20/1/25.
209. Eden to Churchill 26 April 1945, AP 33/9.
210. Eden to Churchill 16 April 1945, AP 20/13/165; Wheeler-Bennett and Nicholls, *Semblance of Peace*, p. 309.
211. See, for example, WP(45)249, CAB 66/64.
212. Eden to Churchill 26 May 1945, AP 20/13/149.
213. R. Butler and M. E. Pelly, eds, *Documents on British Policy Overseas*, series 1, vol. 1 (London, 1984), p. 188.
214. *The Economist*, 21 July 1945.
215. N. Nicolson, ed., *Harold Nicolson: Diaries and Letters 1939–1945* (London, 1967), p. 385.
216. Avon, *The Reckoning*, pp. 444–5.
217. Diary 2 April 1942, AP 20/1/22. See also diary 7 Feb. 1943, AP 20/1/23: 'I replied his position was not the same as R[oosevelt]'s.'
218. Note by Eden 20 Oct. 1975, AP 7/23/590.
219. Dilks, ed., *Cadogan Diaries*, p. 496 (21 Nov. 1942).
220. House of Commons Debates, 5th series, vol. 395, col. 1432.
221. Eden to Churchill 7 April 1945, AP 20/13/113.
222. Barker, *Churchill and Eden*, p. 138.
223. Churchill to Eden 19 Sept. 1943, AP 20/10/592.

CHAPTER 7

1. Eden to H. Macmillan 12 Sept. 1968, AP 23/48/91A.
2. Diary 1 Oct. 1940, AP 20/1/20A.
3. Diary 22 Aug. 1940, AP 20/1/20A.
4. Diary 1 Oct. 1940, AP 20/1/20A.
5. Diary 17 Dec. 1940 and 20 Jan. 1941, AP 20/1/20A and AP 20/1/21.
6. J. Barnes and D. Nicholson, eds, *The Empire at Bay: The Leo Amery Diaries 1929–1945* (London, 1988), p. 669.
7. R. R. James, ed., *Chips: The Diaries of Sir Henry Channon* (London, 1967), p. 293.
8. B. Pimlott, ed., *The Second World War Diary of Hugh Dalton 1940–45* (London, 1986), p. 150.
9. D. Dilks, ed., *The Diaries of Sir Alexander Cadogan 1938–45* (London, 1971), p. 345 (31 Dec. 1940).
10. Dilks, ed., *Cadogan Diaries*, p. 374 (28 April 1941).
11. Dilks, ed., *Cadogan Diaries*, pp. 381–2 (30 May 1941).
12. Dilks, ed., *Cadogan Diaries*, pp. 376, 385 (9 May 1941 and 1 June 1941).
13. K. Young, ed., *The Diaries of Sir Robert Bruce Lockhart 1939–1965* (London, 1980), p. 99.
14. Diary 2 June 1941, AP 20/1/21. Eden even judged Cadogan to be 'very unhelpful': Harvey MS diary, Add. MSS 56397.
15. J. Harvey, ed., *The War Diaries of Oliver Harvey 1941–1945* (London, 1978), p. 50 (8 Oct. 1941).
16. Pimlott, ed., *Dalton War Diary*, p. 161; Dilks, ed., *Cadogan Diaries*, p. 360 (1 March 1941).
17. Pimlott, ed., *Dalton War Diary*, p. 190; see also F. de Guingand, *Operation Victory* (London, 1947), p. 59.
18. R. R. James, *Anthony Eden* (London, 1986), p. 251.
19. Dilks, ed., *Cadogan Diaries*, pp. 361–2 (6 March 1941).

20. Cadogan to Halifax 18 March 1941, Halifax MSS A 4.410.4.16.
21. Eden to R. Lumley 9 May 1941, AP 20/37/48B.
22. See, for example, R. W. Thompson, *Churchill and Morton* (London, 1976), p. 49; Eden's comment on Churchill's draft memoirs dealing with this episode, 18 May 1949, AP 19/4/15B and Eden to Wheeler-Bennett 13 Jan. 1971, AP 33/4.
23. M. Van Creveld, 'Prelude to Disaster: the British Decision to Aid Greece, 1940–41', *Journal of Contemporary History*, 9, 3 (1974), p. 91.
24. Eden to Lumley 9 May 1941, AP 20/37/48B.
25. P. M. Williams, ed., *The Diary of Hugh Gaitskell 1945–1956* (London, 1983), pp. 593–6.
26. Diary 14 Nov. 1944, AP 20/1/24; cf. D. Carlton, *Anthony Eden: A Biography* (London, 1981), p. 184: 'Eden, at least until 1944, appears to have seen the Soviet Union as a genuine ally of liberal democracy and an agent of "progress".'
27. Minute 24 Jan. 1941, FO 371/29500, cited G. Gorodetsky, *Stafford Cripps' Mission to Moscow 1940–42* (Cambridge, 1984), p. 89.
28. M. Kitchen, *British Policy Towards the Soviet Union during the Second World War* (London, 1986), p. 5.
29. Eden to Chamberlain 10 Feb. 1940, PREM 1/408.
30. Eden to Cripps 28 Dec. 1940, FO 371/24845.
31. Eden to Cripps 17 Jan. 1941, cited G. Ross, ed., *The Foreign Office and the Kremlin: British Documents on Anglo-Soviet Relations 1941–45* (Cambridge, 1984), p. 70.
32. Lord Avon, *The Reckoning* (London, 1965), p. 263.
33. Ross, ed., *Foreign Office and Kremlin*, p. 11; Gorodetsky, *Cripps*, p. 110; V. H. Rothwell, *Britain and the Cold War 1941–1947* (London, 1982), p. 77.
34. Eden 15 April 1941, cited Kitchen, *British Policy Towards the Soviet Union*, p. 47.
35. Diary 5 June 1941, AP 20/1/21.
36. Gorodetsky, *Cripps*, p. 170.
37. Diary 24 Sept. 1941, AP 20/1/21.
38. Eden to Churchill 8 July 1941, AP 20/8/797; see also R. Langhorne, ed., *Diplomacy and Intelligence during the Second World War: Essays in Honour of F. H. Hinsley* (Cambridge, 1985), pp. 176–7.
39. Eden to Churchill 9 July 1941, AP 20/8/500.
40. Diary 9 July 1941, AP 20/1/21.
41. Harvey, ed., *War Diaries*, pp. 17–18 (10 July 1941).
42. Harvey, ed., *War Diaries*, pp. 15, 20 (25 June and 16 July 1941).
43. Harvey, ed., *War Diaries*, pp. 19, 24 (12 and 28 July 1941).
44. T. E. Evans, ed., *The Killearn Diaries 1934–1946* (London, 1972), p. 136.
45. J. Colville, *The Churchillians* (London, 1981), p. 163: 'For Churchill to intervene in foreign policy was one thing, but if anybody else did so, or criticised Eden's decisions, his wrath knew no bounds.'
46. J. Colville, *The Fringes of Power: Downing Street Diaries 1939–1955* (London, 1985), p. 408; Pimlott, ed., *Dalton War Diary*, pp. 246–7; Eden to Churchill 13 Aug. 1941, AP 20/8/808; D. Stafford, *Britain and European Resistance 1940–1945* (Toronto, 1983), p. 76.
47. Pimlott, ed., *Dalton War Diary*, p. 354.
48. Yet Eden always found it difficult to confront delicate personal questions head on. Dalton's diary entry for 11 July 1941 (Pimlott, ed., *Dalton War Diary*, p. 250) has the ring of authenticity about it: '[I] ask whether it was true that he thought I interfered in his affairs. He feigns surprise and says, "No, certainly not. What makes you think that?" I say I had gathered this impression.' See also entry for 22

Jan. 1942 (p. 354). As Gladwyn Jebb explained, 'Eden doesn't really like "frank talks". He prefers brooding and exploding to his own officials about the wickedness of outside intermeddlers' (p. 315).

49. Pimlott, ed., *Dalton War Diary*, p. 369.
50. Eden to Churchill 28 Aug. 1941, AP 20/8/530.
51. Diary 5 Sept. 1941, AP 20/1/21.
52. Diary 4 Sept. 1941, AP 20/1/21.
53. War Cabinet 26 Oct. 1941, WP(41)248, CAB 66/19.
54. Harvey, ed., *War Diaries*, p. 57 (27 Oct. 1941).
55. N. Nicolson, ed., *Vita and Harold: The Letters of Vita Sackville-West and Harold Nicolson 1910–1962* (London, 1992), p. 342; Young, ed., *Lockhart Diaries*, p. 117.
56. Ross, ed., *Foreign Office and Kremlin*, p. 77.
57. Diary 14 Nov. 1941, AP 20/1/21.
58. Sir E. L. Woodward, *British Foreign Policy in the Second World War*, vol. 2 (London, 1971), p. 51.
59. Rothwell, *Britain and the Cold War*, p. 85.
60. Diary 14 Nov. 1941, AP 20/1/21.
61. Diary 21 Nov. 1941, AP 20/1/21.
62. Diary 21 July 1941, AP 20/1/21.
63. US Department of State, *Foreign Relations of the United States, 1941*, vol. 1 (Washington, 1958), pp. 188–95; cf. J. G. Winant, *A Letter from Grosvenor Square: An Account of a Stewardship* (London, 1947), pp. 196–7; J. P. Lash, *Roosevelt and Churchill 1939–1941* (London, 1977), p. 490; M. A. Stoler, *The Politics of the Second Front* (Westport, 1977), pp. 21–2; C. Thorne, *Allies of a Kind* (Oxford, 1979), p. 109.
64. Diary 27 Nov. 1941, AP 20/1/21.
65. WP(41)288, 29 Nov. 1941, CAB 66/20.
66. Avon, *The Reckoning*, p. 284.
67. Diary 4 Dec. 1941, AP 20/1/21.
68. Young, ed., *Lockhart Diaries*, p. 130.
69. War Cabinet 4 Dec. 1941, WM(41)124, CAB 65/24.
70. Barnes and Nicholson, eds, *Empire at Bay*, p. 752.
71. Young, ed., *Lockhart Diaries*, p. 130.
72. Cripps diary 30 Nov.–7 Dec. 1941, cited Gorodetsky, *Cripps*, p. 277.
73. Woodward, *British Foreign Policy*, vol. 2, p. 53.
74. Diary 7 Dec. 1941, AP 20/3/3.
75. Law to Harvey 5 Dec. 1941, Harvey MSS Add. MS 56402.
76. Diary 8 Dec. 1941, AP 20/3/3; Harvey, ed., *War Diaries*, p. 80 (28 Dec. 1941): 'He is a little jealous of Winston with all his limelight in Washington but feels, as I do, that he talks too much.'
77. War Cabinet 12 Dec. 1941, WM(41)127, CAB 65/20. Eden later wrote that he accepted the changed decision over aircraft as inevitable, but 'the prospects for my mission looked bleaker still'. (*The Reckoning*, p. 287).
78. Diary 16 Dec. 1941, AP 20/3/3.
79. Diary 17 Dec. 1941, AP 20/3/3.
80. Harvey, ed., *War Diaries*, p. 75 (18 Dec. 1941).
81. Gorodetsky, *Cripps*, p. 280.
82. Harvey, ed., *War Diaries*, p. 77 (18 Dec. 1941).
83. James, ed., *Chips*, p. 315.
84. E. Barker, *Churchill and Eden at War* (London, 1978), p. 233.

85. Eden to Churchill 5 Jan. 1942, FO 371/32874.
86. Halifax diary 8 Jan. 1942, A7.8.19.31.
87. Churchill to Eden 8 Jan. 1942, FO 371/32874.
88. Harvey, ed., *War Diaries*, p. 86 (8 Jan. 1942).
89. C. Ponting, *Churchill* (London, 1994), p. 569; cf. Eden to Halifax 10 Feb. 1942, PREM 3/395, and Pimlott, ed., *Dalton War Diary*, pp. 348–9.
90. V. H. Rothwell, *Anthony Eden: A Political Biography 1931–57* (Manchester, 1992), p. 63.
91. Eden to Halifax 22 Jan. 1942, Halifax MSS A4.410.4.15.
92. Kitchen, *British Policy Towards the Soviet Union*, p. 148; Beaverbrook to Eden 3 March 1942, FO 954/25.
93. Memorandum 28 Jan. 1942, WP(42)48, CAB 66/21; Avon, *The Reckoning*, pp. 317–19.
94. Dilks, ed., *Cadogan Diaries*, p. 437 (24 Feb. 1942).
95. Kitchen, *British Policy Towards the Soviet Union*, p. 53; Langhorne, ed., *Diplomacy and Intelligence*, p. 179.
96. Harvey, ed., *War Diaries*, p. 91 (6 Feb. 1942).
97. Cabinet 6 Feb. 1942, WM(42)17, CAB 65/29.
98. Eden minute 8 Feb. 1942, cited Ross, ed., *Foreign Office and Kremlin*, pp. 89–90.
99. Dilks, ed., *Cadogan Diaries*, p. 437 (24 Feb. 1942).
100. Eden to Churchill 6 March 1942, AP 20/9/47.
101. W. F. Kimball, ed., *Churchill and Roosevelt: The Complete Correspondence*, vol. 1 (London, 1984), p. 394; Churchill to Roosevelt 7 March 1942, FO 954/25.
102. Halifax diary 8 March 1942, Halifax MSS A7.8.19.38.
103. Eden to Churchill 23 March 1942, AP 20/9/52.
104. Diary 20 March 1942, AP 20/1/22.
105. Harvey, ed., *War Diaries*, p. 112 (24 March 1942); Dilks, ed., *Cadogan Diaries*, pp. 442–3 (24 March 1942).
106. Eden to Halifax 26 March 1942, FO 954/25.
107. Dilks, ed., *Cadogan Diaries*, p. 446 (9 April 1942).
108. Ross, ed., *Foreign Office and Kremlin*, p. 22.
109. Eden to Halifax 27 April 1942, Halifax MSS A4.410.4.15.
110. A. J. P Taylor, ed., *Off the Record: W. P. Crozier Political Interviews 1933–1943* (London, 1973), p. 315.
111. Dilks, ed., *Cadogan Diaries*, p. 449 (3 May 1942).
112. Eden to Churchill 28 April 1942, AP 20/9/75.
113. Young, ed., *Lockhart Diaries*, p. 161.
114. J. Charmley, *Duff Cooper: The Authorised Biography* (London, 1986), p. 164; Cooper to Eden 22 April 1942, FO 954/25.
115. Young, ed., *Lockhart Diaries*, p. 179.
116. Simon to Churchill 8 May 1942, Simon MSS 91 fo. 20.
117. Churchill to Eden 14 May 1942, FO 954/25.
118. Diary 1 May 1942, AP 20/1/22.
119. Ross, ed., *Foreign Office and Kremlin*, p. 23.
120. Dilks, ed., *Cadogan Diaries*, p. 450 (5 May 1942) and pp. 453–4 (22–23 May 1942).
121. Dilks, ed., *Cadogan Diaries*, p. 455 (24 May 1942).
122. Beaverbrook to Eden 30 May 1942, AP 33/9.
123. Eden to A. Hardinge 27 May 1942, AP 20/39/90A.
124. Eden to Churchill 3 June 1942, AP 20/9/121.
125. Young, ed., *Lockhart Diaries*, p. 258.

126. Eden to S. Baldwin 19 June 1942, AP 14/2/142.

127. Most notably in R. Lamb, *The Ghosts of Peace 1935–1945* (Salisbury, 1987), *passim*; see also R. Edmonds, *The Big Three: Churchill, Roosevelt and Stalin in Peace and War* (London, 1991), p. 325.

128. Eden to Churchill 17 Feb. 1943, AP 20/10/26.

129. Dilks, ed., *Cadogan Diaries*, p. 551 (11 Aug. 1943).

130. A. Horne, *Macmillan 1894–1956* (London, 1988), p. 218.

131. Avon, *The Reckoning*, pp. 366–7; J. Wheeler-Bennett and A. Nicholls, *The Semblance of Peace: The Political Settlement after the Second World War* (London, 1972), pp. 538–9.

132. Avon, *The Reckoning*, p. 366.

133. Pimlott, ed., *Dalton War Diary*, p. 608.

134. See, for example, Dilks, ed., *Cadogan Diaries*, pp. 531, 650 (17 May 1943 and 24 July 1944).

135. Cranborne to Halifax 10 May 1943, Halifax MSS A4.410.4.19.

136. Harvey, ed., *War Diaries*, p. 210 (19 Jan. 1943).

137. Diary 29 Jan. 1943, AP 20/1/23.

138. Eden to Churchill 16 March 1943, FO 954/26.

139. Avon, *The Reckoning*, p. 373; R. Sherwood, *Roosevelt and Hopkins: An Intimate History* (New York, 1950), p. 715; R. Dallek, *Franklin D. Roosevelt and American Foreign Policy 1932–1945* (New York, 1979), p. 400.

140. Rothwell, *Britain and the Cold War*, pp. 163–4.

141. War Cabinet 19 April 1943, CAB 65/34.

142. P. M. H. Bell, *John Bull and the Bear: British Public Opinion, Foreign Policy and the Soviet Union 1941–1945* (London, 1990), p. 111.

143. Cranborne to Halifax 10 May 1943, Halifax MSS A4.410.4.19.

144. Eden to Churchill 25 Feb. 1944, AP 20/11/99.

145. Young, ed., *Lockhart Diaries*, p. 251.

146. Harvey, ed., *War Diaries*, pp. 286–7 (24 Aug. 1943).

147. Avon, *The Reckoning*, p. 405.

148. Avon, *The Reckoning*, p. 402.

149. Note by Eden for Cadogan 19 Aug. 1943, FO 954/2.

150. Eden to Churchill 19 Aug. 1943, AP 20/10/272.

151. Note by Eden for Cadogan 21 Aug. 1943, FO 954/2; Harvey, ed., *War Diaries*, p. 288 (30 Aug. 1943).

152. Diary 26 Aug. 1943, AP 20/3/3.

153. Eden to Churchill 3 Sept. 1943, AP 20/54/2C.

154. Memorandum 5 Oct. 1943, WP(43)438, CAB 66/41.

155. Harvey, ed., *War Diaries*, p. 304 (6 Oct. 1943).

156. Harvey, ed., *War Diaries*, p. 293 (10 Sept. 1943).

157. H. Macmillan, *War Diaries: Politics and War in the Mediterranean, January 1943– May 1945* (London, 1984), p. 286.

158. Avon, *The Reckoning*, p. 418.

159. Lord Strang, *Home and Abroad* (London, 1956), pp. 199–200.

160. Lord Ismay, *Memoirs* (New York, 1960), p. 327.

161. Harvey, ed., *War Diaries*, pp. 312–13 (24 Oct. 1943); K. Sainsbury, *The Turning Point* (Oxford, 1985), p. 60.

162. Diary 20–21 Oct. 1943, AP 20/3/5.

163. W. A. Harriman and E. Abel, *Special Envoy to Churchill and Stalin 1941–1946* (New York, 1975), p. 242.

Notes

164. Kitchen, *British Policy Towards the Soviet Union*, p. 173.
165. W. Deakin *et al.*, eds, *British Political and Military Strategy in Central, Eastern and Southern Europe in 1944* (London, 1988), p. 212.
166. E. Barker, *British Policy in South-East Europe in the Second World War* (London, 1976), p. 132.
167. Harriman and Abel, *Special Envoy*, pp. 244–5.
168. Sainsbury, *Turning Point*, p. 90.
169. Kitchen, *British Policy Towards the Soviet Union*, p. 168.
170. Eden to Churchill 2 Nov. 1943, FO 954/26.
171. Dilks, ed., *Cadogan Diaries*, p. 580 (29 Nov. 1943).
172. K. Eubank, *Summit at Teheran: The Untold Story* (New York, 1985), p. 372.
173. Diary 28 Nov. 1943, AP 20/3/5.
174. War Cabinet 13 Dec. 1943, CAB 65/36; Kitchen, *British Policy Towards the Soviet Union*, pp. 176–7.
175. See, for example, Barnes and Nicholson, eds, *Empire at Bay*, p. 956.
176. Minute 28 Dec. 1943, FO 371/36590; Avon, *The Reckoning*, p. 435.
177. Dilks, ed., *Cadogan Diaries*, p. 593.
178. Memorandum 11 Jan. 1944, FO 371/39386.
179. Churchill to Eden 16 Jan. 1944, AP 20/12/2.
180. Eden to Churchill 25 Jan. 1944, AP 20/11/21.
181. Avon, *The Reckoning*, p. 439; Minute 3 April 1944, FO 371/43304. See also Harriman and Abel, *Special Envoy*, p. 328.
182. Eden to Churchill 18 Feb. and 5 April 1944, AP 20/11/84 and AP 20/11/206; G. Ross, 'Foreign Office Attitudes to the Soviet Union, 1941–45', *Journal of Contemporary History*, 16, 3 (1981), pp. 526–7; Kitchen, *British Policy Towards the Soviet Union*, p. 205.
183. Eden to Churchill 5 April 1944, AP 20/11/200.
184. Eden to Duff Cooper 25 July 1944, cited Avon, *The Reckoning*, p. 445.
185. Minute 6 July 1944, cited Ross, ed., *Foreign Office and Kremlin*, p. 146.
186. Ross, ed., *Foreign Office and Kremlin*, p. 120.
187. E. Barker, 'Some Factors in British Decision Making over Yugoslavia, 1941–4', in P. Auty and R. Clogg, eds, *British Policy Towards Wartime Resistance in Yugoslavia and Greece* (London, 1975), pp. 22–48.
188. Eden minute 29 May 1944, cited Barker, *British Policy in South-East Europe*, p. 150.
189. Eden to Churchill 17 Dec. 1942, AP 20/9/308.
190. Eden to R. Bruce Lockhart 31 Aug. 1949, AP 16/1/186.
191. Eden to Churchill 9 May 1944, AP 20/11/322.
192. Eden to Churchill 19 May 1944, AP 20/11/351.
193. Wheeler-Bennett and Nicholls, *Semblance of Peace*, p. 557; K. Sainsbury, *Churchill and Roosevelt at War* (New York, 1994), p. 102.
194. WP(44)304, 7 June 1944, CAB 66/51.
195. WP(44)436, 9 Aug. 1944, CAB 66/53; Kitchen, *British Policy Towards the Soviet Union*, p. 200; Ross, ed., *Foreign Office and Kremlin*, p. 45.
196. A. Bryant, *Triumph in the West 1943–1946* (London, 1959), p. 290.
197. Sir E. L. Woodward, *British Foreign Policy in the Second World War*, vol. 5 (London, 1976), pp. 206–7.
198. Meeting with Chiefs of Staff 4 Oct. 1944, FO 954/22.
199. Eden to Montgomery 7 Oct. 1944, AP 23/54/2.
200. Eden to Churchill 8 Aug. 1944, AP 20/11/578; Kitchen, *British Policy Towards the Soviet Union*, pp. 227–8.

201. Eden to Foreign Office 17 Oct. 1944, cited Barker, *British Policy in South-East Europe*, p. 145.

202. Avon, *The Reckoning*, p. 481.

203. Lord Moran, *Winston Churchill: The Struggle for Survival 1940–65* (London, 1966), pp. 193–4.

204. Diary 11 Oct. 1944, AP 20/3/9.

205. N. Nicolson, ed., *Harold Nicolson: Diaries and Letters 1939–1945* (London, 1967), p. 421.

206. Eden to Toby Low 29 March 1945, AP 20/13/254.

207. Harvey, ed., *War Diaries*, pp. 365–6 (11 Nov. 1944); see also p. 344 (12 June 1944).

208. P. Dixon, *Double Diploma: The Life of Sir Pierson Dixon, Don and Diplomat* (London, 1968), p. 129.

209. Churchill minute on Eden to Churchill 3 Jan. 1945, AP 20/13/4.

210. Eden to Churchill 28 Jan. 1945, AP 33/9.

211. Eden to Churchill 28 Jan. 1945, AP 20/13/38.

212. Diary 4 Feb. 1945, AP 20/3/11.

213. Dixon, *Double Diploma*, p. 138; T. M. Campbell and G. C. Herring, eds, *The Diaries of Edward R. Stettinius* (New York, 1975), p. 241.

214. Diary 6 Feb. 1945, AP 20/3/11.

215. Diary 9 Feb. 1945, AP 20/3/11.

216. James, *Eden*, pp. 290–1. According to Roosevelt's Chief of Staff the Yalta agreement on Poland was 'so elastic that the Russians can stretch it all the way from Yalta to Washington without ever technically breaking it'. W. D. Leahy, *I Was There* (New York, 1950), pp. 315–16.

217. Cadogan to Halifax 20 Feb. 1945, Halifax MSS A4.410.4.16.

218. Colville, *Fringes of Power*, p. 560.

219. James, *Eden*, pp. 290–1; N. Tolstoy, *Victims of Yalta* (London, 1977); N. Bethell, *The Last Secret: Forcible Repatriation to Russia 1944–7* (London, 1974).

220. Eden to Churchill 2 Aug. 1944, AP 20/52/17E.

221. P. Dean to Eden 19 Dec. 1974, AP 20/52/388.

222. Eden to Low 15 March 1945, AP 20/42/66.

223. Diary 23 March 1945, AP 20/1/25.

224. Eden to Churchill 24 March 1945, FO 954/26.

225. Halifax to Churchill 22 April 1945, PREM 4/27; Eden to Churchill 22 April 1945, AP 20/13/185. Indeed pressure from Eden, exerted via Stettinius, may have been a factor behind Truman's celebrated reprimand of Molotov on 23 April. W. D. Miscamble, 'Anthony Eden and the Truman–Molotov Conversations, April 1945', *Diplomatic History,* 2, 2 (1978), pp. 167–80.

226. Eden to Churchill 23 April 1945, FO 954/2.

227. Eden to Churchill 4 May 1945, AP 20/13/207.

228. Dilks, ed., *Cadogan Diaries*, p. 737 (1 May 1945); Young, ed., *Lockhart Diaries*, p. 433.

229. Eden to Churchill 12 May 1945, AP 20/13/266.

230. Eden to Churchill 10 July 1945, FO 954/2.

231. Eden to Churchill 12 July 1945, FO 954/2.

232. Sir C. Keeble, *Britain and the Soviet Union 1917–89* (London, 1990), p. 199.

233. Moran, *Churchill*, p. 279.

234. Dilks, ed., *Cadogan Diaries*, p. 769 (22 July 1945).

235. Diary 19 July 1945, AP 20/1/25.

236. Eden to Churchill 17 July 1945, AP 20/13/231.

237. Diary 17 July 1945, AP 20/1/25.
238. Sainsbury, *Churchill and Roosevelt*, pp. 42, 86.
239. Diary 1 Jan. 1968, AP 20/2/13; see also Eden to Wheeler-Bennett 21 Aug. 1968, AP 33/4. When presented with the draft of an NBC broadcast to mark Churchill's eightieth birthday, which contained a tribute to Churchill's perception of the dangers of Soviet imperialism, Eden complained: 'This is almost all untrue and quite nauseating! I simply cannot say any of it. . . . I'm not going to stand up and tell a series of colossal lies!' Note by Eden 22 Nov. 1954, AP 11/10/230B.
240. Vansittart 29 March 1940, cited Kitchen, *British Policy Towards the Soviet Union*, p. 20.
241. Lockhart to Eden 17 Nov. 1943, AP 20/40/88.
242. R. Bruce Lockhart, *Comes the Reckoning* (London, 1947), p. 267.
243. Kitchen, *British Policy Towards the Soviet Union*, is an example of a work which, at least in its conclusion, gives a rather one-sided assessment of Eden's view of the Soviet Union.
244. Sainsbury, *Churchill and Roosevelt*, p. 140. See also K. Sainsbury, 'British Policy and German Unity at the End of the Second World War', *English Historical Review*, XCIV (1979), p. 802.
245. War Cabinet paper March 1943, cited Sainsbury, *Churchill and Roosevelt*, p. 136.
246. As Lieutenant-General Pownall complained in 1941, Eden and the Foreign Office 'think they are dealing with normal people': B. Bond, ed., *Chief of Staff: The Diaries of Lieutenant-General Sir Henry Pownall 1940–1944* (London, 1974), p. 36.
247. Barnes and Nicholson, eds, *Empire at Bay*, p. 718.
248. See, for example, A. Shlaim *et al.*, *British Foreign Secretaries since 1945* (Newton Abbot, 1977), p. 83.
249. Note for Nigel Fisher, 1976, AP 23/32/12G.
250. Harvey, ed., *War Diaries*, p. 385 (7 Aug. 1945).

CHAPTER 8

1. A. Eden, *Full Circle* (London, 1960), p. 266.
2. Lord Avon, *The Reckoning* (London, 1965), p. 91.
3. J. Vincent, ed., *The Crawford Papers* (Manchester, 1984), p. 605.
4. J. Colville, *The Fringes of Power* (London, 1985), p. 56. See also W. Armstrong, ed., *With Malice toward None: A War Diary by Cecil H. King* (London, 1970), p. 18.
5. K. Young, ed., *The Diaries of Sir Robert Bruce Lockhart 1939–1965* (London, 1980), p. 46.
6. Halifax diary 18 March 1940, A7.8.3; Armstrong, ed., *With Malice*, p. 22.
7. R. R. James, ed., *Chips: The Diaries of Sir Henry Channon* (London, 1967), p. 242.
8. Armstrong, ed., *With Malice*, p. 27.
9. Diary 9 May 1940, AP 20/1/20A.
10. Diary 7 Aug., 12 Aug. and 18 May 1940, AP 20/1/20A.
11. Diary 22 Aug. 1940, AP 20/1/20A.
12. C. Stuart, ed., *The Reith Diaries* (London, 1975), p. 265; Eden to Churchill 30 Sept. 1940, AP 20/8/174.
13. Halifax diary 30 Sept. 1940, A7.8.5.
14. Chamberlain diary 24 Sept. 1940, NC 2/24A; J. Charmley, *Churchill: The End of Glory* (London, 1993), pp. 433–4.
15. Diary 1 Oct. 1940, AP 20/1/20A.

16. J. Ramsden, *The Age of Balfour and Baldwin* (London, 1978), p. 375; Harry Crookshank diary 9 Oct. 1940.
17. Diary 14 Aug. 1940, AP 20/1/20A.
18. Salisbury to Eden 20 Feb. 1941, FO 954/4.
19. Diary 30 Sept. 1940, AP 20/1/20A.
20 A. Roberts, *The Holy Fox: A Biography of Lord Halifax* (London, 1991), p. 274.
21. Diary 18–20 Dec. 1940, AP 20/1/20A.
22. Colville, *Fringes of Power*, p. 330.
23. Diary 18 Dec. 1940, AP 20/1/20A.
24. D. Irving, *Churchill's War*, vol. 1 (Bullsbrook, 1987), pp. 477–8.
25. Hankey to Halifax 1 May 1941, Halifax MSS A4.410.5.
26. Diary 14 April 1941, AP 20/1/21.
27. Statement on Lord Moran's book, *Winston Churchill: The Struggle for Survival*, 17 June 1966, AP 19/2/141B.
28. Eden to Churchill 10 Jan. 1941, FO 954/7.
29. J. Harvey, ed., *The War Diaries of Oliver Harvey 1941–1945* (London, 1978), p. 50 (9 Oct. 1941).
30. B. Pimlott, ed., *The Second World War Diary of Hugh Dalton 1940–45* (London, 1986), pp. 141–2.
31. D. Day, *Menzies and Churchill at War* (London, 1986), p. 125; G. Gorodetsky, *Stafford Cripps' Mission to Moscow 1940–42* (Cambridge, 1984), p. 74.
32. Halifax diary 17 Feb. 1941, A7.8.19.
33. Pimlott, ed., *Dalton War Diary*, p. 122.
34. Young, ed., *Lockhart Diaries*, p. 91; Harvey, ed., *War Diaries*, p. 51 (10 Oct. 1941).
35. Day, *Menzies and Churchill*, p. 87; James ed., *Chips*, p. 301.
36. James ed., *Chips*, pp. 302–3.
37. Colville, *Fringes of Power*, p. 412.
38. Diary 10 Oct. and 11 Nov. 1941, AP 20/1/21.
39. Charmley, *Churchill*, p. 456.
40. Diary 14 Nov. 1941, AP 20/1/21.
41. Harvey to Eden 13 Feb. 1942, Harvey MSS, Add. MS 56402.
42. Harvey, ed., *War Diaries*, p. 94 (10 Feb. 1942).
43. Harvey diary (unpublished), 17 Aug. 1941.
44. J. Grigg, *1943: The Victory that Never Was* (London, 1980), p. 35.
45. Harvey, ed., *War Diaries*, p. 102 (27 Feb. 1942).
46. Harvey, ed., *War Diaries*, pp. 97–9 (16–18 Feb. 1942).
47. Harvey, ed., *War Diaries*, pp. 99–100 (19 Feb. 1942).
48. Cranborne to Eden 5 April 1942, AP 20/39/153.
49. Harvey to Eden 13 Feb. 1942, AP 20/39/150.
50. Halifax diary 25 March 1942, A7.8.19.
51. James ed., *Chips*, p. 330.
52. Cranborne to Eden 5 April 1942, AP 20/39/153.
53. Diary 7 April 1942, AP 20/1/22.
54. J. Wheeler-Bennett, ed., *Action This Day: Working with Churchill* (London, 1968), p. 108.
55. Harvey, ed., *War Diaries*, pp. 130–1 (27 May 1942).
56. Harvey, ed., *War Diaries*, p. 143 (24 July 1942).
57. Harvey, ed., *War Diaries*, pp. 149–50 (13 Aug. 1942).
58. R. Pearce, ed., *Patrick Gordon Walker: Political Diaries 1932–1971* (London, 1991), p. 113.

59. Pearce, ed., *Gordon Walker Diaries*, p. 111; T. Jones, *A Diary with Letters* (Oxford, 1954), p. 505; K. Jefferys, ed., *Labour and the Wartime Coalition: From the Diary of James Chuter Ede 1941–1945* (London, 1987), pp. 130–1.

60. Young, ed., *Lockhart Diaries*, p. 186; A.J.P. Taylor, *Beaverbrook* (London, 1972), p. 540.

61. Harvey, ed., *War Diaries*, pp. 151–2 (24 Aug. 1942).

62. Harvey, ed., *War Diaries*, p. 201 (17 Dec. 1942).

63. Harvey, ed., *War Diaries*, pp. 201–3. (17–23 Dec. 1942).

64. Harvey, ed., *War Diaries*, p. 242. (2 April 1943).

65. Harvey, ed., *War Diaries*, pp. 250–1 (26 April 1943); Grigg, *1943*, p. 131.

66. Harvey, ed., *War Diaries*, p. 265 (5 June 1943).

67. Harvey, ed., *War Diaries*, p. 245 (16 April 1943).

68. Harvey, ed., *War Diaries*, pp. 247–8 (22 April 1943).

69. Harvey, ed., *War Diaries*, pp. 250, 257 (25 April and 18 May 1943); Cranborne to Eden 26 April 1943, AP 20/40/25.

70. J. Wheeler-Bennett, *King George VI: His Life and Reign* (London, 1958), pp. 700–2.

71. Thomas to Eden 7 Aug. 1943, AP 20/10/680.

72. Law to Eden 8 Aug. 1943, AP 20/10/680A.

73. Harvey, ed., *War Diaries*, p. 285 (9 Aug. 1943).

74. Young, ed., *Lockhart Diaries*, pp. 290–1.

75. Harvey, ed., *War Diaries*, p. 337 (28 March 1944).

76. Diary 6 March 1944, AP 20/1/24.

77. Diary 16 March 1944, AP 20/1/24.

78. Diary 24 March 1944, AP 20/1/24.

79. Harvey, ed., *War Diaries*, p. 346 (28 June 1944); Young, ed., *Lockhart Diaries*, p. 310.

80. Young, ed., *Lockhart Diaries*, p. 386.

81. Pimlott, ed., *Dalton War Diary*, pp. 707, 722.

82. P. Ziegler, *Diana Cooper* (London, 1981), p. 218.

83. A. Campbell-Johnson, *Sir Anthony Eden: A Biography* (London, 1955), p. 194.

84. Diary 5 July 1945, AP 20/1/25.

85. Harvey to E. Evans 26 Aug. 1945, Evans MSS.

86. Diary 26–27 July 1945, AP 20/1/25.

87. Eden to Lord Birkenhead 29 Jan. 1970, AP 23/9/29.

88. Lockhart to Eden 3 Aug. 1945, AP 20/14/21.

89. Harvey, ed., *War Diaries*, p. 385 (28 July 1945); Young, ed., *Lockhart Diaries*, p. 474.

90. T. Barman, *Diplomatic Correspondent* (London, 1968), p. 196.

91. Young, ed., *Lockhart Diaries*, p. 497.

92. Young, ed., *Lockhart Diaries*, p. 510.

93. Young, ed., *Lockhart Diaries*, p. 511.

94. Young, ed., *Lockhart Diaries*, p. 511.

95. Young, ed., *Lockhart Diaries*, p. 512.

96. Jones, *Diary with Letters*, p. 538.

97. Cranborne to Eden 8 Jan. 1946, AP 20/43/1.

98. H. Dalton, *High Tide and After: Memoirs 1945–1960* (London, 1962), p. 104.

99. Young, ed., *Lockhart Diaries*, p. 523.

100. Eden to Cranborne 15 March 1946, AP 20/43/17A.

101. Young, ed., *Lockhart Diaries*, p. 538.

102. Churchill to Eden 7 April 1946, AP 20/43/25.

103. Eden to Churchill 10 April 1946, AP 20/43/25A.

104. Churchill to Eden 11 April 1946, AP 20/43/27.

Notes

105. M. Gilbert, *Winston S. Churchill*, vol. 8 (London, 1988), p. 228; Churchill seems to have renewed his offer to divide the leadership functions in the summer – but then withdrew it again. Eden to Thomas 14 March 1947, AP 20/14/41.

106. Lockhart to Eden 22 Sept. 1946, AP 20/43/50.

107. Eden to Halifax 8 Aug. 1946, Halifax MSS A4.410.21.2.

108. Young, ed., *Lockhart Diaries*, pp. 570–1.

109. Young, ed., *Lockhart Diaries*, p. 631.

110. J. Stuart, *Within the Fringe* (London, 1967), p. 146.

111. Churchill to Macmillan 9 Feb. 1948, AP 19/1/27E.

112. C. Ponsonby to Eden 14 Dec. 1947, AP 16/1/81.

113. J. Ramsden, *The Making of Conservative Party Policy: The Conservative Research Department since 1929* (London, 1980), p. 144.

114. Lord Moran, *Winston Churchill: The Struggle for Survival 1940–1965* (London, 1966), p. 333.

115. Lockhart to Eden 28 Jan. 1948, AP 16/1/83.

116. Young, ed., *Lockhart Diaries*, p. 696.

117. Young, ed., *Lockhart Diaries*, p. 655.

118. Young, ed., *Lockhart Diaries*, p. 727.

119. J. Bright-Holmes, ed., *Like it Was: The Diaries of Malcolm Muggeridge* (London, 1981), p. 437.

120. Young, ed., *Lockhart Diaries*, p. 720.

121. Young, ed., *Lockhart Diaries*, pp. 519–20; N. Nicolson, ed., *Harold Nicolson: Diaries and Letters 1945–1962* (London, 1968), pp. 35–6.

122. Young, ed., *Lockhart Diaries*, p. 539; M. D. Kandiah, 'The Later Political Career of Frederick Marquis, First Earl of Woolton', University of Exeter Ph.D. (1993), p. 56; J. Ramsden, *The Age of Churchill and Eden, 1940–1957* (London, 1995), p. 96.

123. Diary 11 Aug. 1975, AP 20/2/20.

124. Campbell-Johnson, *Eden*, p. 203.

125. Note of conversation with Archibald James 6 Sept. 1946, Woolton MSS box 21.

126. Salisbury to Emrys Evans 9 Jan. 1947, Evans MSS Add. MS 58240; cf. Cranborne to Thomas 26 April 1946, Cilcennin MSS.

127. Mann to Eden 25 Feb. 1946, AP 20/43/13A.

128. Derby to Beaverbrook 20 Nov. 1946, Derby MSS 920 DER (17) 33.

129. Diary 6 March 1965, AP 20/2/10.

130. Colville, *Fringes of Power*, p. 632.

131. Young, ed., *Lockhart Diaries*, p. 731.

132. J. Colville, *The Churchillians* (London, 1981), pp. 170–1; A. Horne, *Macmillan 1894–1956* (London, 1988), p. 353; Moran, *Churchill*, p. 710; J. Margach, *The Abuse of Power* (London, 1978), p. 105.

133. A. Howard, *Rab* (London, 1987), p. 210.

134. Eden to Salisbury 20 June 1966, AP 23/60/130A; see also diary 17 June 1966, AP 20/2/11.

135. E. Shuckburgh, *Descent to Suez: Diaries 1951–56* (London, 1986), p. 25. Like Sir Geoffrey Howe in 1989, Eden had to be reminded that there was no constitutional basis for this post. Wheeler-Bennett, *George VI*, p. 797.

136. Shuckburgh, *Descent to Suez*, pp. 65–6.

137. Taylor, *Beaverbrook*, p. 602; R. Rayner MP to Butler July 1953, cited Ramsden, *Age of Churchill and Eden*, p. 233.

138. Macmillan diary 7 Dec. 1951, cited Horne, *Macmillan*, p. 352.

139. Shuckburgh, *Descent to Suez*, p. 46.

140. A. Seldon, *Churchill's Indian Summer: The Conservative Government, 1951–55* (London, 1981), pp. 420–1.
141. Crookshank diary 26 Feb. 1953.
142. Note for Nigel Fisher 1 Aug. 1976, AP 23/32/12A.
143. Seldon, *Indian Summer*, p. 524.
144. James, ed., *Chips*, p. 470.
145. P. Gore-Booth, *With Great Truth and Respect* (London, 1974), p. 187.
146. H. Macmillan, *Tides of Fortune 1945–55* (London, 1969), p. 499.
147. D. Acheson, *Present at the Creation* (London, 1970), p. 662.
148. Shuckburgh, *Descent to Suez*, pp. 46–7, 68–9.
149. Moran, *Churchill*, p. 374.
150. Shuckburgh, *Descent to Suez*, p. 41; J. Morgan, ed., *The Backbench Diaries of Richard Crossman* (London, 1981), p. 112.
151. Diary 4 June 1952, AP 20/1/28.
152. Shuckburgh, *Descent to Suez*, p. 61.
153. Shuckburgh, *Descent to Suez*, p. 62.
154. Shuckburgh, *Descent to Suez*, p. 63.
155. Shuckburgh, *Descent to Suez*, p. 66.
156. Diary 23 Jan. 1953, AP 20/1/29.
157. Shuckburgh, *Descent to Suez*, p. 78.
158. R. R. James, *Anthony Eden* (London, 1986), pp. 361–4.
159. Diary 7 April 1968, AP 20/2/13A.
160. Moran, *Churchill*, p. 401.
161. Colville to Clarissa Eden 26 June 1953, AP 20/16/123.
162. Butler to Eden 27 June 1953, AP 20/16/124.
163. Thomas to Eden 29 June 1953, AP 20/16/125.
164. Colville to Clarissa Eden 1 July 1953, AP 20/16/128.
165. Salisbury to Eden 12 July 1953, AP 20/16/137; Moran, *Churchill*, p. 436.
166. Colville, *Fringes of Power*, p. 671.
167. Shuckburgh to Eden 23 July 1953, AP 20/16/140.
168. Shuckburgh, *Descent to Suez*, p. 93.
169. Colville, *Fringes of Power*, p. 673; Moran, *Churchill*, p. 447.
170. Shuckburgh to Eden 25 Aug. 1953, AP 20/16/159A.
171. Thomas to Eden 28 Aug. 1953, AP 20/16/148A, and 9 Sept. 1953, AP 20/16/148.
172. Shuckburgh to Eden 22 Sept. 1953, AP 20/16/152.
173. Salisbury to Eden 6 Sept. 1953, AP 20/16/145, and 24 Sept. 1953, AP 20/16/155.
174. Carr to Eden 20 Sept. 1953, AP 20/16/150.
175. Shuckburgh, *Descent to Suez*, p. 102.
176. Moran, *Churchill*, p. 483.
177. Shuckburgh, *Descent to Suez*, p. 119.
178. Private information.
179. D. R. Thorpe, *Selwyn Lloyd* (London, 1989), p. 175.
180. Woolton diary 14 April 1954.
181. Private information. An American observer noted of the British delegation: 'behind the pride and not too far below the surface you can always get the considerable strain of worry about the Prime Minister's condition and the fact that the leadership of the British Empire still rests in his hands'. R. H. Ferrell, ed., *The Diary of James C. Hagerty* (Bloomington, 1983), p. 78.
182. Carr to Eden 5 Feb. 1954, AP 20/17/163.
183. Shuckburgh, *Descent to Suez*, p. 145.

184. Colville to Eden 28 May 1954, AP 20/17/124.
185. Nutting to Eden 8 June 1954, AP 20/17/135.
186. Churchill to Eden 11 June 1954, AP 20/17/138.
187. Shuckburgh, *Descent to Suez*, p. 221; J. W. Young, 'Churchill's Bid for Peace with Moscow, 1954', *History,* 73, 239 (1988), pp. 434–5.
188. Colville, *Fringes of Power*, p. 697.
189. Churchill to Eden 24 Aug. 1954, AP 20/17/179.
190. Churchill to Eden 30 Aug. 1954, AP 20/17/188.
191. Campbell-Johnson, *Eden*, p. 252.
192. Macmillan, *Tides of Fortune*, p. 550.
193. Eden to Churchill Dec. 1954 (draft), AP 20/17/218.
194. Macmillan diary 14 March 1955, cited Horne, *Macmillan*, p. 354.
195. Colville, *Fringes of Power*, p. 706.
196. Eden, *Full Circle*, p. 265.

CHAPTER 9

1. R. Carr to Eden 21 Oct. 1957, AP 23/16/5.
2. Lord Swinton, *Sixty Years of Power* (London, 1966), p. 162.
3. Lord Home, *The Way the Wind Blows* (London, 1976), p. 69.
4. It should be noted that from 1942 to 1945 Eden was Leader of the Commons while retaining the post of Foreign Secretary. As such he had a wide-ranging association with, though not direct responsibility for, much of the government's domestic business.
5. D. Bardens, *Portrait of a Statesman* (London, 1955), p. 49.
6. A. Campbell-Johnson, *Anthony Eden* (London, 1955), p. 70.
7. D. Carlton, *Anthony Eden* (London, 1981), pp. 20–7. But, as Carlton shows, the views which Eden expressed at this time about the League of Nations are not without interest.
8. Campbell-Johnson, *Eden*, p. 47.
9. Diary 31 Dec. 1930, AP 20/1/10.
10. *The Times*, 6 Oct. 1930.
11. Diary 26 Nov. 1941, AP 20/1/21.
12. Cited R. R. James, *Boothby* (London, 1991), p. 150.
13. W. Rees-Mogg, *Sir Anthony Eden* (London, 1956), p. 49.
14. T. Jones, *A Diary with Letters* (Oxford, 1954), p. 176.
15. See above, pp. 122–3.
16. A. Eden, *Freedom and Order* (London, 1948), p. 48.
17. Halifax to Eden 5 Jan. 1942, FO 954/29A.
18. Churchill to Eden 24 May 1941, AP 20/8/636.
19. B. Pimlott, ed., *The Second World War Diary of Hugh Dalton 1940–45* (London, 1986), p. 142.
20. J. Harvey, ed., *The War Diaries of Oliver Harvey 1941–1945* (London, 1978), p. 47 (29 Sept. 1941).
21. Harvey, ed., *War Diaries*, p. 47 (29 Sept. 1941).
22. Harvey, ed., *War Diaries*, p. 30 (11 Aug. 1941).
23. K. Jefferys, ed., *Labour and the Wartime Coalition: From the Diary of James Chuter Ede 1941–1945* (London, 1987), p. 57.
24. Harvey, ed., *War Diaries*, p. 258 (19 May 1943).
25. Harvey, ed., *War Diaries*, p. 152 (25 Aug. 1942).

Notes

26. K. Young, ed., *The Diaries of Sir Robert Bruce Lockhart 1939–1965* (London, 1980), pp. 326–7.
27. Lord Avon, *The Reckoning* (London, 1965), pp. 453–4. At around the same time Bevin told Field Marshal Smuts that he was prepared to serve under Eden in a post-war coalition: A. Bullock, *Ernest Bevin: Foreign Secretary* (London, 1983), p. 80.
28. N. Nicolson, ed., *Harold Nicolson: Diaries and Letters 1939–1945* (London, 1967), p. 387.
29. Diary 30 March 1940, AP 20/1/20. '[Baldwin] mentioned his view that it would be my duty to take the helm here for the peace and the after war.'
30. Harvey, ed., *War Diaries*, p. 26 (3 Aug. 1941).
31. Churchill to Eden 24 May 1941, AP 20/8/636.
32. Eden to Churchill 20 May 1941, AP 20/8/450.
33. Bardens, *Portrait*, p. 230.
34. Diary 6 Sept. 1942, AP 20/1/22.
35. Harvey, ed., *War Diaries*, pp. 162–3 (26 Sept. 1942).
36. Pimlott, ed., *Dalton War Diary*, pp. 594, 820.
37. J. Barnes and D. Nicholson, eds, *The Empire at Bay* (London, 1988), p. 969.
38. Halifax diary 29 April 1941, Halifax MSS A.7.8.8.
39. Law to Eden 19 Feb. 1942, AP 20/39/152.
40. D. Dilks, ed., *The Diaries of Sir Alexander Cadogan* (London, 1971), p. 497 (24 Nov. 1942).
41. Barnes and Nicholson, eds, *Empire at Bay*, p. 975. See also Pimlott, ed., *Dalton War Diary*, p. 770.
42. Butler to Eden 18 Feb. 1943, AP 20/10/679.
43. For example, at Bristol in October 1944; J. P. L. Thomas to Eden 16 Oct. 1944, AP 20/12/879.
44. Diary 5 Sept. 1941, AP 20/1/21.
45. Note by Eden May 1945, AP 11/4/4.
46. A. Mann to Eden 21 June 1945, AP 20/42/87.
47. Diary 6 June 1945, AP 20/1/25.
48. Young, ed., *Lockhart Diaries*, p. 462.
49. The text is in Eden, *Freedom and Order*, pp. 346–55. 'His voice was more natural. The matter of his speech, too, was excellent and quite free from abuse and extravagant promises.' R. Bruce Lockhart, *Comes the Reckoning* (London, 1947), p. 357.
50. Hankey to Eden 28 June 1945, AP 20/42/54.
51. P. Dixon, *Double Diploma: The Life of Sir Pierson Dixon* (London, 1968), p. 165; A. Clark Kerr to Eden 27 July 1945, AP 20/14/2.
52. Eden to Sir A. Southby 29 Nov. 1945, AP 20/42/146A.
53. Harvey, ed., *War Diaries*, p. 385 (28 July 1945).
54. For a stimulating interpretation of Conservatism in the 1930s, see J. Ramsden, 'A Party for Owners or a Party for Earners', *Transactions of the Royal Historical Society*, 5th series, 37, 1987, pp. 49–63.
55. 'I spoke at every by-election from 1946 to 1951 and we never won one.' A. Eden, *Full Circle* (London, 1960), p. 278.
56. Law to Beaverbrook 1 Jan. 1950, Beaverbrook MSS, BBK C214.
57. Eden to A. Southby 29 Nov. 1945, AP 20/42/146A.
58. P. Addison, *The Road to 1945* (London, 1975), p. 20.
59. D. Dilks to Eden 26 Feb. 1964, AP 33/2.
60. Eden to J. Wheeler-Bennett 13 Nov. 1973, AP 33/4.

Notes

61. H. Macmillan, *Tides of Fortune 1945–55* (London, 1969), p. 46.

62. Eden, *Freedom and Order*, p. 424.

63. Young, ed., *Lockhart Diaries*, p. 516; R. Maudling, *Memoirs* (London, 1978), pp. 43–4.

64. *Freedom and Order* 'has the authentic Eden ring about it. One critic said that he thought it impossible to say so much and say so little'. Bardens, *Portrait*, p. 268.

65. Note by Robert Carr 15 July 1976, AP 23/32/11C.

66. R. Allen to Eden 12 Feb. 1973, AP 23/6/115.

67. The idea is particularly associated with Sir Keith Joseph. See P. Jenkins, *Mrs Thatcher's Revolution* (London, 1987), pp. 61–5.

68. Eden to Churchill 30 July 1945, AP 20/43/41.

69. Eden, *Freedom and Order*, p. 409.

70. Among these was Lord Cranborne. His letter to Eden on 9 Aug. 1946 anticipated much of Eden's speech to the 1946 party conference: AP 20/14/34.

71. Young, ed., *Lockhart Diaries*, p. 562.

72. Eden, *Freedom and Order*, pp. 419–24.

73. Lord Colyton to Eden 5 Sept. 1956, AP 14/4/56.

74. Letter to *The Times*, 8 Jan. 1986.

75. 'A Property Owning Democracy and the Industrial Charter', March 1948, AP 33/1.

76. A. Eden, *Days for Decision* (London, 1949), p. 20.

77. Eden, *Days for Decision*, p. 12.

78. J. Ramsden, *The Making of Conservative Party Policy* (London, 1980), p. 111.

79. Young, ed., *Lockhart Diaries*, pp. 593–4.

80. Thomas to Eden Sept. 1947, AP 16/1/62.

81. Eden, *Days for Decision*, p. 120.

82. For a more positive view of the Conservatives' contribution to their ultimate return to power, see I. Zweiniger-Bargielowska, 'Rationing, Austerity and the Conservative Party Recovery after 1945', *Historical Journal*, 37, 1 (1994), pp. 173–97.

83. Eden to Butler 27 March 1949, quoted in A. Howard, *Rab* (London, 1987), p. 163.

84. Butler to Eden 28 March 1949, AP 16/1/147.

85. R. R. James, *Anthony Eden* (London, 1986), p. xi. Though it is worth noting that on an earlier occasion the same author had described Eden as a 'disaster': *Ambitions and Realities: British Politics 1964–1970* (London, 1972), p. 198.

86. Note by R. Carr 15 July 1976, AP 23/32/11C.

87. Colin Coote, quoted in T. E. B. Howarth, *Prospect and Reality: Great Britain 1945–1955* (London, 1985), p. 140.

88. See, for example, memorandum by Lord Simon for Lord Woolton 27 Feb. 1950, Simon MSS 98 fos 130–1.

89. Young, ed., *Lockhart Diaries*, p. 696.

90. Eden to Thomas 10 Jan 1947, AP 20/14/39.

91. C. Sulzberger, *A Long Row of Candles: Memoirs and Diaries 1934–54* (London, 1969), pp. 422–3.

92. R. Bruce Lockhart, *Friends, Foes and Foreigners* (London, 1957), p. 206.

93. Text of party-political broadcast 19 Oct. 1951, AP 11/7/17.

94. Even so, new council housing formed a substantial part of the government's target of 300 000 houses a year. Local authorities were given limited consent to sell council houses. Little more than 3 000 were sold before the 1955 general election.

95. *The Times*, 13 Nov. 1958.

96. Eden to R. Carr 14 Sept. 1967, AP 33/3.

97. Eden to W. Monckton 1 July 1952, AP 11/10/21A. Carr's maiden speech in the Commons was on the theme of joint consultation in industry.

98. J. Colville, *The Fringes of Power* (London, 1985), p. 652; J. Morgan, ed., *The Backbench Diaries of Richard Crossman* (London, 1981), p. 112.

99. A. Seldon, *Churchill's Indian Summer* (London, 1981), p. 40.

100. Lord Moran, *Winston Churchill: The Struggle for Survival* (London, 1966), p. 550.

101. Seldon, *Indian Summer*, pp. 172–3; 'Operation Robot' by E. A. Berthoud 19 Feb. 1953, AP 20/16/34C; E. Shuckburgh, *Descent to Suez: Diaries 1951–56* (London, 1986), p. 37.

102. Shuckburgh to Berthoud 7 Jan. 1953, AP 20/16/34.

103. Macmillan, *Tides of Fortune*, p. 529.

104. Butler to Eden 6 Feb. 1954, AP 14/3/549.

105. J. Boyd-Carpenter, *Way of Life* (London, 1980), p. 123.

106. Note by R. Carr 15 July 1976, AP 23/32/11C.

107. R. Lamb, *The Failure of the Eden Government* (London, 1987), pp. 5–6.

108. Woolton diary 5 April 1955.

109. Salisbury to Eden 27 March 1955, AP 20/18/59.

110. Eden, *Full Circle*, p. 279.

111. D. Butler, *The British General Election of 1955* (London, 1955), p. 75.

112. Note by R. Carr 15 July 1976, AP 23/32/11C.

113. Seldon, *Indian Summer*, pp. 18–19.

114. Diary 6 Jan. 1955, AP 20/1/31.

115. *The Times*, 4 Jan. 1986.

116. Eden to H. W. Blenkinsop 15 Aug. 1956, AP 20/31/605.

117. Lord Carr, letter to *The Times* 8 Jan. 1986.

118. Eden, *Full Circle*, p. 268.

119. Eden, *Full Circle*, pp. 320–1.

120. Eden to Salisbury 30 Aug. 1955, AP 20/30/564; cf. Eden to Macmillan 30 Aug. 1955, AP 20/20/62: 'we must put the battle against inflation before anything else.'

121. Chandos to Eden n.d. [1955], AP 14/3/817. Back in 1947 Lyttelton had warned Churchill that a future Conservative government would have to reverse Labour's policies through deflation, devaluation and decontrol. In 1951, however, he had been passed over for the Exchequer in favour of Butler: P. Addison, *Churchill on the Home Front 1900–1955* (London, 1992), p. 397.

122. Eden, *Full Circle*, p. 370.

123. M. Dockrill, *British Defence since 1945* (Oxford, 1988), p. 56.

124. Eden, *Full Circle*, p. 371; Minute by Eden 31 May 1956, AP 20/21/106.

125. Eden to Butler 17 Aug. 1955, AP 20/20/48.

126. Eden to Salisbury 30 Aug. 1955, AP 20/30/564; Eden to Macmillan 22 Dec. 1955, AP 20/20/142.

127. Lamb, *Failure*, pp. 8, 46.

128. Eden to Macmillan 22 Dec. 1955, AP 20/20/142.

129. W. Clark, *From Three Worlds* (London, 1986), p. 161.

130. P. Williams, ed., *The Diary of Hugh Gaitskell 1945–1956* (London, 1983), p. 414.

131. Lamb, *Failure*, p. 14.

132. C. Lysaght, *Brendan Bracken* (London, 1979), p. 299; see also R. Cockett, *My Dear Max* (London, 1990), p. 185.

133. Eden, *Full Circle*, p. 323.

134. Minute by Macmillan 8 Feb. 1956, AP 33/8iii.

135. Shuckburgh, *Descent to Suez*, p. 338; A. Horne, *Macmillan 1894–1956* (London, 1988), p. 380; Eden to F. Bishop 12 May 1971, AP 23/10/105.
136. H. Macmillan, *Riding the Storm* (London, 1971), p. 14.
137. Clark, *From Three Worlds*, pp. 162–3.
138. Macmillan to Eden 5 April 1956, PREM 11/1326.
139. Eden to Macmillan 11 April 1956, AP 20/21/64.
140. Eden to O. Poole 4 July 1956, AP 20/31/507.
141. Eden to Head 23 Dec. 1956, AP 20/21/229.
142. Eden to Macmillan 28 Dec. 1956, AP 20/21/234.
143. Note by Eden n.d. (1957), PREM 11/1138 and AP 20/34/4C.
144. Transcript in AP 11/11/5.
145. Eden to J. Wheeler-Bennett 13 Nov. 1973, AP 33/4.
146. A. Cairncross, ed., *The Robert Hall Diaries 1947–53* (London, 1989), pp. 193–4.
147. Addison, *Churchill on Home Front*, p. 387.
148. Interview with Lord Hailsham 10 Dec. 1992.
149. Memorandum by Eden 28 May 1956, AP 20/21/104.
150. Lamb, *Failure*, p. 32.
151. Diary 26 Aug. 1955, AP 20/1/31.

CHAPTER 10

1. Eden to R. Turton 30 June 1969, AP 23/64/25A.
2. 'I had taken the line, I thought rightly, that history was to judge and I did not therefore want to make a running commentary on Suez myself.' Note by Eden 26 Oct. 1967, AP 7/19/3A.
3. House of Lords Debates, 5th series, vol. 243, col. 438.
4. Eden viewed Boothby as 'a dubious character': Eden to P. Emrys-Evans 15 May 1962, AP 23/31/8. See also Eden note on Boothby to Eden 6 July 1944, AP 20/41/7: 'I have *no* belief in this creature's good faith.'
5. Earl of Kilmuir, *Political Adventure* (London, 1964), pp. 186–9; H. Macmillan, *Tides of Fortune* (London, 1969), pp. 156–7, 463–8.
6. Eden to D. Dilks 11 Oct. 1962, AP 33/2.
7. M. Camps, *Britain and the European Community 1955–63* (London, 1964), p. 506; see also A. Shlaim *et al.*, *British Foreign Secretaries since 1945* (Newton Abbot, 1977), p. 102.
8. M. Charlton, *The Price of Victory* (London, 1983), pp. 140–1.
9. N. Fisher, *Harold Macmillan* (London, 1982), pp. 307–8.
10. R. R. James, *Anthony Eden* (London, 1986), p. 350.
11. *The Observer*, 12 Oct. 1986. Yet John Campbell in his biography of Jenkins stresses that, at least until 1957, Jenkins himself subscribed to the traditional view that Commonwealth commitments made it impossible for Britain to join a customs union which imposed a common tariff against the rest of the world. J. Campbell, *Roy Jenkins: A Biography* (London, 1983), p. 50.
12. Eden to Lord Halifax Sept. 1939, AP 7/11/23V. Eden appears to have been sympathetic towards Briand's plans at the time. A. Campbell Johnson, *Anthony Eden: A Biography* (London, 1938), p. 192.
13. J. Barnes and D. Nicholson, eds, *The Empire at Bay: The Leo Amery Diaries 1929–1945* (London, 1988), p. 840.
14. Charlton, *Price of Victory*, pp. 14–15.
15. Eden to Churchill 28 March 1943, FO 954/22.

Notes

16. Memorandum by Eden 8 Nov. 1942, WP(42)516, CAB 66/30.
17. By September Eden was expressing doubts as to whether the idea of leaving British troops in post-war Europe under American command was compatible with the desire to build up a Western European Security Group. Eden to Churchill 10 Sept. 1944, AP 33/9.
18. Eden to E. Grigg 14 July 1943, AP 20/10/667A.
19. Memorandum by Eden 24 March 1944, WP(44)181, CAB 66/48; J. Charmley, 'Duff Cooper and Western European Union, 1944–47', *Review of International Studies*, 11 (1985), p. 55.
20. Eden to N. Bland, L. Oliphant and L. Collier 19 July 1944, FO 954/22.
21. Charmley, 'Duff Cooper', p. 56.
22. K. Larras, 'A Search for Order: Britain and the Origins of a Western European Union, 1944–55', in B. Brivati and H. Jones, eds, *From Reconstruction to Integration: Britain and Europe since 1945* (London, 1993), p. 75.
23. Churchill to Eden 25 Nov. 1944, AP 20/12/546.
24. Emphasis added.
25. Eden to Churchill 29 Nov. 1944, AP 20/11/742.
26. V. H. Rothwell, *Britain and the Cold War 1941–1947* (London, 1982), p. 411.
27. Rothwell, *Britain and the Cold War*, p. 412.
28. Rothwell, *Britain and the Cold War*, p. 413.
29. See J. W. Young, *Britain, France and the Unity of Europe, 1945–51* (Leicester, 1984); G. Warner, 'The Labour Governments and the unity of Western Europe 1945–51', in R. Ovendale, ed., *The Foreign Policy of the British Labour Governments 1945–51* (Leicester, 1984). The argument is conveniently summarized in S. Greenwood, *Britain and European Cooperation since 1945* (Oxford, 1992), pp. 7–29.
30. House of Commons Debates, 5th series, vol. 450, col. 1279.
31. K. Young, ed., *The Diaries of Sir Robert Bruce Lockhart 1939–1965* (London, 1980), p. 632.
32. 'The Hague Congress', AP 33/1/vi.
33. Kilmuir, *Political Adventure*, p. 177; Macmillan, *Tides of Fortune*, pp. 156–7; R. Boothby, *My Yesterday, Your Tomorrow* (London, 1962), pp. 56–7.
34. Boothby to Eden 16 Sept. 1949, AP 16/1/196.
35. *The Times*, 19 Feb. 1949.
36. AP 11/6/172.
37. Harvey to Eden 27 Dec. 1949, AP 16/1/218.
38. Eden to R. Turton 13 Sept. 1969, AP 23/64/34.
39. Charlton, *Price of Victory*, p. 286; Macmillan, *Tides of Fortune*, pp. 196–7.
40. H of C Debs, 5th series, vol. 476, col. 1918.
41. J. Monnet, *Memoirs* (London, 1978), pp. 315–16.
42. Eden to Turton 13 Sept. 1969, AP 23/64/34.
43. U. Kitzinger, *The European Common Market and Community* (New York, 1967), p. 10.
44. A. Eden, *Full Circle* (London, 1960), p. 30.
45. Macmillan, *Tides of Fortune*, p. 217.
46. Salisbury to Eden 14 Aug. 1950, AP 16/1/250.
47. H of C Debs, 5th series, vol. 484, col. 49.
48. Eden to R. Carr 2 July 1969, AP 23/16/50A.
49. R. Bullen and M. Pelly, eds, *Documents on British Policy Overseas* (hereafter *DBPO*), series II, vol. 1, p. xxv.
50. *DBPO*, II, 1, no. 392.
51. *DBPO*, II, 1, no. 395.

52. Macmillan, *Tides of Fortune*, p. 462.
53. CAB 128/23, CC(51)10.
54. CAB 129/48, C(51)32.
55. Charlton, *Price of Victory*, pp. 135–6; 137–8.
56. *DBPO*, II, 1, no. 398.
57. AP 23/58/78.
58. Macmillan, *Tides of Fortune*, p. 461.
59. Charlton, *Price of Victory*, p. 129.
60. Charlton, *Price of Victory*, p. 131.
61. E. Shuckburgh, *Descent to Suez: Diaries 1951–56* (London, 1986), p. 18.
62. I. McDonald, *A Man of the Times* (London, 1976), p. 132.
63. This question is fully examined in J. W. Young, 'Churchill's "No" to Europe: the "Rejection" of European Union by Churchill's Post-War Government, 1951–52', *Historical Journal*, 28, 4 (1985), and H. J. Yasamee, 'Anthony Eden and Europe, November 1951', *Foreign and Commonwealth Office Occasional Papers*, i (1987).
64. Yasamee, 'Eden and Europe', p. 43.
65. Charlton, *Price of Victory*, p. 148.
66. A. Nutting, *Europe Will Not Wait* (London, 1960), pp. 40–1.
67. Boothby, *My Yesterday*, p. 83; Lord Boothby, *Recollections of a Rebel* (London, 1978), pp. 220–1.
68. Diary 6 April 1969, AP 20/2/14.
69. Eden to Lord Aldington 2 May 1964, AP 23/4/87A.
70. *The Times*, 29 Nov. 1951.
71. *DBPO*, II, 1, no. 404.
72. *DBPO*, II, 1, no. 421.
73. *DBPO*, II, 1, no. 423.
74. '. . . much evidence that no one in Britain wished to join E.D.C. and that the six were perfectly happy with this'. Massigli to Eden 30 June 1969, AP 20/50/21A.
75. '. . . didn't leave any doubt about this. . . . Britain's participation in the formation of Europe was thus set out. It was derisory.' P. H. Spaak, *Combats Inachevés* (Brussels, 1969), vol. 2, pp. 47–8.
76. *DBPO*, II, 1, no. 406.
77. *DBPO*, II, 1, no. 402, calendar i.
78. Note by Eden on Boothby to Eden 15 Dec. 1951, AP 20/50/8.
79. Eden, *Full Circle*, p. 33.
80. J. Morgan, ed., *The Backbench Diaries of Richard Crossman* (London, 1981), p. 40.
81. *DBPO*, II, 1, no. 409.
82. Eisenhower to Eden 8 Dec. 1951, AP 20/15/1.
83. *DBPO*, II, 1, no. 414.
84. *DBPO*, II, 1, no. 416.
85. *New York Times*, 12 Jan. 1952; Charlton, *Price of Victory*, p. 160; J.W. Young, ed., *The Foreign Policy of Churchill's Peacetime Administration 1951–1955* (Leicester, 1988), p. 85.
86. Eden to Salisbury 12 March 1969, AP 23/60/152A.
87. Macmillan to Eden 21 Dec. 1951, AP 20/15/2.
88. *DBPO*, II, 1, no. 424. In his memoirs Macmillan struck a different note. The Permanent Under-Secretary's paper was 'based upon a complete misapprehension of the reality and strength of the movement for European unity. It indeed treated the whole of these efforts with a certain amount of levity and contempt': Macmillan, *Tides of Fortune*, p. 468.

89. *DBPO*, II, 1, no. 424, calendar.
90. *DBPO*, II, 1, no. 424.
91. *DBPO*, II, 1, no. 428.
92. *DBPO*, II, 1, no. 429.
93. *DBPO*, II, 1, no. 430.
94. Cabinet 12 March 1952, CAB 128/24, CC(52)29.
95. Macmillan, *Tides of Fortune*, p. 472; Macmillan to Lord Chandos 22 May 1969, AP 23/17/136B.
96. *DBPO*, II, 1, no. 437.
97. J. Colville, *The Fringes of Power* (London, 1985), p. 649.
98. P. H. Spaak, *The Continuing Battle* (London, 1971), p. 226; Monnet, *Memoirs*, p. 378.
99. H of C Debs, 5th series, vol. 499, col. 24.
100. CAB 129/52, C(52)189.
101. *DBPO*, II, 1, no. 463.
102. Cabinet 22 Feb. 1954, AP 20/54/7.
103. Monnet, *Memoirs*, p. 378.
104. Greenwood, *Britain and European Cooperation*, p. 58.
105. *DBPO*, II, 1, no. 466.
106. *DBPO*, II, 1, no. 505.
107. Greenwood, *Britain and European Cooperation*, p. 74.
108. Greenwood, *Britain and European Cooperation*, p. 59.
109. Cabinet 13 March 1952, CAB 128/24, CC(52)30.
110. Charlton, *Price of Victory*, p. 157.
111. Note by D. Dilks 2 Aug. 1962, AP 20/50/3.
112. *DBPO*, II, 1, no. 520.
113. Eden to R. Carr 2 July 1969, AP 23/16/50A.
114. Record of conversation 16 Dec. 1952, Papiers Massigli, Quai d'Orsay, 54, fos 228–34. I am grateful to Philip Bell for this reference.
115. Eden to Churchill 26 Feb. 1953, AP 20/16/50.
116. Eden, *Full Circle*, p. 58.
117. Shuckburgh, *Descent to Suez*, pp. 125–6.
118. Eden to Churchill 20 Feb 1954, AP 20/17/51.
119. Cabinet 27 Aug. 1954, AP 20/54/10.
120. Eden, *Full Circle*, p. 151.
121. Charlton, *Price of Victory*, p. 163; Nutting, *Europe Will Not Wait*, p. 71.
122. Jebb to Eden 30 Aug. 1954, AP 20/17/25.
123. M. Dockrill and J. W. Young, eds, *British Foreign Policy 1945–56* (London, 1989), p. 168.
124. Diary 5 Sept. 1954, AP 20/1/30.
125. Eden, *Full Circle*, p. 160.
126. Diary 1 Jan. 1957, AP 20/2/5.
127. Dockrill and Young, eds, *British Foreign Policy*, pp. 156–7.
128. N. Beloff, *Transit of Britain* (London, 1973), p. 89. See also L. B. Pearson, *Memoirs 1948–1957: The International Years* (London, 1974), p. 89.
129. Nutting, *Europe Will Not Wait*, p. 71.
130. Lord Gladwyn, *Memoirs* (London, 1972), p. 273.
131. Diary 15 Sept. 1954, AP 20/1/30.
132. Nutting, *Europe Will Not Wait*, p. 73.

133. Dockrill and Young, eds, *British Foreign Policy*, pp. 166–7; S. Dockrill, *Britain's Policy for West German Rearmament 1950–1955* (Cambridge, 1991), pp. 147–8.
134. Lloyd to Eden 4 Oct. 1954, AP 20/17/27.
135. Spaak, *Continuing Battle*, pp. 187–8.
136. H of C Debs, 5th series, vol. 533, col. 636.
137. J. Amery to Eden 5 Oct. 1954, AP 14/3/512.
138. Boothby to Eden 13 July 1953, AP 20/16/135.
139. H of C Debs, 5th series, vol. 555, col. 1674.
140. M. Ceadel, 'British Parties and the European Situation 1950–1957', in E. Di Nolfo, ed., *Power in Europe? II: Great Britain, France, Germany and Italy and the Origins of the EEC 1952–1957* (Berlin, 1992), p. 325.
141. Diary 23 Feb. 1969, AP 20/2/14.
142. R. Lamb, *The Failure of the Eden Government* (London, 1987), pp. 72–3.
143. Sir H. Ellis-Rees to Macmillan 11 Oct. 1955, cited Lamb, *Failure*, p. 81.
144. J. Turner, *Macmillan* (London, 1994), p. 98.
145. Spaak, *Continuing Battle*, pp. 233–4.
146. Notes for biographer 12 Nov. 1968 and 6 June 1968, AP 33/7ii. One recent study of this period has emphasized that 'ministers appear only infrequently in the story. . . . Eden hardly appears in this story.' S. Burgess and G. Edwards, 'The Six plus One: British policy-making and the question of European economic integration, 1955', *International Affairs*, 64, 3 (1988), pp. 394, 413.
147. Note for biographer 6 June 1968, AP 33/7ii.
148. Cabinets 14 and 18 Sept. 1956, CAB 128/30, CC(56)65 and 66.
149. H of C Debs, 5th series, vol. 561, cols 37–8.
150. Eden to R. Carr 26 May 1969, AP 23/16/49.
151. Cabinet 8 Jan. 1957, CAB 128/30, CC(57)3. See also Lloyd to Eden 15 Nov. 1974, AP 23/44/116A.
152. S. Lloyd, *Suez 1956: A Personal Account* (London, 1978), p. 236.
153. Note by Eden, Jan. 1957, PREM 11/1138.
154. See, for example, H. McRae, 'Still Learning the Lessons of Suez', *The Independent*, 25 Nov. 1993.
155. D. Dutton, 'Anticipating Maastricht: the Conservative Party and Britain's First Application to Join the European Community', *Contemporary Record*, 7, 3 (1993).
156. Eden to Chandos 5 Oct. 1962, AP 23/17/61A.
157. Draft speech, AP 20/50/108.
158. Strang to Eden 26 Sept. 1962, AP 20/50/23.
159. V. Rothwell, *Anthony Eden: A Political Biography 1931–57* (Manchester, 1992), p. 112.
160. B. Pimlott, 'E.D.C. and E.E.C.', AP 26/20/3A.
161. Morgan, ed., *Crossman Backbench Diaries*, p. 951.
162. R. Bullen, 'Britain and Europe 1950–57', in Nolfo, *Power in Europe*, p. 501.
163. Shuckburgh to Eden 1 May 1969, AP 24/61/13.
164. Eden to Massigli 20 June 1969, AP 23/50/21.
165. *DBPO*, II, 1, no. 501.
166. Boothby to Eden 11 July 1966, AP 20/50/55.
167. Boothby to Eden 20 April 1955, AP 14/3/796.
168. J. W. Young, *Britain and European Unity, 1945–1992* (London, 1993), p. 53.
169. Macmillan, *Tides of Fortune*, pp. 480–1.
170. Turner, *Macmillan*, p. 91.
171. Eden to Chandos 12 Oct. 1962, AP 23/17/62A.

Notes

CHAPTER 11

1. Article by Eden, 'Soviet Foreign Policy: the Chameleon and the Bear', 6 Oct. 1949, AP 33/1/6.
2. *Documents on British Policy Overseas*, series I, vol. 1, no. 226.
3. W. S. Churchill, *Triumph and Tragedy* (London, 1954), pp. 546–7.
4. T. H. Anderson, *The United States, Great Britain and the Cold War 1944–1947* (London, 1981), p. 84.
5. R. Ovendale, 'Britain, the U.S.A. and the European Cold War, *1945–8*', *History*, 67 (1982), pp. 221–2.
6. *DBPO*, I, 1, no. 102.
7. Diary 26 July 1945, AP 20/1/25.
8. Diary 28 July 1945, AP 20/1/25.
9. A. Bullock, *Ernest Bevin: Foreign Secretary* (London, 1983), p. 80.
10. Diary 1 Aug. 1945, AP 20/1/25.
11. K. Young, ed., *The Diaries of Sir Robert Bruce Lockhart 1939–1965* (London, 1980), p. 484.
12. Young, ed., *Lockhart Diaries*, p. 498.
13. T. M. Campbell and G. C. Herring, eds, *The Diaries of Edward R. Stettinius* (New York, 1975), p. 423.
14. P. Dixon, *Double Diploma: The Life of Sir Pierson Dixon, Don and Diplomat* (London, 1968), p. 182.
15. Diary 10 Aug. 1945, AP 20/1/25.
16. J. F. Byrnes, *Speaking Frankly* (London, 1948), p. 79.
17. House of Commons Debates, 5th series, vol. 413, col. 321.
18. Lord Avon, *Facing the Dictators* (London, 1962), p. 486. Eden's distaste for partisan politics may have been one of the factors which prompted calls for more vigorous opposition from a critical meeting of the 1922 Committee in November 1945.
19. Young, ed., *Lockhart Diaries*, p. 498; Byrnes, *Speaking Frankly*, p. 79.
20. R. Ovendale, ed., *The Foreign Policy of the British Labour Governments 1945–1951* (Leicester, 1984), p. 2.
21. Dixon, *Double Diploma*, p. 179. Sir Frank Roberts has also testified that he acted as an unofficial liaison between the Labour Foreign Secretary and his Conservative predecessor: Ovendale, ed., *Foreign Policy of British Labour Governments*, p. 23.
22. B. Jones, *The Russia Complex* (Manchester, 1977), p. 117.
23. Diary 19 Oct. 1945, AP 20/1/25.
24. H of C Debs, 5th series, vol. 416, cols 609–19.
25. Dixon, *Double Diploma*, p. 199.
26. Young, ed., *Lockhart Diaries*, p. 511.
27. Diary 25 Dec. 1945, AP 20/1/25. See also B. Pimlott, ed., *The Political Diary of Hugh Dalton* (London, 1986), p. 367.
28. Eden to Halifax 17 Jan. 1946, Halifax MSS A4.410.4.15.
29. Young, ed., *Lockhart Diaries*, p. 526.
30. H of C Debs, 5th series, vol. 419, cols 1340–2.
31. H of C Debs, 5th series, vol. 420, col. 236.
32. Cranborne to Eden 13 March 1946, AP 20/43/17; Eden to Cranborne 15 March 1946, AP 20/43/17A; see also N. Nicolson, ed., *Harold Nicolson: Diaries and Letters 1945–1962* (London, 1968), p. 63.
33. N. Mansergh, *The Transfer of Power*, vol. 9 (London, 1980), p. 190; N. Tiratsoo, ed., *The Attlee Years* (London, 1991), p. 159.

34. Young, ed., *Lockhart Diaries*, pp. 570–1.
35. Eden to J. P. L. Thomas 26 Dec. 1946, AP 20/14/38.
36. A reference to the American Liberal, Henry Wallace, who had just had to resign his post as President Truman's Secretary of Commerce.
37. A. Eden, *Freedom and Order* (London, 1947), pp. 413–17; D. Carlton, *Anthony Eden* (London, 1981), pp. 269–70.
38. R. R. James, *Anthony Eden* (London, 1986), pp. 319–20.
39. H of C Debs, 5th series, vol. 457, col. 21.
40. Young, ed., *Lockhart Diaries*, p. 670.
41. Eden, 'Soviet Foreign Policy: the Chameleon and the Bear', 6 Oct. 1949, AP 33/1/6.
42. A. Eden, *Days for Decision* (London, 1949), p. 24.
43. Eden, *Days for Decision*, p. 184.
44. H of C Debs, 5th series, vol. 472, cols 50–1.
45. H of C Debs, 5th series, vol. 477, cols 578–86.
46. Note for biographer 5 Sept. 1973, AP 33/7.
47. Diary note 1951, AP 20/1/27.
48. A. Cairncross, *Years of Recovery* (London, 1985), p. 500.
49. Carlton, *Eden*, p. 291.
50. D. Reynolds, 'Eden the Diplomatist, 1931–56: Suezide of a Statesman?' *History*, 74, 240 (1989), p. 75; Eden visited Canada, New Zealand, Australia and India between January and March 1949.
51. A. Eden, 'Anglo-American Relations', *Everybodys*, 16 June 1951, AP 33/1/6.
52. Carlton, *Eden*, p. 299.
53. Eden to O. Franks 10 Sept. 1951, AP 16/1/303A.
54. A. Nutting, *No End of a Lesson: The Story of Suez* (London, 1967), p. 19.
55. A. Adamthwaite, in J. W. Young, ed., *The Foreign Policy of Churchill's Peacetime Administration, 1951–1955* (Leicester, 1988), p. 10.
56. A. Shlaim, P. Jones and K. Sainsbury, *British Foreign Secretaries since 1945* (Newton Abbot, 1977), p. 91.
57. Shlaim *et al.*, *British Foreign Secretaries*, p. 108.
58. See above pp. 279–313.
59. F. Roberts, *Dealing with Dictators: The Destruction and Revival of Europe 1930–70* (London, 1991), p. 163. One of Eden's most persistent critics for his lack of long-term planning both during the war and after 1951 was the Foreign Office official, Gladwyn Jebb. Jebb, of course, became a committed supporter of European integration and was also closely associated with the 'Russia Committee' established early in 1946 to co-ordinate Foreign Office policy for containing Soviet expansionism. On his return to office Eden soon allowed the committee to lapse. Interview with Lord Gladwyn, 22 July 1993.
60. H of C Debs, 5th series, vol. 494, cols 34–53.
61. Young, ed., *Churchill's Peacetime*, p. 13.
62. J. Morgan, ed., *The Backbench Diaries of Richard Crossman* (London, 1981), p. 37.
63. H. Morrison, *An Autobiography* (London, 1960), p. 273.
64. I. McDonald, *A Man of the Times* (London, 1976), p. 133.
65. A. Nutting, in H. van Thal, ed., *The Prime Ministers: From Lord John Russell to Edward Heath* (London, 1975), p. 336.
66. H. Macmillan, *Tides of Fortune 1945–55* (London, 1969), pp. 486–7.
67. E. Shuckburgh, *Descent to Suez: Diaries 1951–56* (London, 1986), p. 27; see also p. 14.
68. Private information.
69. Van Thal, ed., *Prime Ministers*, p. 341.

Notes

70. C. L. Sulzberger, *A Long Row of Candles: Memoirs and Diaries 1934–54* (London, 1969), p. 629.
71. A. Seldon, *Churchill's Indian Summer: The Conservative Government, 1951–55* (London, 1981), p. 380.
72. V. Bartlett, *And Now, Tomorrow* (London, 1960), p. 204.
73. McDonald, *Man of the Times*, pp. 75–6.
74. Shuckburgh, *Descent to Suez*, p. 19; Seldon, *Indian Summer*, p. 409.
75. B. Lapping, *End of Empire* (London, 1985), p. 255.
76. Paper by Eden 18 June 1952, C(52)292, CAB 129/53.
77. C(52)202, CAB 129/53.
78. Lord Kilmuir, *Political Adventure* (London, 1964), p. 193.
79. Seldon, *Indian Summer*, p. 85.
80. P. M. Williams, ed., *The Diary of Hugh Gaitskell 1945–1956* (London, 1983), p. 307.
81. Shuckburgh, *Descent to Suez*, p. 25.
82. Lord Moran, *Winston Churchill: The Struggle for Survival 1940–1965* (London, 1966), pp. 358–9.
83. Simonds to Eden 22 Dec. 1952, AP 16/1/311A.
84. Diary 4 March 1952, AP 20/1/28.
85. H of C Debs, 5th series, vol. 494, col. 34.
86. D. Acheson, *Present at the Creation: My Years in the State Department* (London, 1970), pp. 594–606.
87. Shuckburgh, *Descent to Suez*, p. 32.
88. Macmillan diary 17 Jan. 1952, cited A. Horne, *Macmillan 1894–1956* (London, 1988), p. 347.
89. Morgan, ed., *Crossman Backbench Diaries*, p. 70.
90. D. S. McLellan, *Dean Acheson: The State Department Years* (New York, 1976), p. 390.
91. Diary 1957, AP 20/2/5.
92. A. Eden, *Full Circle* (London, 1960), p. 200; Acheson, *Present at the Creation*, p. 578.
93. See correspondence in AP 23. After one meeting in February 1971, Eden noted: 'We think so much alike' (diary 5 Feb. 1971, AP 20/2/16). But at least one observer believed that Acheson had 'nothing but contempt' for Eden and only saw him 'for old times' sake': C. King, *The Cecil King Diary 1965–1970* (London, 1972), p. 319.
94. *Newsweek*, 14 April 1952.
95. Acheson, *Present at the Creation*, p. 605.
96. H. Brandon, *Special Relationships* (London, 1988), p. 69.
97. L. Pearson, *Memoirs 1948–1957: The International Years* (London, 1974), p. 78.
98. Pearson, *International Years*, p. 328.
99. Diary 4 June 1952, AP 20/1/28.
100. Cabinet 24 June 1952, CC(52)63, CAB 128/25.
101. Morgan, ed., *Crossman Backbench Diaries*, p. 114.
102. Minute by Eden 1 Sept. 1952, cited M. Dockrill and J. W. Young, eds, *British Foreign Policy 1945–56* (London, 1989), p. 138.
103. Eden to Foreign Office 23 Nov. 1952, cited Young, ed., *Churchill's Peacetime*, p. 221.
104. Private information.
105. Shuckburgh, *Descent to Suez*, pp. 53–4.
106. Shuckburgh, *Descent to Suez*, p. 57.
107. Carlton, *Eden*, p. 323.

108. D. Eisenhower, *Mandate for Change 1953–1956* (New York, 1963), p. 142.

109. J. Wheeler-Bennett and A. Nicholls, *The Semblance of Peace: The Political Settlement after the Second World War* (London, 1972), pp. 619–20.

110. Eden to Wheeler-Bennett 22 July 1970, AP 33/4.

111. Eden to Churchill 21 Nov. 1952, PREM 11/323.

112. Eden, *Full Circle*, p. 63.

113. D. Kunz, *The Economic Diplomacy of the Suez Crisis* (Chapel Hill, 1991), p. 29.

114. D. Neff, *Warriors at Suez: Eisenhower takes America into the Middle East* (New York, 1981), p. 145.

115. L. Mosley, *Dulles* (London, 1978), p. 353.

116. Shuckburgh, *Descent to Suez*, p. 55.

117. R. Makins to Eden 9 Jan. 1953, AP 20/16/23.

118. R. H. Ferrell, ed., *The Eisenhower Diaries* (New York, 1981), p. 223.

119. Diary 4 March 1953, AP 20/1/29.

120. Eden to Churchill 9 March 1953, PREM 11/666.

121. Eisenhower to Eden 16 March 1953, AP 20/16/25A; Eisenhower, *Mandate*, p. 152.

122. Eden to Eisenhower 1 April 1953, AP 20/16/26A.

123. Eden, *Full Circle*, p. 49.

124. Eden minute 21 March 1953, cited V. H. Rothwell, *Anthony Eden: A Political Biography 1931–57* (Manchester, 1992), p. 136.

125. Eden to Churchill 28 March 1953, AP 20/16/57.

126. Diary 2 April 1953, AP 20/1/29.

127. Diary 3 April 1953, AP 20/1/29.

128. J. Colville, *The Fringes of Power: Downing Street Diaries 1939–1955* (London, 1985), p. 673.

129. A. Nutting, *Europe Will Not Wait* (London, 1960), p. 50; D. R. Thorpe, *Selwyn Lloyd* (London, 1989), p. 171.

130. H of C Debs, 5th series, vol. 515, col. 897.

131. Eden to Churchill 9 Nov. 1953, AP 20/16/81.

132. M. S. Fish, 'After Stalin's Death: the Anglo-American Debate over a New Cold War', *Diplomatic History*, 10 (1986), p. 354.

133. Diary 1 Oct. 1953, AP 20/1/29A.

134. Diary 1 Oct. 1953, AP 20/1/29A.

135. Diary 19 Oct. 1953, AP 20/1/29A.

136. Draft paper submitted to Churchill *c.* 19 Nov. 1953, AP 20/16/85A.

137. Eden to Churchill 25 Nov. 1953, AP 20/16/87.

138. C(53)330, CAB 129/64; J. Cable, *The Geneva Conference of 1954 on Indochina* (London, 1986), p. 34.

139. McDonald, *Man of the Times*, p. 134.

140. Private information.

141. Diary 6 Dec. 1953, AP 20/1/29.

142. Diary, 10 Dec. 1953, AP 20/1/29.

143. J. W. Young, 'Churchill, the Russians and the Western Alliance: The Three-Power Conference at Bermuda, December 1953', *English Historical Review*, 401 (1986), p. 906; Sulzberger, *Long Row of Candles*, p. 781; diary 6 Dec. 1953, AP 20/1/29.

144. McDonald, *Man of the Times*, p. 135.

145. Colville, *Fringes of Power*, p. 683. From the Moscow embassy William Hayter confirmed Eisenhower's analysis: 'The girl inside the dress is the same girl, exercising the same profession. . . . We have yet to see the sight of a good

Stalinist who feels that his master's heritage has been betrayed by Malenkov.' (Hayter to Eden 1 Jan. 1954, AP 20/17/14.)

146. Dockrill and Young, eds, *British Foreign Policy,* p. 145.
147. K. Ruane, 'Anthony Eden, British Diplomacy and the Origins of the Geneva Conference of 1954', *Historical Journal,* 37,1 (1994), pp. 156–7.
148. Memorandum for cabinet 11 Jan. 1954, C(54)13, CAB 129/65.
149. Ruane, 'Origins of Geneva Conference', pp. 159–60.
150. Eden to Churchill 25 Nov. 1953, AP 20/16/87.
151. This episode is well covered in Ruane, 'Origins of Geneva Conference'.
152. Carlton, *Eden,* p. 339.
153. Morgan, ed., *Crossman Backbench Diaries,* p. 289.
154. Diary 25 Jan. 1954, AP 20/1/30.
155. Eden to Lloyd 6 Feb. 1954, AP 20/17/10A.
156. Shuckburgh, *Descent to Suez,* p. 133.
157. Private information.
158. Cable, *Geneva Conference,* p. 57; Young, ed., *Churchill's Peacetime,* p. 244; T. Hoopes, *The Devil and John Foster Dulles* (London, 1974), pp. 215–16.
159. Shuckburgh, *Descent to Suez,* p. 189.
160. FO 371/112053; Cable, *Geneva Conference,* pp. 58–9.
161. Sulzberger, *Long Row of Candles,* p. 852; Hoopes, *Devil and Dulles,* p. 216.
162. McDonald, *Man of the Times,* p. 137.
163. G. C. Herring and R. H. Immerman, 'Eisenhower, Dulles and Dienbienphu: "The Day We Didn't Go to War" Revisited', *Journal of American History,* 71, 2 (1984), p. 360.
164. Ferrell, ed., *Eisenhower Diaries,* pp. 279–80.
165. Eden to Churchill 28 April 1954, AP 20/17/34D.
166. Ruane, 'Origins of Geneva Conference', p. 154.
167. Cable, *Geneva Conference,* p. 3. This judgement perhaps does less than justice to Eden's own subsequent resolution of the crisis over German rearmament.
168. Eden to Lloyd 21 May 1954, AP 20/17/15A.
169. Diary 3 Feb. 1974, AP 20/2/19.
170. Diary 29 April 1954, AP 20/1/30.
171. Eden, *Full Circle,* p. 127.
172. Eisenhower, *Mandate,* p. 355.
173. Shuckburgh, *Descent to Suez,* p. 186.
174. Shuckburgh, *Descent to Suez,* p. 193.
175. T. B. Millar, ed., *Australian Foreign Minister: The Diaries of R. G. Casey 1951–60* (London, 1972), p. 157. For further tributes see Cable, *Geneva Conference,* pp. 33, 64, and H. Trevelyan, *Worlds Apart* (London, 1971), p. 78.
176. Diary 3 May 1954, AP 20/1/30.
177. Diary 4 May 1954, AP 20/1/30; Shuckburgh, *Descent to Suez,* p. 191.
178. Eden to Churchill 5 May 1954, AP 20/17/35A.
179. Eden to Salisbury 16 May 1954, AP 20/17/118A.
180. Salisbury to Eden 9 May 1954, AP 20/17/118.
181. Makins to Eden 21 May 1954, AP 20/17/18A.
182. Young, ed., *Churchill's Peacetime,* p. 253.
183. Eden to Makins 10 June 1954, AP 14/3/654A.
184. Eden, *Full Circle,* p. 131; Cable, *Geneva Conference,* p. 105; R. Dingman, 'John Foster Dulles and the Creation of the South-East Asia Treaty Organisation in 1954', *International History Review,* 11, 3 (1989), p. 460.

185. Moran, *Churchill*, p. 558.

186. Makins to Eden 18 June 1954, AP 20/17/19.

187. Moran, *Churchill*, p. 564; see also Colville, *Fringes of Power*, p. 694.

188. Moran, *Churchill*, p. 566; see also Seldon, *Indian Summer*, pp. 392–3, and P. G. Boyle, ed., *The Churchill–Eisenhower Correspondence, 1953–1955* (Chapel Hill, 1990), p. 154.

189. Private information.

190. J. W. Young, 'Churchill's Bid for Peace with Moscow, 1954', *History,* 73, 239 (1988), pp. 435–6; Moran, *Churchill*, p. 575; J. Wheeler-Bennett, ed., *Action This Day: Working with Churchill* (London, 1968), pp. 135–6.

191. Young, 'Churchill's Bid', pp. 437–9; cabinets 7, 8 and 9 July 1954, CC(54)47–9, CAB 128/27.

192. Salisbury to Eden 15 July 1954, AP 20/17/144.

193. O. Lyttelton to Eden 16 July 1954, AP 20/17/146.

194. Moran, *Churchill*, p. 576; Colville, *Fringes of Power*, pp. 699–700.

195. McDonald, *Man of the Times*, p. 138.

196. Diary 31 Dec. 1954, AP 20/1/30.

197. Macmillan diary 20 July 1954, cited Macmillan, *Tides of Fortune*, p. 537.

198. Eden to Churchill 22 July 1954, AP 20/17/79.

199. Cabinet 23 July 1954, CC(54)52, CAB 128/27.

200. Eden to Churchill 30 July 1954, AP 20/17/82.

201. Diary 1 Aug. 1954, AP 20/1/30; Salisbury to Eden 20 Aug. 1954, AP 20/17/176.

202. C. J. Bartlett, *'The Special Relationship': A Political History of Anglo-American Relations since 1945* (London, 1992), p. 71.

203. Dulles to Eden 3 Oct. 1954, AP 14/3/587.

204. Acheson to Eden 1 Nov. 1954, AP 14/3/506A.

205. Nutting, in van Thal, ed., *Prime Ministers*, p. 329.

206. Dingman, 'South-East Asia Treaty Organisation', pp. 465–6.

207. Eden to Makins 20 Jan. 1955, cited Dockrill and Young, eds, *British Foreign Policy,* pp. 182–3.

208. Millar, ed., *Australian Foreign Minister*, p. 206.

209. Eden note *c.* 19 April 1955, cited Dockrill and Young, eds, *British Foreign Policy,* p. 187.

210. M. Charlton, *The Price of Victory* (London, 1983), p. 172.

211. Shuckburgh, *Descent to Suez*, p. 187.

212. Cabinet 4 Oct. 1955, CC(55)34, CAB 128/29.

213. W. S. Lucas, in A. Deighton, ed., *Britain and the First Cold War* (London, 1990), pp. 254–5.

214. Young, ed., *Churchill's Peacetime*, p. 167.

215. Eden to Wheeler-Bennett 18 Aug. 1970, AP 33/4.

216. M. Heikal, *Nasser: The Cairo Documents* (London, 1972), p. 12.

217. D. Eisenhower, *Waging Peace 1956–1961* (London, 1966), pp. 22–3; R. Ovendale, *The Origins of the Arab–Israeli Wars* (London, 1984), p. 136.

218. D. R. Devereux, *The Formulation of British Defence Policy Towards the Middle East 1948–56* (London, 1990), pp. 106–7.

219. Devereux, *British Defence Policy,* pp. 24–5.

220. Amery to Eden 26 May 1951, AP 14/3/56.

221. J. Barnes and D. Nicholson, eds, *The Empire at Bay: The Leo Amery Diaries 1929–1945* (London, 1988), p. 1063.

222. Copy by A. Grant, Eden's research assistant, AP 33/3.

223. Eden, 'Defence of the Middle East' 13 April 1945, WP(45)256, CAB 66/65.
224. Shuckburgh, *Descent to Suez*, p. 210.
225. Cabinets 14 and 18 Feb. 1952, CC(52)17,18,19, CAB 128/24.
226. Colville, *Fringes of Power*, p. 643.
227. Churchill to Eden March 1952, cited Young, ed., *Churchill's Peacetime*, p. 138.
228. Cabinet 1 April 1952, CC(52)35, CAB 128/24.
229. Cabinet 4 April 1952, CC(52)37, CAB 128/24.
230. See, for example, Lloyd to Eden 6 March 1953, AP 14/3/354.
231. Macmillan to Eden 28 Oct. 1952, AP 20/44/185.
232. Colville, *Fringes of Power*, p. 645.
233. Shuckburgh, *Descent to Suez*, p. 75.
234. Diary 9 Feb. 1953, AP 20/1/29.
235. Cabinet 11 Feb. 1953, CC(53)9, CAB 128/26.
236. Buchan-Hepburn to Eden 11 Feb. 1953, AP 11/10/97.
237. Cabinet 11 Feb. 1953, CC(53)10, CAB 128/26.
238. Shuckburgh, *Descent to Suez*, p. 77.
239. Eden, 'Egypt: the Alternatives', 16 Feb. 1953, C(53)65, CAB 129/59.
240. Eden to Hankey 25 Feb. 1953, PREM 11/636.
241. Eden, *Full Circle*, p. 198.
242. Memorandum 5 Aug. 1952, C(52)276, CAB 129/54.
243. Diary 17 Dec. 1954, AP 20/1/30.
244. Eden to Churchill 11 April and 1 Dec. 1953, AP 20/16/61, 88.
245. Pearson, *International Years*, p. 187; Eden to Foreign Office 6 March 1953, cited Devereux, *British Defence Policy*, pp. 130–1.
246. Diary 9 March 1953, AP 20/1/29. See above pp. 335–6.
247. Memorandum by Dulles, cited K. Love, *Suez: The Twice-Fought War* (London, 1969), p. 194.
248. Devereux, *British Defence Policy*, p. 156.
249. Cabinet 15 Oct. 1953, CC(53)38, CAB 128/26.
250. S. I. Troen and M. Shemesh, eds, *The Suez–Sinai Crisis 1956* (London, 1990), pp. 112–15; Lapping, *End of Empire*, p. 255.
251. Barnes and Nicholson, eds, *Empire at Bay*, p. 1064.
252. Sulzberger, *Long Row of Candles*, p. 782; draft Eden to Churchill 30 Nov. 1953, AP 20/16/158.
253. Churchill to Eden 11 Dec. 1953, AP 20/16/21; Shuckburgh, *Descent to Suez*, p. 118.
254. Eden to Churchill 21 Dec. 1953, PREM 11/484; Eden, *Full Circle*, pp. 256–7.
255. Briefing paper 15 Jan. 1954, cited Devereux, *British Defence Policy*, p. 158.
256. Eden, 'Middle Eastern Policy', 7 Jan. 1954, cited Deighton, ed., *Britain and the First Cold War*, p. 262.
257. Diary 22 Jan. 1954, AP 20/1/54.
258. Lloyd to Eden 3 Feb. 1954, AP 20/17/10.
259. Shuckburgh, *Descent to Suez*, p. 151.
260. Boyle, ed., *Churchill–Eisenhower Correspondence*, p. 148.
261. Cabinet 22 June 1954, CC(54)43, CAB 128/27.
262. Eden to Churchill 21 June 1954, PREM 11/702.
263. Shuckburgh, *Descent to Suez*, p. 15.
264. Heikal, *Cairo Documents*, p. 79.

Notes

CHAPTER 12

1. Diary Jan. 1957, AP 20/2/5.
2. Eden to R. Blake 13 Dec. 1959, AP 33/3.
3. K. Kyle, *Suez* (London, 1991), and W. S. Lucas, *Divided We Stand: Britain, the United States and the Suez Crisis* (London, 1991), are outstanding. But see also D. Kunz, *The Economic Diplomacy of the Suez Crisis* (Chapel Hill, 1991), D. Carlton, Britain and the Suez Crisis (Oxford, 1988) and W. R. Louis and R. Owen, eds, *Suez 1956: The Crisis and its Consequences* (Oxford, 1989). H. Thomas, *The Suez Affair* (London, 1967), remains valuable as it is based on the (largely unattributed) testimony of key participants.
4. Correspondence with R. Lacey in AP 23/42. It is a curious commentary on the White Paper on Open Government that this innocuous file, having been open to inspection, has now been reclosed at the request of the Cabinet Office. P. Ziegler, ed., *From Shore to Shore: The Final Years; The Diaries of Earl Mountbatten of Burma 1953–1979* (London, 1989), pp. 340–1.
5. The most obvious example is the British copy of the Protocol of Sèvres. When Eden was questioned on the issue of the burning of files, he declined to comment: Lord Kennet to Eden 21 June 1971, AP 23/10/108A.
6. A. Deighton, ed., *Britain and the First Cold War* (London, 1990), pp. 261–2.
7. M. Copeland, *The Game of Nations* (London, 1969), p. 78.
8. Eden to Churchill 21 Feb. 1955, AP 20/23/3.
9. Cabinet 15 March 1955, CM(55)24, CAB 128/28.
10. Eden to Makins 24 March 1955, AP 20/23/28.
11. Eden to H. Beeley 31 March 1955, FO 371/121282.
12. Eden to Eisenhower 17 July 1955, AP 20/23/64.
13. Makins to Foreign Office 19 Aug. 1955, AP 20/23/73.
14. Eden to Macmillan 19 Aug. 1955, AP 20/20/58.
15. Diary 30 Aug. 1955, AP 20/1/31.
16. Diary 3 Oct. 1955, AP 20/1/31.
17. Shuckburgh to Macmillan 23 Sept. 1955, cited Louis and Owen, eds, *Suez*, p. 106.
18. Eden to Macmillan 27 Sept. 1955, AP 20/22/156.
19. Eden's comments on Stewart (Ankara) to Foreign Office 14 Oct. 1955, AP 20/23/113.
20. Cabinet 20 Oct. 1955, CM(55)36, CAB 128/29.
21. Memorandum of Dulles–Macmillan conversation 3 Oct. 1955, cited Kunz, *Economic Diplomacy,* p. 46.
22. Macmillan to Stewart 20 Oct. 1955 , AP 20/23/122.
23. Cabinet 20 Oct. 1955, CM(55)36, CAB 128/29.
24. Eden to Eisenhower 26 Nov. 1955, FO 371/113739.
25. Eden to Macmillan 6 Nov. 1955, AP 20/23/129; Duke to Macmillan 18 Nov. 1955, AP 20/23/154.
26. Cabinet 24 Nov. 1955, CM(55)43, CAB 128/29.
27. Eden to Eisenhower 17 Nov. 1955, AP 20/27/29.
28. Record of meeting 9 Nov. 1955, AP 20/23/136.
29. Against this Eden minuted, 'I fear they will be very reluctant before the election' – a sentiment he would have been wise to remember the following year.
30. Macmillan to Eden 25 Nov. 1955, AP 20/23/163.
31. Eden to Makins 20 Nov. 1955, AP 20/22/270.
32. Against this Eden simply minuted 'No'.

33. Macmillan to Dulles 25 Nov. 1955, AP 20/23/164B, and Dulles to Macmillan 6 Dec. 1955, AP 20/23/179.
34. Eden denied this. K. Love, *Suez: The Twice-Fought War* (London, 1969), p. 201; H. Trevelyan, *The Middle East in Revolution* (London, 1970), pp. 56–7.
35. Cabinet 20 Oct. 1955, CM(55)36, CAB 128/29; see also Eden to Trevelyan 27 Oct. 1955, PREM 11/859.
36. William Clark diary 29 Nov. 1955.
37. R. Lamb, *The Failure of the Eden Government* (London, 1987), p. 173.
38. Eden to Nehru 9 Dec. 1955, AP 20/22/321.
39. S. Adams, *First Hand Report* (London, 1962), p. 196.
40. Eden to Nehru 9 Dec. 1955, AP 20/22/321.
41. S. I. Troen and M. Shemesh, eds, *The Suez–Sinai Crisis 1956: Retrospective and Reappraisal* (London, 1990), p. 104; Eden later commented that 'Foster was incurable in this conviction which had done untold harm to his relations with the French and ourselves': diary 31 July 1970, AP 20/2/15.
42. Eden to Makins 28 Dec. 1955, AP 20/22/343.
43. Dulles to Hoover 16 Dec. 1955, cited Kunz, *Economic Diplomacy,* p. 56.
44. Eden to Eisenhower 16 Jan. 1956, AP 20/27/37.
45. E. Shuckburgh, *Descent to Suez: Diaries 1951–56* (London, 1986), p. 326.
46. Private information.
47. Shuckburgh, *Descent to Suez*, p. 327.
48. Shuckburgh, *Descent to Suez*, p. 332.
49. Eden to Lloyd and Butler 31 Jan. 1956, AP 20/24/69; private information.
50. Eisenhower diary 9 Feb. 1956, cited T. Petersen, 'Anglo-American Rivalry in the Middle East: the Struggle for the Buraimi Oasis, 1952–57', *International History Review*, 14, 1 (1992), pp. 86–7.
51. Diary Jan. 1957, AP 20/2/5.
52. Lloyd to Eden 2 March 1956, AP 20/24/140; Trevelyan, *Middle East in Revolution*, pp. 64–5.
53. A. Nutting, *No End of a Lesson* (London, 1967), p. 29; Nutting claimed that he 'spent the evening and half of the night' after Glubb's dismissal arguing with Eden (p. 18). Eden expended some energy trying to use his engagement diary to show that Nutting had not even stayed for dinner. Unfortunately Eden's information related to Sunday, 4 March: note for biographer 27 Nov. 1970, AP 33/7.
54. Duke to Foreign Office 2 March 1956, AP 20/32/66.
55. Shuckburgh, *Descent to Suez*, p. 341.
56. Eden to Duke 2 March 1956, AP 33/6.
57. Lloyd to Eden 3 March 1956, AP 20/24/145.
58. Lamb, *Failure*, p. 191; T. Royle, *Glubb Pasha* (London, 1993 paperback edn), p. 465.
59. M. Heikal, *Cutting the Lion's Tail: Suez through Egyptian Eyes* (London, 1986), p. 97.
60. Shuckburgh, *Descent to Suez*, p. 340; see also p. 344: 'Yesterday the P.M., in order to placate the party, told the Foreign Affairs Committee of his plans (a) to get the US Government to join the Bagdad Pact (which they won't . . .) and (b) to get the King of Iraq to persuade the King of Jordan to come back into the fold.'
61. Woolton diary 24 Oct. 1955.
62. William Clark diary 24 Oct. 1955.
63. William Clark diary 2 Jan. 1956; Eden to J. Langford-Holt 11 April 1956, AP 20/31/272.
64. Lord Kilmuir, *Political Adventure* (London, 1964), p. 257; see also W. Clark, *From Three Worlds* (London, 1986), p. 155.

65. Shuckburgh, *Descent to Suez*, pp. 330–1.

66. For the origins of this piece see D. Hart-Davis, *The House the Berrys built: Inside The Telegraph* (London, 1991 paperback edn), p. 214.

67. Diary 17 Feb. 1957, AP 20/2/5.

68. Shuckburgh, *Descent to Suez*, p. 337.

69. Eden to Eisenhower 5 March 1956, PREM 11/1177.

70. Cabinet 6 March 1956, CM(56)19, CAB 128/30.

71. Eden to Lloyd 6 March 1956, AP 20/24/156.

72. Clark, *From Three Worlds*, p. 162.

73. J. Morgan, ed., *The Backbench Diaries of Richard Crossman* (London, 1981), p. 475. After this speech the journalist Ian Waller predicted that by 1957 Macmillan would replace Eden in Downing Street.

74. Shuckburgh, *Descent to Suez*, p. 345.

75. Lloyd to Eden 7 March 1956, AP 20/24/159.

76. Lloyd to Eden 7 March 1956, AP 20/24/161.

77. Eisenhower to Eden 10 March 1956, AP 20/27/50.

78. Shuckburgh, *Descent to Suez*, p. 346.

79. Eden to Eisenhower 15 March 1956, PREM 11/1177; R. H. Ferrell, ed., *The Eisenhower Diaries* (New York, 1981), p. 321.

80. Macmillan diary 16 March 1956, H. Macmillan, *Riding the Storm 1956–1959* (London, 1971), p. 93.

81. Cabinet 21 March 1956, CM(56)24, CAB 128/30; S. Lloyd, *Suez 1956: A Personal Account* (London, 1978), p. 60.

82. Eden to Eisenhower 19 March 1956, AP 33/6.

83. Dulles memorandum 28 March 1956, cited G. Warner, 'The United States and the Suez Crisis', *International Affairs*, 67, 2 (1991), p. 307; Kunz, *Economic Diplomacy*, pp. 64–5.

84. O'Connor memorandum 10 April 1956, cited Warner, 'United States and Suez', p. 304.

85. Eisenhower to Eden 5 April 1956, AP 33/6.

86. L. Pearson, *The International Years: Memoirs 1948–1957* (London, 1974), p. 225.

87. Lloyd to Eden 13 March 1956, AP 20/24/182.

88. Eden to Lloyd 18 March 1956, AP 20/21/46.

89. Lloyd, *Suez*, pp. 68–9.

90. Cabinet 17 July 1956, CM(56)50, CAB 128/30.

91. Undated (1957) memorandum by G. Millard, AP 20/46/2.

92. A. Eden, *Full Circle* (London, 1960), p. 422.

93. Louis and Owen, eds, *Suez*, p. 110.

94. P. M. Williams, ed., *The Diary of Hugh Gaitskell 1945–1956* (London, 1983), p. 553.

95. William Clark testimony, 'Suez 1956: Neither War nor Peace at 10 Downing Street', BBC Radio Three broadcast (1979).

96. Cabinet 27 July 1956, CM(56)54, CAB 128/30. Despite public statements to the contrary, the government's legal advisers did not regard Nasser's action as illegal, or at least as justifying military intervention. The exception was the Lord Chancellor, Lord Kilmuir, who sought to justify the use of force in self-defence under Article 51 of the UN Charter. But international law was not an area where Kilmuir could claim any particular expertise (R. V. F. Heuston, *Lives of the Lord Chancellors 1940–1970* (Oxford, 1987), p. 170). As Sir Gerald Fitzmaurice, legal adviser to the Foreign Office, put it: 'Whatever illegalities the Egyptians may have committed . . . these do not in any way . . . justify forcible action on our part' (Louis and Owen, eds,

Notes

Suez, p. 114). On 5 November Fitzmaurice took steps to distance himself from the allied intervention (Kunz, *Economic Diplomacy*, p. 129).

97. I. McDonald, *A Man of the Times* (London, 1976), pp. 142–4; I. McDonald, *The History of the Times: Struggles in War and Peace 1939–1966* (London, 1984), p. 261.
98. Lord Butler, *The Art of the Possible* (London, 1971), pp. 188–9.
99. 'I have often been struck by the similarities between Suez and the Rhineland. In both instances the dictator was acting within his own territory while breaking an international agreement': Eden to P. Emrys-Evans 6 Aug. 1963, AP 23/31/12A.
100. Eden to Lord Lambton 16 June 1967, AP 23/43/91.
101. Lloyd, *Suez*, p. 66.
102. K. Young, *Sir Alec Douglas-Home* (London, 1970), p. 91.
103. Warner, 'United States and Suez', p. 309.
104. Eden to Lloyd 29 July 1956, AP 20/21/163.
105. Thomas, *Suez Affair*, p. 88.
106. Kirkpatrick to Makins 10 Sept. 1956, FO 800/740; cf. Shuckburgh, *Descent to Suez*, p. 360: '[Kirkpatrick said] that in two years' time Nasser will have deprived us of our oil, the sterling area fallen apart, no European defence possible, unemployment and unrest in the UK and our standard of living reduced to that of the Yugoslavs or Egyptians.'
107. D. Neff, *Warriors at Suez: Eisenhower Takes America into the Middle East* (New York, 1981), p. 277.
108. House of Commons Debates, 5th series, vol. 557, col. 1701.
109. H of C Debs, 5th series, vol. 557, col. 1613.
110. H of C Debs, 5th series, vol. 557, col. 1660.
111. H of C Debs, 5th series, vol. 557, col. 1617.
112. Williams, ed., *Gaitskell Diary*, p. 568.
113. Eden to A. Hodge 2 July 1959, AP 33/3; it seems to have been Douglas Jay and John Hynd who, on learning of the government's military plans, persuaded Gaitskell to include this passage: D. Jay, *Change and Fortune: A Political Record* (London, 1980), p. 254.
114. J. Griffiths, *Pages from Memory* (London, 1969), p. 151; T. Benn, *Years of Hope: Diaries, Letters and Papers 1940–1962* (London, 1994), pp. 203–4. See also Morgan, ed., *Crossman Backbench Diaries*, p. 508: 'According to [Barbara Castle], Hugh Gaitskell had let the whole Movement down by supporting Eden . . . in the debate at the beginning of August.'
115. Lord Birkenhead, *Walter Monckton: The Life of Viscount Monckton of Brenchley* (London, 1969), p. 307.
116. H. Macmillan, *Tides of Fortune 1945–55* (London, 1969), p. 567.
117. Chiefs of Staff meeting 27 July 1956, cited A. Gorst and W. S. Lucas, 'Suez 1956: Strategy and the Diplomatic Process', *Journal of Strategic Studies*, 11, 4 (1988), p. 400.
118. R. Fullick and G. Powell, *Suez: The Double War* (London, 1979), p. 73.
119. Heikal, *Cutting the Lion's Tail*, p. 119.
120. G. Millard, 'Memorandum on relations between the United Kingdom, the United States and France in the months following Egyptian nationalisation of the Suez Canal Company in 1956', 21 Oct. 1957, FO 800/728.
121. Back in 1930 Eden had insisted: 'We are not content to leave the protection of a vital artery, the jugular vein of the British Empire, to the goodwill of the people of Egypt.' M. Foot and M. Jones, *Guilty Men, 1957* (London, 1957), p. 56.

122. D. Eisenhower, *Waging Peace: The White House Years 1956–1961* (London, 1966), p. 37.

123. S. Ambrose, *Eisenhower: The President 1952–1969* (London, 1984), p. 330.

124. Memorandum of Eisenhower–Dulles conversation 7 Sept. 1956, cited Peterson, 'Anglo-American Rivalry', pp. 87–8.

125. Eden, *Full Circle*, p. 458.

126. Diary Jan. 1957, AP 20/2/5.

127. Diary note 1966, AP 20/2/11; '[Winthrop Aldrich] said that the problem had been the deviousness of Foster Dulles': note for biographer 14 July 1971, AP 33/7iii.

128. R. G. Casey, cited in J. Cable, *The Geneva Conference of 1954 on Indochina* (London, 1986), p. 51.

129. D. Dimbleby and D. Reynolds, *An Ocean Apart* (London, 1988), p. 206.

130. H. Finer, *Dulles over Suez: The Theory and Practice of his Diplomacy* (London, 1964), p. 70.

131. W. R. Louis and H. Bull, eds, *The Special Relationship: Anglo-American Relations since 1945* (Oxford, 1986), p. 278.

132. Makins to Lloyd 30 July 1956, FO 371/119080.

133. Lloyd, *Suez*, p. 42.

134. Macmillan, *Riding the Storm*, p. 110.

135. R. Murphy, *Diplomat among Warriors* (London, 1964), p. 465.

136. Murphy, *Diplomat among Warriors*, p. 463.

137. Macmillan, *Riding the Storm*, p. 105.

138. Eisenhower to Eden 31 July 1956, PREM 11/1098.

139. Note by Dulles 1 Aug. 1956, PREM 11/1098.

140. Record of Eden–Pineau meeting 30 July 1956, cited Gorst and Lucas, 'Suez 1956', p. 403; Egypt Committee 30 July 1956, CAB 134/1216.

141. Macmillan, *Riding the Storm*, p. 106; A. Horne, *Macmillan 1894–1956* (London, 1988), p. 398.

142. W. Aldrich, 'The Suez Crisis: A Footnote to History', *Foreign Affairs*, 45, 3 (1967), p. 543.

143. R. R. James, *Anthony Eden* (London, 1986), p. 475; L. Mosley, *Dulles: A Biography of Eleanor, Allen and John Foster Dulles and their Family Network* (London, 1978), p. 412.

144. D. R. Thorpe, *Selwyn Lloyd* (London, 1989), p. 219.

145. McDonald, *Man of the Times*, p. 145.

146. Kyle, *Suez*, p. 182; Kunz, *Economic Diplomacy*, p. 84.

147. Eisenhower, *Waging Peace*, p. 44.

148. Egypt Committee 30 July 1956, PREM 11/1098.

149. Eden to Eisenhower 5 Aug. 1956, PREM 11/1098.

150. A. Gorst and W. Scott Lucas, 'The Other Collusion: Operation Straggle and Anglo-American Intervention in Syria, 1955–56', *Intelligence and National Security*, 3 (1988), pp. 576–95.

151. Note by Eden 1957, AP 20/34/4D; see also diary Jan. 1957, AP 20/2/5.

152. J. Aitken to author 30 March 1993; T. Bower, *The Perfect English Spy: Sir Dick White and the Secret War 1935–90* (London, 1995), pp. 191 ff.

153. Egypt Committee 9 Aug. 1956, CAB 134/1216.

154. A. Nutting, *Nasser* (London, 1972), p. 150.

155. Macmillan diary 18 Aug. 1956, cited Horne, *Macmillan 1894–1956*, p. 408.

156. G. Wyndham-Goldie, *Facing the Nation: Television and Politics, 1936–76* (London,

1977), p. 177; M. Cockerell, *Live from Number 10: The Inside Story of Prime Ministers and Television* (London, 1988), p. 45.

157. Clark, *From Three Worlds*, p. 170.

158. L. Behrens to C. Davies 15 Aug. 1956, Clement Davies MSS J/23/12.

159. Kyle, *Suez*, p. 190.

160. L. Epstein, *British Politics in the Suez Crisis* (London, 1964), p. 142.

161. Williams, ed., *Gaitskell Diary*, p. 574.

162. Eden to Gaitskell 12 Aug. 1956, AP 20/31/596.

163. Egypt Committee 14 Aug. 1956, CAB 134/1216.

164. F. R. MacKenzie, 'Eden, Suez and the BBC: A Re-assessment', *The Listener*, 18 Dec. 1969, pp. 841–3; T. Shaw, 'Eden and the BBC during the Suez Crisis: A Myth Re-examined', *Twentieth Century British History*, 3, 3 (1995), pp. 320–43; F. Bishop to Eden 20 Feb. 1969, AP 20/51/7.

165. Eden to Cadogan 16 Aug. 1956, AP 20/31/612.

166. Note for biographer 5 Sept. 1973, AP 33/7.

167. Adams, *First Hand Report*, p. 202.

168. Diary 21 April 1975, AP 20/2/20; undated note for biographer AP 33/7i; photocopied note AP 23/58/78.

169. Diary 15 Aug. 1956, AP 20/1/32.

170. Millard, 'Memorandum on relations', FO 800/728.

171. Eden to Lord Boyd 19 July 1966, AP 23/11/38.

172. Note for biographer n.d., AP 33/7i; Butler to Eden 19 May 1969, AP 23/13/24; Eden to Salisbury 26 July 1966, AP 23/60/134A.

173. Note for biographer 9 Sept. 1968, AP 33/7. For the misgivings of Iain Macleod, Minister of Labour, see R.Shepherd, *Iain Macleod* (London, 1994), pp. 116–21.

174. Salisbury to Eden 2 Aug. 1956, AP 20/33/4.

175. Egypt Committee 2 and 3 Aug. 1956, CAB 134/1216; Horne, *Macmillan 1894–1956*, pp. 400–1. Compare Eden's comment in 1953 that 'any negotiations now [with Israel] would ruin our hopes of securing the cooperation of the Arabs in arrangements for the defence of the Middle East': Eden to Churchill 23 March 1953, AP 20/16/56.

176. Egypt Committee 7 Aug. 1956, CAB 134/1216; Horne, *Macmillan 1894–1956*, p. 404.

177. Horne, *Macmillan 1894–1956*, p. 406.

178. Macmillan diary 24 Aug. 1956, cited Horne, *Macmillan 1894–1956*, p. 410.

179. Lucas, *Divided We Stand*, p. 176.

180. Clark, *From Three Worlds*, p. 178.

181. Home to Eden 22 Aug. 1956, PREM 11/1152.

182. Macmillan diary 24 Aug. 1956, cited Horne, *Macmillan 1894–1956*, p. 410.

183. Home to Eden 24 Aug. 1956, PREM 11/1152.

184. Lennox-Boyd to Eden 24 Aug. 1956, PREM 11/1152.

185. Salisbury to Eden 24 Aug. 1956, PREM 11/1152, and Salisbury to Eden, n.d., AP 20/33/4A.

186. N. Brook to Eden 25 Aug. 1956, PREM 11/1152.

187. Macmillan diary 9 Aug. 1956, cited Horne, *Macmillan 1894–1956*, p. 405; Clark, *From Three Worlds*, p. 173.

188. Clark, *From Three Worlds*, pp. 170–1.

189. Clark, *From Three Worlds*, p. 172.

190. Eden to Lloyd 26 Aug. 1956, AP 20/21/181; Clark, *From Three Worlds*, p. 181.

191. Cabinet 28 Aug. 1956, CM(56)62, CAB 128/30.

192. Menzies to Eden 9 Sept. 1956, AP 23/51/1.

193. Eden to Monckton 26 Aug. 1956, AP 20/21/179.
194. Lloyd, *Suez*, p. 129. In early September Dulles declared: 'Every day that goes by without some outbreak is a gain, and I just keep trying to buy that day. I don't know anything to do but keep improvising': E. J. Hughes, *The Ordeal of Power: A Political Memoir of the Eisenhower Years* (London, 1963), p. 178.
195. Note for biographer 23 Aug. 1976, AP 33/7iii.
196. Eden, *Full Circle*, pp. 479, 481: 'It provided a means of working with the United States. I was prepared to lean over backwards to achieve this. . . . Close cooperation with the United States had been a guiding principle throughout my political life.' No wonder Eden approved of Finer, *Dulles over Suez*, p. 224: 'Eden was tempted by the possibilities as coloured by Dulles and he fell. He fell from the most honourable motive a man could have in the circumstances: to maintain what had been the keystone of Churchill's British policy since one of the darkest moments of World War Two.'
197. Note by H. Beeley 31 Aug. 1956, FO 371/119128.
198. Cabinet 11 Sept. 1956, CM(56)64, CAB 128/30.
199. Millard, 'Memorandum on relations', FO 800/728; cf. Eden, *Full Circle*, p. 484: 'Such cynicism towards allies destroys true partnership. It leaves only the choice of parting or a master and vassal relationship in foreign policy.'
200. Eisenhower to Eden 3 Sept. 1956, AP 20/27/78.
201. See, for example, Kyle, *Suez*, p. 224.
202. Clark, *From Three Worlds*, p. 183.
203. S. Ambrose and R. Immerman, *Ike's Spies: Eisenhower and the Espionage Establishment* (New York, 1981), p. 240.
204. A. Atherton, in Troen and Shemesh, eds, *The Suez–Sinai Crisis*, p. 270.
205. Eden to Eisenhower 6 Sept. 1956, PREM 11/1177.
206. Eisenhower, *Waging Peace*, p. 50; *Foreign Relations of the United States [FRUS]*, *1955–1957* vol. 16 (Washington, 1990), pp. 403–4.
207. Eisenhower to Eden 8 Sept. 1956, AP 20/27/80.
208. In the event the delays inherent in the diplomatic process obliged a reluctant Eden to accept the replacement of Plan Musketeer by Musketeer Revise in early September. The latter, which reverted to the idea of an invasion at Port Said, at least offered the advantage of allowing the military option to be left open for a further month. 'I was not at first much enamoured of [the plan]', noted Eden on 7 September, 'but in discussion, and as Chiefs of Staff amplified it, became more reconciled' (diary 7 Sept. 1956, AP 20/1/32). The Egypt Committee adopted Musketeer Revise on 10 September.
209. Diary Jan. 1957, AP 20/2/5.
210. Makins to Lloyd 9 Sept. 1956, FO 800/740.
211. Makins to Lloyd 8 Sept. 1956, AP 20/25/34.
212. Millard, 'Memorandum on relations', FO 800/728.
213. Cabinet 11 Sept. 1956, CM(56)64, CAB 128/30.
214. 'Economic and financial measures in the event of war with Egypt'; Bridges to Macmillan 7 Sept. 1956, T236/4188.
215. Cabinet 11 Sept. 1956, CM(56)64, CAB 128/30.
216. Lloyd, *Suez*, p. 134.
217. H of C Debs, 5th series, vol. 558, cols 2–15.
218. Eden, *Full Circle*, p. 483.
219. H of C Debs, 5th series, vol. 558, cols 304–5.

220. N. Nicolson, ed., *Harold Nicolson: Diaries and Letters 1945–1962* (London, 1968), p. 310.
221. Williams, ed., *Gaitskell Diary*, p. 606.
222. Egypt Committee 17 Sept. 1956, CAB 134/1216.
223. Macmillan, *Riding the Storm*, p. 128.
224. J. Devine (Eden's constituency agent) to Chloe Otto 23 Sept. 1956, AP 11/2/107.
225. Eden to Churchill 21 Sept. 1956, AP 20/31/673.
226. Eden to Lloyd 21 Sept. 1956, AP 20/21/192. The epitaph on SCUA was contained in an exchange of letters between Lloyd and Dulles in mid-October. Lloyd said he was 'deeply disappointed to find how far apart our conceptions of the purpose of the Users' Association now are'. In a letter which crossed with Lloyd's, Dulles wrote: 'Our idea, made clear from the beginning, is that [SCUA] was to be a means of practical working cooperation with the Egyptian authorities which would seek to establish *de facto* international participation in the control of the Canal' (Millard, 'Memorandum on relations', FO 800/728).
227. Eden to Lloyd 24 Sept. 1956, AP 20/21/195 (offering advice for Lloyd's broadcast).
228. Eden to Makins 1 Oct. 1956, AP 20/25/49.
229. Millard, 'Memorandum on relations', FO 800/728; Finer, *Dulles over Suez*, p. 263. Dulles had been warned on 19 September that Macmillan and Salisbury still regarded military action as the only satisfactory solution and that the way might be prepared in about a month after the British had 'arranged' a 'grave incident' in Egypt (*FRUS, 1955–1957*, vol. 16, pp. 521–2).
230. Clark, *From Three Worlds*, p. 192.
231. Eden to Macmillan 23 Sept. 1956, AP 20/25/44.
232. Macmillan to Eden 26 Sept. 1956, PREM 11/1102.
233. Note by Macmillan 26 Sept. 1956, PREM 11/1102.
234. Horne, *Macmillan 1894–1956*, p. 422; Kunz, *Economic Diplomacy*, pp. 103, 106–7; H. J. Dooley, 'Great Britain's "Last Battle" in the Middle East: Notes on Cabinet Planning during the Suez Crisis of 1956', *International History Review*, 11, 3 (1989), p. 506.
235. Makins to Lloyd 2 Oct. 1956, PREM 11/1174; Millard, 'Memorandum on relations', FO 800/728.
236. Makins to Eden 3 Oct. 1956, AP 20/25/51; McDonald, *Man of the Times*, p. 149; note for biographer 31 July 1970, AP 33/7iii.
237. Dimbleby and Reynolds, *Ocean Apart*, p. 213.
238. Notes in 1957 diary, AP 20/2/5.
239. Nutting, *Nasser*, p. 162.
240. Diary 4 Oct. 1970, AP 20/2/15.
241. N. Brook to Churchill 4 Oct. 1956, cited M. Gilbert, *Winston S. Churchill*, vol. 8 (London, 1988), pp. 1213–14.
242. Dixon to Foreign Office 6 Oct. 1956, AP 20/25/68.
243. Lloyd to Eden 11 Oct. 1956, AP 20/25/83; B. Urquhart, *Hammarskjöld* (London, 1972), p. 167.
244. Lloyd, *Suez*, p. 151.
245. Millard, 'Memorandum on relations', FO 800/728.
246. Lloyd to Eden 14 Oct. 1956, AP 20/25/100.
247. Eden to Lloyd 14 Oct. 1956, PREM 11/1102.

Notes

CHAPTER 13

1. Note by Eden in preparation of *Full Circle*, AP 7/10/72A.
2. G. Mollet, *Bilan et Perspectives Socialistes* (Paris, 1960), p. 31.
3. A. Nutting, *No End of a Lesson* (London, 1967), p. 32.
4. Eden to Boothby 2 Oct. 1956, AP 20/31/678; Lord Boothby, *My Yesterday, Your Tomorrow* (London, 1962), p. 65.
5. Lord Gladwyn, *Memoirs* (London, 1972), p. 282.
6. Cabinet 18 Oct. 1956, CM(56)71, CAB 128/30.
7. Cabinet 23 Oct. 1956, CM(56)72, CAB 128/30.
8. D. Logan, 'Collusion at Suez', *Financial Times*, 8 Nov. 1986.
9. Emphasis added. 'Memorandum on relations between the United Kingdom, the United States and France in the months following Egyptian nationalisation of the Suez Canal Company in 1956', 21 Oct. 1957 (drawn up by Guy Millard), FO 800/728.
10. R. Lamb, *The Failure of the Eden Government* (London, 1987), p. 243.
11. Cabinet 25 Oct. 1956, CM(56)74, CAB 128/30.
12. H. Wilson, *The Chariot of Israel: Britain, America and the State of Israel* (London, 1981), p. 267.
13. H. J. Dooley, 'Great Britain's "Last Battle" in the Middle East: Notes on Cabinet Planning during the Suez Crisis of 1956', *International History Review*, 11, 3 (1989), p. 511.
14. N. Nicolson, ed., *Vita and Harold: The Letters of Vita Sackville-West and Harold Nicolson 1910–1962* (London, 1992), p. 421.
15. E. Monroe, 'Suez Secrets: the Jigsaw Completed', *Observer*, 24 July 1966.
16. See Macmillan's evasive reply when challenged on this issue on television by Robert McKenzie, *The Listener*, 13 May 1971.
17. P. Hennessy, 'The Scars of Suez', *The Listener*, 5 Feb. 1989.
18. Diary 12 Feb. 1957, AP 20/2/5; according to his obituary in the *New York Times* of 15 January 1977 Eden, in a 1967 interview with Alden Whitman, on the understanding that nothing would appear in print before his death, 'acknowledged secret dealings with the French and "intimations" of the Israeli attack'.
19. 'Suez', undated note by R. Blake, AP 33/3i.
20. S. Lloyd, *Suez 1956: A Personal Account* (London, 1978), pp. 247, 249.
21. House of Commons Debates, 5th series, vol. 597, col. 1068. On 24 December 1956 Boothby wrote to *The Times* to defend the government against the charge of collusion: 'If the British and French . . . had decided . . . to attack Egypt, for the purposes of getting rid of Nasser and occupying the Suez Canal, and to use Israel as an instrument of their policy by giving her the green light for an invasion of the Sinai Peninsula in order to justify their own intervention . . . , they would have been guilty of collusion in the accepted sense of the word.' As Geoffrey Warner has commented, 'one could hardly improve upon this as a concise statement of what the British and French governments actually did': G. Warner, '"Collusion" and the Suez Crisis of 1956', *International Affairs*, 55 (1979), pp. 238–9.
22. D. R. Thorpe, *Selwyn Lloyd* (London, 1989), p. 265.
23. L. D. Epstein, *British Politics in the Suez Crisis* (London, 1964), pp. 46–7.
24. *The Times*, 12 Oct. 1956.
25. D. Carlton, *Britain and the Suez Crisis* (Oxford, 1988), p. 58.
26. Lord Carr, 'Living with Anthony', BBC Radio Four broadcast (1990).
27. R. R. James, *Anthony Eden* (London, 1986), p. 366; see also R. R. James in S. I. Troen

and M. Shemesh, eds, *The Suez–Sinai Crisis 1956: Retrospective and Reappraisal* (London, 1990), p. 108.

28. Eden to Lloyd 9 Oct. 1956, AP 20/25/77.

29. Eden to W. Aldrich 3 May 1967, AP 23/5/10B.

30. W. Clark, *From Three Worlds* (London, 1986), p. 160. Clark came to believe that Eden was 'mad, literally mad, and that he went so the day his temperature rose to 105°' (p. 209).

31. H. Thomas, *The Suez Affair* (London, 1967), p. 127.

32. Interview Lord Gladwyn with author 22 July 1993; draft of N. Fisher, *The Tory Leaders: Their Struggle for Power* (London, 1977), AP 23/32/6, and note by R. Carr 15 July 1976, AP 23/32/11C. Benzedrine, an amphetamine, was commonly prescribed in the 1950s. Its side effects can include insomnia, irritability and euphoria.

33. K. Kyle, *Suez* (London, 1991), p. 557.

34. Diary 14 Aug. and 7 Sept. 1956, AP 20/1/32.

35. Diary 21 Aug. 1956, AP 20/1/32.

36. Note for cabinet of 9 Jan. 1957, AP 20/33/11.

37. Engagement diary, AP 20/30/1.

38. Lord Butler, *The Art of Memory: Friends in Perspective* (London, 1982), p. 100; cf. Thomas, *Suez Affair*, p. 34.

39. M. Foot and M. Jones, *Guilty Men*, 1957 (London, 1957), pp. 234–5.

40. Eden to D. Dilks 25 July 1966, AP 33/2.

41. T. Peterson, 'Anglo-American Rivalry in the Middle East: the Struggle for the Buraimi Oasis, 1952–57', *International History Review*, 14, 1 (1992), p. 89; see also R. Blake, in J. Mackintosh, ed., *British Prime Ministers in the Twentieth Century*, vol. 2 (London, 1978), p. 113. As Vita Sackville-West noted at the time, 'Granted that A. E. may have lost his wits, surely his entire Cabinet can't have?': N. Nicolson, ed., *Harold Nicolson: Diaries and Letters 1945–1962* (London, 1968), p. 313.

42. R. Carr to N. Fisher 15 July 1976, AP 23/32/11C.

43. Note by Eden on Foreign Office to Paris 15 Oct. 1955, AP 20/23/117.

44. T. Robertson, *Crisis: The Inside Story of the Suez Conspiracy* (New York, 1965), p. 161.

45. Eden to Lloyd 29 July 1956, AP 20/21/162.

46. Eden to Foreign Office 26 Sept. 1956, PREM 11/1102; Clark, *From Three Worlds*, p. 192.

47. B. Urquhart, *Hammarskjöld* (London, 1972), p. 167.

48. Millard, 'Memorandum on relations', FO 800/728.

49. I. McDonald, *A Man of the Times* (London, 1976), p. 147.

50. Diary Jan. 1957, AP 20/2/5.

51. E. Kedourie, 'The Entanglements of Suez', *Times Literary Supplement*, 30 Nov. 1979, p. 69.

52. Cabinet 23 Oct. 1956, CM(56)72, CAB 128/30.

53. C. Pineau, *1956 Suez* (Paris, 1976), p. 135.

54. A. Eden, *Full Circle* (London, 1960), pp. 512–13.

55. Diary Jan. 1957, AP 20/2/5. Emphasis added.

56. Cabinet 18 Oct. 1956, CM(56)71, CAB 128/30.

57. Hailsham to Eden 10 Oct. 1956, AP 14/4/97.

58. Cabinet 25 Oct. 1956, CM(56)74, CAB 128/30; for Buchan-Hepburn's assessment of the divisions of opinion inside the cabinet on the use of force on 23 October, see J. Ramsden, *The Age of Churchill and Eden* (London, 1995), p. 308.

59. Lord Hailsham, *A Sparrow's Flight* (London, 1990), p. 288; interview Lord Hailsham, with author 10 December 1992.

60. H. Brandon, *Special Relationships* (London, 1988), p. 127; H. Trevelyan, *The Middle East in Revolution* (London, 1970), p. 127.

61. A. Gorst and W. S. Lucas, 'Suez 1956: Strategy and the Diplomatic Process', *Journal of Strategic Studies*, 11, 4 (1988), p. 429.

62. A. Horne, *Macmillan 1894–1956* (London, 1988), p. 431.

63. Salisbury to Eden 31 Oct. 1956, AP 20/33/6; note passed by Salisbury to Lord Halifax during 1957 House of Lords debate on Suez, Halifax MSS A2.278.64.

64. J. Colville, *The Fringes of Power: Downing Street Diaries 1939–1955* (London, 1985), p. 718.

65. Colville, *Fringes of Power*, p. 724.

66. W. R. Louis and H. Bull, eds, *The Special Relationship: Anglo-American Relations since 1945* (Oxford, 1986), p. 279; D. Eisenhower, *Waging Peace: The White House Years 1956–1961* (London, 1966), p. 73.

67. Cabinet 30 Oct. 1956, CM(56)75, CAB 128/30.

68. Clark, *From Three Worlds*, p. 198.

69. Nicolson, ed., *Vita and Harold*, pp. 420–1.

70. Eisenhower to Eden 30 Oct. 1956, AP 20/27/90.

71. Eden to Eisenhower 30 Oct. 1956, AP 20/27/89.

72. Eisenhower to Eden 30 Oct. 1956, AP 20/27/92.

73. Thorpe, *Lloyd*, p. 226.

74. Note for talk with Kenneth Harris, Oct. 1972, AP 33/7.

75. P. Dixon, *'Double Diploma': The Life of Sir Pierson Dixon, Don and Diplomat* (London, 1968), p. 275.

76. R. Griffith, ed., *Ike's Letters to a Friend 1941–1958* (Lawrence, Kansas, 1984), p. 176.

77. Note by D. Dilks 26 July 1967, AP 33/6.

78. H. Macmillan, *Riding the Storm 1956–1959* (London, 1971), p. 157.

79. Eden may even have believed that America's preoccupation with electoral politics required Britain to take the lead during 1956 on the world stage on behalf of the Western alliance: see diary 25 Feb. 1956, AP 20/1/32.

80. Eisenhower, *Waging Peace*, p. 84.

81. Millard, 'Memorandum on relations', FO 800/728.

82. D. Neff, *Warriors at Suez: Eisenhower takes America into the Middle East* (New York, 1981), p. 385.

83. Eisenhower, *Waging Peace*, p. 85.

84. W. S. Lucas, *Divided We Stand: Britain, the United States and the Suez Crisis* (London, 1991), p. 270.

85. See, for example, Pineau, *1956 Suez*, p. 195, citing Dulles's views in 1957 and Richard Nixon in *The Times*, 28 Jan. 1987, suggesting that Eisenhower came to view Suez as 'his major foreign policy mistake'; see also C. L. Sulzberger, *An Age of Mediocrity: Memoirs and Diaries 1963–1972* (New York, 1973), p. 40, quoting Eisenhower in 1967: 'The British attitude then was understandable in terms of their resentment. But they did it all wrong.'

86. Eisenhower to Eden 8 Sept. 1956, AP 20/27/80.

87. Lloyd, *Suez*, p. 219; Lloyd to Eden 17 Nov. 1956, AP 20/25/166; cf. Eisenhower–Dulles conversation 12 Nov. 1956, *Foreign Relations of the United States [FRUS]*, 1955–1957, vol. 16 (Washington, 1990), pp. 1112–14.

88. L. Pearson, *The International Years: Memoirs 1948–1957* (London, 1974), p. 238.

89. Cabinet 2 Nov. 1956, CM(56)77, CAB 128/30.

90. H of C Debs, 5th series, vol. 558, col. 1710.
91. H of C Debs, 5th series, vol. 558, col. 1454.
92. Lord Kilmuir, *Political Adventure* (London, 1964), p. 274.
93. J. Morgan, ed., *The Backbench Diaries of Richard Crossman* (London, 1981), p. 533.
94. T. Benn, *Years of Hope: Diaries, Letters and Papers 1940–1962* (London, 1994), p. 207.
95. Benn, *Years of Hope*, p. 200.
96. M. Cockerell, *Live from Number 10: The Inside Story of Prime Ministers and Television* (London, 1988), p. 49; J. Margach, *The Abuse of Power* (London, 1978), p. 112.
97. Morgan, ed., *Crossman Backbench Diaries*, p. 539. On the evening of 6 November, after Eden had announced a ceasefire, Crossman received a telephone call from Alec Spearman, a leading Tory critic of Suez: 'Country before Party, Dick. I beg you to stop the "Eden must go" campaign, since this will make it impossible to get rid of him' (p. 542).
98. Kyle, *Suez*, p. 88.
99. Nicolson, ed., *Nicolson Diaries and Letters*, p. 319.
100. Louis and Bull, eds, *Special Relationship*, p. 276.
101. P. Gore-Booth, *With Great Truth and Respect* (London, 1974), p. 229; P. Hennessy and M. Laity, 'Suez – What the Papers Say', *Contemporary Record*, 1, 1 (1987), p. 5. P. Hennessy, *Whitehall* (London, 1990), pp. 165–7.
102. Gladwyn, *Memoirs*, p. 282; interview Lord Gladwyn with author 22 July 1993.
103. W. Hayter, *A Double Life* (London, 1974), p. 150.
104. Hayter, *Double Life*, p. 153.
105. Dixon, *Double Diploma*, p. 278.
106. P. Ziegler, *Mountbatten: The Official Biography* (London, 1985), p. 545.
107. *Observer*, 4 Nov. 1956; R. Cockett, 'The *Observer* and the Suez Crisis', *Contemporary Record*, 5, 1 (1991), pp. 9–31.
108. Morgan, ed., *Crossman Backbench Diaries*, p. 544.
109. Dixon to Foreign Office 5 Nov. 1956, cited Lamb, *Failure*, p. 266.
110. Note for biographer 9 June 1975, AP 33/7iii.
111. R. Hankey to Eden 12 Oct. 1976, AP 23/35/26 and note by Eden Oct. 1976, AP 23/35/26A. The opinion quoted was that of Paul Gore-Booth.
112. Morgan, ed., *Crossman Backbench Diaries*, p. 534.
113. T. B. Millar, ed., *Australian Foreign Minister: The Diaries of R. G. Casey 1951–60* (London, 1972), p. 250; K. Love, *Suez: The Twice-Fought War* (London, 1969), p. 420; C. H. King, *Strictly Personal* (London, 1969), p. 131.
114. Lloyd, *Suez*, p. 206.
115. Cabinet 4 Nov. 1956, CM(56)79, CAB 128/30.
116. Lord Butler, *The Art of the Possible* (London, 1971), p. 193; Lord Hailes to Eden 17 Oct. 1971, AP 23/37/74; Lord Boyd to Eden 7 April 1972, AP 33/6; Cabinet Office to Eden 22 March 1972, AP 20/32/79.
117. James, *Eden*, p. 567.
118. Diary Jan. 1957, AP 20/2/5.
119. Note for biographer 18 April 1972, AP 33/7i; M. Dayan, *Story of My Life* (London, 1976), p. 209.
120. Home to Eden 5 Nov. 1956, cited Hennessy, 'Scars', p. 8.
121. Eden to Eisenhower 5 Nov. 1956, FO 800/726.
122. Eden to A. Hodge 27 Oct. 1959, AP 33/3.
123. Eden, *Full Circle*, pp. 554, 557.

124. Dixon, *Double Diploma*, p. 272.
125. Dixon to Eden 5 Nov. 1956, PREM 11/1105.
126. Norman Brook's comments on a draft of Eden, *Full Circle*, 14 May 1958, AP 23/56/15B.
127. Note from the Prime Minister's office to Eden 24 Aug. 1959, AP 23/56/46A.
128. Cabinet 6 Nov. 1956, CM(56)80, CAB 128/30.
129. Memorandum by Chiefs of Staff 8 Nov. 1956, CAB 134/1217; PREM 11/1163.
130. Eden, *Full Circle*, p. 556.
131. Millard, 'Memorandum on relations', FO 800/728.
132. Brook's comments on a draft of Eden, *Full Circle*, 14 May 1958, AP 23/56/15b.
133. Lloyd, *Suez*, p. 209.
134. Thomas, *Suez Affair*, p. 146.
135. Kyle, *Suez*, p. 464; D. Kunz, 'Did Macmillan Lie over Suez?', *Spectator*, 3 Nov. 1990.
136. Lamb, *Failure*, p. 282.
137. Hennessy and Laity, 'What the Papers Say', p. 5.
138. Pineau's notoriously unreliable memoirs suggest that pressure from Eisenhower was the main reason for Eden's capitulation (*1956 Suez*, pp. 175–6). In later years Eden sought to confirm that the President did not apply personal pressure and in particular that he did not telephone him until the decision to cease fire had already been taken (Eden to Eisenhower 17 June 1964 and Eisenhower to Eden 23 June 1964, AP 23/29/22). But according to Mollet 'there was previously on November 5th a more important telephone conversation between you and Eisenhower and it was during that conversation, he said, that you yielded to American pressure and obtained in exchange certain definite promises which Mollet did not specify'. P. Winkler to Eden 22 Oct. 1959, AP 7/19/9.
139. Morgan, ed., *Crossman Backbench Diaries*, p. 541.
140. Lamb, *Failure*, p. 276.
141. Diary 24 March 1975, AP 20/2/20.
142. Report of meeting 9 Nov. 1956, AP 20/56/10.
143. Dixon, *Double Diploma*, p. 274; cf. Eden to Salisbury 14 Dec. 1959, AP 23/60/41A: 'I came to the conclusion that the Americans behaved much worse after the cease-fire than before and that it was in their obstinate failure to cooperate then that the missed opportunities were most conspicuous.'
144. Eden, 'Eisenhower and the telephone', diary 18 Feb. 1957, AP 20/2/5.
145. Eden to Mollet 6 Nov. 1956, PREM 11/1105.
146. S. Adams, *First Hand Report* (London, 1962), p. 209.
147. Eden, 'Eisenhower and the telephone', diary 18 Feb. 1957, AP 20/2/5; Lucas, *Divided We Stand*, pp. 299–300.
148. Note by Eden, AP 23/58/78.
149. Hoover 'couldn't stand' Eden: R. Murphy, *Diplomat among Warriors* (London, 1964), p. 468.
150. W. S. Lucas, 'Suez, the Americans and the Overthrow of Anthony Eden', *LSE Quarterly*, 1, 3 (1987), pp. 230–1.
151. Lucas, 'Overthrow of Anthony Eden', pp. 232–3.
152. Ben-Gurion diary 1 Dec. 1956, cited Troen and Shemesh, eds, *Suez–Sinai Crisis*, p. 324.
153. Diary Jan. 1957, AP 20/2/5.
154. Cabinet 7 Nov. 1956, CM(56)81, CAB 128/30.
155. Eden to K. Adenauer 17 Nov. 1956, AP 20/25/163.
156. Boothby to Eden 9 Nov. 1956, AP 14/4/38.

157. Lloyd to Eden 12 Nov. 1956, AP 20/25/141.

158. Lloyd to Eden 14 Nov. 1956, AP 20/25/148.

159. Egypt Committee 15 Nov. 1956, CAB 134/1216.

160. Lloyd to Eden 18 Nov. 1956, AP 20/25/167.

161. Lloyd to Eden 17 Nov. 1956, AP 20/25/163.

162. Lloyd to Eden 17 Nov. 1956, AP20/25/163.

163. Macmillan, *Riding the Storm*, p. 169.

164. Harcourt to Rowan 19 Nov. 1956, cited D. Kunz, *The Economic Diplomacy of the Suez Crisis* (Chapel Hill, 1991), p. 140.

165. M. Amory, ed., *The Letters of Ann Fleming* (London, 1985), pp. 188–9; cf. A. Montague Brown, *Long Sunset* (London, 1995), p. 212. Eden might have been less enthusiastic about his Jamaican retreat had he known of the longstanding affair between Ann Fleming and the Labour leader, Hugh Gaitskell.

166. D. Carlton, *Anthony Eden: A Biography* (London, 1981), pp. 457–62; Lucas, 'Overthrow of Anthony Eden', pp. 227ff. Humphrey also acted as a channel of communication, especially with Butler. See also *FRUS, 1955–1957*, vol. 16, pp. 1115–16, 1150–3.

167. Macmillan diary 3 Feb. 1957, cited Horne, *Macmillan 1894–1956*, p. 453; cf. Macmillan, *Riding the Storm*, p. 180: 'I was deeply shocked, for I had not been at all prepared for this sudden and tragic end to the adventure on which we had set out so gaily some twenty months before.'

168. Cabinet 20 Nov. 1956, CM(56)85, CAB 128/30.

169. Eden to F. Bishop 1 Dec. 1956, PREM 11/1826.

170. Lamb, *Failure*, p. 289.

171. Millard, 'Memorandum on relations', FO 800/728.

172. Lloyd, *Suez*, p. 232; note by Lloyd May 1958, FO 800/728.

173. Cabinet 20 Nov. 1956, CM(56)85, CAB 128/30.

174. Butler to Eden 1 Dec. 1956, cited Lamb, *Failure*, p. 298.

175. R. Allen to Eden 5 Dec. 1956, AP 20/25/210.

176. R. Allen to Eden 7 Dec. 1956, AP 20/25/230.

177. Eden to Butler and Lloyd 3 Dec. 1956, AP 20/25/204.

178. For Eden's own draft, see Eden to Bishop 10 Dec. 1956, AP 20/25/242.

179. Butler to Eden 12 Dec. 1956, AP 20/25/252.

180. Nicolson, ed., *Nicolson Diaries and Letters*, p. 323.

181. A. Boyle, *Poor Dear Brendan: The Quest for Brendan Bracken* (London, 1974), p. 338.

182. Eden to M. Adeane 16 Dec. 1956, AP 20/31/810; Eden to Lord Montgomery 18 Dec. 1956, AP 20/31/819.

183. Eden to Lloyd 22 Dec. 1956, AP 20/21/227.

184. Kilmuir, *Political Adventure*, p. 281.

185. J. Morrison to Eden 19 Dec. 1956, AP 20/19/69.

186. Morgan, ed., *Crossman Backbench Diaries*, p. 557.

187. Sir Henry Studholme to Eden 18 Dec. 1956, AP 14/4/198A.

188. H of C Debs, 5th series, vol. 562, col. 1518. Sir Donald Logan later declared: 'I felt that at that moment his attempt to justify his intervention to separate the forces simply exploded.' Hennessy, *Whitehall*, p. 167.

189. Kilmuir, *Political Adventure*, p. 283.

190. Buchan-Hepburn to Eden 28 Dec. 1956, AP 20/33/7. By this date at least one minister had confessed that his 'confidence and respect for the man [had]

evaporated totally'. Stuart to Butler 13 Dec. 1956, cited Ramsden, *Churchill and Eden*, p. 317.

191. Coleraine to C. Eden 5 Jan. 1957, AP 20/33/19.

192. Eden to Lord Iliffe 1 Jan. 1957, AP 14/4/116B.

193. Note by Prime Minister 5 Jan. 1957, AP 33/8iii; see also note for biographer 30 March 1960, AP 33/7ii.

194. Diary 3 Jan. 1957, AP 20/2/5.

195. Millard, 'Memorandum on relations', FO 800/728.

196. *The Listener*, 13 May 1971.

197. 'Memorandum for the record' 12 Dec. 1956, cited Kunz, *Economic Diplomacy*, p. 156.

198. R. Davenport-Hines, *The Macmillans* (London, 1992), p. 266; Viscount Tonypandy, *George Thomas, Mr Speaker* (London, 1985), p. 75.

199. Bracken to Beaverbrook 7 Dec. 1956, Beaverbrook MSS BBK C58; compare Macmillan's efforts – 'so blatant as to be embarrassing' – to be seen at Eden's side during the party conference in October: note for biographer 30 March 1960, AP 33/7ii.

200. The argument is based on a somewhat cryptic letter from Sir Horace Evans to Butler on 14 Jan. 1957, Butler MSS G31. If Evans was party to a conspiracy to remove the premier from office by exaggerating the severity of his condition, he sustained the deception with great tenacity. Eden was forbidden, for reasons of health, to attend the annual Remembrance Day service at the Cenotaph for each of the next five years.

201. Eden to Halifax 16 Jan. 1957, Halifax MSS A4.410.21.3.

202. Note by Eden 11 Jan. 1957, AP 20/33/12A. Though Eden was annoyed by statements in the memoirs of Macmillan and Kilmuir that the Queen had not asked for his advice about his successor, these accounts were technically correct, as Eden came to realize when he rediscovered the note he had dictated two days after his audience: note for biographer 27 Nov. 1970, AP 23/2/20A; Eden to M. Adeane 13 Jan. 1972, AP 23/2/23.

203. Note by R. Carr 15 July 1976, AP 23/32/11C.

204. Diary 3 Jan. 1957, AP 20/2/5.

205. 'Secret and personal' note by Eden, Jan. 1957, PREM 11/1138.

206. Eden to Lloyd 27 Jan. 1957, cited Thorpe, *Lloyd*, p. 259.

207. Diary 15 Sept. 1975, AP 20/2/20.

208. Note for Eden, *Full Circle*, AP 7/10/72A.

209. See, for example, Eden, *Full Circle*, p. 579: 'The insidious appeal of appeasement leads to a deadly reckoning.'

210. Brook to Eden 14 May 1958, AP 23/56/15A.

211. Nicolson, ed., *Nicolson Diaries and Letters*, p. 312.

212. Kedourie, 'Entanglements', p. 67.

213. W. R. Louis and R. Owen, eds, *Suez 1956: The Crisis and its Consequences* (Oxford, 1989), pp. 338–9.

214. Kyle, *Suez*, pp. 258, 534.

215. V. Bogdanor and R. Skidelsky, *The Age of Affluence 1951–1964* (London, 1970), p. 178.

216. Home to Eden 10 Jan. 1957, AP 9/1/113.

217. R. Blake, in Mackintosh, ed., *British Prime Ministers*, p. 74.

218. R. Bowie, in Louis and Owen, eds, *Suez*, p. 213.

219. M. Gilbert, *Winston S. Churchill*, vol. 8 (London, 1988), p. 1234. For a more

sympathetic analysis of Dulles's conduct see W. R. Louis, 'Dulles, Suez and the British', in R. H. Immerman, ed., *John Foster Dulles and the Diplomacy of the Cold War* (Princeton, 1990), pp. 133–58.

220. Note for biographer 23 Aug. 1976, AP 33/7iii.

CHAPTER 14

1. Lord Chandos, *Memoirs* (London, 1962), p. 291.
2. Note on meeting of 22 Jan. 1957, AP 11/11/5.
3. C. Sulzberger, *The Last of the Giants* (New York, 1970), p. 405.
4. E. Shuckburgh, *Descent to Suez: Diaries 1951–56* (London, 1986), p. 9.
5. P. Williams, ed., *The Diary of Hugh Gaitskell 1945–1956* (London, 1983), p. 421.
6. Shuckburgh, *Descent to Suez*, p. 9.
7. Diary 20 Jan. 1951, AP 20/1/27.
8. R. R. James, *Boothby* (London, 1991), p. 112.
9. Diary 25 Sept. 1941, AP 20/1/21.
10. Diary 14 Aug. 1942, AP 20/1/22.
11. Lord Avon, *Another World 1897–1917* (London, 1976), p. 148.
12. Eden to Marjorie Eden 15 Nov. 1918, AP 22/2/93.
13. V. Lawford, 'Three Ministers', *Cornhill Magazine*, 1010 (1956–7), p. 90.
14. K. Young, ed., *The Diaries of Sir Robert Bruce Lockhart 1939–1965* (London, 1980), p. 386.
15. A. Campbell Johnson, *Anthony Eden: A Biography* (London, 1938), p. 9.
16. E. Kelen, *Peace in their Time: Men who Led us in and out of War 1914–1945* (London, 1964), p. 337; cf. Lawford, 'Three Ministers', p. 75.
17. Lawford, 'Three Ministers', p. 73.
18. G. Hagglof, *Diplomat* (London, 1972), p. 79.
19. Young, ed., *Lockhart Diaries*, pp. 307, 374.
20. L. Mosley, *Dulles: A Biography of Eleanor, Allen and John Foster Dulles and their Family Network* (London, 1978), p. 382; cf. Shuckburgh, *Descent to Suez*, p. 14.
21. B. Pimlott, ed., *The Second World War Diary of Hugh Dalton 1940–45* (London, 1986), p. 594, citing the opinion of George Harvie-Watt.
22. Pimlott, ed., *Dalton War Diary*, p. 722.
23. R. R. James, ed., *Chips: The Diaries of Sir Henry Channon* (London, 1967), p. 302.
24. J. Colville, *The Fringes of Power: Downing Street Diaries 1939–1955* (London, 1985), p. 319.
25. Pimlott, ed., *Dalton War Diary*, p. 263.
26. Pimlott, ed., *Dalton War Diary*, p. 369.
27. D. Bardens, *Portrait of a Statesman: The Personal Life Story of Sir Anthony Eden* (London, 1955), p. 81.
28. W. Clark, *From Three Worlds* (London, 1986), p. 160; cf. C. R. Coote, *Editorial* (London, 1965), p. 277: 'all that glittered was not gold. I was astonished to find that in the departments over which he had presided, he had the reputation of succumbing to fits of irritability.'
29. T. Eden, *The Tribulations of a Baronet* (London, 1933), p. 22.
30. Lawford, 'Three Ministers', p. 82.
31. Shuckburgh, *Descent to Suez*, p. 73.
32. 'People talk of Anthony's impatience and his moments of petulence and irritability, but generally these soon vanish, the cloud is gone and the sun is shining

again': Lord Moran, *Winston Churchill: The Struggle for Survival 1940–1965* (London, 1966), p. 675.

33. Clark, *From Three Worlds*, p. 164.
34. Clark, *From Three Worlds*, p. 165.
35. Lord Boothby, *My Yesterday, Your Tomorrow* (London, 1962), p. 58.
36. James, ed., *Chips*, p. 301.
37. Sulzberger, *Last of the Giants*, p. 638.
38. Macmillan diary 5 July 1956, cited A. Horne, *Macmillan 1894–1956* (London, 1988), p. 390.
39. Lord Harvey, 'Did you Hear That?', *The Listener*, 14 April 1955.
40. Hagglof, *Diplomat*, p. 79.
41. Diary 19 July 1941, AP 20/1/21.
42. 'Eden: the Iron as Well as the Smile', *New York Times Magazine*, 4 April 1954.
43. R. Bruce Lockhart, *Comes the Reckoning* (London, 1947), p. 109.
44. Churchill well expressed the difference of approach between himself and Eden. 'Groping among these fogs of verbiage', he complained of Foreign Office telegrams which 'simply record endless arguments on either side which may be great fun for the juniors of the Department to digest, but are simply unreadable by those who have to take the decisions': Churchill to Eden 26 May 1944, AP 20/12/291B.
45. Clark, *From Three Worlds*, p. 162.
46. Young, ed., *Lockhart Diaries*, p. 271.
47. *New York Times Magazine*, 4 April 1954.
48. Law to Eden 14 Feb. 1975, AP 20/52/89.
49. Young, ed., *Lockhart Diaries*, p. 379.
50. R. Bruce Lockhart, *Friends, Foes and Foreigners* (London, 1957), p. 200.
51. Shuckburgh, *Descent to Suez*, p. 226.
52. Bardens, *Portrait*, p. 103.
53. Young, ed., *Lockhart Diaries*, p. 557.
54. J. Connell, *Wavell: Scholar and Soldier* (London, 1964), p. 354. Compare J. Cable, *The Geneva Conference of 1954 on Indochina* (London, 1986), p. 74: 'For the preparation of a speech in the House of Commons he needed the assistance of half a dozen officials for hours on end. At Geneva something so simple as "I propose that we should now adjourn our meeting and reassemble at 3 p. m. on . . ." had to be typed, in triple spacing on blue crested paper using the special machine with extra large characters, for him to read out.'
55. Horne, *Macmillan 1894–1956*, p. 375. According to Gladwyn Jebb, 'Eden lives in a constant whirl, people rushing in with papers and doors opening and shutting in all directions, like the best French farce': Pimlott, ed., *Dalton War Diary*, p. 343.
56. Horne, *Macmillan 1894–1956*, p. 376.
57. R. R. James, *Anthony Eden* (London, 1986), p. 59; J. R. St John, 'The Rt. Hon. Anthony Eden', Nov. 1947, AP 33/6.
58. Young, ed., *Lockhart Diaries*, p. 386.
59. J. G. Winant, *A Letter from Grosvenor Square: An Account of a Stewardship* (London, 1947), pp. 67–8.
60. H. van Thal, ed., *The Prime Ministers: From Lord John Russell to Edward Heath* (London, 1975), p. 336; cf. Young, ed., *Lockhart Diaries*, p. 498: 'After luncheon everyone else . . . was content to rest and sleep. Not Anthony. . . . Even when we sat down . . . Anthony could not keep still, but jumped up every few minutes to chase away his hens who were invading his grass plots.'
61. Lawford, 'Three Ministers', p. 85; cf. D. Dilks, ed., *The Diaries of Sir Alexander*

Cadogan 1938–1945 (London, 1971), p. 592: 'He doesn't like it if one sends him too many papers, but is always ready to complain if *one* slips past him.'

62. Lawford, 'Three Ministers', p. 86.
63. P. M. H. Bell, 'Some French Diplomats and the British, *c.* 1940–1955: Aperçus and idées reçues', *Franco-British Studies*, 14 (1992), p. 44; R. Massigli, *Une Comédie des Erreurs, 1943–1956* (Paris, 1978), p. 67.
64. N. Nicolson, ed., *Harold Nicolson: Diaries and Letters 1939–1945* (London, 1967), p. 344.
65. Young, ed., *Lockhart Diaries*, pp. 299–300.
66. Young, ed., *Lockhart Diaries*, p. 510.
67. Colville, *Fringes of Power*, p. 697.
68. Moran, *Churchill*, p. 572.
69. M. Gilbert, *Winston S. Churchill*, vol. 8 (London, 1988), p. 1062.
70. Young, ed., *Lockhart Diaries*, p. 656.
71. Diary 24 April 1930, AP 20/1/10.
72. Young, ed., *Lockhart Diaries*, p. 462.
73. Diary entry end 1929, AP 20/1/9.
74. Diary 24 Aug. 1929, AP 20/1/9.
75. A. Cave Brown, *The Secret Servant: The Life of Sir Stewart Menzies, Churchill's Spymaster* (London, 1987), pp. 490–1; P. Wright, *Spycatcher* (New York, 1987), p. 373.
76. Young, ed., *Lockhart Diaries*, p. 512.
77. G. Payn and S. Morley, eds, *The Noel Coward Diaries* (London, 1982), p. 68.
78. Young, ed., *Lockhart Diaries*, p. 633.
79. Young, ed., *Lockhart Diaries*, p. 697.
80. Clark, *From Three Worlds*, p. 156.
81. Diary 5 and 11 Sept. 1976, AP 20/2/21.
82. Earl of Swinton, *Sixty Years of Power* (London, 1966), p. 163.
83. W. S. Churchill, *The Gathering Storm* (London, 1948), p. 188.
84. Diary 8 Oct. 1941, AP 20/1/21.
85. *Birmingham Mail*, 23 Oct. 1951.
86. Diary 2 Aug. 1941, AP 20/1/21.
87. Diary 28 Sept. 1943, AP 20/1/23.
88. J. Charmley, *Churchill: The End of Glory* (London, 1993), p. 492.
89. Young, ed., *Lockhart Diaries*, p. 490; cf. Ivor Thomas to Thomas Jones 14 Aug. 1942, cited P. Addison, *The Road to 1945: British Politics and the Second World War* (London, 1975), pp. 208–9: 'The Conservative Party regard Eden as a weak figure who has got his present position only by birth, the right school, good looks and luck.'
90. Eden to Churchill 26 Feb. 1942, AP 20/9/35.
91. Lord Butler, *The Art of the Possible* (London, 1971), p. 165.
92. T. Evans, ed., *The Killearn Diaries 1934–1946* (London, 1972), pp. 270–1.
93. Young, ed., *Lockhart Diaries*, p. 727.
94. D. Neff, *Warriors at Suez: Eisenhower Takes America into the Middle East* (New York, 1981), p. 184.
95. H. Brandon, *Special Relationships* (London, 1988), p. 135.
96. E. Barker, *Churchill and Eden at War* (London, 1978), pp. 125–6.
97. Eden to Eisenhower 18 July 1956, AP 20/27/69.
98. Harold Macmillan seemed to share the illusion. He wrote to congratulate Eden on his achievements in Washington in January 1956: 'You have put Anglo-American

relations back where they ought to be. We are not one of many foreign countries. We are allies and more than allies – blood brothers. In my short time I have tried to get Foster Dulles towards this concept. You have crowned the work.' Macmillan to Eden 8 Feb. 1956, AP 33/8iii.

99. Clark, *From Three Worlds*, p. 149; cf. P.-H. Spaak, *The Continuing Battle* (London, 1971), p. 83, quoting Eden in 1944: '[The Americans] are averse to taking the initiative, but, on the other hand, they do not like things to be done without them.'

100. Mosley, *Dulles*, p. 359.

101. Young, ed., *Churchill's Peacetime*, p. 13.

102. Draft of speech, [?] 1962, AP 20/50/108.

103. Moran, *Churchill*, p. 627.

104. Moran, *Churchill*, p. 501.

105. Gilbert, *Churchill*, vol. 8, p. 970: '[After a disastrous speech by Churchill in April 1954, Eden] told me [Anthony Montague-Browne] . . . that two Ministers had advised him not to salvage the Prime Minister in his winding up speech. He, to his everlasting credit, rejected this advice.'

106. Shuckburgh, *Descent to Suez*, p. 93.

107. J. Boyd-Carpenter, *Way of Life* (London, 1980), p. 122.

108. J. Morgan, ed., *The Backbench Diaries of Richard Crossman* (London, 1981), p. 416; André Laguerre, 'Forecast on Eden as Prime Minister', *Life*, 25 April 1955.

109. Shuckburgh, *Descent to Suez*, p. 284.

110. M. Cockerell, *Live from Number 10: The Inside Story of Prime Ministers and Television* (London, 1988), pp. 36–40.

111. Lord Kilmuir, *Political Adventure* (London, 1964), p. 257; cf. Payn and Morley, eds, *Coward Diaries*, p. 308: 'Anthony Eden's popularity has spluttered away like a blob of fat in a frying-pan.'

112. Horne, *Macmillan 1894–1956*, pp. 371–2.

113. Shuckburgh, *Descent to Suez*, p. 315.

114. P. Ziegler, *Diana Cooper* (London, 1981), pp. 288–9.

115. Shuckburgh, *Descent to Suez*, p. 324.

116. Boyd-Carpenter, *Way of Life*, p. 125.

117. A. Horne, *Macmillan 1957–1986* (London, 1989), p. 13.

118. J. Bright-Holmes, ed., *Like It Was: The Diaries of Malcolm Muggeridge* (London, 1981), p. 479; Kilmuir, *Political Adventure*, p. 308; Lord Hailsham, *A Sparrow's Flight* (London, 1990), p. 296.

119. P. Walker, *Staying Power* (London, 1991), p. 19.

120. Diary 19 Oct. 1940, AP 20/3/1.

121. *The Times*, 2 June 1956.

122. P. Hennessy and M. Laity, 'Suez – What the Papers Say', *Contemporary Record*, 1, 1 (1987), p. 3.

123. J. Charmley, *Churchill's Grand Alliance* (London, 1995), p. 349.

Bibliographical note

I NCOMPARABLY the most important single source for this book has been the private papers of Sir Anthony Eden held in the Special Collections Department of the Library of the University of Birmingham (Avon Papers). A further substantial collection of Eden's private office papers as Foreign Secretary is contained in the FO 954 series. The originals of these are held in Birmingham with copies at the Public Record Office. As the ownership of this collection was the subject of lengthy and vigorous discussion in the last years of his life between Eden and the Cabinet Office, the contents of the two versions are not identical. That held by the University of Birmingham is the more complete.

Details of other unpublished papers, private and official, used in the preparation of this work may be found in the notes.

Eden's published memoirs are among the most important to come from his generation of British politicians. They were published in three volumes: *Full Circle* (London, 1960), *Facing the Dictators* (London, 1962) and *The Reckoning* (London, 1965). In 1976, the last full year of his life, Eden published a slim volume covering his early years including his service in the First World War. *Another World* (London, 1976) met with considerable critical acclaim.

Eden's career covers a period of British history which has already been well-trodden by historians. It would be profitless to recapitulate here all the titles of books and articles referred to in the notes, but my indebtedness to the scholarship of others should be apparent from even a cursory examination of those notes. There is, however, an obligation upon me to list works which have been of the greatest value in the preparation and writing of this book.

As I have explained elsewhere, I have sought to complement rather than replace the work of Eden's earlier biographers. Robert Rhodes James, *Anthony Eden* (London, 1986) and David Carlton, *Anthony Eden: a Biography* (London, 1981) are outstanding. Though they reach strikingly different conclusions, each remains an indispensable source for any student of Eden's career.

At various stages in his life a number of figures aspired to the role of Eden's Boswell. The following published diaries are particularly important. John Harvey (ed.), *The Diplomatic Diaries of Oliver Harvey, 1937–1940* (London, 1970) and *The War Diaries of Oliver Harvey, 1941–1945* (London, 1978); Kenneth Young (ed.), *The Diaries of Sir Robert Bruce Lockhart, 1939–1965* (London, 1980); Piers Dixon, *Double Diploma: The Life of Sir Pierson Dixon, Don and Diplomat* (London, 1968); Evelyn Shuckburgh, *Descent to Suez : Diaries 1951–56* (London, 1986); William Clark, *From Three Worlds* (London, 1986).

The most comprehensive study of Eden's career before 1939 is A. R. Peters, *Anthony Eden at the Foreign Office, 1931–1938* (Aldershot, 1986). Among works which focus on the Second World War Elisabeth Barker, *Churchill and Eden at War* (London, 1978) is particularly perceptive. Much valuable information can also be gleaned from the pages of the official history of Britain's wartime foreign policy: Sir Llewellyn Woodward, *British Foreign Policy in the Second World War* (five volumes, London, 1970–76). The on-going theme in Eden's career of Anglo-American relations has been perceptively discussed in John Charmley, *Churchill's Grand Alliance: the Anglo-American Special Relationship, 1940–57* (London, 1995). The literature on the Suez Crisis would justify a bibliographical essay in its own right, but Keith Kyle, *Suez* (London, 1991) and W. Scott Lucas, *Divided we Stand: Britain, the United States and the Suez Crisis* (London, 1991) are unlikely to be superseded in the foreseeable future. Richard Lamb's *The Failure of the Eden Government* (London, 1987) remains the only substantial work devoted specifically to Eden's premiership. It is vigorously argued though many of its judgments are unlikely to stand the test of time. It should be supplemented by John Ramsden's thoughtful and elegant *The Age of Churchill and Eden, 1940–1957* (London, 1995). The last mentioned is indispensable for Eden's career as a Conservative politician.

Index

Index

Index

Index

Index

Index

Index

Index

Index

Index

Index

Index

Index

Index

Index

Index